FACILITIES SITE PIPING SYSTEMS HANDBOOK

ABOUT THE AUTHOR

Michael Frankel, CPD, is a recognized and often-quoted national and international authority on the design of a wide variety of basic and advanced facility piping systems and consultant on all phases of plumbing design and engineering. A native of Brooklyn, New York, he received his engineering degree from the City University of New York and has more than 45 years of experience as a designer and department head. He is an active member of the American Society of Sanitary Engineering and the American Society of Plumbing Engineers (ASPE), and is past president of the New Jersey chapter of ASPE. Mr. Frankel is a member of the National Fire Protection Association (NFPA) and principal voting member of the Technical Committee on Piping Systems for NFPA 99, the Code for Health Care Facilities.

He is a prolific writer and a frequent contributor to *Plumbing Systems and Design* magazine and *PM Engineering* magazine, and also the author of McGraw-Hill's *Facility Piping Systems Handbook,* now in its Third Edition (2010). Other accomplishments include writing the revised plumbing section for *Architectural Graphic Standards* for the American Institute of Architects and authoring multiple chapters for major handbooks such as *Piping Handbook, Practical Plumbing Engineering,* and *Handbook of Utilities and Services for Buildings.* Mr. Frankel has been a popular speaker at past ASPE conventions and symposiums and various local chapters of ASPE.

FACILITIES SITE PIPING SYSTEMS HANDBOOK

Michael Frankel, CPD

McGraw Hill

New York Chicago San Francisco Lisbon London Madrid
Mexico City Milan New Delhi San Juan Seoul
Singapore Sydney Toronto

The McGraw-Hill Companies

1 2 3 4 5 6 7 8 9 0 LPI/LPI 1 8 7 6 5 4 3 2

ISBN 978-0-07-176027-0
MHID 0-07-176027-X

Sponsoring Editor
Larry S. Hager

Editing Supervisor
Stephen M. Smith

Production Supervisor
Pamela A. Pelton

Acquisitions Coordinator
Bridget L. Thoreson

Project Manager
Anupriya Tyagi,
Cenveo Publisher Services

Copy Editor
Patti Scott

Proofreader
Meera Abraham,
Cenveo Publisher Services

Indexer
ARC Films, Inc.

Art Director, Cover
Jeff Weeks

Composition
Cenveo Publisher Services

This book is printed on acid-free paper.

This handbook is dedicated to all the people who work to preserve the future of this planet through the use of self-sustaining material and all those concerned with the conservation of water.

CONTENTS

Chapter 4. Water Treatment 4.1

Chapter 5. Desalination 5.1

Chapter 6. Site Domestic Water Service 6.1

Chapter 7. Site Fire Protection 7.1

Chapter 8. Site Fuel Gas Systems 8.1

Chapter 9. Fats, Oils, and Grease Interceptors, and Motor Oil Separation Units 9.1

Chapter 10. pH Neutralization Systems 10.1

Chapter 11. Infectious and Biological Waste Drainage Systems 11.1

Chapter 12. Nuclear Waste 12.1

Chapter 13. Industrial Waste 13.1

Chapter 14. Fire Suppression Water Drainage 14.1

Chapter 15. Volatile Liquids: Treatment and Disposal 15.1

Chapter 16. Stormwater Harvesting and Reuse 16.1

Chapter 17. Stormwater Drainage and Disposal 17.1

Chapter 18. Flow in Ditches and Open Channels 18.1

Chapter 19. Sanitary Gravity Flow 19.1

PREFACE

This handbook is a companion book to my *Facility Piping Systems Handbook* (McGraw-Hill). It is intended to be a design guide for the disposal of various kinds of waterborne waste, the supply of water and natural gas for facilities, and other site-related facility work that a facilities engineer is responsible for. It concentrates on the site work and piping outside the walls of the building. It will be particularly useful to mechanical design engineers in addressing the issues of many different water and gas supplies, waste production and disposal, and other important miscellaneous systems.

The disposal of all waterborne waste from facilities has always presented a potential problem because of the large number of possible pollutants as well as applicable codes and standards. Numerous pollutants are produced as by-products of many industrial processes, and other pollutants are generated or otherwise required to be disposed of by the various industrial, commercial, and other types of end users. Because of the myriad number of waterborne discharges possible from different types of facilities, this book concentrates on elements of the production and discharge of such pollutants, only from smaller-scale commercial, industrial, and chemical facilities. Methods are discussed that assist in the measurement and transport of these wastes. For the waste that is intended to be discharged into public sewers, local treatment is discussed that conforms with the requirements and approval of local authorities. If treatment is not permitted, the waste will be stored on site in a limited and approved manner by the operator of each facility until collected by a disposal company.

A definition of *wastewater* for this handbook is nonfecal waterborne wastes from commercial buildings, industrial plants, and institutions, including surface water and stormwater. Special attention is given to informing the design engineer and others who are responsible for identifying potential and undesirable waterborne contaminants in a water supply. Also covered are the waste streams of facilities whose release of pollutants could be detrimental to both the environment and public health. One part of this book concentrates on identifying undesirable waste characteristics rather than actual treatment, except for the rare instances where the contaminants are well known and are intended or required to be treated on site prior to discharge into the environment. Large-scale facilities, municipal waste and water treatment, sanitary discharge, local facility treatment, and plumbing and sanitary codes are outside the scope of this handbook.

All wastewater legislation falls into two categories. The first category pertains to the environment. Legislation and related regulations relative to wastewater discharges are necessary to protect the environment, health, and welfare of the public. These requirements act as a safeguard for various conditions recognized as being unique by local authorities. They also preserve cultural and historic environments. The second category consists of codes and standards that have been developed to protect the health and safety of the workplace and the clients who use the facilities, and to regulate the dimensional accuracy and suitability of construction materials so that installers and users remain safe.

Numerous codes have been written by federal, state, and local officials that can be applied to facilities addressed in this book. A multitude of local codes are used for specific subjects, most of which will be revised periodically to reflect local conditions and improvements. An effort is made to present many of the codes, standards, and other legislation closely associated with wastewater. Where possible, the reader is directed to the sources that originate the codes for further study. The ultimate discharge to the environment, where allowable, is also discussed.

There is confusion about all the federal codes and standards that apply to various waste discharges. In addition, most states and local governments have written additional codes patterned after federal programs. An overview is given of the federal, state, and local legislation and regulations, which are constantly changing, for waste streams. Many public administrators, engineers, and other code writers recognize that, along with the actual writing of the environmental regulations, there is a need to recognize the concerns and importance of the cost and results of strictly adhering to all aspects of such codes. This is also discussed.

Because of the increasing use of potable water, chapters are included on the conservation of water. Finding a supplemental supply of water for such uses as drinking, agricultural, and industrial water is imperative. The use of harvested precipitation and reuse of graywater as a substitute for potable water are discussed. A chapter is included to inform the design professional of which nationally accepted building rating and assessment organizations are available and the goals they seek to accomplish.

The other part of this handbook evaluates the quality of the water supply for commercial and industrial use. The various methods necessary to provide the quality of water used for potable and other needs in various processes and facilities are discussed. The use of freshwater throughout the world has continued to rise significantly. Groundwater overpumping has resulted in the lowering of water tables and saltwater intrusion. In most parts of the world freshwater is being used faster than it can be replenished by rain and snowfall. Industrial demands are also increasing; some means must be found to reduce this amount and to improve the quality of water used for manufacturing purposes.

Michael Frankel, CPD

CHAPTER 1
WATER WELLS

This chapter discusses only the drilling of a well used for providing a water supply to a facility from outside the normal municipal or public water system. The size of the pump is outside the scope of this handbook.

GENERAL

When the supply of public or municipal water is not easily available, it is necessary to find another source. The first choice is a well, because it provides the most likely source of water that is generally potable and also suitable for almost any purpose.

A *well* is a vertical hole or shaft, excavated or bored through the earth for the purpose of bringing water to the surface from an underground aquifer. A major consideration is that construction of a well must protect the overall quality of the groundwater. Wells are often classified as deep or shallow. A shallow well is considered to be about 50 to 100 ft (15 to 30 m) deep. Deep wells are generally considered a superior source of water because the water is less susceptible to surface contamination and the depth of the aquifer usually fluctuates less than the depth of water in a shallow well.

CODES AND STANDARDS

1. The Safe Drinking Water Act
2. Local requirements

WATER WELL SYSTEM COMPONENTS

The total well system consists of the well pump and the well structure, which is made up of the borehole and casing, intake section, grout, and distribution pipe that brings the water to the surface. Additional components on the surface include an aboveground pumping arrangement that supplies adequate water pressure to a facility and a softening arrangement if the water is too hard.

Well Pumps

The selection of the pump depends on many factors, including the following:

1. Flow rate and total head
2. Well diameter
3. Depth of the well
4. Straightness of the well
5. Presence of sand in the water
6. Duration of the pumping cycle

The well pump is the motive power used to bring water in the aquifer to the surface. Three types of well pumps are generally used: vertical turbine, jet, and submersible pumps.

Vertical Turbine Pumps. Vertical turbine pump systems consist of an impeller suspended at the bottom of the well in the aquifer and a centrifugal pump with the motor mounted on the surface over the borehole connected to the impeller. The impeller is enclosed in a bowl and connected to the motor by a long shaft, positioned by bearings in the discharge pipe. This type of pump is well suited for larger flow rates, deep wells, and high discharge heads. A typical vertical pump installation is illustrated in Fig. 1.1.

Jet Pumps. Jet pumps are centrifugal pumps that use the flow of water through a special fitting to create a partial vacuum at the bottom of the well, which draws additional amounts of water into the discharge pipe. This is a two-pipe system: The first pipe supplies water down into the well, and the second pipe discharges the supply and the additional water to the surface for use. The disadvantages of jet pumps are that since a portion of the discharge must be pumped back down the well, the jet pump is generally used for smaller flow rates. In addition, a larger casing is usually required. The advantage is that the pump system is located on the surface for ease of maintenance. The pump could be located over the well or in a remote location. A simplified, remote deep well jet pump installation is illustrated in Fig. 1.2.

Submersible Pumps. The submersible pump is a small-diameter, totally self-contained centrifugal impeller in a housing that is close-coupled to an electric motor. It is placed inside the casing near the bottom of the well in the aquifer. The pump is supported by the discharge pipe and available in a wide range of flow rates and pressures. The electrical supply is attached to the discharge pipe. The submersible pump must be installed lower than the expected lowest drawdown level of the well. Advantages are that aboveground installation enclosures are avoided, the pumps can be installed in wells that have crooked casings, and the pumps make little noise and are easy to maintain. An installation of a total submersible pump system is illustrated in Figs. 1.3 and 1.4.

Casing

The purpose of the casing is both to maintain the borehole opening by keeping the loose and unstable soil from falling in and to prevent contamination. It is a thin-walled cylinder placed in the borehole of the well, usually as it is drilled. It acts as a lining and extends from the ground surface to the bottom of the well where the intake screen is located. It helps to withstand shifts in the earth and to prevent cave-ins. It also helps to eliminate seepage of contaminants into the well from undesirable aquifers. Casing material is available in steel, fiberglass, or plastic.

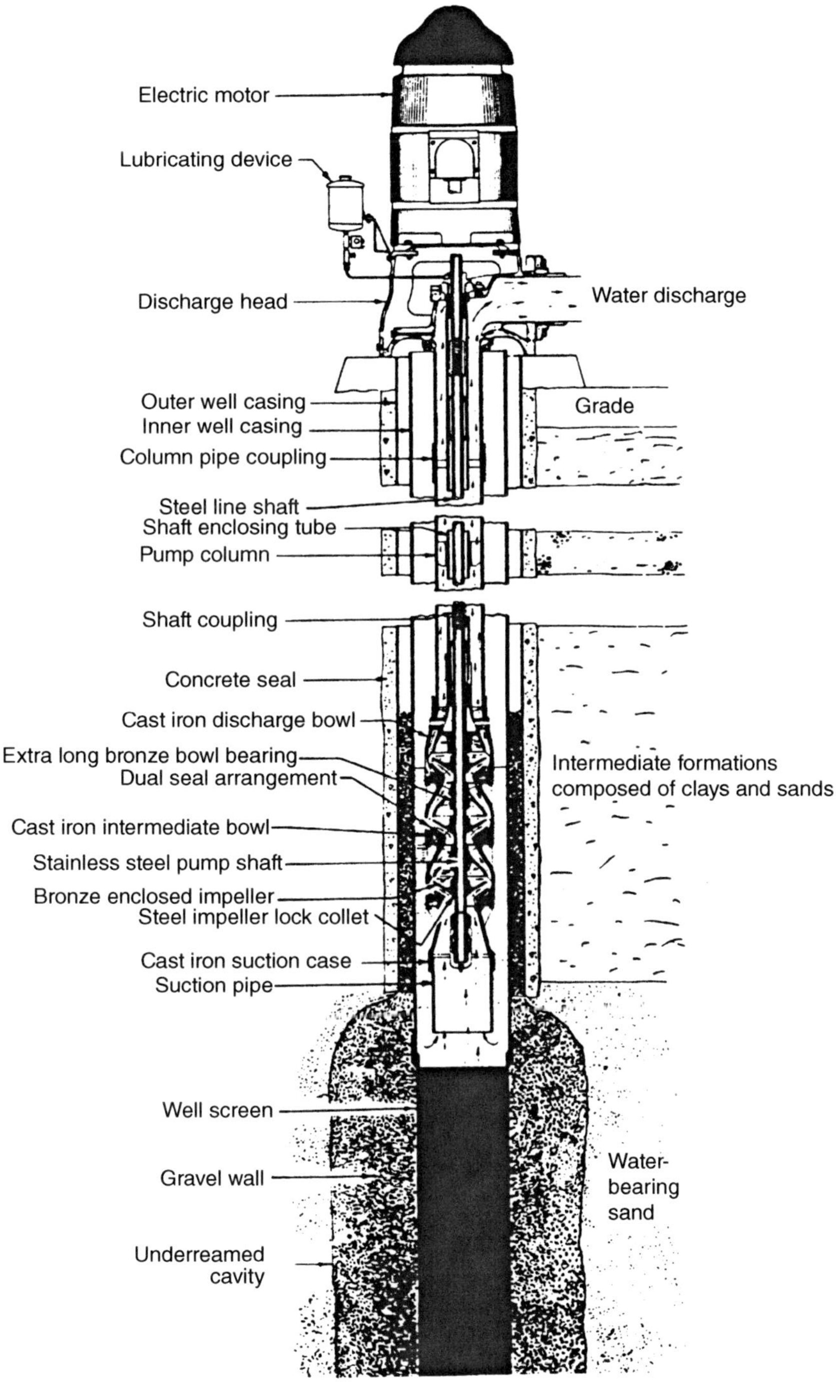

FIGURE 1.1 Typical vertical pump installation.

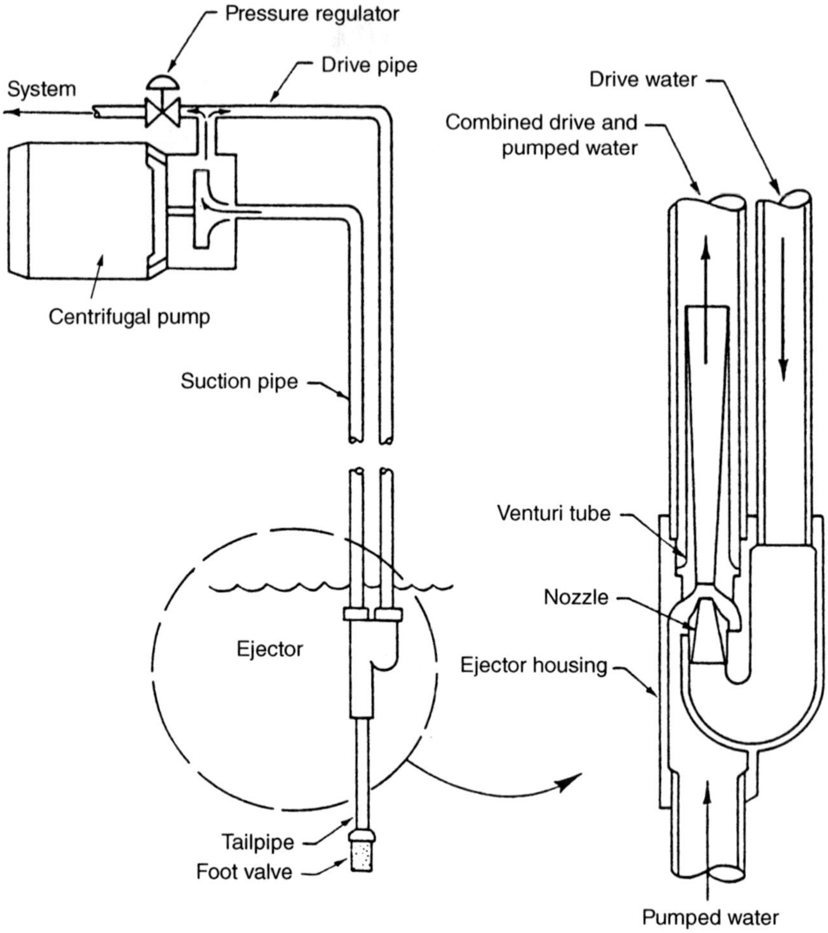

FIGURE 1.2 Schematic deep jet pump installation.

The casing is subject to physical and chemical forces. The physical forces are tension, column loading, and collapse pressures. The chemical forces are those from contaminants in the soil and groundwater that cause corrosion and degradation of the casing.

Plastics are used primarily in short, small-diameter wells; fiberglass is primarily used for highly corrosive waters; and steel, the strongest material, is most commonly used. Cable tool drilling requires the use of steel casing.

The casing diameter is generally based on yield. The recommended diameters of wells based on yield are listed in Table 1.1. Another consideration in selecting the casing size is to keep the maximum flow through the smallest part of the casing at or below a generally accepted velocity of 3 ft/s (1 m/s). Another common practice is to use casing two sizes larger than the discharge pipe required by the pump selected.

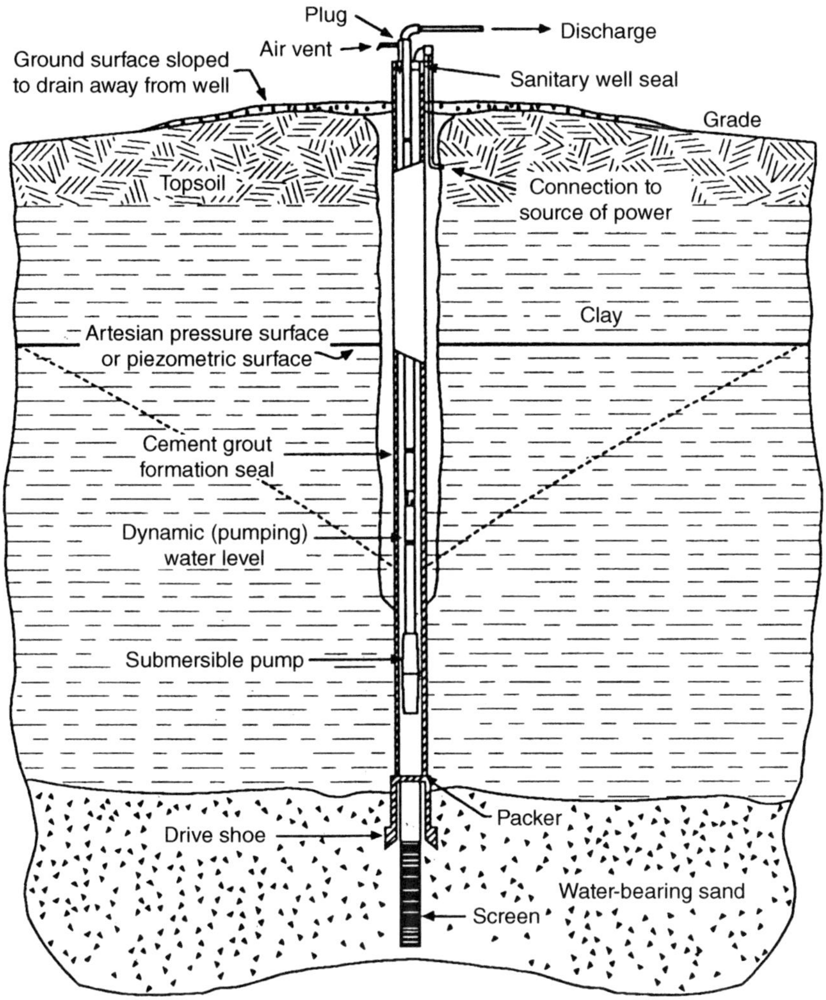

FIGURE 1.3 Belowground submersible well pump installation.

Well Intakes

The intake is regarded by many as the heart of the well. It is installed at the bottom of the well pipe to allow water from the aquifer to enter the well pipe. Two of the most common intake constructions are the tube type, often called *naturally developed,* and the gravel-packed type.

The *well screen,* or *intake,* is a perforated device that allows water from the aquifer to enter the well while preventing sand from entering. It is connected to the end of the casing. Without the screen, many wells would not be capable of producing a useful flow of water. Typical well screens are slotted, louvered, and wire-wrapped. The wire-wrapped

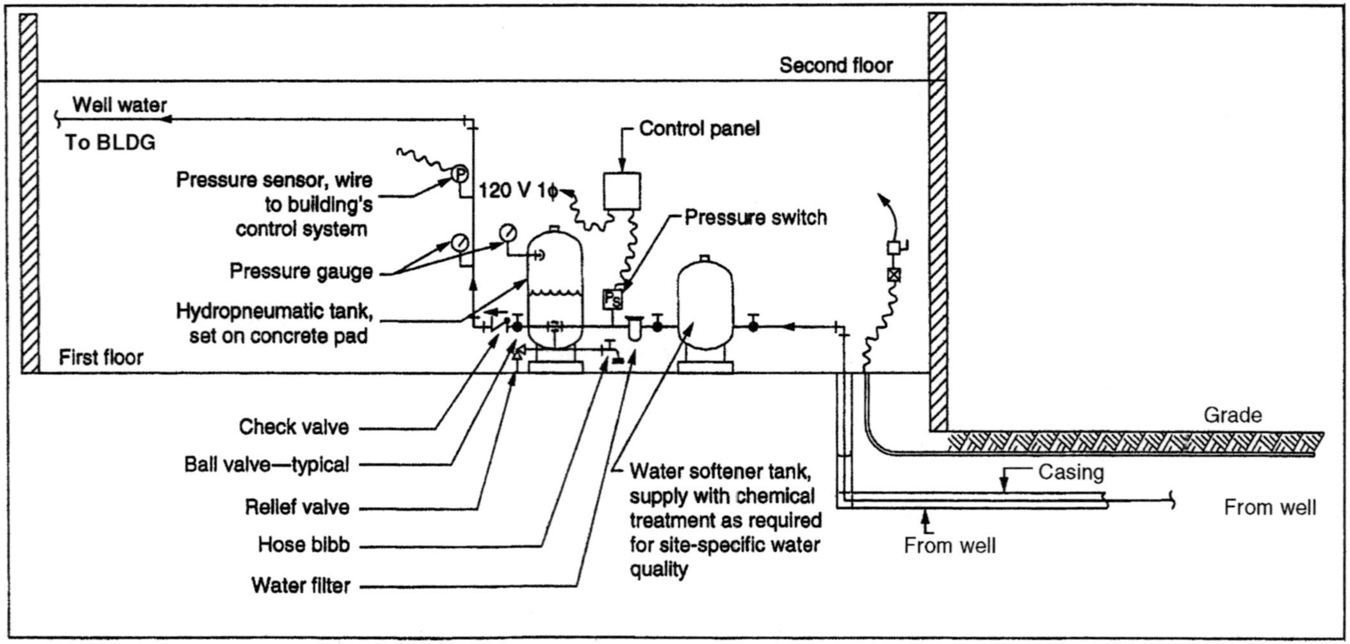

FIGURE 1.4 Typical aboveground submersible pump installation.

TABLE 1.1 Recommended Well Diameters for Wells Based on Yield

Anticipated well yield, gpm	Nominal pump bowl size		Optimum well casing size		Smallest well casing size, in
	in	mm	in	DN	
<100	4	100	6	150	5 ID
75–175	5	125	8	200	6 ID
150–400	6	150	10	250	8 ID
350–650	8	200	12	300	10 ID
600–900	10	250	14	350	12 ID
850–1300	12	300	16	400	14 OD
1200–1800	14	350	20	500	16 OD
1600–3000	16	400	24	600	20 OD

screen is well suited for large-diameter, high-yield wells. Wells drilled in consolidated rock are called *naturally developed* because the fractured rock acts as an intake and there is an absence of sand. There is little need for a screen intake in this case. Intakes are commonly manufactured from plastic, bronze, steel, stainless steel, and Monel metal. The length of the screen is determined by the type of aquifer (confined or unconfined), the diameter of the casing, and the expected flow rate. For large flow rates, lengths of 30 to 40 ft (9 to 12 m) are not uncommon. Generally accepted criteria limit the velocity of water passing through the screen to about 2 to 4 in/s (50 to 100 mm/s). It is recommended that the intake be located at the lowest practical portion of the aquifer so that the drawdown can be as great as possible.

Gravel Packing

Gravel packing is a common method of enhancing the flow of water by reducing the intake of sand particles at the well intake. This procedure replaces material around the screen with gravel or coarse sand packed around the intake to maximize yield and prevent clogging of the intake openings by fine sand in the aquifer.

Grouting and Sealing the Well Casing

Grout is a material used both to fill the annular space between the casing and the borehole and to seal the well at the surface. This material must be able to be pumped for underground and capable of hardening sufficiently to form a seal. The material used most often is a slurry of plain concrete or a mixture of concrete and up to 8 percent bentonite. Other additives are available that resist water intrusion, add strength, and resist corrosion. The concrete-bentonite mixture has less shrinkage than plain concrete. Grouting and sealing are done for the following reasons:

1. To prevent seepage of polluted surface water down into the well along the outside of the casing
2. To seal out water of unsuitable quality from the strata above the aquifer
3. To stabilize and secure the casing in the drilled hole
4. To form a protective barrier around the casing, thereby increasing its life and protecting it against corrosion

It is not necessary to place grout down to the lowest level of the well, but only to the lowest level necessary to keep out pollutants of concern.

WELL DRILLING

The primary methods used to drill a water well depend on the expected borehole depth, access to the site, purity of the water required, and often the driller's preference.

Other methods, such as digging wells by hand, using water jets (jetted wells), and using an earth auger turned by hand or light machines (bored wells), are rarely used to obtain sufficient water for any facility. The digging of wells by hand and earth auger is outside the scope of this book.

Cable Tool Drilling

The cable tool rig is often referred to as the "yo-yo" method. This system is based on the simple lifting and dropping action of the tool string in the borehole. Gravity, the weight of the drill bit, and the natural rotation of the drill string break up and cut through the underground formations. The *tool string* consists of the rope and socket that connects the string to the surface engine, drilling jars (optional), the drill stem, which serves as a guide to the string, and the drill bit. Water is added to produce a slurry mixture of the cuttings at the borehole bottom. All this is connected at the surface to a derrick that lifts and drops the entire string.

Cuttings from the borehole must be removed periodically. This is done by means of a *bailer,* which is a 10- to 20-ft (3- to 6-m) section of pipe with a check valve that holds the cuttings along with the slurry. The tool string is raised clear of the hole, and the bailer is lowered for this purpose.

Although the cable tool method is not used very often, the advantages are the low cost of the entire rig as a whole, the simplicity and sturdiness of the design, and the ability of the drilling rig to move over rough terrain. It is also able to collect qualitative data on the water-bearing characteristics of various strata as the casing is being inserted. Disadvantages are the relatively slow drilling speed and the need for a casing to be inserted when drilling through unconsolidated formations.

Hollow Rod or Hydraulic Percussion

This method is similar to jet drilling except the water is pumped down through the annular space, and the cuttings are carried up through the pipe by the reciprocating motion and a series of valves that keep the water from flowing downward.

Rotary Drilling

The rotary drilling method is used to drill approximately 80 percent of all water wells. The *rotary drilling* process bores a hole by means of a rapidly rotating drill bit that has a constant downward pressure exerted on it. The rotary drilling system consists of the rig, a pump to circulate or create the drilling fluid, the drill stem, and a drill bit. The rig is made up of a derrick and hoist, a motor, and a removable, rotating table placed directly over the borehole through which the drill stem passes. The drill stem is composed of the drill collar, which connects the bit to the pipe; the drill pipe, which connects the bit to the surface; and the Kelly fitting, which is a fitting at the top of the pipe that imparts rotary motion to the stem from the table.

The use of drilling fluids is required in rotary drilling. Drilling liquid or air performs the following four primary functions:

1. It removes cuttings produced by the bit.
2. It transports cuttings up to the surface. The fluids can be interrupted for information about the geophysical features of the borehole.
3. It maintains borehole stability.
4. It cools the drill bit.

Rotary drilling is separated into two primary categories based on the method used to drill the borehole—liquid (or hydraulic rotary) and air rotary drilling.

Hydraulic Rotary. This method can be used to drill in most formations. It functions by using a rotary drill along with the pumping of the drilling fluid, which is generally drilling mud, referred to in the field simply as *mud*. This could be any type of mixture. The filter cake that forms around the length of the borehole wall from the mud flow assists in retaining soft formations and provides a protective layer that resists the corrosive effects of the drilling mud.

Air Rotary. This method is mostly used to drill in consolidated formations. Often mud or water is also used in the unconsolidated strata before bedrock is encountered. The same types of tools are used as for drilling. The difference is that compressed air, usually around 50 psig (340 kPa), is used to depths of about 250 ft (80 m).

Reverse Circulation Rotary. In this method, the flow of drilling fluid is reversed from that of conventional drilling. It is commonly used when the completion of the well is to be done by gravel packing.

WATER WELL DESIGN AND CONSTRUCTION

The design of a well includes criteria for present and future flow rates, the purpose of the well, and its expected useful life.

The drilling or construction of a well requires prior planning. Criteria such as the probable depth of water-bearing formations and the required permits and approvals of various agencies must be obtained. It must be understood that as the drilling of the well progresses, plans and methods may have to be altered or modified based on actual formations and drilling conditions encountered.

Several issues should be considered in the design of a well:

1. Minimum desired output (yield) of the well and the conditions under which the flow is desired.
2. Proposed diameter of the well casing.
3. Out-of-plumb borehole. A common maximum figure is 1 well diameter per 100 ft (33 m).
4. Out-of-straight borehole. Generally accepted requirements should be limited to a dimension that would permit passage of a 33-ft-long (10-m) blank, ½ in (13 mm) less in diameter than the casing inside diameter. Straightness is more important than being plumb.
5. Screen characteristics, including size and type of openings, length, and net open area.
6. Casing material, type, and weight.
7. Cleaning and disinfection.

WELL DEVELOPMENT AND TESTING

Every type of drilling causes some kind of disturbance to the aquifer by clogging the pores of the formation where the aquifer exists. After the borehole is completed, the casing is inserted, the intake is attached, and the gravel packing is installed (if desired). The next step is to have it "developed." Development is the process that removes finer material from the natural formation around the intake, enlarging it and leaving only larger gravel and stones around the screen. The larger pores allow the water in the aquifer to flow into the intake at maximum capacity. Development is the last stage of construction and is regarded as much of an art as a science. The experience of the driller is an important factor. After development, the well is capable of providing water.

During the drilling of the well, constant analysis of the drilling fluid used to find an aquifer should be made as the bit passes through each formation. There are no set standards to recognize water-bearing strata. The experience of the well driller is important in this respect. Often, a pilot well drilled for this purpose is highly recommended.

When the borehole is complete, the casing shall be installed (if not previously done), the water level in the well accurately measured, gravel packing done (if desired), the intake connected, and the well grouted. The last step in the construction process is disinfection, where a strong chlorine solution is introduced into the well. It is also recommended that intermediate sterilization be performed during the drilling operation.

Development

There is no single, effective method; rather, several are available, each with its own effectiveness. Most of the methods discussed are used in combination with others. The most common methods include overpumping, surging, air lifting, air surging, and high-velocity jetting.

Overpumping. This is a process in which the capacity of the well is greatly exceeded by pumping. All backflow-preventing devices are in place. This process cleans out the fine materials and begins to collapse the aquifer strata material around the well. A separate pump is usually selected because one impeller would be excessively worn by the additional sand brought into the pump by the large volume of water.

Rawhiding. This is a variation of overpumping, except that all backflow-preventing devices are removed. The pump is started and stopped, and when the water in the discharge pipe falls to the bottom of the borehole, it loosens the aquifer material adjacent to the well screen. Starting and stopping the pump is done many times.

Surging. This process uses a plunger-like tool that is moved up and down inside the well and intake. A tight-fitting insert called a *surge block* creates an alternating vacuum and pressure that force water in and out of the intake.

Air Lifting. In this process an air rotary drill string is placed at the bottom of the well without the bit. Compressed air is pumped down the well, allowing the air pressure to loosen the material around the bottom of the borehole. This method often results in too great a flow out of the well. This method has many drawbacks and is used primarily in large-production wells at the end of the development process.

Air Surging. This method is a variation of air lifting that uses two types of air rotary drills, one within the other. The inner one carries air and the outer one carries water to the surface. A valve is alternately opened and closed after compressed air is allowed to enter the well, resulting in a surge of water that provides the desired results.

High-Velocity Jetting. Generally regarded as the most effective method of development, jetting uses multiple nozzles to spray water at 150 to 300 ft/s (50 to 100 m/s) through the intake while it is raised and lowered. Often, the pump will be turned on to take some water to the surface.

WATER LEVEL MEASUREMENT

The *static water level* is the level that the water reaches in a well under atmospheric conditions when the well is not being pumped. The *dynamic water level* is the elevation to which water falls in the well during pumping at a given flow rate. The distance between the static level and the dynamic level is the *drawdown.* This level will change depending on the pump flow rate and time. When water is pumped from a well, the level of water adjacent to the intake creates an inverted cone of depression of the water level in the aquifer. This is illustrated in Fig. 1.5.

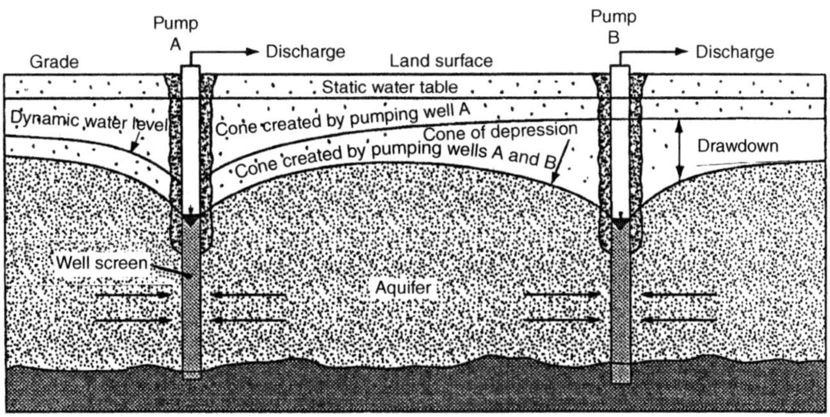

FIGURE 1.5 Cone of depression.

To determine drawdown, it is important that this level be checked often over a relatively long time at various flow rates. This test shall be recorded in a log. No calculation of capacity should be accepted unless it has been made for at least several hours at hourly intervals. For practical purposes, the pumping level is considered established when three test levels are the same for a minimum of 3 h. Often, a 24-h test is conducted. A common method used to describe the yield is to express the discharge capacity in relation to the drawdown distance. This relationship is called the *specific capacity* of the well and is expressed as gallons per minute per foot squared (gpm/ft^2) or liters per meter squared (L/m^2).

TESTING FOR YIELD

Following development of the well, the next procedure is to test the well by measuring the flow rate of pumping and the level of water at the various pumping stages. Testing will determine the quantity of water, or yield, that can be produced in a given time and the acceptable drawdown level. Testing will provide a basis to estimate the water supply quantity available from the well and to determine the type of pump to be used. *Yield* is the average flow available over a period of time.

Many states require at least two wells. A complete backup system must be installed so that the maximum daily demand will be met with one pump out of service. It is considered good practice to alternate the primary operation of the pumps so that equal wear will be obtained.

LOCATION OF WELLS

An initial estimate in setting the location of wells relative to one another may be made by using one of the two rules of thumb:

1. For two wells, if the aquifer is more than 340 ft (102 m) or less in thickness, the distance between the wells should be at least twice the thickness of the aquifer.
2. For more than two wells, the distance between the wells should be at least 240 ft (75 m) regardless of the thickness of the aquifers.

In either case, the degree of interference must be evaluated.

The character of the aquifer and the construction of the well each will affect the yield. In determining the yield, the result that pumping has on the aquifer shall also be determined. The following are definitions of different types of yield:

1. *Safe yield* is the quantity of water that can be withdrawn annually without the ultimate depletion of the aquifer.
2. *Permissive sustained yield* is the maximum rate at which water can be withdrawn from an aquifer on a continuing basis without developing undesirable results.
3. *Maximum sustained yield* is the maximum rate at which water can be withdrawn from the aquifer on a continuing basis.

The yield can be effectively increased by lengthening the intake and changing the type or number of openings in the intake screen. Another method is to make the diameter of the borehole larger, but studies have shown that doubling the diameter increases the yield by only 10 percent. In most cases, this is not an economical solution.

DISINFECTION

All newly drilled wells, as well as wells that have been repaired, should be thoroughly cleaned of soil, grease, and scum and disinfected before being put into service. The most common disinfectant is a 65 percent solution of calcium hypochlorite. As an alternate, common household bleach or sodium hypochlorite may be used.

The recommended practice is to prepare a liquid that will achieve 100 mg/L of available chlorine at the static water level. An additional 150 gal (570 L) of solution is then added to the well to force water into the surrounding formation. The well should be pumped until the odor of chlorine is apparent and then shut down. After that, the solution should be kept in place between 4 and 24 h—the longer the better. Finally, the well is pumped until a sample analysis for coliform bacteria is completed and approved.

WELL SYSTEM DESIGN CONSIDERATIONS

The total well system takes water from the well and distributes it in a manner that is useful to a facility, almost always as a supply to a water storage tank from which the water will be withdrawn. The purpose of the tank is to provide a storage capacity so that the well pump can be shut down for a time. If the tank is large, it can be filled by the well pump at a lower flow rate than the high instantaneous flow rate that might be required by the facility.

For very small systems similar in capacity to those in private residence systems, the well water is routed into a small hydropneumatic tank. This tank has an airspace that is compressed by the incoming water. When the volume of water in the tank compresses the air to a high enough level, the pump stops. For large systems, water should be supplied to ground-level or elevated water tanks to prevent the well pumps from running continuously.

The *pump head*, or total dynamic head, is calculated by adding the distance between the impeller and the surface, the friction loss in the pipe and fittings, and the height from 1 ft (0.3 m) above the inlet into the elevated storage tank (if one is used) or the amount of pressure required in the hydropneumatic tank. If the system is used to directly supply a specific piece of equipment, the value for the pressure required by the equipment is used.

The discharge pipe material is the same as that used for normal plumbing at the facility.

BIBLIOGRAPHY

Corbit, R. A., ed., *Standard Handbook of Environmental Engineering*, McGraw-Hill, New York, 1989.

Design Manual, Johns-Manville Transit Pipe.

Mackenzie, L. D., *Water and Wastewater Engineering*, McGraw-Hill, New York, 2010.

National Water Well Association, *Design and Construction of Water Wells*, Van Nostrand Reinhold, New York, 1988.

CHAPTER 2
GRAYWATER

INTRODUCTION

This chapter is written to acquaint the reader with the general parameters of graywater and is not intended to be used as a basis of design for specific purposes. Consultation with applicable local code bodies for compliance is mandatory. Tables of contaminants and treatment are given to acquaint the reader with available criteria.

DEFINITIONS

Graywater is defined as untreated wastewater from both commercial and domestic activities.

Residential systems are those that receive wastewater from an individual residence, specifically bathing, showering, laundry waste, and lavatory sinks that will be recycled into water available for useful purposes.

Commercial systems are those that treat wastewater from business activities where no manufacturing, assembly, or food processing is involved. Health care facilities that discharge waste from laundries, hotels, and other facilities that would discharge high amounts of high-strength commercial cleaners are also considered graywater. Under some circumstances, clean heating, ventilating, and air conditioning (HVAC) condensate and clean industrial waste could be included. This is an increasing resource mandated by many cities. This chapter discusses only graywater. Harvested rainwater used for this purpose is discussed in Chap. 16.

Blackwater is defined as water originating from toilets, urinals, dishwashers, kitchen sinks, garbage disposers, and diaper washing, and it is excluded from reuse consideration. Blackwater treatment and reuse is outside the scope of this book because of both the very costly treatment associated with this category and the reluctance of an authority having jurisdiction (AHJ) to recognize this class of wastewater. Blackwater would be very costly and would be considered for reuse only if there were an extreme or total lack of other kinds of water. Many in this industry are concerned about the inevitable water shortage and when this will become a fact of life leading to the acceptance of the reuse of all wastewater.

Reclaimed water is often referred to as *recycled*, secondary and tertiary treated municipal water obtained directly from a publicly owned treatment works (POTW). This water is filtered and chemically treated to within stringent standards. It is almost exclusively used for municipal landscape irrigation, particularly in arid regions such as the southwest United States. This wastewater contains residual disinfectants and is distributed through off-color purple pipes. The Health Canada guidelines address domestic graywater reuse as "reclaimed water." Reclaimed water is outside the scope of this book.

GENERAL

In recent years, concerns over the dwindling reserves of groundwater and overloaded or costly water treatment plants have generated great interest in the reuse and recycling of graywater, both domestically and for use in facilities.

All people remotely associated with the water industry are aware of the looming lack of water availability and quality issues. The increasing population growth is putting a strain on the total available water supplies, diminishing potable water supplies in particular. This is particularly true in arid portions of the globe and some parts of the United States.

Recycling graywater can stretch a water budget during the hot summer months. Graywater, which is recycled shower, bath, and laundry water, can be used to keep thirsty plants alive, but some precautions must be followed. If graywater is used where human contact is possible and it has not been disinfected, it could be contaminated. A careful, commonsense approach to the use of graywater can virtually eliminate potential hazards. Municipal reclaimed water is well established and represents one source for use in irrigation, toilet and urinal flushing, decorative fountains, and other nonpotable uses. The same concept has been recently applied inside residences and commercial buildings, particularly in Australia where a number of wastewater technologies exist in the market today.

A lack of knowledge and experience with modern graywater systems has led to some concerns over potential health and environmental risks. If properly designed, a graywater system will lower sewage costs and at the same time will reduce water usage for irrigation. If desired, the water quality can be increased to better than that supplied from municipal sources. Using this approach, less water enters the city sewage system, saving the government from building new treatment plants. Some jurisdictions demand highly intensive treatment systems for legal reuse of graywater, and others encourage the untreated use of graywater. Despite many obstacles, graywater is often used for irrigation both legally and illegally.

We must consider all sources of water, regardless of their quality, and investigate the feasibility of recovery and reuse. Most graywater would be used for agricultural purposes. It is very rare that any surface water source is pure enough to serve directly as a source of potable water for individual facilities. If one is available, it must be approved by the local authorities. To supplement the available supply, we must first consider all sources of water, regardless of quality, and investigate the feasibility of recovery and reuse. Water that is recycled from earlier uses is a very large resource in irrigation and other nonpotable water uses. The following is true of all water sources:

1. All water supplies, regardless of source, contain contaminants.
2. These contaminants can be classified into specific chemical classes.
3. Treatment technologies exist to reduce the concentration of contaminants in each category.

CODES, STANDARDS, AND REGULATIONS

At this time, there are no federal regulations addressing graywater reuse. However, many states or other governing bodies have promulgated regulations covering residential and/or commercial graywater reuse. To protect human health, enforceable state regulations are in place influencing treatment, system design, installation, and operation. These regulations must be read carefully. They often vary from one location within a state to another.

The major regional codes are in the process of writing green code supplements. The International Association of Plumbing and Mechanical Officials (IAPMO), e.g., has established requirements for green building and mechanical efficiency in Supplement 502.

A part of the Section 502 supplement specifically covers allowable parameters of contaminants for graywater systems, including the minimum distance of the tank and irrigation field from building structures. The following excerpt is from IAPMO 502: "Graywater shall be permitted to be diverted away from a sewer or private sewage disposal system, and to discharge to a subsurface irrigation or subsoil distribution system. The graywater shall be permitted to discharge to a mulch basin for a single family or multiple family residential dwellings. Gray water shall not be used to irrigate root crops or food intended for human consumption that come in contact with the soil."

Graywater systems cannot be connected to any potable, reclaimed, or on-site treated nonpotable supply. Potable sources can be used for makeup water in a nonpressurized storage tank, provided that the connection is protected by an approved air gap. The graywater supply system must be located only on the property that discharges the graywater.

Numerous other parts of Section 502 regulations include graywater components, soil absorption, and the design of irrigation fields. This standard must be read when anything that uses this model code is designed. That document also makes compliance with the local authority having jurisdiction mandatory, to require adherence to more stringent requirements. The supplement also shows how to replace some potable water, earn Leadership in Energy and Environmental Design (LEED) points, and do it safely.

NSF International (National Sanitation Foundation) has developed four standards that will "establish minimum materials, design, construction and performance requirements for onsite residential and commercial water reuse treatment systems." This requires testing the water proposed to be used for a 6-month duration. The four standards are

1. NSF/ANSI Standard 40, Residential Wastewater Treatment Systems, NSF International/American National Standards Institute
2. NSF/ANSI Standard 245, Wastewater Treatment Systems—Nitrogen Reduction, NSF International/American National Standards Institute
3. NSF/ANSI Standard 350, On-Site Residential and Commercial Water Reuse Systems
4. NSF/ANSI Standard 350-1, On-Site Residential and Commercial Graywater Treatment Systems for Subsurface Discharge

Model codes, such as the Green Supplement of the International Plumbing Code and the Uniform Plumbing Code, have added provisions for the applicability and acceptance of on-site treated nonpotable reuse. The recent acknowledgment of numerous alternative water sources, such as graywater, is a milestone to permit water from nonpotable sources to be used for such purposes as irrigation, ornamental water features, and indoor plumbing applications.

The acceptable quality of reuse water for on-site applications is determined by local and state regulations, not federal. This has resulted in a wide range of varying criteria and approvals. Clearly there is a large disparity and no agreement among the differing authorities about the regulations or even definitions. Table 2.1 summarizes the state's graywater regulations for the 30 states where they exist at the time of this writing. Table 2.2 summarizes the standards appearing in NSF/ANSI 350 and 350-1.

TABLE 2.1 Existing Codes as of February 2010

State	Agency	Title/chapter	Residential rules?	Flow limits, gpd	Commercial rules?
Alabama	Alabama Department of Public Health, Environmental Services	Onsite Sewage Disposal Subdivision	Yes	N/A	Yes
Arizona	Arizona Department of Environmental Quality	R18-9-701 ADEQ Administration Code	Yes	N/A	Yes
Arkansas	Arkansas Department of Health Sanitary Division	Alternate Systems Manual	Yes	N/A	Yes
California	California Building Standards Commission	California Plumbing Code (Title 24, Part 5, Chapter 16A)	Yes	250 gpd	No
Colorado	Colorado Department of Health and Environment	Guidelines on Individual Sewage Disposal Systems	Yes	N/A	Yes
Connecticut	Connecticut Department of Public Health	Public Health Code Regulations and Technical Standards	Yes	N/A	Yes
Florida	Florida Department of Health Bureau of Onsite Sewage Programs	Public Health Chapter 381.0065; Florida Administrative Code 64E-6	Yes	75	Yes
Georgia	Georgia Department of Human Resources, Division of Public Health	Department of Human Resources Chapter 290-5-26	Yes	> 500	Yes
Hawaii	Hawaii Department of Health, Wastewater Branch	Title 11 Department of Health, Chapter 62 Wastewater Systems	Yes	150	Yes
Idaho	Idaho Division of Environmental Quality	IDAPA 16.01.03 Rules for Individual/ Subsurface Sewage Disposal Systems	Yes	N/A	No
Kentucky	Kentucky Department of Public Health, Protection and Safety Division	902 KY Administrative Regulations 10:085	No	N/A	No
Maine	Maine Department of Human Services, Bureau of Health, Division of Health Engineering	1509.0 Separated Laundry Disposal Systems	Yes	N/A	No
Maryland	Maryland Department of the Environment, Water Management Administration	Innovative and Alternative Program	No	N/A	No
Massachusetts	Massachusetts Department of Environmental Protection	310 CMR 15.000, Title 5: Innovative and Alternative Systems	Yes	10,000	No

TABLE 2.1 Existing Codes as of February 2010 (*Continued*)

State	Agency	Title/chapter	Residential rules?	Flow limits, gpd	Commercial rules?
Michigan	Michigan Department of Environmental Quality, Hearth and Water	Act 421 P.A, Acceptable Alternative Graywater Systems	Yes	N/A	Yes
Minnesota	Minnesota Pollution Control Agency/Water Quality	Chapter 7080.9010 Alternative/ Experimental	Yes	1000	Yes
Montana	Department of Environmental Quality	DEQ 4	No	N/A	No
Nevada	Department of Human Resources	R129-98 Sewage Disposal Administrative Code	Yes	N/A	Yes
New Jersey	Department of Environmental Protection	NJ Administrative Code 7:9A	Yes	75% of blackwater	Yes
New Mexico	Environment Department	House Bill 114	Yes	250	No
New York	Department of Health	10NYCRR Appendix 75-A	Yes	75	No
North Dakota	ND Department of Health, Environmental Health Division	Ch. 62.03-16-01 Individual Sewage Treatment	Yes	N/A	Yes
Oregon	Department of Environmental Quality, Water Quality Division	OR Administrative Rules Ch. 918 Division 790	Yes	N/A	Yes
Pennsylvania	Department of Environmental Protection, Bureau of Water Quality	Title 25 Environmental Protection Ch.73	Yes	N/A	Yes
Rhode Island	Department of Environmental Management	Ch. 12-120-002	No	N/A	No
South Dakota	Department of Environment and Natural Resources	Administrative Rule: ARSD 74:53 Water Supply and Treatment Systems. Specifically, sections 74:63:01:38 and 74:53:01:19	Yes	Minimum of 25gpd/ person	Yes
Texas	Texas Commission on Environmental Quality	Existing rules under 30 TAC, Ch. 285, Subchapter H. Currently working on a set of rules to fulfill HB2661	No	N/A	No
Washington	Department of Ecology	Publication No. 98-37 WQ; Section E1	Yes	N/A	Yes
Wisconsin	Department of Commerce	Chapter Comm 82, WI Admin Code; Specifically Comm 82.34, 82.40, 82.41, and 82.70	No	N/A	No

Bahman Sheikh, Ph.D., P.E., "White Paper on Graywater."

TABLE 2.2 Scope of NSF/ANSI Standards 350 and 350-1

NSF/ANSI Standard 350: On-Site Residential and Commercial Water Reuse Treatment Systems	
Building types	Residential, up to 1500 gallons per day Commercial, more than 1500 gallons per day and all capacities of commercial laundry water
Influent types	Combined black and graywater Graywater Bathing water only Laundry water only
Effluent uses	Nonpotable applications, such as surface and subsurface irrigation and toilet and urinal flushing
Ratings	Two classifications that vary slightly in effluent quality: • Class R: single-family residential • Class C: multifamily and commercial Systems are further described based on the type of influent (combined, graywater, bathing only, laundry only).
NSF/ANSI Standard 350-1: On-Site Residential and Commercial Graywater Treatment Systems for Subsurface Discharge	
Building types	Residential, up to 1500 gallons per day Commercial, more than 1500 gallons per day and all capacities of commercial laundry water
Influent types	Combined black and graywater Graywater Bathing water only Laundry water only
Effluent uses	Subsurface irrigation only
Ratings	Single effluent quality with no classifications Systems are further described based on the type of influent (graywater, bathing only, laundry only).

GRAYWATER BACKGROUND

Graywater reuse is viewed by the "green-leaning" layperson as the panacea for water shortages, groundwater depletion, surface water contamination, and all our other water woes. But graywater reuse is seen by many public health officials, as well as some water utilities, as a threat to the health and safety of the users themselves and their neighbors, particularly with regard to microorganism contaminants.

No cases of any disease caused by exposure to graywater have been documented, although systematic research on this public health issue is virtually nonexistent. While this does not prove that there has never been such a case, the fact remains that untreated graywater is wastewater with microbial concentrations far in excess of levels established in treated drinking, bathing, and irrigation water standards for recycled water. This water must be adequately treated if human contact is possible.

SOURCES

With local and regional water shortages, the use of graywater has pushed conservation to the forefront of people's minds. Graywater sources include the following:

1. Domestic water from residential use
2. Water from nonresidential use
3. Condensate from cooling activities (air conditioners, refrigerators, etc.)

Domestic Residential Graywater Sources

While we tend to think of *domestic activities* as those resulting from a single-family household, they can be expanded to include the activities from apartments, condominiums, multifamily houses, vacation homes and any residential dwelling. Domestic activities can also result from nonresidential locations: institutions such as schools and nursing homes; commercial establishments such as restaurants, hotels; and industrial facilities.

Given that the graywater is specifically from hand-washing, bathing and showering, condensate discharge, and laundry activities, virtually all residential dwellings should produce similar per capita quantities and qualities of water. The quantity and quality of wastewater generated in institutions, commercial establishments, and industrial facilities should be similar with regard to hand-washing activities. The presence and size of bathing and showering, laundry, and cooling/HVAC in other than domestic facilities will vary considerably from one location to another.

For collected graywater from lavatories, showers, laundry, etc., the health risk is quite low when it is used for subsurface irrigation, although a study of potential soil impact should be run prior to the installation of the system. If graywater is used for flushing fixtures or any other indoor application, one must ensure the proper disinfection and treatment of the graywater. The potential for contamination by human, animal, or food waste—and other substances—is very real.

Census data indicate that the average U.S. household size is 2.7 people, and an American Water Works Association (AWWA) study determined the median daily per capita flow rate to be 54 to 67 gallons/day (gpd) [204 to 253 liters/day (L/day)] for residential applications.

From one estimate, the profile of all indoor household wastewater generation per person per day in the United States is approximated in Table 2.3. Of the per capita wastewater

TABLE 2.3 Household Waste Generated per Person per Day

Wastewater source	Gallons per person per day*
Laundry	13.1
Bath	1.0
Shower	10.2
Hand washing	9.5
Toilet flushing	16.2
Dishwasher	0.8
Leakage	8.3
Other	1.4

*Based on average per capita flow rate of 60.5 gpd.

TABLE 2.4 Approximate Water Use of Standard Appliances

U.S. clothes washing machine (top-loading)	30 gal per cycle
European (front-loading) clothes washer	10 gal per cycle
Dishwasher	3–5 gal per cycle
Low-flow showerhead (per shower)	3–7 gal per average use
Other sink use (shaving, hand washing, etc.)	1–5 gal per average use

"generation," 8.3 percent is represented by leakage from broken pipes and defective faucets. In comparison, the approximation of water use from appliances per use is listed in Table 2.4. Not all water discharged from domestic sources is suitable for graywater use. Of all the water generated by domestic sources only a portion is suitable for use as graywater. This portion is listed in Table 2.5.

TABLE 2.5 Domestic Water Suitable to Use as Graywater

Wastewater source	Gallons per person per day	Percent of total
Laundry	13.1	38.8
Bath	1.0	3.0
Shower	10.2	30.2
Hand washing	9.5	28.1
Total	33.8	

Out of an estimated total of 60.5 gpd (2.35 L/day)/person generated from all domestic sources, Table 2.5 shows that a little more than one-half, or 33.8 gal (128 L), of the indoor household wastewater generated per person per day can be defined as suitable for graywater use.

Yet another source, IAPMO supplement Section 502, provides the following for domestic graywater generation:

1. The building occupants are calculated by assuming that the first bedroom has two occupants and each additional bedroom has one occupant. Other authorities have calculated that there are two occupants per bedroom.
2. The graywater volume generated by each occupant is calculated as follows:
   a. Shower and bathtub lavatory, 25 gpd per occupant
   b. Laundry, 15 gpd per occupant

The following precautions are recommended:

1. Graywater should never be used for direct consumption.
2. Graywater should not be used directly on anything that may be eaten.
3. Graywater should not be sprayed, allowed to puddle, or run off the property.
4. Use only water from clothes washing, bathing, or the bathroom sink. Do not use water that has come in contact with soiled diapers, meat or poultry, or anyone with an infectious disease.

TABLE 2.6 Nonresidential Potable Water Estimated Usage, Percent, by Building Type

	Hospitals	Schools	Hotels	Office buildings	Restaurants
Landscape irrigation	10	38	21	10	3
Restrooms, showers	30	43	23	39	30
Laundry	5	3	10	—	—
Cleaning, sanitation	5	1	12	—	—
Cooling, HVAC	30	4	5	37	2
Kitchen	5	6	15	2	50
Miscellaneous	15	5	14	12	13

Extracted from Mechanical Contractors Education and Research Foundation (MCERF), "Water-Efficient Technologies for Mechanical Contractors."

Whether its called *reuse*, *cascading*, or *recycle*, the use of water that has already been used in another application brings added challenges along with potential savings in water and cost. Water purification and decontamination systems are available that can be used for applications where potable water is required, such as drinking, washing, and showering. They are costly.

Nonresidential Graywater Sources

Given that most of the graywater is generated as a result of human activity, there should be little difference between residential and nonresidential graywater quality. However, the relative quantities of nonresidential graywater produced vary considerably as a function of the individual facility. Table 2.6 lists nonresidential estimated water usage by building type.

Industrial wastewater condensate from cooling activities is referred to as water from a "point source" and is site-specific. Potential sources from air conditioning include clean condensate discharge from HVAC cooling units, condensate from freezer and refrigeration units in food supermarkets, and other similar equipment. There are a number of manufacturers of systems designed to recover water from the humidity in the air by utilizing condensing technologies and cooling tower discharges.

Industrial wastewater represents an excellent source of recoverable water; however, wastewater qualities are so extremely variable, and the topic is so complex, that they are not discussed in this book. In addition, the ultimate utilization of most industrial recovered wastewater is that it is sent back into the process from which it was obtained. Normally, industrial wastewater is not permitted to be used.

Harvested Rainwater, Including Snowmelt

Harvested rainwater and snowmelt are discussed in Chap. 16.

WATER CONTAMINANTS

Although the specific chemical parameters in graywater result from the particular incoming water source (as well as from chemicals introduced during the activities), and can vary considerably from one household to another, Table 2.7 summarizes and describes the ranges of

TABLE 2.7 Water Contaminants by Class

Class	Examples
Suspended solids	Dirt, clay, colloidal materials, silt dust, insoluble metal oxides, and hydroxides
Dissolved organics	Trihalomethanes, synthetic organic chemicals, humic acids, fulvic acids
Dissolved ionics (salts)	Heavy metals, silica, arsenic, nitrate, chlorides, sulfates
Microorganisms	Bacteria, viruses, protozoan cysts, fungi, algae, molds, yeast cells
Gases	Hydrogen sulfide, methane, radon, carbon dioxide

all significant contaminants by classes. Although many general blackwater contaminants are close to those of graywater, the bacterial concentrations are 100 to 1000 times those of graywater, making them generally unsuitable except for unusual conditions.

Studies have found that all water supplies contain contaminants brought into domestic use that are too numerous to count. In addition, soaps, detergents, fabric softeners, and other products are introduced from hands, faces, bodies, and fabrics. Table 2.8 lists chemical contaminants and concentrations of contaminants in typical domestic graywater.

TABLE 2.8 Domestic Graywater Chemical Contaminants and Concentrations

Ingredient	Concentration, mg/L
Sunscreen	1,500
Moisturizer	1,000
Toothpaste	3,250
Deodorant	1,000
Na_2SO_4	3,500
$NaHCO_3$	2,500
Na_4PO_4	3,900
Clay	5,000
Vegetable oil	700
Shampoo or hand soap	7,200
Laundry detergent	15,000
Boric acid	140
Lactic acid	2,800

In the past, lakes and rivers were clean and supported a variety of plant and animal life. As the waters started to accumulate organic pollution from agriculture, industry, and other present-day practices, these primary pollutants decomposed into the water. The measurement of this pollution is the *biochemical oxygen demand* (BOD) and the *chemical oxygen demand* (COD). Secondary pollution, such as algae, started to grow due to the primary pollutants. The principal secondary pollutant elements are nitrogen, phosphorus, and potassium. Contaminants typically found in surface water before treatment are listed in Table 2.9.

For graywater use in landscape irrigation, Table 2.10 from ANSI/NSF 350 provides an excellent set of effluent contaminant levels acceptable for water used for irrigation. Class R is water from residential sources, and class C is water from commercial sources. Canadian guidelines for treated water used to flush toilets and urinals are listed in Table 2.11. NSF 350

TABLE 2.9 Significant Water Contaminants in Surface Water

Parameter	Unit	Range
BOD (biochemical oxygen demand)	mg/L	60–300
TSS (total suspended solids)	mg/L	30–185
E. Coli (bacteria)	CFU*/100 ML	80–30,000
Fecal coliform (bacteria)	CFU*/100 ML	50–100,000
P_{total} (phosphorus)	mg/L	1–15
N_{total} (nitrogen)	mg/L	4–30
pH (acidity/alkalinity)	units	5–10

*Colony-forming units.

TABLE 2.10 Acceptable Effluent Contaminant Levels for Irrigation Purposes

	Class R		Class C	
Measure	Test average	Single sample maximum	Test average	Single sample maximum
BOD_5, mg/L	10	25	10	25
TSS, mg/L	10	30	10	30
Turbidity, NTU	5	10	2	5
E. coli‡, MPN*/100 mL	14	240	2.2	200
pH, units	6.0–9.0	N/A†	6.0–9.0	N/A
Storage vessel disinfection, mg/L	≥ 0.5–≤ 2.5§	N/A	≥ 0.5–≤ 2.5³	N/A
Color	MR	N/A	MR¶	N/A
Odor	Nonoffensive	N/A	Nonoffensive	N/A
Oily film and foam	Nondetectable	Nondetectable	Nondetectable	Nondetectable

*Most probable number.
†N/A = not applicable.
‡Calculated as geometric mean.
§Values are for chlorine disinfectant residual only. Other disinfectants can be used, as described below, in which case appropriate measurements shall be made to ensure adequate disinfection.
¶MR = measured and reported only.
Source: ANSI/NSF 350

TABLE 2.11 Canadian Recommended Levels for Water Treatment Effluent Used for Toilet Flushing

		Water quality parameters	
Parameter	Unit	Median	Maximum
BOD*	mg/L	≤ 10	≤ 20
TSS†	mg/L	≤ 10	≤ 20
Turbidity†	NTU	≤ 2	≤ 5
Escherichia coli‡	CFU/100 mL	Not detected	≤ 200
Thermotolerant coliforms‡	CFU/100 mL	Not detected	≤ 200
Total chlorine residual§	mg/L	≤ 0.5	

*Unless otherwise noted, recommended quality limits apply to the reclaimed water at the point of discharge from the treatment facility or treatment unit. BOD_5 = 5-day biochemical oxygen demand; TSS = total suspended solids; NTU = nephelometric turbidity unit; CFU = colony-forming unit.
†Measured prior to disinfection point. Only one of TSS and turbidity needs to be monitored in a given system.
‡Only one of *Es. coli* and thermotolerant coliforms needs to be monitored in a given system.
§Measured at the point where the treated effluent enters the distribution/plumbing system.

TABLE 2.12 NSF Graywater Test Concentrations (Combined Laundry and Bathing Water)

Parameter	Required range
Total suspended solids (TSS)	80–160 mg/L
Five-day biochemical oxygen demand (BOD_5)	130–180 mg/L
Temperature	25–35°C
pH	6.5–8
Turbidity	50–100 NTU
Total phosphorous P	1–3 mg/L
Total Kjeldahl nitrogen N	3–5 mg/L
Chemical oxygen demand	250–400 mg/L
Total organic carbon	50–100 mg/L
Total coliforms	10^3–10^4 CFU/100 mL
E.coli	10^2–10^3 CFU/100 mL

Source: NSF/ANSI

uses an artificial influent supply using test water containing combined concentrations of combined bathing water and residential laundry water assumed to be in the test water. These concentrations are shown in Table 2.12. Recommended treatment levels for use in effluent contaminants for use in classes R and C are shown in Table 2.13.

For graywater use, California's graywater guide recommends that laundry detergents be selected to minimize the effect on plants. Specifically, liquid (not powder) detergents are recommended. Other substances to avoid include

1. Bleaches
2. Softener additives
3. Boron

TABLE 2.13 NSF 350 Recommendations for Water Treatment Effluent

	Class R		Class C	
Parameter	Overall test average	Single sample maximum	Overall test average	Single sample maximum
$CBOD_5$, mg/L	10	25	10	25
TSS, mg/L	10	30	10	30
Turbidity, NTU	5	10	2	5
E.coli,* (MPN/100 mL)	14	240	2.2	200
pH, SU	6–9	NA	6–9	NA
Storage vessel disinfection, mg/L†	≥ 0.5–≤ 2.5	NA	≥ 0.5–≤ 2.5	NA
Color	MR‡	NA	MR	NA
Odor	Nonoffensive	NA	Nonoffensive	NA
Oily film and foam	Nondetectable	Nondetectable	Nondetectable	Nondetectable
Energy consumption	MR	NA	MR	NA

*MPN = most probable number
†As chlorine. Other disinfectants can be used.
‡MR = Measured and reported only.
Source: NSF/ANSI

4. Borax
5. Peroxygen of sodium perborate
6. Petroleum distillate
7. Axylbenzene
8. Sodium hypochlorate

If a water softener is present, a demand-initiated regeneration" model should be used.

However, graywater is used for other nonpotable water uses such as equipment cooling, car washing, and various other purposes. Surface water quality is not generally as uniform as groundwater. This variability of surface water can be a challenge for industrial water applications.

The largest discrepancy and confusion in the industry stems from how to use alternative water sources. There is a large deviation in standards among various localities. Inspectors and code officials think it best to avoid using graywater because they are very conservative and feel it best to avoid the risk of something happening. Green building rating systems give points for the use of alternative water sources, but they do not show how to use the graywater correctly.

TREATMENT TECHNOLOGIES

To protect human health, enforceable regulations are in place that mandate the treatment, system design, installation, and operation. These regulations vary from one location to another.

The key to water reuse is to have the necessary treatment technologies available to remove the pollutants efficiently to an acceptable level and to completely remove those that are undesirable. There is no single technology that will sufficiently and economically remove all classes of contaminants, although there are treatment technologies that collectively are capable of effectively reducing the concentration of virtually any contaminant down to acceptable levels for any water reuse requirement, or to meet any standard. The largest problem is cost.

Most graywater is easier to treat and recycle than blackwater because of the lower level of contaminants. All graywater contains microorganisms, and the health hazards associated with graywater from multiple dwellings should be considered differently from those obtained from a single dwelling unit.

Basic Suspended Solids Removal

Turbidity is a general term used to describe the optical opacity of water containing any form of insoluble suspended solids matter. Different methods are used to remove suspended solids, including bacteria, and dissolved solids. The engineer is required to know the desired quality of the water to be reused, the flow rate, and the daily volume of water expected in order to determine the type and size of the filtration system. It is generally required that a chlorine residual be maintained in any public building. Treatment methods are outlined below.

Settlement Tanks. In a settling tank, solids and large particles will settle to the bottom, while grease, oils, and small particles will float. The remaining liquid will be further filtered before being reused. A settling tank also allows hot water to cool before reuse. The tank should be large enough to hold twice the expected daily flow plus 40 percent, to allow for sludge accumulation and surge loading. A tank is specifically designed to allow settling; but do not confuse the tank used to treat graywater with the conventional use of a septic tank.

Graywater coming out of a storage tank contains little or no oxygen. Graywater from an aerobic tank will contain more oxygen, which is better for irrigation purposes. An electrical pump or aerator added to a tank can create an aerobic environment. Aerobic conditions allow some decomposition of wastes in the tank, and may help minimize sludge buildup and blockages in the system. Aerobic tanks will need to be pumped out every 3 to 5 years for removal of sludge.

Several types of tanks may be suitable for settling or storage of small quantities of graywater. In addition to metal, polyethylene, fiberglass, or wooden tanks that are commonly used, consider using plastic garbage cans or 55-gal drums for domestic, residential use.

Graywater Intended for Cartridge Filters

The type of filter required for a graywater system depends largely upon the amount to be filtered and the type of contaminants present. A drain filter is an easy and inexpensive way to filter lint and hair out of bath or laundry water. A simple cloth bag tied over the end of a hose or pipe may be sufficient for irrigating outdoor or similar applications.

A cartridge filter is inserted into a housing. The filter itself could be any thin, porous material, constructed of plastic, paper, or membrane-like material, with a pore size selected capable of capturing the size of desired suspended solids inside the filter. Filter cartridges are generally of the wound or pleated type. The wound fiber type is used most often. It is intended to be replaced when the pressure drop through the filter exceeds approximately 10 psig (78 kPa). Offered in many different designs and capable of solids removal down to the submicrometer level, filter cartridges provide a wide variety of excellent choices to the knowledgeable design engineer. They are typically used at flow rates of 5 gallons per minute (gpm) (14 L/min). An illustration of a typical filter cartridge and housing unit is given in Fig. 16.2.

Media Bed Filter. These large filters consist of housings where a single-medium or multimedia material, such as sand, anthracite, activated carbon granules, or garnet, is added. This material captures suspended solids inside the bed until it is taken off line and backwashed. Often, piping arrangements will allow duplex units to be backwashed separately while the other is in operation. After backwashing, the media bed could be reused. The discharge is routed to a sewer. This type of filter is usually used in the range of 5 to 20 gpm (78 to 156 kPa). Refer to an illustration of a granular carbon filter in Fig. 4.3.

Carbon and Ceramic Block Filters. Similar in nature to the cartridge filter, the medium in the carbon or ceramic block filter consists of specially prepared activated carbon in the housing or solid porous ceramic material. The advantage of carbon block cartridge filters is that they will adsorb organic contaminants and certain gases. Ceramic cartridges can be cleaned and reused, but the activated charcoal filters must be replaced. They are typically used at flow rates of 5 gpm (14 L/min).

Membrane Technologies

Membrane technologies is a general term for removing contaminants from feedwater by means of a thin, porous barrier called a membrane. It is based on a process known as *cross-flow filtration* that allows for continuous treatment of input liquid streams. In this type of process, the pressurized raw water inlet flows parallel to a porous semipermeable

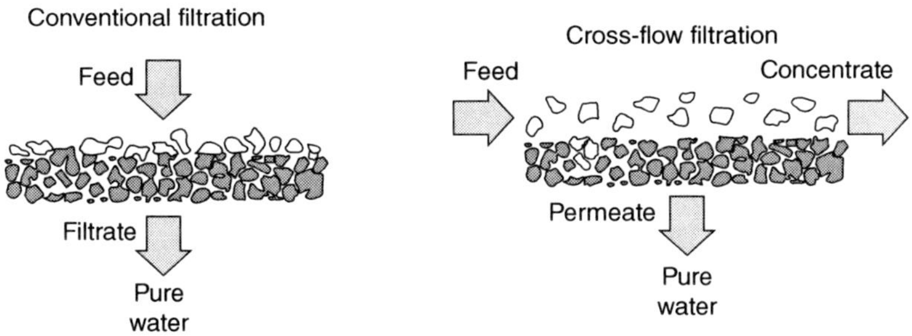

FIGURE 2.1 Conventional versus cross-flow filtration.

membrane filter. Because this system is pressurized, water is forced through the filter membrane. The clean permeate passes through the semipermeable membrane, and the unwanted particles remain in the concentrate stream that is discharged to the sewer. The difference between conventional flow and cross-flow operation is illustrated in Fig. 2.1. For a complete configuration of the reverse-osmosis operation, refer to Figs. 4.13 to 4.16.

Four pressure-driven membrane technologies are in use today: They are microfiltration (MF), ultrafiltration (UF), nanofiltration (NF), and reverse osmosis (RO). These technologies behave differently from regular filters in that the feed stream is continuously pumped at a high flow rate across the surface of the filter membrane. A portion of this stream is forced through the membrane, producing clean water (permeate), and the remainder (concentrate) has the undesired particulates. The concentrated contaminant remaining in the other stream exits the housing containing the membranes on a continuous basis and is discharged to the sanitary system. Because the concentrate stream is continuously removing contaminants, this technology requires only occasional backwashing and cleaning.

Microfiltration is the most often used technology designed for suspended solids removal. There are systems available to remove particulate contaminants down to the submicrometer range, including bacteria. Properly designed and specified, they are capable of operating at virtually all flow rates.

Dissolved Organics Removal

Activated Carbon. Activated carbon removal utilizes a specially prepared granular carbon medium capable of absorbing dissolved organic contaminants and certain gases. It is very effective in removing many taste and odor contaminants, including chlorine, and is usually installed in housings similar to media or cartridge filters. The activated carbon usually requires replacement once or twice a year, depending on service.

Special resin adsorbents are also available for organics removal. They are designed for a particular removal function, such as humic acids, and require regeneration and/or replacement.

Ultrafiltration. Ultrafiltration is a membrane technology with smaller pores than MF, and UF is capable of removing dissolved organics. Instead of adsorbing the contaminants, it continuously removes them as part-of-the concentrate stream.

Dissolved Salts Removal. Most graywater contains relatively high concentrations of salts, both from the incoming water supply and from soluble contaminants resulting from activities inside a facility. These can include both benign and potentially hazardous compounds. The most practical technology for salts reduction is RO (or possibly NF). These technologies are often designed to be used at a point of use, usually at a single tap point. Point-of-use systems are in common use throughout the United States today.

Disinfection. Microorganisms are bacteria and viruses. Disinfection can be achieved by removal using a filter with the proper pore size (0.2 μm), destroyed by disinfectant or at least inactivated to keep them from reproducing by maintaining a residual of disinfectant. This is usually accomplished by chlorination, filtration, ultraviolet (UV) radiation, or ozone generation. An important consideration is that disinfection material must remain as a recommended residual for private homes, and it is generally mandatory when graywater is used for public buildings.

Two chemicals used to disinfect water are chlorine and iodine, with chlorine being more common. Not only is it readily available (as household liquid bleach or at swimming pool supply houses) and relatively inexpensive, but also it is stable in storage and will, in time, vaporize from the water after disinfection. Organic material in graywater may combine with chlorine and reduce the amount available for disinfection. For this reason, a filter or settling tank before the disinfection point may be advisable.

One method of disinfection is to use ozone for water entering a storage tank followed by feeding a low concentration of sodium hypochlorite (bleach) to maintain a residual of 0.3 to 0.5 mg/L free chlorine. Another allows the use of chlorine which requires contact time and the establishment of residual chlorine. A range of 0.5 to 1.0 ppm is a generally acceptable residual for typical water. Several devices are available commercially that dispense appropriate amounts of iodine or chlorine (in solid or liquid form) to a graywater system.

Ozone will inactivate all microorganisms much more effectively than chlorine and will also break down dissolved organic molecules to a certain extent. This allows activated charcoal to be more effective.

Many experts favor a multibarrier approach using disinfection in conjunction with a separate technology such as UF.

Gases

In general, graywater does not contain noxious gases. It has been found that bacterial action in water that is stored may produce gases such as hydrogen sulfide that is both unpleasant and dangerous. This possibility makes it imperative that the gases be removed. Activated carbon adsorption is usually effective in removing these gases.

TREATMENT THAT RESULTS IN POTABLE WATER

With purification, there are filtering methods for the water to become potable or near potable. Numerous systems are based on the following principles:

1. Mechanical systems include sand filtration, lava filtration, and systems based on UV radiation.
2. Biological systems include constructed wetlands and treatment ponds, biological reactors, activated sludge systems, bioreactors, aerobic and anaerobic biofilters, and submerged aerobic filters.
3. Direct "hard" mechanical processes include distillation, membrane filtration, and reverse osmosis.

4. To purify graywater into potable water, several of the above systems are usually combined to work as combinations in two or three stages. Sometimes tertiary purification is also added. Some municipalities recycle a certain amount of graywater using a high standard of treatment; that graywater is utilized for irrigation and other purposes.

APPLICATION

Regardless of the water source, or its degree of contamination, it is technically possible to have the water suitably treated for virtually any application, including potable. It is often not desirable or cost-effective to make water completely free of any existing contaminant outside of that used for laboratory use. There are such a variety of treatments available for use that, it is strongly recommended that a competent individual or company be hired to assist in the design of the treatment system. For larger nonpotable systems, refer to Fig. 16.7.

When one is investigating graywater reuse possibilities, the following fundamental questions must be answered:

1. What specific contaminants, and their concentrations, can be expected from the source?
2. What is the estimated quantity of graywater from each fixture and building type?
3. What are the potential reuse possibilities in a given application?
4. What treatment technologies (if any) will be required?
5. What will be required in the way of infrastructure (storage, distribution, etc.)?
6. What are the design engineering requirements?
7. What are the regulatory requirements?
8. What are the economic factors?

Domestic graywater can be recycled directly within the home, garden, or company and used immediately or processed and stored. Recycled water of this kind is never safe to drink, but a number of stages of filtration and microbial digestion can be used to provide water for washing or flushing toilets operated with flush valves. It is important not to use graywater in flush tanks as there may be stoppages in the flushing mechanism. Some graywater may be applied directly from the sink to the garden field, receiving further treatment from soil life and plant roots. Given that graywater may contain nutrients and pathogens and is often discharged warm, it is very important not to store it before use in irrigation unless it is properly treated first.

Plant specialists warn that graywater, when used for irrigation, should not be used on vegetables, seedlings, container plants, or acid-loving plants such as azaleas and rhododendrons. Graywater should be rotated with freshwater to leach out any harmful buildup. Chlorine bleach may damage plants, especially if it touches the foliage. Biodegradable soaps appear to have the least harmful effects.

The simplest use of a graywater system without purification is to divert the water directly to a garden. Regulations are very different between county and region, but common guidelines for safe usage include not storing graywater for more than 24 h, ensuring that it cannot pool. It is essential that graywater be diverted to the sewer when garden-unfriendly products are being used.

Water recycling without purification is used in certain agricultural applications, such as tree nurseries, and dwellings for application where potable water is not required, such as for garden and land irrigation and toilet flushing. It could also be used in dwellings where graywater is fairly clean to begin with and has not been polluted with nondegradable chemicals such as nonnatural soap. Graywater from the shower or bath is generally good-quality water for the garden. The soap dilution levels are actually good for the garden, and they are a wetting agent as well. When laundry graywater is diverted to the garden, the laundry products used must be carefully chosen to ensure that phosphate and salt levels are low and the pH level is neutral. The engineer has to make this determination with consultation from authorities.

Graywater Diversion Systems

Graywater diversion systems can be both designed for new homes and retrofitted to many existing homes for flushing toilets. When systems are fully designed, manufactured, and installed to relevant standards, the diversion systems tend to be highly efficient, effective, and safe for simple applications where potable water is not required. Diversion systems can be as basic as running outlet hoses from a washing machine out of a window and out to a garden, or they can be designed as a permanent part of a building plumbing system. Fully engineered systems incorporate a sump pump and surge tanks to deliver the water for subsurface irrigation.

VALVES IN GRAYWATER USE

The use of treated graywater has been found to have high levels of disinfectant chemicals that will degrade conventional materials. For example, rubber compounds, when used for strictly potable water, would be expected to last for many years, but are now subjected to new exposure issues. Stainless steel, copper, brass, and many plastics will also be vulnerable.

In the past, the valve-damaging component was the increased level of chlorine as a disinfectant. The modern treatment uses chloramines as a more persistent chemical that does not break down as rapidly. Elastomeric components are especially susceptible to damage. The water attacks the cross-linked structure, and cracks appear at the surface.

Users should watch carefully for any signs of corrosion, leaks, or any changes in treatment. An inspection program should be implemented. Finally, users should communicate findings of any in-place service issues to the manufacturer.

PLANNING A SYSTEM

The following items should be evaluated when planning a new system:

1. Make certain that a graywater system is useful for the actual situation expected.
2. Make certain that the system is applicable for the house (based on the number of bedrooms) or project regardless of who occupies the house or facility.
3. Take an inventory of the graywater sources and the number of uses that the occupants have for them.
4. Determine how many gallons (liters) that appliances use per cycle and that a fixture uses per day, using previously discussed water use tables.

TABLE 2.14 Design Considerations

A. Overall site plan	B. Regulatory requirements

• Topography
• Slope
• Soil conditions
• Ledge
• Groundwater
• Buildings
• Utilities
• Property lines
• Vegetation
• Wetlands
	• Plumbing code
	• State regulations
	• Reuse Regulations
	• Effluent discharge limitations
	  • BOD
	  • NO_3
	  • *E.coli*
	  • Other
	• Land application regulations
	• Permit question
	  • Local board of health
	  • State environmental regulations
	  • Building code requirements
	  • Monitoring Requirements
	  • Test data for approvals

C. Design information	**D. New system specifications**

• Existing treatment facilities
  • Septic tank
  • Leach field
  • Cesspool
  • Other
• Influent quality and quantity
  • Number of bedrooms
  • Number of persons regularly in residence
  • Occupancy (i.e., on a year-round or seasonal basis)
• Type of appliances and Fittings
  • Flow rates
  • Dishwashers
  • Washing machines
  • Bathtubs, showers
• Evaporation rates
• Temperature data
• Rainfall data
• Effluent reuse goals
	• Propose treatment train
	• Technical specifications
	• Effluent treatment character—typical transformation
	• Reduction in pollution
	• Production of fertilizer
	• Reuse opportunities
	• Optimum use of existing Facilities
	  • Existing pipes
	  • Dispersion beds
	  • New use of existing Equipment
	• Potential for savings
	  • Water
	  • Fees—water and sewage
	• Overall improvement evaluation

5. Table 2.14 is provided for reference which indicates most design considerations. Not all considerations will be applicable for all individual projects.

6. The most important first consideration is to consult the local authorities.

BIBLIOGRAPHY

American Society of Sanitary Engineering, "Reclaimed Water: Sources and Treatment Methods," *Plumbing Standards*, May 2010.

Bruursema, T., "Keeping It Sustainable," *NSF Magazine*, April 2009.

Cartwright, P. S., "Graywater Systems and Design," Paper delivered at 2010 American Society of Plumbing Engineers (ASPE) Convention, Tampa, FL, October 28.

NSF International, techstreet.com/nsfgate.html.

New Mexico State University, "Safe Use of Household Graywater," January 2008.

Phillips, N., "Graywater: Reduce, Reuse, Recycle," *Water Technology Magazine,* May 2011.

Reynolds, R., "Quantifying Quality," *PM Magazine*.

Strutzenberg, R., "Engineering Challenges of Reclaimed Water," *Valve Magazine*, September 2011.

CHAPTER 3
GROUNDWATER MONITORING WELLS

Groundwater monitoring wells are required to detect contamination in groundwater. The objective of groundwater monitoring is to collect data that most accurately represent the chemistry of the water in the aquifer being monitored. Typically, groundwater is monitored because it must be done to comply with federal or state regulations. Monitoring is often required as a means to ensure compliance with operating permits, with sampling systems employed for laboratory analysis. Polluters face extremely large cleanup costs as well as legal costs associated with noncompliance. It is essential that sampling methods stand up to any challenge. It has been estimated that many dollars have been wasted each year on costly samples that have been biased by the sampling equipment or process used.

Methods for recording defensible groundwater samples are well documented although not widely known. This has resulted most often through lack of knowledge rather than by design. Two of the crucial elements for obtaining representative groundwater samples involve the manner in which the samples are lifted and the materials to which the samples are exposed. Where the monitoring requires great accuracy, improper lift devices, materials, and procedures can physically alter or contaminate the water samples. In this regard, selection of appropriate sampling equipment is critical to the success of the monitoring program. Appropriate sampling equipment is comprised of material that remains inert in the presence of waterborne chemicals. The most accurate of the samplers is the permanent pneumatic bladder pump. This pump uses pneumatic pressure to squeeze the water in the internal bladder, forcing water out through the discharge line. It never contacts the air that might contaminate the sample. The electrical, submersible pump, on the other hand, vaporizes water for a brief time and alters the sample chemistry. The EPA recommends the use of SS 316 for overall chemical inertness.

It is very unlikely that the design engineer who reads this book is knowledgeable enough to design a complicated monitoring system that requires extreme accuracy. A competent firm should be retained to do this. A less complicated and less accurate method should be used as an alternative to obtain samples that will discover the existence of and identify potential chemicals in the groundwater. In its simplest form, the method would use a portable sampling system such as a hand-lifted bailer. This consists of a sample collector that is nothing more than a special can lowered into the well. The can is lifted to the surface with the sample which is transferred into a sterile container and delivered to a laboratory. This sample only takes water from the surface of the aquifer. Because the integrity of samples depends in large part on well construction, the drilling methods and materials used in the construction must have no influence on the later quality of groundwater samples brought to the surface.

CODES AND STANDARDS

1. ASTM D 5092.
2. American Society for Testing and Materials, Standard Guide for the Development of Groundwater Monitoring Wells in Granular Aquifers.
3. Resource Conservation and Recovery Act, Ground Water Monitoring Technical Enforcement Guidance Document.
4. Environmental Protection Agency.

DRILLING METHODS

The most critical aspect of the drilling method selected is that it will not have any effect on groundwater quality near the borehole. The drilling methods must not use liquids. Cable tool, dual-tube, and reverse circulation air rotary methods are acceptable. A typical monitoring well is illustrated in Fig. 3.1. Another method of well drilling is to use direct-push drilling where a hydraulically powered hammer pushes a drill directly into the soil.

CASING

The material selected for the casing must be structurally strong, easy to handle, and durable enough to resist the long-term subsurface environment. Durability depends on groundwater and soil chemistry.

The casing material must be capable of resisting any degradation, adsorption, and reaction to the possible contamination products being measured. If it is for water only, it should not contribute any chemicals of its own.

Polyvinyl chloride (PVC), polyethylene (PE), and other plastic materials are structurally strong and resistant to normal groundwater conditions, but should not be used when solvents might be encountered. Fluoropolymer materials such as Teflon are chemically inert, but their high cost and low tensile strength make them poor choices. PVC is used most often because solvents are rarely encountered. Stainless steel type 316 is also considered acceptable for many applications.

GROUT

The most common materials are straight (neat) cement or cement and bentonite mixtures. A surface seal of cement must be placed so that surface runoff is prevented from running down the well.

INTAKES

If the intake screen and filter pack allow sediment into the well, increased sampling time and cost will result because of the additional filtering required. Factory-slotted casings and continuous slot wire-wound intakes are preferred. For formations less than 50 ft (15 m),

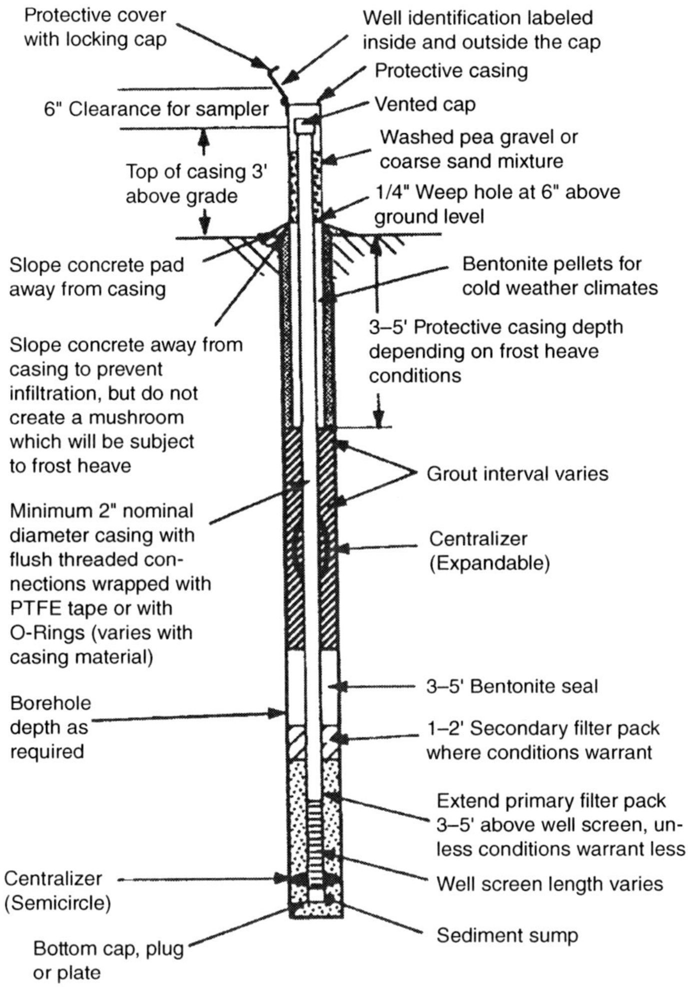

FIGURE 3.1 Typical monitoring well components.

screen lengths are recommended. For larger formations, proportionally longer lengths can be used. Filter packs generally consist of medium sand to fine gravel with a thickness of 2 to 3 in (50 to 75 mm) surrounding the intake. They should extend about 3 ft (1 m) above the intake to allow for settlement.

WELL DEVELOPMENT

The most effective and popular development technique is surging, although hydraulic jetting and overpumping can be used. Well development is not effective in clay formations or in wells that straddle the water table, such as those drilled to monitor petroleum hydrocarbons.

FINISHING THE MONITOR WELL

A surface seal of cement or concrete may be required by local codes to extend some distance beyond the borehole. Two types of well completion, sometimes called *finishes*, are common: the aboveground and flush-to-ground completions. In the aboveground type, a protective casing is installed around the well casing. It should be either painted or manufactured from a corrosion-resistant material. It is important to drill a drainage hole in the protective casing to drain any water that might accumulate and possibly freeze in colder climates. If no enclosure is installed around the monitor well, bumper guards are recommended if the well is in an area well traveled by vehicles.

For a flush-to-ground completion, a small protective box the size of a small meter is set into the casing with the top flush with the grade. This type is typically used in urban areas, parking lots, and streets where an aboveground completion is impractical. It must be watertight.

BIBLIOGRAPHY

ASTM D 5092.

American Society for Testing and Materials, Standard Guide for the Development of Groundwater Monitoring Wells in Granular Aquifers.

Environmental Protection Agency.

Resource Conservation and Recovery Act, Ground Water Monitoring Technical Enforcement Guidance Document.

CHAPTER 4
WATER TREATMENT

GENERAL

Where there is an abundant source of surface water adjacent to a facility, the location of that water may make it economically feasible to be used for various purposes in the facility. For this water to be used, information will be furnished that an engineer requires to make an informed decision about the presence and removal of impurities, which will result in selection of the treatment appropriate to provide the quality of water necessary for end use. This chapter describes the following:

1. Filtration and straining of water to remove suspended solids
2. Various water treatment methods used to remove and replace various impurities
3. Water-conditioning additives used to neutralize impurities and inhibit corrosion to a level at or below limits desired by the end user
4. Water softening for potable use
5. Removal of natural radiation

Information is presented on basic water chemistry, impurities found in water, water analysis, and impurity measurement. This information includes treatments that may be used for a variety of water uses, but not all may apply to any single project. Water treatment for this handbook is intended to include only the removal and/or replacement of undesired impurities in water with more desirable impurities. A brief description of water softening for potable water is given.

Potable water supplied from municipal sources and municipal treatment of wastewater is outside the scope of this handbook. Water treatment for graywater is discussed in Chap. 2. Water treatment for stormwater that is to be reused is discussed in Chap. 16.

Definitions

1. *Filtration* means the separation, retention, and removal of suspended and colloidal particles by mechanical capture and adsorption from fluids by passage through a porous medium. The resultant water is not pure. In very general terms, a filter is used to remove and retain particles 45 μm (325-mesh) and smaller. Filters retaining material below 1000 daltons are called *nanofilters*. Often, different manufacturers' nomenclature of filter categories becomes blurred.

2. *Membrane filters*
   a. Membrane filters use a semipermeable membrane for removing solutes, such as dissolved ions from a solution. Reverse osmosis (RO) is a continuous, steady-state process that uses a semipermeable membrane which allows the product water to diffuse, or pass through, but is impervious to the waste solute that is discarded. Membrane-type filters are generally selected for particles between 45 and 10 μm.
   b. In general, ultrafilters and nanofilters are membranes used to concentrate and purify fluids and to remove particulate contaminants, and microfilters are used to clarify a solution for applications where quantitative retention is not required. In addition, membrane filters are used to sterilize various solutions that cannot use heat due to the loss of biological activity after exposure to elevated temperatures.
   c. The membrane filter is still the most effective in the separation process referred to as *microfiltration.* When the size of particulates falls below about 0.02 μm, the removal process is called *ultrafiltration.* The molecular weight of any compound is measured in daltons. 1 dalton = 1/12 × mass of a carbon atom. Ultrafilters retain material ranging in size from 1,000,000 to 1000 daltons while allowing water to pass through. Some filter media measure passage through the filter by molecular weight for separation of one compound from another.
3. *Depth filters* can be of two types, granular and preformed. Preformed depth filters are composed of fibrous or sintered materials that have a random pore structure. Granular depth filters have either a graded or a consistent density of granular media and typically are long. Graded granular filters have layers of media that become progressively denser through the matrix as water flows through them. Constant-density granular filters have the same size filter media or openings throughout the matrix. Depth-type filters are generally used to remove particulate sizes ranging between 10 and 0.02 μm.
4. *Precoat (septum) filters* generally use diatomaceous earth as the filter medium. They are depth-type filters used to clarify water when the solids concentration is low.
5. A *screen filter* is a single, thin layer of a material that has a symmetric arrangement of openings or passages called *pores*. These pores trap all particles larger than the pore size on the surface of the filter. This process is called *sieving*, or *size exclusion*, and is the classic filtration method. Screen filtration is essentially absolute because any particle larger than the pore size cannot pass through. Examples of screen filters are woven metal, nylon, and Dacron mesh. Cast polymeric membranes are used where the smallest pores are required for submicron and macromolecular separations.
6. A *surface filter* is thicker than the screen filter and constructed from thick or multiple layers of filter media, often glass or polymeric fibers. When the water passes through a surface filter, particles larger than the spaces within the fiber matrix are retained, primarily on the surface. Smaller particles are trapped within the matrix, giving this type of filter the properties of both a screen filter and a depth filter.
7. *Granular filters* are larger units generally used to remove suspended particles larger than 10 μm. Examples of granular filter media are single-medium or multimedia sand and activated charcoal contained in a vessel, tank, or column and also septum filters. Deep-bed filters are designed to remove coarse suspended particulates larger than 10 μm. They are pressure-type filters that use either multigraded sand or multimedia in the filter. Particulate removal on the order of 98 percent should be expected.
8. *Bridging* is a condition that occurs as particles captured by direct interception form a particle mat across the filter medium. By partially blocking the filter pores, this bridge or filter cake may produce a smaller filter pore structure that will aid in particle capture.

9. *Cartridge filters* are relatively small and generally used to clarify a previously filtered stream of water containing suspended particles smaller than 10 μm. When the media are plugged, the cartridge is replaced. Another type of cartridge filter is the capsule filter, in which the filter media are contained in a sealed housing. When plugged, the entire capsule is discarded. Commonly used materials for cartridge filters are paper, cloth, polymetric fibers, and various combinations of these.
10. A *hydrophilic filter* is one that has an affinity for water; it can be wetted with almost any liquid. A *hydrophobic filter* is one that cannot be wetted by an aqueous solution.
11. *Adsorption* is the attraction to and adherence of a particle to the surface of the filter medium. Adsorption can occur even if the pore is larger than the particle. This attraction is due to a variety of surface chemical forces between the particle and filter medium.
12. *Dezincification* is a condition associated with brass alloys in which zinc ions diffuse out of solution, leaving porous copper behind.
13. *Stress crack corrosion* is a process in which the formation and propagation of cracks are accelerated by tensile stresses in a chemically reactive environment.
14. *Raw water* is the untreated water obtained from the initial source. Potable water and surface water could be considered raw water for producing pure water applications.
15. *Surface water* is water that comes from sources such as lakes and rivers and as groundwater when it originates from streams or other underground aquifers.
16. Calcium and magnesium are the hardness twins, and anything said about one applies to the other. Hardness was the invention of the softener industry. Water dissolves virtually all minerals contacted as it passes through the soil.
17. *Water conditioning* is intended to mean the addition of chemicals to water for the purposes of inhibiting corrosion and neutralizing undesirable impurities.
18. *Sieving* can also be referred to as *screening* or *straining*.
19. Alpha radiation is low level, capable of being stopped by skin or paper.
20. Beta radiation is comprised of fast-moving particles and can be stopped by thin aluminum.

Initial Filtering

If the supply is from surface water, a coarse or fine screen is usually placed at the intake to keep out fish and other large debris. Coarse screens are usually ½-in-diameter (13-mm) bars with a clear opening of 1 to 3 in (25 to 75 mm). Fine screens could have openings of approximately 3/8 in^2 (6 mm^2), with the water velocity through the screen limited to about 2 ft/s (0.6 m/s). If the quantity of water is small enough, basket strainers can be used. In climates where freezing may occur, the inlet should be placed far enough below the low water level to prevent freezing.

Clarification

After initial filtering, clarification is required to obtain water that meets certain standards for proposed use. The selection of the clarifier is based on the volume of water to be treated and the final quality desired. If the volume of water is small and the raw water is not very turbid, filters may be used.

Biological Control

To control microorganism fouling of the system, they must be destroyed, if possible, or at least inactivated (to keep them from reproducing) and then removed from the water stream. This is usually accomplished by chlorination, filtration, uv radiation, ozone generation, and special adsorbents.

Chlorination is the least costly and most common method. The action of chlorine requires a reasonable contact time and the establishment of a residual chlorine amount. A range of 0.5 to 1.0 ppm is a generally acceptable residual for water typically used.

CODES AND STANDARDS

The codes and standards that apply to various systems include these:

1. Potable water treatment shall comply with the 1986 Safe Drinking Water Act and Amendments.
2. There are various standards for nonpotable water based on ultimate use.

BASIC WATER CHEMISTRY

Water to be treated is known variously as *raw water*, *feedwater*, or *source water*. Water that has been treated is known as *treated water*, *product water*, or *solute*. Impurities that dissolve in water are called *electrolytes*. The dissolved impurities dissociate (separate) to form negatively and positively charged atoms called ions. The positively charged atoms are called *cations* because they migrate to the cathode electrode; the negatively charged atoms are called *anions* because they migrate to the anode electrode. The ions in solution act almost independently. For example, magnesium sulfate dissociates to form positive magnesium ions and negative sulfate ions. Ions are often referred to as *salts* in reverse-osmosis water production.

When a compound dissolves in water, the molecules of the compound separate, disperse among the molecules of water, and then are held in suspension. When the solubility limit is reached, those same compounds become suspended solids because it is no longer possible for them to be held in suspension. There is no chemical reaction. Dissolved materials cannot be removed by filtration.

Some types of liquids cannot be dissolved, such as oil. They break down into extremely small particles and then disperse into water, but do not dissolve. These liquids are called *immiscible liquids*, and the resulting small particles are known as *colloids*. Colloidal material is composed of suspended particles at the upper end of the size range for ions and molecules, from 0.001 to 1.0 μm. Other liquids, such as oil and grease, cannot separate into smaller particles and become dispersed, but simply remain in suspension. If they are lighter than water, they float on top; if heavier, they sink below the surface.

All acid compounds in water chemistry consist of hydrogen combined with an acid radical. Since the acid radical moves around as a unit, it is convenient to view an acid radical as an integral anion unit. When a metal radical and an acid radical combine, they form a class of chemicals called *salts*. When a metal cation and a hydroxide anion combine, a *base* results.

The valence of any element is the measure of its chemical combining power compared to that of a hydrogen atom, which has the assigned valence of 1. An element with a valence of 2 can replace two hydrogen atoms in a compound. When atoms combine to form compounds, a cation atom must combine with an anion atom; in addition, the valences of the two basic elements must be equal in order to form the compound.

Most of the basic chemical reactions in water treatment consist of rearranging cation and anion atoms using their valences. As can be seen in Table 4.1, hydrogen has a valence of 1 and sulfate has a valence of 2. When combined to form sulfuric acid, two hydrogen atoms are required to form the combination, resulting in the formula H_2SO_4. The chief exception is the case where carbonates and bicarbonates are destroyed by heating or aeration, giving off carbon dioxide.

The term *equivalent weight* is the weight, in pounds (kilograms), of any element listed that could combine with 1 lb (2.2 kg) of hydrogen. Since the valence of an element is proportional to its combining power, the equivalent weight is based on its valence and is shown in Table 4.1.

WATER IMPURITIES

Natural or raw water is never pure. Water picks up impurities falling through the air, when it comes into contact with the ground or, when percolated through the earth, minerals found in the soil. It also contains dissolved gases and dust picked up by rain, snow, and hail or by surface water in contact with the air above the water level.

Suspended Matter (Particulates)

Turbidity. *Turbidity* is a general term used to describe the optical opacity of water containing any form of insoluble matter, or suspended solids, suspended in water. *Color* is often used to describe turbidity and may be used when referring to water containing decaying vegetation. However, turbidity is used most often when referring to mineral particulates, such as silt, that are usually the most plentiful substances in the water. Other commonly occurring impurities are liquids, such as oil, and the residue caused by decaying vegetation. Coarse particles that settle rapidly when water is standing are referred to as *sediment*; fine particles that remain in suspension are called *silt*.

The mechanical properties of the particles suspended in the water stream must also be considered. At one extreme are solid, undeformable particles such as sand or quartz, and at the other extreme are gelatinous or deformable materials such as synthetic colloids and bacteria. Because they can deform, they are more likely than hard particles of the same size to pass through a filter.

Microorganisms. Microorganisms are bacteria and viruses. They are living forms of particulate matter. Their unusual physiology allows them to grow and multiply in water containing only trace levels of nutrients. The presence of these nutrients in untreated water is an indicator of the presence of microorganisms (if the temperature is favorable for their growth). Although microorganisms are suspended solids, the treatment required for their removal or neutralization puts them in a separate category.

Pyrogens cause fever, and pathogenic organisms (such as *Legionella*) cause diseases of any kind. Endotoxins, which are fragments derived from the cell walls of gram-negative bacteria, are considered the most important and widely occurring group of pyrogens. Other organic growths include algae (a primitive form of plant life), fungi (plants that lack the chlorophyll required for photosynthesis), and bacteria that exhibit both plant and

TABLE 4.1 Important Elements, Acid Radicals, and Acids in Water Chemistry

Element	Symbol	Atomic weight	Valence	Equivalent weight
Aluminum	Al	27.0	3	9.0
Barium	Ba	137.4	2	68.70
Calcium	Ca	40.1	2	20.05
Carbon	C	12.0	Variable	—
Chlorine	Cl	35.46	Variable	35.46
Fluorine	F	19.0	1	19.0
Iron (ferrous)	Fe″	55.8	2	27.9
Iron (ferric)	Fe‴	55.8	3	18.6
Hydrogen	H	1.0	1	1.0
Magnesium	Mg	24.3	2	12.15
Nitrogen	N	14.0	Variable	—
Potassium	K	39.1	1	39.1
Oxygen	O	16.0	2	8.00
Phosphorus	P	31.02	Variable	—
Sodium	Na	23.0	1	23.0
Sulfur	S	32.0	Variable	—
Silicon	Si	28.06	4	7.01
Acid radicals	**Formula**	**Molecular weight**	**Valence**	**Equivalent weight**
Bicarbonate	HCO_3	61.0	1	61.0
Carbonate	CO_3	60.0	2	30.0
Chloride	Cl	35.46	1	35.46
Nitrate	NO_3	62.0	1	62.0
Hydroxide	OH	17.0	1	17.0
Phosphate	PO_4	95.0	3	31.66
Sulfite	SO_3	80.0	2	40.0
Sulfate	SO_4	96.06	2	48.03

Acid	Formula	Molecular weight	Equivalent weight
Carbonic acid	H_2CO_3	62.0	31.0
Hydrochloric acid	HCl	36.46	36.46
Phosphoric acid	H_3PO_4	98.0	32.67
Sulfuric acid	H_2SO_4	98.1	49.05
Sulfurous acid	H_2SO_3	82.1	41.05
Miscellaneous compounds	**Formula**	**Molecular weight**	**Equivalent weight**
Aluminum hydroxide	$Al(OH)_3$	78.0	26.0
Calcium bicarbonate	$Ca(HCO_3)_2$	162	81.0
Calcium carbonate	$CaCO_3$	100	50.0
Calcium sulfate	$CaSO_4$	136	68.0
Carbon dioxide	CO_2	44.0	22.0
Ferric hydroxide	$Fe(OH)_3$	107	35.6
Magnesium carbonate	$MgCO_3$	84.3	42.1
Magnesium hydroxide	$Mg(OH)_2$	58.3	29.1
Magnesium sulfate	$MgSO_4$	120	60.1
Sodium sulfate	Na_2SO_4	142	71.0

animal characteristics. Bacteria are further subdivided into slime bacteria, which secrete slime; iron bacteria, which thrive on iron; sulfate-reducing bacteria, which live by consuming sulfate and converting it to hydrogen sulfide gas; and nitrifying bacteria, which use ammonia and whose by-product results in the formation of nitric acid.

Several methods of measurement are used, including viable count assays, direct count epifluorescent microscopy, scanning electron microscopy, and biochemical techniques. The most common means of measuring bacterial contamination is the viable count method, in which the water being measured is passed through a sterile nutrient medium, and the number of colonies appearing on the medium is counted after a period of time is allowed for growth. These are called *colony-forming units* (CFUs). Endotoxins are measured in endotoxin units per milliliter (EU/mL). A popular form of measurement for endotoxins is the limulus amoebocyte lysate (LAL) test, in which a blood extract of the horseshoe crab becomes turbid in the presence of bacterial endotoxins. This detection technique uses optical density (turbidity level) measured over a period of time.

Other Organisms. This form of impurity is also applied to larger living things such as clams, mussels, and their larvae. They tend to clog water inlets of salt water and freshwater and may also find their way into the piping system of a facility. See the section on nonpotable water systems for further discussion.

Dissolved Minerals and Organic Substances

Organic Substances. Dissolved organic substances typically found in water include both humanmade and natural substances. Humanmade impurities include herbicides, pesticides, trihalomethanes, surfactants, and detergents. Naturally occurring impurities include lignins, tannins, humic and fulvic acid, and other biodecomposition products.

Alkalinity. All natural water contains some alkalinity. *Alkalinity* is a measure of the quantity of dissolved earth minerals in water, and it reflects the water's ability to neutralize acids. It is mainly the sum of the carbonate, bicarbonate, and hydroxide ions in water, with borate, phosphate, and silicate ions partially contributing to the total. It is reported as the parts per million (ppm) equivalent of calcium carbonate. Alkalinity is regarded as the most important characteristic in determining the scale-forming tendency of water.

Alkalinity is measured using two endpoint indicators. The phenolphthalein alkalinity (or P alkalinity) measures the strong alkali in the solution. The methyl orange alkalinity (or M alkalinity) measures all the alkalinity present in the solution. M alkalinity is often called *total alkalinity* because it also includes P alkalinity. Alkalinity is not a measure of pH.

Iron. The most common form of iron is ferrous bicarbonate. It is also considered a form of hardness of water. Iron causes problems with many ion exchange resins.

Magnesium. The most common forms are magnesium carbonate, magnesium bicarbonate, and magnesium chloride. These impurities tend to deposit scale on surfaces they come in contact with.

Silica. The three most common kinds of silica are soluble, colloidal, and particulate. Soluble silica is often referred to as *reactive silica*, and colloidal silica is sometimes called *nonreactive* or *polymeric*. The most common form in solution is silicon oxide, and it is found in suspension as a fine colloid. These impurities tend to deposit a scale on surfaces they come in contact with and form a gelatinous mass on reverse-osmosis (RO) membranes.

Sodium and Potassium. Both elements form similar salts, with the three most common being sodium or potassium chloride, sodium or potassium carbonate, and sodium or potassium bicarbonate.

Chlorides and Sulfates. The most common forms are dissolved salts of sodium, potassium, calcium, and magnesium. These impurities tend to deposit a scale on surfaces they come in contact with.

Hardness. *Hardness* is the total amount of calcium, magnesium, iron, and other metallic elements that contribute to the "hard" feel of water. Carbonate, sulfate, and chloride salts of these elements are responsible for most of the scaling deposited on pipe walls and boilers. Generally accepted practice limits the term *hardness* to include only calcium and magnesium. Often water is characterized in general terms by the amount of hardness, expressed in milligrams per liter as $CaCO_3$ (calcium carbonate), as follows:

Soft	0 to 75 mg/L
Moderate	76 to 150 mg/L
Hard	151 to 300 mg/L
Very hard	Over 300 mg/L

Trace Elements. Trace elements are present in very small quantities and are considered problems only if the amount is above an accepted level for the purpose for which the water is to be used. Examples are lead, cadmium, copper, barium, silver, lithium, zinc, chromium, mercury, arsenic, and selenium.

Dissolved Gases. The most common dissolved gases in natural raw water are oxygen, carbon dioxide, nitrogen, radon, and hydrogen sulfide. In addition, water obtained from a potable water supply usually has chlorine and fluorides present, added for public health. Of increasing concern is the presence of radon gas in many water supplies obtained from wells.

Oxygen is the basic factor in the corrosion process and must be present for the corrosion of metals. Its removal or reduction reduces the corrosiveness of the water.

For chlorine, no pretreatment is usually necessary for feedwater containing less than 1 part per million (ppm). When more than 1 ppm of chlorine is present, an activated carbon filter is recommended.

Radon is a by-product of uranium that rises to the surface of the ground because of its light weight and is released into the atmosphere. During this process, some radon becomes dissolved in groundwater. The main concern arises when this type of radiation is ingested into the body. Radiation is measured by its half-life, which indicates how fast the radiation is dissipated.

Radon (Rn) is an isotope that naturally occurs in three forms: Rn-219, Rn-220, and Rn- 222.

1. Rn-219 is an alpha-type radiation and has a half-life of 4 s.
2. Rn-220 is an alpha-type radiation and has a half-life of 56 s.
3. Rn-222 is an alpha-type radiation and has a half-life of 3.8 days.

Radon is easily removed using granular carbon, which will hold it long enough for it to decay.

Water Analysis and Impurity Measurement

General. Analyzing a water sample is the process of finding the quantities of various impurities present. To accomplish this, the quantities must be presented in a logical and understandable manner to allow for easy and practical interpretation. It is of utmost importance that the initial analysis of incoming water be accurate and contain a worst-case scenario and that the desired output quality be established prior to the selection of any treatment system.

The most accurate analysis of a water sample is done by laboratories specializing in this type of work. Sterile containers must be used, and several samples must be taken over a period of time to ensure that peak readings and average values will be obtained. There are also many field tests that are not as accurate as laboratory tests but give an acceptable accuracy for all but pure water.

The results of the analysis can be expressed in many ways. A common method is to report the concentration of ions in solution by the weight of an element or compound per liter of water, expressed as milligrams per liter (mg/L) of water. Another method is expressed in parts per million (ppm). It can be expressed either by the weight of an impurity compared to the weight of water, abbreviated w/w (weight to weight), or by the volume of the impurity compared to the volume of water, abbreviated v/v. Other units are also used, such as grains per gallon (gpg) and equivalents per million (epm). The unit mg/L differs from ppm in expressing a proportion in weight per volume. This finds specific use in the analysis of saline waters. Where the specific gravity of a liquid is around 1, the mg/L and ppm are equal. *Grains per gallon* (gpg) is a term used in the discussion of ion exchange equipment capabilities, where 1 gpg = 17.1 ppm.

As previously explained, compounds break down into ions when dissolved. Although chemists can measure the amount of each ion present in a sample, it is not practical to find the total amount of each compound that actually went into solution. In practice, the actual method of analysis measures ions only. Using the ionic form of measurement when reporting impurities makes it easier to interpret the results.

To further simplify reporting, it is desirable to reduce all ions present in solution to a common denominator. The common denominator is calcium carbonate. This is accomplished by comparing the equivalent weight of all ions present and expressing them as the ppm anion and cation equivalent of calcium carbonate. The main reason for this is that the molecular weight of calcium carbonate is 100 and its equivalent weight is 40. This method of expression is a widely accepted, but not universal, standard for reporting water analysis. Table 4.2 presents the conversion factors used for major impurities. Figure 4.1 illustrates a

TABLE 4.2 Converting ppm of Impurities to ppm of Calcium Carbonate

Cations	Ionic ppm multiplier	Anions	Ionic ppm multiplier
Hydrogen	50.00	Hydroxide	2.94
Ammonium	2.78	Chloride	1.41
Sodium	2.18	Bicarbonate	0.82
Potassium	1.28	Nitrate	0.81
Magnesium	4.10	Bisulfate	0.52
Calcium	2.50	Carbonate	1.67
Ferrous	1.79	Sulfate	1.04
Ferric	2.69		
Cupric	1.57	Other	
Zinc	1.53		
Aluminum	5.55	Carbon dioxide	2.27
Chromic	2.89	Silica	1.67

WATER ANALYSIS REPORT

Sample No. 605 | Collected ______
For ABC Co. | Analyzed ______
| Reported ______

	Ion		epm	ppm as $CaCO_3$
Cations	Calcium as Ca	62 ppm	3.10	155
	Magnesium as Mg	31 ppm	2.54	127
	Sodium and potassium as Na	38 ppm	1.64	83
	Total cations		7.28	365
Anions	Bicarbonate as HCO_3	250 ppm	4.10	205
	Carbonate as CO_3	0 ppm	0	0
	Hydroxide as OH	0 ppm	0	0
	Chloride as Cl	11 ppm	0.31	15
	Sulfate as SO_4	138 ppm	2.87	145
	Nitrate as NO_3	____ ppm		
	Total anions		7.28	365

Silica as SiO_2	5 ppm	Total hardness	282 ppm $CaCO_3$
Iron as Fe_2O_3	1.2 ppm	Methyl orange alkalinity	205 ppm $CaCO_3$
Total dissolved solids	536 ppm		
Suspended solids (weight)	5 ppm	Phenolphthalein alkalinity	0 ppm $CaCO_3$
Chloroform-extractable matter	____ ppm		
Turbidity (after shaking)	5 ppm	pH 7.7	Color ______
Carbon dioxide as CO_2	10 ppm	Sp conductance ______ µmhos	

FIGURE 4.1 Typical water analysis report.

typical water analysis report indicating impurities in parts per million, equivalents useful in calculating reacting chemicals, and a comparison of positive and negative ions.

pH. When alkalines (bases) are mixed in water, hydroxyl ions result. In a mixture of acid and water, hydrogen ions result. *pH* is a measurement of the hydrogen ion concentration of a solution. Since the balance of hydroxyl (cation) and hydrogen (anion) ions must be constant, a change in one ion concentration produces a corresponding change in the other. The pH value is calculated from the logarithmic reciprocal of the hydrogen ion concentration in water. The pH scale ranges from 0 to 14, with 0 being acid, 14 being alkaline, and 7.0 being neutral. A change of 1 unit represents a 10-fold increase (or decrease) in strength. pH is not a measure of alkalinity. pH is further discussed in Chap. 10.

Specific Resistance. *Specific resistance* is a measure of the amount of electrolytes in water. It measures the ability of 1 cm^3 of a sample solution at a given temperature to resist the flow of an electric current. It is based on the activity of the compounds, i.e., ionized salts, dissolved in water and varies with the temperature of the water. It is the most practical method of measuring impurities from a given sample. Resistance is given in ohms; pure water has

TABLE 4.3 Resistivity and Conductivity Conversion

Gr/gal* as $CaCO_3$	ppm as $CaCO_3$	ppm NaCl	Conductivity, μmho/cm	Resistivity, MΩ/cm
99.3	1700	2000	3860	0.00026
74.5	1275	1500	2930	0.00034
49.6	850	1000	1990	0.00050
24.8	425	500	1020	0.00099
9.93	170	200	415	0.0024
7.45	127.5	150	315	0.0032
4.96	85.0	100	210	0.0048
2.48	42.5	50	105	0.0095
0.992	17.0	20	42.7	0.023
0.742	12.7	15	32.1	0.031
0.496	8.5	10	21.4	0.047
0.248	4.25	5.0	10.8	0.093
0.099	1.70	2.0	4.35	0.23
0.074	1.27	1.5	3.28	0.30
0.048	0.85	1.00	2.21	0.45
0.025	0.42	0.50	1.13	0.88
0.0099	0.17	0.20	0.49	2.05
0.0076	0.13	0.15	0.38	2.65
0.0050	0.085	0.10	0.27	3.70
0.0025	0.042	0.05	0.16	6.15
0.00099	0.017	0.02	0.098	10.2
0.00070	0.012	0.015	0.087	11.5
0.00047	0.008	0.010	0.076	13.1
0.00023	0.004	0.005	0.066	15.2
0.00012	0.002	0.002	0.059	16.9
0.00006	0.001	0.001	0.057	17.6
None	None	None	0.055	18.3†

*Gr/gal = 17.1 ppm ($CaCO_3$).
†Theoretical maximum.

a resistance of 18.3 megaohms (MΩ). Resistivity and conductivity conversions are given in Table 4.3. Specific conductance in actual practice is normally measured by probes suspended in the stream of water.

Total Suspended Solids. This figure is the sum of all the suspended material found in a water sample, and it is commonly measured in parts per million, weight per weight (ppm, w/w), or milligrams per liter (mg/L). For all practical purposes, these two forms of measurement are equal (1 ppm = 1 mg/L).

The most common method of reporting is the *nephelometric turbidity unit* (NTU). The NTU measures the color of a beam of light passed through a water sample. A common standard for potable water is the Standard Method for the Examination of Water and Wastewater by the American Public Health Service, which uses formazin as the standard for producing a known volume of turbidity. The standard color scale to which it is compared is derived from the *platinum cobalt unit* (PCU). Other methods used less frequently are the comparator tube determination using formazin, called the *formazin turbidity unit* (FTU), and the original test, the *Jackson turbidity unit* (JTU), named after the man who developed the standard candle used to compare the color of the candlelight through the sample to a color standard. The most accurate method of measuring solids is the *gravimetric method,* in which a known quantity of water is evaporated and the resulting solids are weighed.

The most effective devices for removing turbidity are filters and strainers. The equipment chosen to accomplish this task depends for the most part on the size and type of solids to be retained. Other factors include materials composing the device, the nature of the raw water, flow rate requirements, particle removal target, initial and operating costs, and maintenance requirements.

Total Dissolved Solids (TDS). Often referred to as *dissolved inorganics* and *mineral salts, TDS* is the sum of all the dissolved minerals, including chlorides, sulfates, and carbonates. Dissolved solids contribute to scale deposit and corrosion of piping and equipment. When dissolved in water, mineral salts form positively charged ions, mostly sodium and calcium, and negatively charged ions, mostly chlorides and sulfates.

Total Organic Carbon (TOC). *TOC* is a measurement of the carbon level in water and is widely used to determine the level of contamination of water by organic compounds. These compounds contribute to corrosion, cause problems in manufacturing, and usually indicate the presence of endotoxins in water for pharmaceutical use.

The measurement is generally complicated and dependent on the expected level. For high levels, the organic compound is first converted to carbon dioxide which is then measured by infrared absorption. Gas stripping is required to remove other forms of carbon ions from dissolved mineral compounds. For parts per billion (ppb) levels, photolytic oxidation is used and the resultant carbon dioxide is measured.

Silt Density Index (SDI). The *SDI* is a measure of the fouling potential of a feedwater source. Since colloids and other solids can be any size in the submicrometer range, there is no direct method to measure their concentration in feedwater. The SDI is found by passing the feedwater through a 0.45-μm rated Millipore filter at 30 psi (207 kPa) and then applying the following formula:

$$\text{SDI} = \frac{1 - \frac{t_1}{t_2} \times 100}{T} \tag{4.1}$$

where t_1 = initial time, in seconds, needed to collect 500-mL sample of water through freshwater 0.45-μm filter, 47 mm in diameter
t_2 = time in seconds to filter and collect a second 500-mL sample after exposing the same filter as above for 15 min to the flow of feedwater
T = total test time, in minutes, typically 15 min; for high SDI, T may be higher

To obtain an accurate test, the filter should not have become more than approximately 74 percent plugged by the end of the elapsed time. If this figure is exceeded, the test should be repeated using a shorter overall elapsed time. A Millipore filter is the only membrane currently approved by the American Society for Testing and Materials (ASTM) for determining the SDI. The higher the SDI, the greater the potential for fouling.

Many manufacturers of RO cartridges recommend allowable SDI figures for feedwater. For hollow-fiber modules the maximum SDI is 3, and for spiral-wound modules the maximum SDI is 4. For continuous deionization, an SDI of 4 or less is recommended. In practice, when water has an SDI greater than 4, a 4-μm depth prefilter is recommended. In addition to the 4-μm filter, an additional l-μm filter is recommended downstream. The use of a 4-μm filter in the feedwater stream is always recommended as a precaution against fouling regardless of the potential SDI.

Deposits and Corrosion

The contaminants previously discussed will cause piping system fouling by depositing materials on the walls of the pipe, thereby reducing the efficiency of the system and reducing the thickness of the pipe wall by corrosion, which will eventually cause failure of the piping system. Following is a brief discussion of these categories and treatment methods as they apply to most systems.

Deposits

Scale and Sludge. *Scale* is a solid deposit on the walls of a pipe resulting from precipitation of dissolved solids in the fluid stream. Scale reduces heat transfer and interferes with the flow of water by increasing the friction of the fluid against the walls of the pipe. Boiler scale consists of calcium, magnesium, iron, and silica minerals. It is prevented by pretreating water before it enters the boiler to remove much of the scale-forming ingredients or by adding effective chemicals to feedwater to adjust pH, prevent corrosion, and prevent deposits from occurring.

Sludge is a sticky and adherent deposit in the feedwater, which results from the settling out of suspended matter from several sources. One source is an excess of iron, generally iron oxide (rust) and iron carbonate (a corrosion product). Other sources are mud, dirt, and clay that tend to collect and adhere in areas of low circulation. Sludge is prevented by filtering the incoming feedwater and adding chemical dispersants to keep the solids in suspension. They are removed by blowdown. Mud, dirt, and clay are rarely encountered except where the feedwater comes from surface sources.

Condenser scale deposits consist of calcium carbonate, calcium sulfate, or silica minerals when their concentrations exceed their solubility or their pH exceeds saturation, causing these minerals to come out of solution. This is prevented by controlling the pH, diluting the circulating water to prevent concentration, and adding chemicals to inhibit and prevent scale formation.

Biological Fouling. Microbiological fouling is caused by the growth of bacteria, algae, fungi, and other organisms. Their growth is helped by a favorable water temperature, favorable pH, and the presence of oxygen and food. Slime is the buildup of microbes and their waste products and also a deposit of dust and other suspended matter.

Microbial control consists of sterilization, disinfection, and sanitation. *Sterilization* is defined as the lethal disruption of all bacteria, molds, and yeasts and the elimination of biofilm and spores. Numerically, it is a 12 log reduction in bacteria. Disinfection is a 6 log reduction of microbials. Sanitation is generally considered to be the killing of the vegetative organisms, and it minimizes the presence of bacteria and endotoxins. Numerically, it is a 3 log reduction in bacteria.

There are different methods of controlling biofouling. The method selected depends on the intended use of the treated water and the proposed materials of the system components. Chemicals, ultraviolet radiation, heat, filtering, and ozone are the most commonly used.

Corrosion. *Corrosion* is the loss and eventual failure of metals and alloys from electrochemical reaction between water and the pipe material. It is separated into two basic types: general and localized. General corrosion describes the potential dissolution of pipe over its entire exposed surface. Localized corrosion affects only a small area of the pipe surface.

General Corrosion. This is a breakdown of the pipe material at a uniform rate over its entire surface by direct chemical attack. It is caused by the loss of the protective passive film that forms on the surface of the pipe coupled with a chemical reaction occurring between the pipe material and the chemical in the fluid.

GALVANIC CORROSION. This type of corrosion occurs in a liquid medium (called an electrolyte) when a more active metal (anode) and a less active metal (cathode) come in contact with each other and form an electrode potential. When this occurs, the more active (noble) metal will tend to dissolve in the electrolyte and go into solution.

INTERGRANULAR CORROSION. This occurs in the pipe wall when material in the grain boundary of some alloys is less resistant to the corroding agent than the grains themselves, and the bonds between the grains are destroyed.

EROSION CORROSION. This is caused by a wearing away of the pipe wall, usually as a result of excessive fluid velocity or constant wearing away by solids in the water striking the walls of the pipe.

DEZINCIFICATION. This condition associated with brass alloys in which zinc ions diffuse out of solution, leaving porous copper behind.

STRESS CRACK CORROSION. In this process, the formation and propagation of cracks are accelerated by tensile stresses in a chemically reactive environment.

Localized Corrosion. This takes place on small areas of the surface, usually at high rates, and takes various forms. Stress corrosion cracking is the physical deterioration and cracking of the pipe wall caused by a combination of high operating temperature, tensile stress on the pipe, and chemicals in the fluid stream. Pitting is characterized by deep penetration of the metal at small areas of the surface, concentrating in small cells, without affecting the entire surface. Crevice attack corrosion occurs at junctions between surfaces (often called *crud traps*) where a crack exists that allows an accumulation of a corroding agent.

Predicting Scale Formation and Corrosion Tendencies—General

A common and costly water-caused problem is the formation and deposit of mineral scale. Although scale deposits may contain a complex mixture of mineral salts, the primary constituent is calcium carbonate.

Most salts are more soluble in hot water than in cold water. Calcium and magnesium salts, however, dissolve more readily in cold water than in hot. As a result, they tend to deposit on surfaces when there is a rise in temperature. The following primary factors affect this tendency:

1. Alkalinity
2. Hardness (calcium)
3. pH
4. Total amount of dissolved solids
5. Temperature

The pH value for scale formation reflects the ratio of bicarbonate to carbonate alkalinity. A higher pH value indicates a greater carbonate content of the water and the increased tendency of calcium and magnesium carbonates to precipitate out of solution. Also, the higher the temperature, the greater the tendency of dissolved solids to precipitate out of solution because of their property of inverse solubility.

Langelier Saturation Index. In the 1930s, W. F. Langelier studied the primary factors that affected the tendency of water to form deposits of mineral scale on heat transfer equipment. As a result of this work, the Langelier saturation index (LSI) was created. The index is based on numerical values given to the factors that affect deposits.

The LSI is actually a calcium carbonate saturation index. It is based on the assumption that water with a scaling tendency will tend to deposit a corrosion-inhibiting film of calcium carbonate and thus be less corrosive. A water with nonscaling tendency will tend to

TABLE 4.4 Prediction of Water Tendencies by the Langelier Index

Langelier saturation index	Tendency of water
2.0	Scale-forming and for practical purposes noncorrosive
0.5	Slightly corrosive and scale-forming
0.0	Balanced, but pitting corrosion possible
–0.5	Slightly corrosive and nonscale-forming
–2.0	Serious corrosion

dissolve protective films and thus be more corrosive. The interpretation of the LSI is based on numerical values given in Table 4.4. The LSI is calculated as follows:

$$\text{LSI} = \text{pH} - \text{pHs} \tag{4.2}$$

where LSI = Langelier saturation index
pH = value obtained from testing the water in question
pHs = value calculated for the pH of saturation for the calcium carbonate present in the water in question

The most accurate method for calculating pHs is to use the following formula:

$$\text{pHs} = 9.3 + A + B - (C + D) \tag{4.3}$$

The numerical values of A, B, C, and D for substitution into Eq. (4.3) are found in Table 4.5. A more empirical method to find pHs is to use the chart in Fig. 4.2.

TABLE 4.5 Numerical Values for Substitution in Eq. (4.3) to find the pHs of Saturation of Water

Total solids	A	Ca as $CaCO_3$	C	M. alkalinity	D
50–330	0.1	10–11	0.6	10–11	1.0
400–1000	0.2	12–13	0.7	12–13	1.1
Temp, °F	B	14–17	0.8	14–17	1.2
		18–22	0.9	18–22	1.3
32–34	2.6	23–27	1.0	23–27	1.4
36–42	2.5	28–34	1.1	28–35	1.5
44–48	2.4	35–43	1.2	36–44	1.6
50–56	2.3	44–55	1.3	45–55	1.7
58–62	2.2	56–69	1.4	56–69	1.8
64–70	2.1	70–87	1.5	70–88	1.9
72–80	2.0	88–110	1.6	89–110	2.0
82–88	1.9	111–138	1.7	111–139	2.1
90–98	1.8	139–174	1.8	140–176	2.2
100–110	1.7	175–220	1.9	177–220	2.3
112–122	1.6	230–270	2.0	230–270	2.4
124–132	1.5	280–340	2.1	280–340	2.5
134–146	1.4	350–430	2.2	350–440	2.6
148–160	1.3	440–550	2.3	450–550	2.7
162–178	1.2	560–690	2.4	560–690	2.8
178–194	1.1	700–870	2.5	700–880	2.9
194–210	1.0	880–1000	2.6	890–1000	3.0

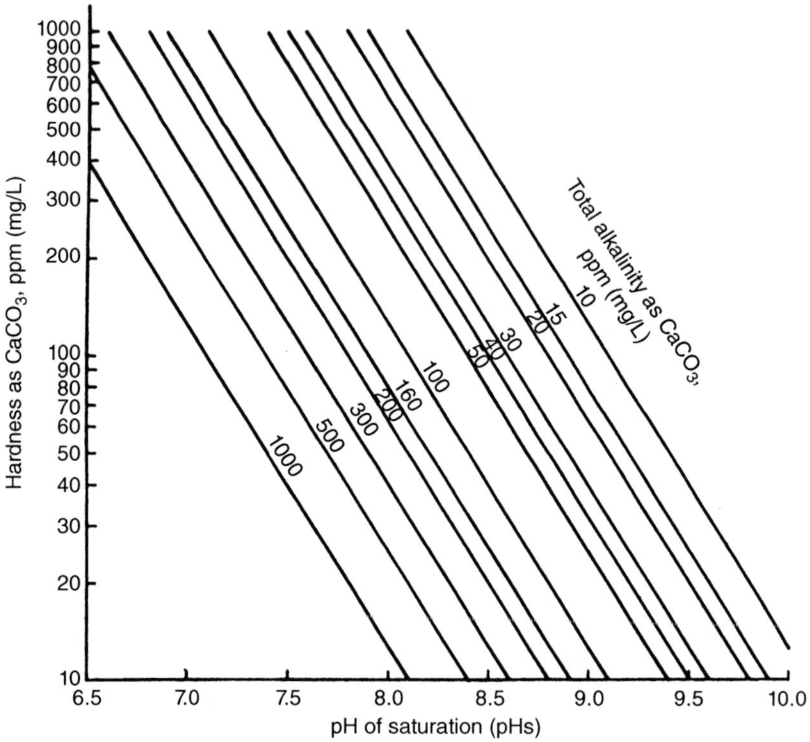

FIGURE 4.2 pH of saturation for water.

Ryznar Stability Index. Also referred to as simply the *stability index* (SI), the Ryznar index (RI) is an empirical method used to predict scale-forming tendencies of water. RI is calculated from the following formula, using the same definitions used for Eq. (4.2):

$$RI = 2pHs - pH \qquad (4.4)$$

RI is always positive. When it falls below 6.0, scale formation is possible and becomes more probable with decreasing RI. For an interpretation of RI, refer to Table 4.6.

TABLE 4.6 Prediction of Water Tendencies by the Ryznar Index

Ryznar index	Tendency of water
4.0–5.0	Heavy scale
5.0–6.0	Light scale
6.0–7.0	Little scale or corrosion
7.0–7.5	Significant corrosion
7.5–9.0	Heavy corrosion
9.0 and higher	Intolerable corrosion

Aggressiveness Index

Developed by the Environmental Protection Agency (EPA), the aggressiveness index (AI) is used rarely as a guideline to find the corrosive tendency of potable water and is given for information only. AI is calculated as follows:

$$AI = pH + \log 10\ (\text{alkalinity} - \text{hardness}) \tag{4.5}$$

Values less than 10 indicate aggressive water, values between 10 and 12 indicate moderately aggressive water, and values greater than 12 indicate nonaggressive water.

FILTERS AND STRAINERS

Filter Categories

Filters are divided into two general classes, depending on their filter media: granular and preformed. Granular filters are depth-type filters using individual grains in layers, such as sand charcoal or combinations of these, as a filter medium. Preformed filters can be either screen, surface, or depth type, ranging in thickness from a single thin membrane element to a thick filter mat. Often, filter elements are contained within a housing called a *cartridge*.

Screen, surface, and depth filtration are the three broad categories of filters. A depth filter relies on the density and thickness of the filter layers to mechanically trap the particles, and it will retain relatively large quantities of them. Depth filtration occurs on the surface and throughout all or part of the filter medium as the water passes through a complex network of flow channels. The particles are retained by random adsorption and mechanical entrapment. An oil skimmer removes oil from water by means of a moving medium, such as a mop, belt, or disk, that causes oil to be retained on the medium by surface tension. Oil and grease are removed from the skimmer by wiper blades or pinch rollers. The skimmers must be operated at an elevated temperature to keep the oils in a fluid state. For more demanding situations, skimmers could be an effective and cost-effective pretreatment.

Filter Ratings

Filters and strainers are rated in several ways. Absolute and nominal ratings are based on the size of the particle that the filter is expected to capture and retain. Particles are measured in micrometers (microns), which is 1/1,000,000 m (1/25,000 in) and abbreviated μm. This rating is a single number called the *micrometer (micron) rating*. The micrometer rating of a filter or strainer can be absolute or nominal. These ratings are often misunderstood, and this is an area of confusion in the filtration industry. Another method is called the *beta rating*, which is based on actual particle counts of different particle sizes of both the influent and effluent liquid stream. The beta rating is considered the most accurate rating measurement of a filter. Refer to Table 4.7 for the relationship between beta value and percent removal efficiency.

TABLE 4.7 Relationship Between Beta Value and Removal Efficiency

Beta ratio	Removal efficiency %
1	0
2	50
10	90
100	99
1,000	99.9
5,000	99.98
10,000	99.99

Efficiency is a measure of particle removal. It indicates what percent of particles above a certain size will be retained. For absolute rated filters, the rated pore size indicates 100 percent removal and is based on the log reduction values associated with bacterial retention testing. Because the pore size of some filters is not well defined, it is not possible to assign those filters an absolute rating. Instead they are given a nominal pore rating, which indicates the particle size above which a predictable percentage of particulates will be retained. As an example, a nominally rated 1.0-m depth filter will remove 90 to 95 percent of all particles 1.0 μm or larger. For a surface filter of the same rating, the efficiency would be 99.99 percent.

An absolute micrometer rating indicates the smallest size particle that the filter will capture; no particles of that diameter or larger will pass through the filter. The absolute rating generally depends on sieving, since the capture of particles by adsorption is never ensured. Since absolute ratings are generally unrealistic for most services, nominal ratings are the most common method used to rate filters. One exception is in pharmaceutical service, where absolute ratings are required to ensure that all particulates of a certain size are removed.

Nominal ratings allow the filter rating to consider particles retained by adsorption. The nominal rating has no generally accepted definition in the industry, and there are no industry standards. As defined by the American National Standards Institute (ANSI), the nominal rating is an arbitrary micrometer value indicated by the filter manufacturer. Due to its lack of reproducibility, this rating is depreciated. The ambiguity of this rating method makes it difficult to achieve reliable and consistent results. Many manufacturers use different methods to rate their filters, e.g., expressing the results gravimetrically, which does not represent the particle size and number in the effluent stream. Some have specific test conditions that do not represent the actual conditions for which the filters will be used. These test conditions may use fine or coarse particles such as AC test dust, latex beads, carbon fines, or bacteria. The nominal rating can be used as a guideline, provided that the micrometer rating includes the percent removal efficiency rating of that micrometer size.

Void volume of preformed fibrous media is the ratio of pore area to the fiber diameter of the filter media. If all other factors are equal, the medium with the greatest void volume will have the longest life and lowest initial clean pressure drop per unit thickness. Factors such as strength, compressibility of the fiber material under pressure (which reduces void volume), cost, and compatibility of the media with the water contaminants being removed should all be considered when selecting a filter for a specific application.

Component Description

General. The filter media type, particle size, and specific gravity are primarily selected for particle retention capability and ease of restratification of the media in the filter tank after backwash. A typical multimedia filter arrangement has a top bed of anthracite, a middle layer of sand, and a bottom layer of garnet. An additional layer of gravel may sometimes be used as an underlayer on the bottom of the unit to support the media bed. Anthracite usually has 1.1-mm grain size with a specific gravity of not less than 1.4. Typical depth is usually between 8 and 12 in. Silica sand has an effective size between 0.35 and 0.50 mm with a specific gravity of 2.6. The sand layer is usually between 8 and 12 in deep. Garnet or ilmenite typically has 0.2-mm grain size and 4.2 specific gravity. Ilmenite is commonly substituted for garnet of the same grain size, has a specific gravity of 4.5, and has a typical depth of about 4 in. The gravel depth can range from 6 to 24 in, but usually falls between 10 and 18 in.

Gravity Filters. Roughing-type gravity filters are used for initial treatment (pretreatment) of surface water, and they are made to handle large flows economically and to remove large particles. These types of filters are most often rectangular, concrete tanks. Conventional gravity filters usually operate on pretreated water and are housed in steel tanks. Gravity filters are usually rated at 2 to 4 gallons per minute per square foot (gpm/ft^2) of cross-sectional bed area. Aluminum or ferric sulfate (which is composed of multivalent ions) could be added as a coagulant to the feedwater to neutralize the surface charge of the colloids, thus making their removal easier. The resulting flocculated aggregates removed by the filter are discharged to drain. This type of filter is used to treat large volumes of water for process use and is outside the scope of this handbook.

Activated Carbon Filters. Activated carbon is a depth-type granular filter that depends on adsorption to separate contaminants from the feedwater. It is discussed in this chapter because it falls in the granular filter category and conforms to the definition for filtration. It is not specifically used for particulate removal but rather for reduction of free chlorine, removal of TOCs, soluble organics, and trihalomethanes. Chlorine removal is necessary to protect some RO membranes and ion exchange resins from attack.

The most common type of activated carbon for water treatment is the granular form, often called *activated charcoal*, which is the only one discussed here. Activated carbon is manufactured by grinding the raw material into uniform sizes, adding a binder if necessary, and reducing (burning) the mixture in the presence of steam. This creates skeletal granules with a very large network of micro- and macropores which thereby become activated. Bituminous-based coal has the highest bed density and therefore a larger number of pores available for attraction of contaminants. In addition, the bituminous product is more amenable to reactivation. Because of these attributes, this is the material most often used. Granular carbon is usually 8- to 30-mesh, with an effective size of 0.9 mm and a uniformity coefficient of 1.8.

Activated carbon media are carbon granules that have been processed from raw high-carbon materials, such as lignite and bituminous coal, wood, peat, and coconut husks. They are available in powdered and granular forms. Another rarely used method is to have powdered carbon added to a process stream and filtered out downstream. This method is not discussed.

The granules are placed within a housing, and the feedwater to be treated is passed through the granules. The organic removal capacity of the media depends on the diffusion rate of the organic molecules through the pores of the media, the surface area of the media, the pore size, and the source method used to manufacture the carbon. The characteristics of organic impurities also play a role in removal. The impurities' molecular weight, polarity, pH, temperature, and concentration are important factors in the rate of adsorption. This filter is much more effective in stripping compounds of low molecular weight because those of high molecular weight tend to be poorly adsorbed.

One problem with activated carbon filters is the growth of bacteria due to trapped organics, mostly because of the removal of chlorine. This can be controlled by frequent cleaning of the filter with steam, hot water [180°F (80°C) or higher], or dilute caustic soda backwash.

The activated carbon media gradually lose their adsorptive capacity and have to be periodically replaced or reactivated. Small amounts of media cannot be economically reactivated. Coal-based media can be activated, but all other types are usually replaced. Reactivation is accomplished by heating the media to a temperature of 1600 to 1800°F. Expect a 10 to 20 percent loss of media during reactivation. After treatment, the carbon will lose some of its smaller adsorption pores, reducing its capacity to retain trace level contaminants. If this is an important factor, the media must be replaced with virgin material rather than reactivated. The indication for replacement is an excessive pressure drop through the unit found through periodic testing of the water quality.

A generally accepted conservative method to select the size of the filter uses a figure of 5 gpm/ft^3 of media. Use manufacturers' specifications to select the housing size, amount, and type of filter media for specific applications and flow rates. Activated carbon is used to remove dissolved, nonionic organics such as residual chlorine disinfectants, trihalomethanes, and chlorimines, as well as a major portion of naturally occurring dissolved organic material, from municipal water supplies. The nonionic organics tend to coat ion exchange resins and all types of membranes.

System designers are reluctant to use the activated carbon filter in the generation of potable water because of the development of significant levels of bacteria in the unit itself. This can be controlled by periodically sanitizing with pure steam or hot water with a temperature greater than (180°F) (80°C). The need for sanitizing can be determined only by testing the water. Because of this sanitizing, the interior of the filter housing should be lined or coated. When potable water is used as feedwater, stainless steel housings should be avoided because of the possible chloride stress corrosion and chloride pitting resulting from the chlorine in the feedwater. A typical detail of a large-scale granular activated carbon filter is illustrated in Fig. 4.3.

Pressure Filters. Pressure filters are the most common types used for general utility and service systems. They usually operate under normal water pressure, but may require pumps to overcome excessive friction loss through the unit if the available water pressure is too

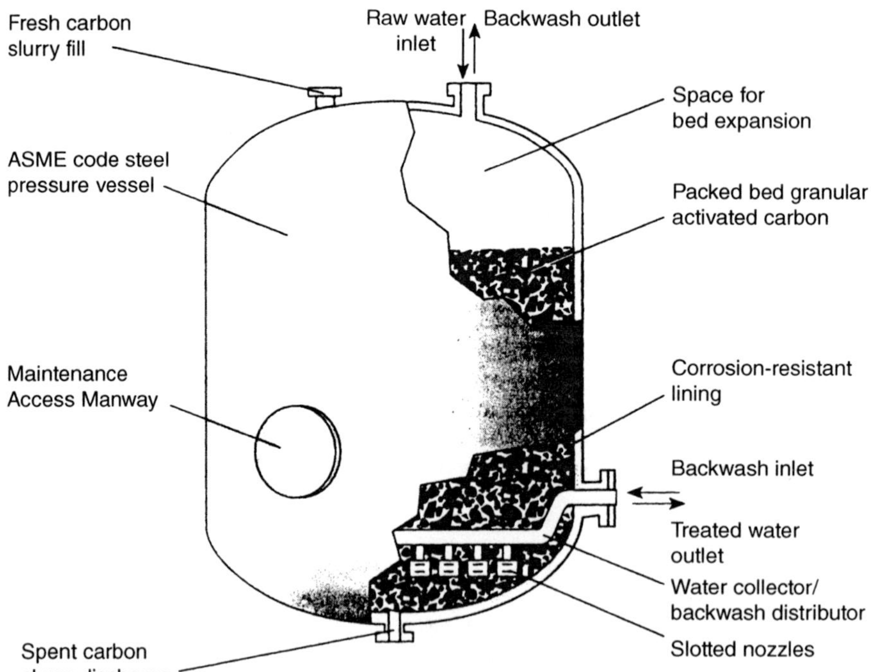

FIGURE 4.3 Schematic detail of typical large-scale granular activated carbon filter.

low. Commonly called high-rate or rapid sand filters, different designs are available that offer flow rates higher than those available for gravity designs. Three different kinds of granular filters are used: single-medium filters with sand media; dual-media filters consisting of a top layer of anthracite and a bottom layer of sand; and multimedia filters consisting of a three-layer bed of anthracite on top, sand in the center, and garnet on the bottom.

Deep-Bed Granular Sand and Multimedia Filtration. Deep-bed sand filters consist of a tank containing either silica or garnet sand of constant size (grade), or layers of a multimedia type consisting of a variety of graded material such as anthracite, silica sand, garnet sand, and quartz. This type of filter is most often used as a prefilter to remove larger suspended solids to extend the duty cycle of finer filters downstream. It has a relatively large retention capacity of solids and removes particles 10 μ and larger. During normal operation, the raw water to be treated enters at one end (or the top) of the unit, the suspended solids adhere to the media, and the clear water collects at the other end or on the bottom.

Sand filters are either gravity or pressure type. If the tank is atmospheric and water flows through the unit with no assistance from pumps, it is a gravity filter. If the filter is in line and uses the pressure of the water supply to force the water through the filter, it is a pressure type. The pressurized filter is the most commonly used because of its smaller size and higher flow rate. A typical pressure sand filter is illustrated in Fig. 4.4.

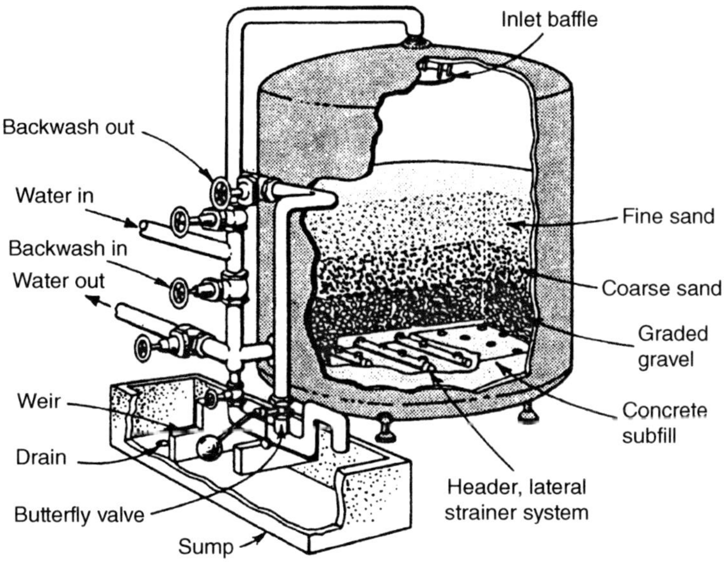

FIGURE 4.4 Typical pressure sand filter.

The multimedia filter achieves a more uniform distribution of filter media throughout the bed and is considered a more effective type of filter. A typical multimedia filter for laboratory use consists of a top layer of anthracite having a 1.1-mm-diameter grain size and 1.5 specific gravity, a middle layer of sand having a 0.5-mm-diameter grain size and a specific gravity of 2.6, and a bottom layer of garnet having a 0.2-mm-diameter grain size and 4.2 specific gravity. Normal operational flow rate is from 6 to 15 gpm/ft^2 of bed area.

Backwashing is required to clean the filter, with the effluent discharged to the sanitary drainage system. A backwash flow rate of 10 to 15 gpm/ft^2 is generally required for effective cleaning.

Cartridge Depth Filters. Cartridge depth filters consist of comparatively thick, replaceable, preformed filter media that are placed inside a housing. The particle size retained by a depth filter is not precisely defined because of the random nature of the fiber matrix. The advantage of the depth filter is that it has superior particle load because the filter is capable of holding particles throughout its entire matrix.

Precoat (Septum) Filters. One of the most frequent uses of filters is to clarify water for swimming pools. A *septum* is a thin, porous membrane with relatively large pores that has little resistance to the flow of water. The filter itself consists of a tank containing one or more vertical filter leaves or plates (septa) onto which a thin coating of filter aid is evenly deposited to form a filter surface, or precoat. The precoat could be pearlite or diatomaceous earth. The most commonly used material is diatomaceous earth. As the feedwater flows through the filter, it passes through the filter precoat first, which traps the particulates. Although pressurized filters are available, most precoat filters use a pump to create a vacuum to pull the water through the filter, with the inlet water level controlled by a float valve. When the filter coat becomes plugged, it must be replaced. This is determined by excessive pressure loss through the filter. The filter is cleaned by backwashing, and the filter aid along with trapped solids is discharged to the sewer. Care must be used in redepositing the new filter aid to ensure uniformity. Generally accepted filter aid coverage is 0.1 lb/ft^2 of filter septum surface. The diatomaceous earth media usually have a dry weight of 8 to 10 lb/ft^3 (128 to 160 kg/m^3) and range from 5 to 64 μ in size.

Filter Selection Criteria

The advantage of the granular filter is its low initial cost. The advantage of the multimedia filter is the consistent distribution of the various size media after backwashing, thereby increasing the time between backwash cycles. Tests have shown that the multimedia filter is more effective in removing particulates than the single-medium sand filter. It is often cost effective to prefilter a liquid containing large amounts of particulates with a high solid capacity sand filter before putting it through a cartridge filter.

Sand filters can be obtained with either vertical or horizontal tanks, depending on the headroom available. The horizontal tank has a larger filter bed area, which requires a higher backflow rate. Compartmentalization of the vessel may offset this requirement somewhat. The key to operation for this type of filter is the arrangement of the filter media, best arranged from coarse to fine in the direction of water flow. In addition, activated carbon added to the filter bed will remove excessive odor and bad taste, although this is usually done in a separate filter. Problems have resulted from failure of the interior lining of the filter due to abrasion of the lining by the sand during backwash. This abrasion process creates iron corrosion products that can pass through the filter and interfere with downstream purification processes.

It is important to operate these units at the face velocities recommended by the manufacturer to prevent the forcing of the impurities through the filter. This phenomenon is called *breakthrough*.

Activated carbon used for pretreatment should allow for a face velocity of about 2 to 4 gpm/ft^2 (1.4 to 2.8 L/min/m^2) of cross-sectional area. A minimum depth of 24 in (600 mm) is recommended for most feedwater streams, but the actual depth should be selected to achieve the recommended contact time with the flowing water. These two figures should be adjusted depending on the quantity and type of contaminants. Tests have shown that 1 g of activated carbon is capable of removing 1 g of residual chlorine. Because of anticipated breakthrough of organic material, the media should be replaced approximately every 6 months.

Filter Media. Depth filter media vary and consist of wound fiber (such as cotton, polypropylene, or rayon) or resin-bonded laminates (such as cellulose, acrylic rayon, or fiberglass). Filter cartridges are of the wound or pleated type. The wound fiber type is used most often.

Wound filters have the same effect as a stack of woven cloth, with the filter wound around a round mandrel instead of flat. The characteristics that affect filter operation are the type of fiber, fiber diameter, and cartridge-winding techniques such as yarn tension, winding pattern, and spacing.

Fibers are available in two types, staple and monofilament. Staple fibers such as cotton and wool are spun into a thread or yarn, which is then wound onto the mandrel. Staple fibers have a fuzzy surface, or nap, that provides a high surface area for adsorption in addition to a long, random path for particle interception.

Monofilament fibers, such as those of polypropylene and other plastics, are not spun but are manufactured in any desired diameter. These fibers must be texturized in some fashion to make them suitable for filter media. They can also be produced in smaller diameters and spun into "yarn."

Pleated filters are generally made from resin-bonded materials, with the pore size determined by the type and size of the fiber and the binding method used. Pleated cartridges usually depend more on sieving than do the wound type, and can better remove particles of a specific size.

Specific Filter Materials

Polysulfone. Polysulfone is a membrane that is basically hydrophilic and has excellent flow rates, low extractability, broad chemical resistance, high mechanical strength, and heat resistance, permitting a variety of sterilization methods.

Nylon. Nylon is a hydrophilic membrane with generally very high flow rates, high tensile strength, low extractables, and limited chemical resistance.

PTFE. PTFE is a naturally hydrophobic membrane often laminated to a polypropylene support for added strength. It has excellent chemical and heat resistance.

Acrylic Copolymer. Acrylic copolymer is a membrane that is basically hydrophilic and has excellent flow rates, low extractables, and a low differential pressure.

Polypropylene. Polypropylene is naturally hydrophobic and is available as a membrane that is chemically inert, has a broad pH stability and high flow rate, and is considered very durable.

Glass. Glass fibers are usually manufactured from borosilicate glass. Thicker fibers are spun and thinner fibers are made into mats with the addition of a binder. They typically have low differential pressures, good wet strength, and high dirt-holding capacity.

Membrane Filter Testing

An important feature of a membrane filtration system is its ability to be tested before and after filtration runs. Testing can detect a damaged membrane, ineffective seals, or a system leak that may result in passage of contaminants that the filter is designed to trap. These tests are commonly called *integrity tests*. Testing before and after a run will ensure that the entire system is intact, thereby validating the process. Prior to testing, cleaning to remove large-scale contamination (and sterilizing the filter and apparatus, if necessary, to ensure elimination of microbial contamination) is required.

The type of test selected is dependent on the specific filter chosen. However, if the previous history of a specific filter is not available, the only accurate method of testing the filter is to place it in service and run an on-site fouling and compatibility test.

Air Permeability Test. An air permeability test is normally used to test wound cartridges. It is a simple, nondestructive test that correlates well with filter performance, and it is considered more revealing than micrometer rating.

Bubble Point Test. Membrane filters have discrete, uniform passages from one side of the membrane to the other which, in effect, are fine uniform capillaries. The bubble point test is based on the fact that a liquid is held in these capillarylike structures by surface tension, and the minimum pressure required to force this liquid out of the capillary space is a measure of the capillary diameter. The pressure required is inversely proportional to the largest pore size. After the filter is wetted, air pressure upstream of the filter is increased.

There are two widely used variations of the bubble point test. The first is the visual test. For this variation, the downstream side is watched for the appearance of bubbles, which indicate that the air is passing through the capillaries. The pressure that produces a steady, continuous stream of bubbles is the bubble point pressure. The second variation is the monitored method, where a pressure drop will occur as the gas begins to flow through the filter.

It is not necessary to determine the exact pressure of a given filter to prove its integrity. If the pressure exceeds the minimum point determined by the manufacturer of the filter, its integrity is assured. The bubble point test is also used to test the integrity of the filter cartridge.

Diffusion Test. In a high-volume system where a large volume of water must be displaced before bubbles can be detected, a diffusion test should be performed instead of the bubble point test. This test is based on the fact that in a wetted membrane filter under pressure, air flows through the water-filled pores of the filter at a differential pressure below the bubble point pressure by a diffusion process following Fick's law. In small filters, the flow of air is very slow. But in a large filter it is significant and can be measured to perform a sensitive filter integrity test. In a wetted filter, a constant air pressure is applied at approximately 80 percent of the bubble point pressure established for that particular filter.

There are two widely used variations of the diffusion test. The first is the forward flow test, which relies on direct measurement of the diffusive gas flow rate. This flow rate is measured either by instruments placed in the gas flow upstream of the filter or by calculating the volume of airflow according to the rate of flow of displaced water downstream of the filter. The second method, called the *pressure decay* method, calculates the loss of diffusion gas pressure from a known volume of gas over a period of time.

Water Breakthrough Test. This is similar to the bubble test, except that water is used instead of air. Water pressure is increased on the upstream side of the filter, and the pressure that results in a steady stream of water downstream of the filter is recorded. The breakthrough pressure must be correlated to empirical data on contaminant retention from the manufacturer to be a valid test.

Water Intrusion Test. Also called the *water pressure integrity test*, this is often used for hydrophobic filters (which resist wetting by water). This test requires that the filter be wetted by an alcohol/water mixture. Water pressure is applied upstream of the filter, and the pressure decay is measured. Care must be taken to test with water from the same source because of variations in surface tension. When used as a vent filter, the membrane must be dried before being placed back in service.

Filter Selection Considerations

The mechanical properties of the particles suspended in the water stream must also be considered. At one extreme are solid, undeformable particles such as sand or quartz, and at the other extreme are gelatinous or deformable materials such as synthetic colloids and bacteria. Because they can deform, they are more likely than hard particles of the same size to pass through a filter.

Some filter materials may leach substances into the fluid as it is processed, thereby affecting its purity. Such substances, called *extractables*, can be minimized by preflushing. There is a test for plastics conforming to USP class VI that is used to ensure that there will be no adverse reaction of body fluids to extractables from filter housing or media materials.

Granular Filter Flow Rates

A single medium filter is usually operated at a flow rate of 3 to 4 gpm/ft^2 (130 to 170 L/m^2) of cross-sectional area. A dual-media filter is operated at a flow rate of about 6 gpm/ft^2. A multimedia filter has a typical flow rate of 6 to 15 gpm/ft^2 (250 to 700 L/m^2). Another consideration is the face velocity of the feedwater, which is the velocity of the feedwater through the surface layer area of the vessel. The manufacturer establishes a face velocity, based on actual tests, that cannot be exceeded.

Filters for radiation should be limited to a flow of 1.5 gpm/ft^2 (5.7 L/m^2). The filter can be backwashed. Careful placement of the filter shall locate it well away from any living quarters and shield it with concrete. The slightly radioactive filters can be disposed of as recommended and as required by the authority having jurisdiction (AHJ).

Backwashing

Eventually, the suspended particles removed from the water will accumulate in the top layer of the filter medium and obstruct the flow of water through the filter. These solids are removed by backwashing, in which clean water flows backward through the filter bed and is discharged to drain. The backwash flow and volume should be capable of expanding the filter media by 50 percent to permit complete dislodging of the trapped particles. The optimum rate is determined by the manufacturer of the equipment and is usually in the range of 10 to 15 gpm/ft^2 (470 to 700 L/m^2) of filter area for multimedia filters. For single- and dual-medium filters, the rate is lower than that of multimedia filters. A backwash pump provides the pressure needed if the water pressure is not high enough, and the typical length of backwash time is between 5 and 15 min. Where surface water is being filtered and to speed the cleaning process, compressed air mixed with the backwash water may be used to effectively remove algae and other organisms. This is called *air scouring*. The need to backwash is indicated when generally the pressure drop falls to between 7 and 11 psi (49 to 77 kPa) above the clean pressure value.

During the backwash cycle, the action of the water in the single-medium type filter distributes the coarsest sand at the bottom and the finest at the top. This distribution will result in the accumulation of a majority of particulates removed from the raw water in the fine sand layer on the top where the raw water enters the filter. This quickly reduces its effectiveness, thereby requiring frequent backwash.

Seals

Seals are used to prevent the feedwater from bypassing the filter medium. They must be reliable enough to withstand repeated changing of the cartridge. The most common sealing system uses a double open-ended cartridge with a top compression spring on the cartridge. This provides a knife-edge seal on a seat at the top of the cartridge. Single open-ended cartridges, most often used in single-cartridge housings, use a piston-type O-ring seal contained within the cartridge itself. The material of the seal must be compatible with the type of fluid being processed, the fluid temperature, and the proposed use.

Housings

Housing selection for filter cartridges is based on the proposed application. Although housings most often contain a single cartridge, units that hold multiple filters are available. General industrial requirements differ greatly from the sanitary requirements of the pharmaceutical and food service industries. Manufacturers provide filters and housings of standard diameter and length.

There are no industry standards for general industrial housings. The material can range from plastic to stainless steel. For safety, pressurized housings for compressed gas and high-pressure liquid applications must be ASME code-stamped.

Housings come in three basic styles: tee, in-line, and L-shaped. The tee style offers easy filter changing, and it is easier to fit into an existing pipeline. It usually has the highest pressure drop of the three styles. The in-line style offers the lowest pressure drop of the three styles. The L-shaped style is used most often for multiple-cartridge installations in industrial applications.

Filter Selection Considerations

The mechanical properties of the particles suspended in the water stream must also be considered. At one extreme are solid, undeformable particles such as sand or quartz, and at the other extreme are gelatinous or deformable materials such as synthetic colloids and bacteria. Because they can deform, they are more likely than hard particles of the same size to pass through a filter.

Some filter materials may leach substances into the fluid as it is processed, thereby affecting its purity. Such substances, called extractables, can be minimized by preflushing. A test for plastics conforming to USP class VI is used to ensure that there will be no adverse reaction of body fluids to extractables from filter housing or media materials.

A wide variety of filter options are available to accomplish separation, and many of these options have overlapping effectiveness. Choosing the appropriate method will often involve tradeoffs between performance and value. The following factors should be taken into consideration when selecting a filter based on the quality of the feedwater and the intended service.

1. The first step to decide the suitability of any filter is to obtain an accurate analysis of the feedwater and to determine the size and quantity of the particulates to be removed.
2. It should be decided whether absolute or nominal filter ratings will be used. For filtering service, most filters are selected by using nominal ratings, with the intended use of the water dictating the various degrees of removal.
3. If the particles are deformable, the wound depth filter is more efficient in removing them. They tend to be extruded through a thin, solid, pleated filter. For hard particles, either wound or pleated can be used.

4. The selection of a pleated or wound filter should be based on the type and quantity of solids to be removed. The pleated filter has about 16 times the effective filtration area of a similar wound filter and so is much more effective in removing smaller particulates captured on the surface. For larger particulates or for a large amount of dirt to be held in the filter, the depth-type filter is more effective. Pleated filters are usually more costly than wound filters.
5. Depth filters have high dirt-holding capacity and are used most often as prefilters for clarification of water prior to treatment by water purification equipment. They are less costly than other types of filters.
6. The removal efficiency of the filter must be known. This value is typically specified by the end user and often based on past experience. Since the micrometer rating alone is not a fair representation of filter efficiency, the degree of removal must also be specified. A rating of 90 percent efficiency at 5 μm means that 90 percent of the particles of that size will be retained on the filter medium.
7. The flow rate must be determined. This is also called the *flow density*, which relates flow rate to the unit of filter area. As the flow rate increases, the contaminant capacity and service life of the filter decrease. Manufacturers have established flow rates for filters. Housings containing multiple cartridges can be selected to lower the flow rate of individual cartridges.
8. The pressure drop across the filter is another consideration. The smaller the pore diameter, the higher the pressure drop across the filter. If the ratio of the initial clean pressure drop through the filter to the total available pressure is high, unacceptably low flow will quickly result even when the particulate-holding capacity is not fully realized. Pressure requirements across the filter from new to plugged condition must also be considered, with the filter selected based on the highest allowable pressure drop. Accepted practice is to have a maximum of 80 percent filter plugging. If the pressure drop is too high or if the amount of plugging is not acceptable, the size of the filter can be increased or multiple filters used. Housings that hold multiple filters are used to reduce the pressure drop when available water pressure is low.
9. If the filter is intended to be an initial filter in the system, the beta rating should be obtained from the manufacturer to determine its removal efficiency and performance characteristics. Ratings for a minimum of five or more particle sizes should be provided. If the filter is used in an intermediate location, the particle sizes will be reduced and the nominal rating of specific particulate sizes will allow for efficient selection.
10. The maximum temperature that the filter media can withstand will determine if they can be used for service where they must be sterilized and if the media are acceptable for sterilization using high-temperature water.
11. The maximum differential pressure is the difference in pressure between the inlet and outlet water pressure beyond which the filter will fail structurally, and it is established by the manufacturer.
12. The effective filtration area is the measure of the usable area of the filter media independent of the material or type.
13. The toxicity and pyrogenicity of the filter materials must be compatible with the intended use of the product water.
14. Compatibility with any chemicals intended to be used for sterilization that will come in contact with the filter must be considered.
15. Extractables from the filter media must be reduced to acceptable limits to prevent problems with any downstream process. The method used most often to control and reduce extractables is to preflush the filters prior to use.

16. Some filter materials may leach substances into the fluid as it is processed, thereby affecting its purity. Such substances, called extractables, can be minimized by pre-flushing. A test for plastics conforming to USP class VI is used to ensure that there will be no adverse reaction of body fluids to extractables from filter housing or media materials.

STRAINERS

General

Straining is coarse type of filtration that uses a closed vessel with a cleanable screen or mesh inside, generally used to remove and retain foreign particles larger than 45 mm (325-mesh) from flowing fluids while allowing the remainder of the fluid to pass through. If a particle is visible to the naked eye, a strainer should be chosen to remove it from the liquid stream. If the device retains particles finer than 45 mm (325-mesh), it is generally considered a filter. The difference between filters and strainers is one of semantics; a strainer could be considered a coarse filter.

Types of Strainers. Types of strainers include Y and basket (or bucket) types. Basket strainers are available in self-cleaning models. All types are available with a large variety of jointing methods for insertion into a pipeline, including soldered types available for copper piping. They are available as off-the-shelf models capable of meeting any reasonable need. Manufacturers can construct units for special conditions.

The Y-type strainer, illustrated in Fig. 4.5, is compact in design and is considered for use where space is at a premium. Frequent cleaning is often required. Its construction makes it a good choice for high pressure applications and for gases such as steam, natural gas, and compressed air where pressures are higher and amount of dirt present is low. It has a smaller dirt-holding capacity than similar sizes of basket strainers. It is installed in a pipeline with its strainer element in the down position, and can be positioned either horizontally or vertically. Very often, a valved pipe will be put in the removable end of the strainer so that the accumulated debris can be easily blown out while keeping the line in service.

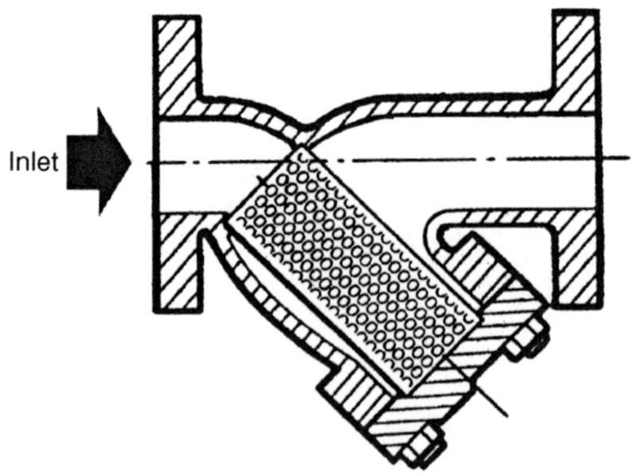

FIGURE 4.5 Y-type strainer.

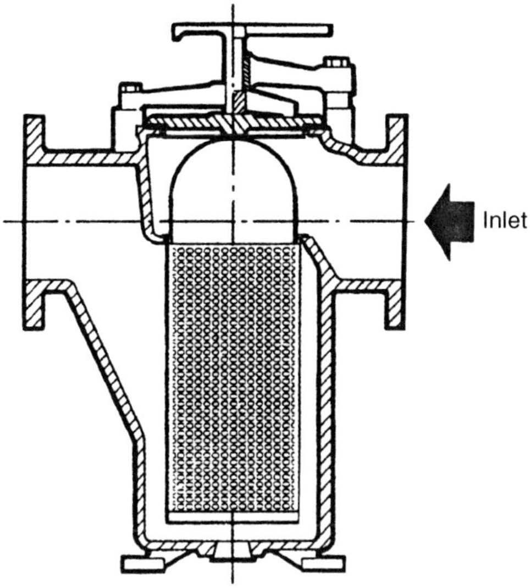

FIGURE 4.6 Basket-type strainer.

The basket strainer, illustrated in Fig. 4.6, gets its name from the upright, perforated basket used to trap particles. It is installed upright in the strainer body, and the top of the strainer must be removed for cleaning. Because of its large size, it has the ability to store large quantities of dirt and so has a lower pressure loss than a similar sized Y strainer. The relationship between mesh and opening size is given in Table 4.8.

Materials of Construction

Strainer Bodies. The materials used most commonly for strainer bodies are cast iron, bronze, carbon steel, stainless steel, and plastic.

Because of its low initial cost, cast iron is the most popular strainer body. It is used in systems where the pressure and temperature of the water are not high and the system is not subject to high thermal or mechanical shock. Cast iron is mostly used for larger potable water lines and many nonpotable water systems in addition to a variety of other product and process uses. A bronze body is preferred for brackish, saline, and seawater service. It is also often used for potable water services in smaller sizes. Its cost is about double that of cast iron.

Carbon steel bodies are used where moderately high temperature and pressure conditions are encountered, and where resistance to high thermal and mechanical shock is required. Carbon steel components are the materials of choice where fire hazards exist, such as in the petroleum and petrochemical industries. Their cost is equal to that of bronze. For high pressures and temperatures over 1000°F, chromemoly steel is usually specified for bodies.

Stainless steel is the preferred body, basket, and screen material for the pharmaceutical, food processing, and chemical industries because of its resistance to corrosion and contamination and ease of cleaning. Stainless steel costs about 4 times as much as cast iron.

Plastic strainer bodies are available in all the materials used for pipelines. Baskets and screens of all metallic materials are also available.

TABLE 4.8 Relationship between Mesh and Opening Size

Mesh size	Particle diameter, μm
Strainer	
4	5205
8	2487
10	1923
14	1307
18	1000
20	840
25	710
30	590
35	500
40	420
45	350
50	297
60	250
70	210
80	177
100	149
120	125
140	105
170	88
200	74
230	62
270	53
Filter	
325	44
400	37
550	25
800	15
1250	10

Basket and Screen Construction and Materials. The actual collection and retaining of the debris, dirt, and other particles in all strainers is done by the basket or screen that is placed inside the body of the strainer. The size of the openings through a screen is referred to as the *mesh*, and the size of the openings through a basket is referred to as the *perforations*.

The term *mesh* describes a screen that uses a woven wire cloth manufactured from the material chosen for the intended service. The most common material for all applications is stainless steel. Mesh screens are generally available in standard sizes from 20- to 200-mesh, with a variable wire size used depending on the mesh. The mesh size does not indicate the particle size retention, since the size of the mesh opening is determined by the diameter and number of wires per inch. For example, 100-mesh means 100 vertical and 100 horizontal strands of wire per inch. For critical applications, a screen should be selected on the basis of particle retention capability, not mesh size. Conventional corrosion treatment of feedwater for boilers and cooling water systems consists of pH control and chemical corrosion inhibitors. Dissolved gases are removed by deaeration.

Standard perforated screens are manufactured from a light- or heavy-gauge sheet metal available in standard sizes generally ranging in diameter from 1/32 to ¼ in. The baskets should be assembled by either welding or brazing. The strongest are of welded construction. Solder is much weaker and more easily broken. If a basket strainer is required for fine straining, it is common practice to add a wire mesh liner inside the perforated bucket since the screen alone is too weak to provide the necessary mechanical strength.

The material to be used depends on both the intended water service and the body into which the strainer is installed. The most commonly used basket materials are brass and stainless steel. Cast iron strainer bodies commonly use baskets of brass and stainless steel, depending on the intended service. Bronze bodies usually require Monel metal baskets because of the severe service required. Stainless steel baskets are used with stainless steel bodies.

The baskets are made from sheet metal with a wide variety of diameter perforations. A "Y" strainer is generally furnished with 1/16-in perforations in sizes up to 4 in, 1/8-in perforations in larger sizes for liquid service, and 3/64-in perforations for steam service. If finer straining is required, a wire mesh screen fitted inside a basket is used. In this case, generally accepted practice limits the perforations in the basket to 50 percent of the wall area in order not to lose strength. In addition, this combination usually provides the best ratio of maximum flow rate with adequate strength. The mesh and basket should be an integral unit, with the mesh fastened to the basket at both the top and the bottom to prevent any debris from bypassing the unit.

By generally accepted practice, the open-area/perforation ratio should be about 4 to 1 to avoid excessive pressure drop through the unit. A smaller ratio will require frequent cleaning. Additional strainer basket area can be obtained by using a pleated basket. If finer filtration is required after the strainer has been in service, a mesh liner can be added inside the basket. If the size particle to be removed is known, the perforations should be slightly smaller.

Fluid streams may contain iron or steel particles that are small enough to pass through the finest screens. If this is a problem, a strong magnet capable of lifting several times its own weight should be suspended in the basket. The magnet should be installed so that all the water passes over it. This magnet should be encased in an inert material to prevent corrosion.

Baskets, especially when full, are not capable of withstanding the same pressure as the body. A particular phenomenon called *runaway buildup* can occur, in which the dirt builds up and plugs the mesh or perforations, thereby reducing the free area. The pressure in the strainer increases slowly at first, but faster and faster over time. The water velocity and pressure inside the basket escalate quickly, which causes the resultant flow to stop or be reduced to a trickle. This full-line pressure can burst the basket.

Covers. A cover is provided in order to remove and clean the basket. The most common type is bolted; the bolts must be loosened and removed to provide access to the basket. This type of cover is the strongest and should be used for high-pressure applications. However, its removal is time consuming. Another type is the clamping yoke, in which threaded, tee-shaped handles are used to secure the cover to the body. Often, the cover is attached to the body with a hinge mechanism, making it very easy to remove. This type of cover is more expensive than the bolted type.

Another type of strainer is the automatic type, which does not require manual cleaning. A rotating, circular screen is used as the basket. The water inlet is directed to the inside of the basket. A rotating backwash inlet inside the basket uses the differential pressure between the atmosphere and line pressure to produce a localized reverse flow across only a portion of the basket, thereby allowing continuous cleaning. This type of strainer is appropriate for large consumers of water such as raw water inlets from rivers and lakes used for cooling and process. Automatic strainers are available to 60-in (1500-mm) size.

Design Considerations and Selection Criteria

When a strainer is selected, the four main considerations are its physical size, friction loss through the unit, price, and ease of cleaning. The viscosity and specific gravity of the fluid and the degree of perforation of the basket (or size mesh) all influence the pressure drop. In many cases, the size of the unit is a consideration if it is to fit into an existing space. Generally accepted practice limits the pressure drop to a maximum of 2 psi (13.8 kPa). Another general rule is to have a minimum 4-to-1 open-area ratio of the perforations in the basket. To reduce the friction loss, a strainer, one or more sizes larger than the pipeline into which it is installed should be selected. In some cases, the basket may become large and unwieldy when filled with debris. If this is a possibility, several smaller units in parallel should be considered. Typical pressure drop through a Y-type strainer is given in Fig. 4.7,

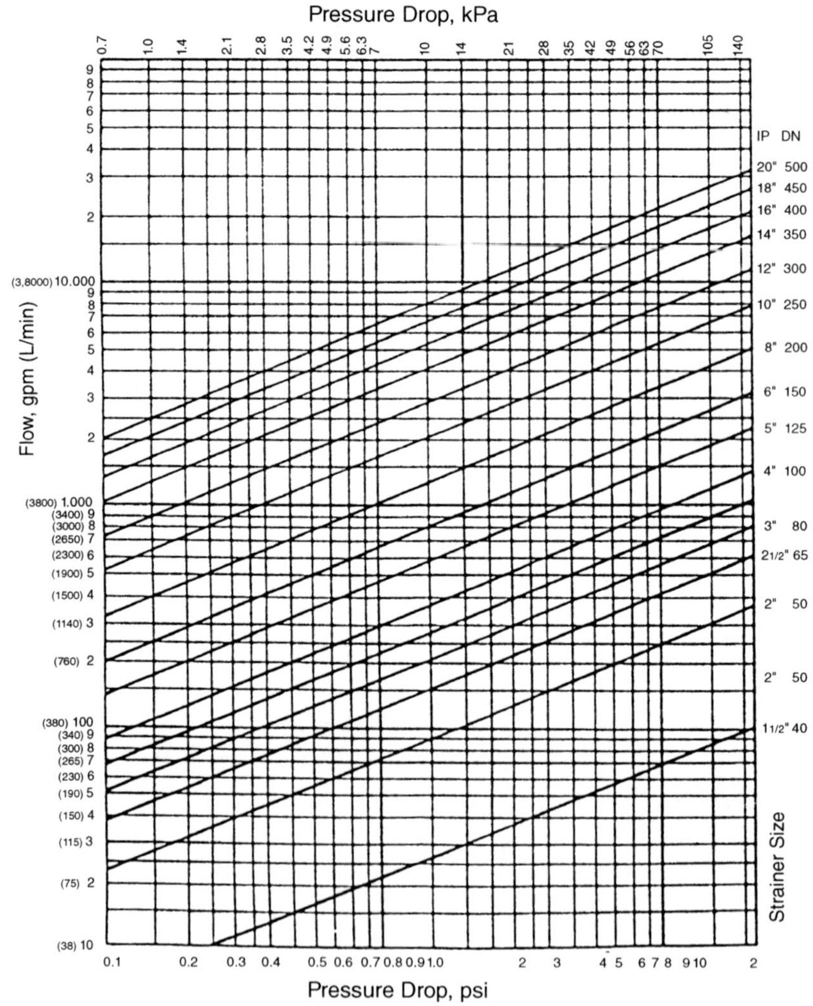

FIGURE 4.7 Pressure drop through Y-type stainers. Specific gravity of one.

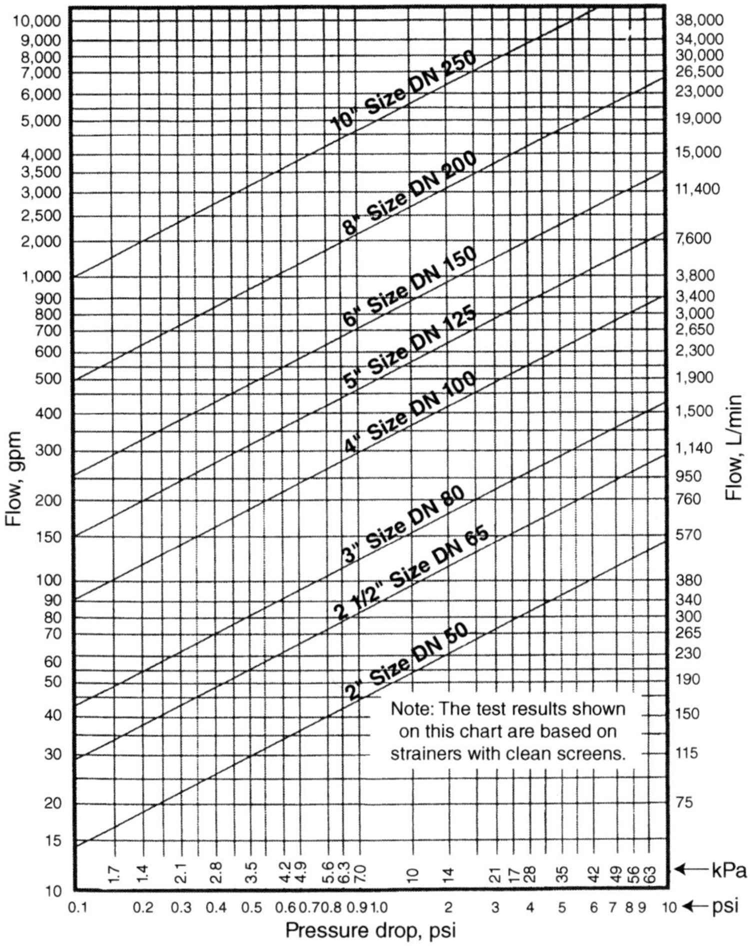

FIGURE 4.8 Presure drop through basket strainers (1/8 perforations). Specific gravity of one.

and for a basket strainer in Fig. 4.8. When the specific gravity of the liquid being strained is other than one, multiply the pressure drop found in the appropriate chart by the specific gravity of the new liquid.

Another consideration is the length of time required between cleanings. Past experience with the specific application (if available) should give a good idea of the size and amount of debris expected. Examination of the total suspended solids present in the water analysis allows calculation of the amount of debris to be expected over a period of time. This, along with any specific preference, should provide good guidelines for the proposed size of the unit to be selected.

If continuous operation of the units is critical, duplex strainers should be used. This allows one to be cleaned while the other is used, permitting full use of the system. Another method is to install two strainers in line, the first with larger openings to trap large particles and the second with a finer mesh or smaller openings for small particles. This would prevent the loading up and frequent cleaning of the strainer in a one-strainer system by spreading out the cleaning load to two strainers.

A number of methods are used to divert flow between strainers for cleaning. The most common is a multiport plug valve. For large pipe, the plug valve may get too large. For larger strainers, sliding gate valves that have synchronized disks operated by handwheels to divert flow are often used. Another method is to use synchronized butterfly valves. The actual method selected depends on the size, available friction loss, available room, and cost.

Basket strainers are usually large units with a high initial cost. In general, Y strainers have a lower initial cost for the same size unit, smaller dirt-holding capacity, and larger friction loss because of their smaller bodies. It is good practice to allow higher-pressure safety margins due to the possibility that the strainers will be placed in a position to receive water hammer shocks and water slugs. There must be enough room around the unit(s) for the necessary frequent cleaning and laydown of the basket.

Other Technology and Component Description

Aeration. Aeration is a gas-transfer process in which water is brought into contact with air for the purpose of transferring volatile substances to or from the raw water. It is used most often to remove undesirable gases such as carbon dioxide, hydrogen sulfide, and methane. Aeration, by introducing oxygen, is also used to remove iron and manganese and to lower the amount of volatile organic compounds (VOCs) from groundwater. The following are criteria for its use:

Reduction of carbon dioxide by 90 percent is obtained by near saturation with oxygen. Carbon dioxide dissolved in groundwater will consume lime in the lime-soda softening process without any accompanying softening. Generally accepted practice indicates that aeration is not economical for carbon dioxide concentrations in water of less than 10 mg/L. The cost of the lime saved should be compared to the cost of purchasing and operating the proposed aerator.

Aeration will partially remove VOCs from raw water by oxidation, making them insoluble. They would then be coagulated and removed from the water.

Aeration alone could be used for the removal of hydrogen sulfide in concentrations of 2 mg/L or less. Above this level, it could be used in conjunction with chlorination, which oxidizes hydrogen sulfide.

Iron and manganese can be removed by aeration if this cannot be done by other methods. These metals are oxidized to form insoluble hydroxides which precipitate out at proper pH levels. They can then be removed by settlement or filtered out of the water stream. Each 1 ppm of dissolved oxygen will oxidize about 7 ppm of iron or manganese.

Several types of aerators are commonly used: waterfall, diffusion (or bubble), and mechanical. The most commonly used system in utility water treatment is the waterfall.

Waterfall aerators are made in several types: cascade, spray nozzle, and multiple tray. They operate by having the raw water enter the unit from the top and fall by gravity to the bottom, using various methods to evenly distribute and disperse the water throughout the unit. The most common waterfall aerator is the naturally ventilated, multiple-tray type. This unit consists of a series of trays, one above the other, with perforated, slot, or mesh bottoms. The trays are filled with 2 to 6 in (40 to 140 mm) of a medium such as coke, stone, or ceramic balls to improve water distribution and gas transfer as well as provide catalytic oxidation in the medium. The vertical openings between trays range from 12 to 30 in (300 to 750 mm) depending on the number of trays required. Water loading on the trays is usually in the range of 10 to 20 gpm/ft^2 (450 to 900 L/m^2). Efficiency can be improved by the use of enclosures and forced-air blowers to provide counterflow ventilation.

The simplest type of aerator is the diffusion type which bubbles compressed air up through the water tank. The large volume of air required limits the diffusion aerator to smaller flows of water. Air requirements vary from 0.1 to 0.2 scfm (0.17 to 0.34 m^3) per gallon of water aerated. A detention time is necessary, ranging from 10 to 30 minutes. Advantages of this type of aerator are the freedom from cold-weather operating problems, very low head loss, and the possibility of using this process for chemical mixing.

The mechanical aerator consists of an open impeller operating on the water surface of a tank. It is not as efficient as either of the two systems previously described, and so longer retention times are necessary. This unit is also free from cold weather problems and can be used for chemical mixing.

Clarification

Clarification is a process to reduce or remove turbidity, silt, and sediment present in the raw water supply. The water can be treated with chemicals or filtered, depending on the amount of impurities present and the volume of water to be treated. If treated with chemicals, time is required for the solids to settle out of suspension. The chemical treatment process is usually reserved for large volumes of water.

Suspended and colloidal particles will normally range from 0.1 to 10 μm in size. They stay apart because negatively charged ionized matter is absorbed on their surfaces, and thus they repel one an other. A measure of the charge that surrounds the colloid is known as the *zeta potential.* These negative charges are reduced by the use of positively charged chemicals called *coagulants.* Coagulants, through a chemical reaction, reduce the zeta potential and allow the colloids to cluster to form a larger, jellylike mass called *floc.* In a process called *flocculation,* colloidal particles, bacteria, and organic matter are mechanically brought together into larger and heavier particles. These particles are now too heavy to remain suspended, and so they settle out of the water to the bottom. This process is called *sedimentation.*

Clarification of large volumes of water is done in a basin. First, the coagulant(s) must be introduced with a strong action in order to completely disperse and mix the chemicals with the incoming water. Often, some of the sludge produced by flocculation is mixed with the coagulant. After coagulation is accomplished, the mixture is then gently mixed in order to allow the now larger particles to settle rapidly to the bottom of the tank. This settling process is called *sedimentation.* The particles and chemicals that settle out are called *sludge,* which must be removed from the tank and disposed of. Most clarifier designs do both processes in the same operating unit.

The chemical coagulant used most frequently is aluminum sulfate, called *alum* or *filter alum.* Other coagulants used often are sodium aluminate and potash alum. Iron coagulants include ferric sulfate, ferris sulfate, and ferric chloride. Organic polyelectrolytes, which are high-molecular-weight polymers, are also employed in low dosages to increase the effectiveness of treatment.

It is usually a matter of trial and error to find the correct amount of alum. This is so because the amount depends on the size and quantity of suspended solids of the raw water, retention time prior to flocculation, water temperature, and the amount of mixing. All of these conditions change from day to day.

As with almost every process, there is more than one method appropriate to achieve a desired goal when the demand for water is not very large. One alternative flocculation method is to produce a finer floc which can be removed by an appropriate filter system. By eliminating a large settling basin, a high-quality water can be produced more quickly with smaller equipment.

Deaeration

Dissolved gases in the water supply can produce corrosion and pitting. They must be removed prior to using the water for most purposes. These gases are oxygen, carbon dioxide, hydrogen sulfide, and ammonia. A common use of deaeration is in the treatment of boiler feedwater.

Although carbon dioxide and hydrogen sulfide can be removed by aeration, in many cases it is easier and less expensive to remove them by deaeration. This process operates on the principle of raising the water temperature to the saturation point for the existing pressure. There are two types of deaerators, steam and vacuum. When heated water is needed, e.g., for boilers, the steam type is preferred. When cold water is required, the vacuum type is used.

Steam deaerators break up water into a spray or thin film, then sweep the steam across and through it to force out the dissolved gases. Using this method, oxygen can be reduced to near the limit of detectability. Designs fall broadly into spray, tray, and combination units. Typical deaerators have a heating and deaeration section and a storage section for hot, deaerated water. Often, a separate tank is provided to hold an additional 10-mm supply of deaerated water. The direction of steam may be crossflow, downflow, or countercurrent. The majority of the steam condenses in the first section of the unit. The remaining mixture of noncondensable gases is discharged to the atmosphere through a vent condenser.

Vacuum deaerators use a steam jet or mechanical vacuum pumps to develop the required vacuum used to draw off the unwanted gases. The vessel has a packing material inside, and the inlet water is introduced to the top of the unit and passed down through this packing. Deaerated water is stored at the bottom of this vessel. The steam or vapor vacuum located at the top of the unit discharges the unwanted gases. The vacuum unit is far less efficient than the steam one, and is most often used in demineralizer systems to reduce the chemical operating cost and the demineralizer size, and to protect anion exchange resins from possible oxidation damage.

Decarbonation

Decarbonation is rarely used but should be considered if the bicarbonate level in the feedwater is in the range of 14 to 20 mg/L or higher. Decarbonation is usually accomplished by the use of filtered air counterflowing through the water stream and stripping out the carbon dioxide.

Dealkalizing

Dealkalizing is a process that reduces the alkalinity of feedwater. This can be done either with or without the use of acid regenerant. Without the acid, the regenerant used is salt, and the process is often called *salt splitting*. The salt-splitting process exchanges all bicarbonate, sulfate, and nitrate anions for chloride anions. For best results, it is recommended that deionized water be used as feedwater. Where hard, alkaline, low-sodium water is available, the use of a weak acid resin should be considered.

The entire regeneration cycle is similar to that previously described for a water softener. It is not uncommon to use the same salt and regenerant piping to accommodate both the softener and the dealkalizer. Some caustic soda may be added (1 part caustic soda to 9 parts salt) to reduce leakage of alkalinity and carbon dioxide. Since this can cause hardness leakage from the dealkalizer, a filter downstream of the processed water is necessary.

A weak acid resin can also be used. The process transfers the alkaline salts of calcium and magnesium to the weak acid resin. This process should include degasification if required by the product water. The weak acid process operates at a very high utilization factor, near the theoretically required amount. Hydrochloric acid is preferred for regeneration rather than sulfuric acid. This process is very sensitive to the flow rate, temperature, and contaminant level of the feedwater. These changing conditions must be considered by the manufacturer in the design of the process. The entire regeneration cycle is similar to that of cation ion exchange column.

Ion Exchange and Removal

Ion exchange is the basic process where by specific ions in a feedwater stream are transferred onto an exchange medium called *resin* and exchanged for different ions of equal charge. When the ion exchange process is used to treat water only for removal of hardness, it is generally known as *water softening*. Water softening is an ion exchange process that reduces or removes the dissolved impurities that cause hardness in water. This is commonly done by passing the raw water through an ion exchange medium. When the ion exchange process is used to treat water for the removal of ions to produce industrially usable water, it is often referred to as *deionization* or *demineralization*.

The deionization or demineralization process uses different types of resin to exchange first anions and then cations, which will result in the removal of all ions from feedwater when carried to completion. When all the ionic components devolved in water are removed by ion exchange, the water is said to be deionized or demineralized. The ion exchange process is also used to remove dissolved inorganics. Water softening only exchanges some types of ions for others less detrimental for the intended end use of the water. Table 4.9 lists the common anions and cations typically found in water. Ion exchange will not remove significant amounts of organics, bacteria, pyrogens, or turbidity.

Regenerable Ion Exchange. Regenerable ion exchange is a batch process where by ions in raw water are transferred onto a resin medium in exchange for other ions bonded to that medium as the raw water percolates through them. This is accomplished by having the ions in the raw water adsorbed onto a bed of exchange resins and replaced with an equivalent amount of another ion of the same charge. This action continues until the medium has reached its exchange capacity, where it is no longer capable of exchanging ions. Water softening and deionization are the two most common ion exchange processes.

There are two general types of deionizers, working and polishing. The working type is used for the initial removal of the bulk of ions from feedwater or as an ion exchange process alone (such as hardness removal) if the purification is a single process. A polishing type is used to purify feedwater after an initial run through a working ion exchange system.

TABLE 4.9 Comparison of Reverse-Osmosis Polymers

	Thin-film composite	Polyamide	Cellulose acetate	Cellulose triacetate	Polysulfone
pH stability	2–12	4–11	2–8	4–7.5	3–11
Chlorine tolerance	Fair–poor	Poor	Good	Fair–good	Good
Biological resistance	Good	Good	Poor	Fair–good	Good
Temperature limit for stability, °F (°C)	122 (45)	95 (35)	95 (35)	86 (30)	95 (35)
Typical rejection ionic species, percent	> 90	> 90	90	90	90
Flux	High	Low	Low–medium	Low–medium	High

Resins. Resin exchange media include natural inorganic aluminum silicates (sometimes called zeolites or greensands), bentonite clay, synthetic gelatinous resins, and synthetic organic resins. Most commercial processes use the synthetic resins and residential process use zeolites. Resins are graded by purity and consistency of bead size.

Resin is manufactured in the form of a large number of spherical beads, typically of about 0.4-mm diameter. These beads have weakly bonded ions present on their surface that are used for the exchange process. Because the process must exchange ions of the same charge, ion exchange resins are composed of either anion or cation exchange resins. Manufacturers are constantly making new resins for different ion removal purposes.

Traditional deionization exchanges cations with H^+ ions (acids) and anions with OH^- ions (bases). Although it is not 100 percent effective, the end result of these two exchange processes combines the ions remaining in the feedwater to create deionized or demineralized water as the end product.

There are a large number of ion exchange resins available. Each has been formulated to obtain optimum performance for different impurities. The affinity for specific ions in solution is termed the *selectivity coefficient.* The number of charges (valence) available on a particular ionic medium is a major factor in the selection of specific resins to remove the desired impurities, and is based on an analysis of raw water. These resins are contained in a vessel often referred to as a *column.* The actual resin bed could be supported by a mat of graded gravel, screen-wrapped pipe, or perforated plates, which also act to evenly distribute feedwater over the entire resin bed. The size of the resin beads in the vessel also creates an effective depth filter. This filtering action leads to fouling and unpredictable operating runs because of an accumulation of particulates.

Anion resins could be either strong or weak bases. A common anion resin is divinyl benzene, a gelatinous bead. Anion resin type 1 premium has a very close tolerance of bead size. Anion resin type 1 regular is generally used for maximum silica reduction. Resin type 2 is used most often unless type 1 is specifically requested. There is a difference in cost and capacity between the two resins. In general, the higher cost of the type 1 resin is considered acceptable in order to obtain a more efficient and longer-lasting resin. Weak base exchangers are not effective in the removal of carbon dioxide or silica. They remove strong acids more by adsorption than by ion exchange. The end result is the same, and the efficiency of weak base regeneration for acid salt removal is far superior to that of the strong base material doing the same job. Thus weak base units are superior when the feedwater is high in sulfates and chlorides.

The two cation exchange resins used most often are strong or weak acids. Strong cation resins remove all cations regardless of the anion with which they are associated. These resins have a moderate exchange capacity and require a strong acid regenerant such as hydrochloric or sulfuric acid.

The deionization process can be arranged as either a two-step (dual-bed) or single-step (mixed-bed) process. In the dual-bed process, one vessel contains the anion exchange resins and the second vessel, the cation exchange resin. In the mixed-bed unit, a single vessel contains a mixture of both resins. The dual-bed arrangement produces water that is less pure than that produced by a mixed bed, but the dual bed has a greater removal capacity. A typical mixed bed contains 40 percent cation resins and 60 percent anion resins. Dual beds are easier to regenerate. It is not uncommon to have a dual-bed exchanger, often referred to as a *working exchanger*, installed in front of a mixed bed to remove the bulk of the impurities, and then have the mixed bed, often called a *polishing exchanger*, further purify the water to the desired high level. A typical single-bed ion exchange unit is illustrated in Fig. 4.9. A typical dual-bed ion exchange unit is illustrated in Fig. 4.10. A typical mixed-bed ion exchange unit is illustrated in Fig. 4.11. The piping and valve arrangement for different manufacturers may be different.

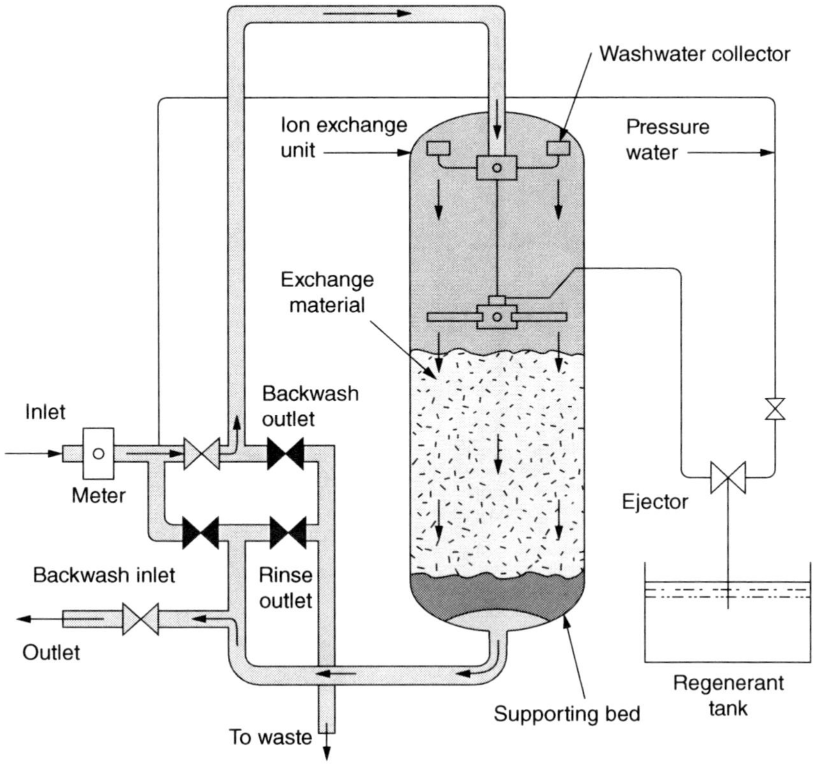

FIGURE 4.9 Typical single-bed ion exchanger.

Regeneration Cycle. The ion exchange process is reversible. As the water continues to pass through the ion exchange resin beds, the number of ions on the resin beads available for exchange declines with time and gradually becomes exhausted. This starts first at the water entry to the vessel and progresses down the bed. When the resin has reached the limit of exchange, the bed is said to have reached its exchange capacity. It is then necessary to take the column out of service to be regenerated.

Regeneration, which is the reverse of deionization, is the term used for the replacement of the ions removed by the feedwater. Regeneration generally consists of three steps: (1) backwashing, (2) application of regenerating solution, and (3) rinsing. Regeneration can be performed either concurrently (in the same direction as the flow of feedwater) or countercurrently (in the opposite direction to the flow of feedwater). All the water used for regeneration must be routed to a drain of adequate size. In addition, the acid and caustic must be neutralized prior to discharge into a public sewer system. It is common practice to combine the acid and caustic waste streams to neutralize the effluent to the greatest extent possible. Additional acid or caustic may have to be added to the final effluent to produce a pH acceptable to the local authorities.

Backwashing is a countercurrent operation that accomplishes two purposes. The first is to remove any particulates that have accumulated in the resin bed and on the beads.

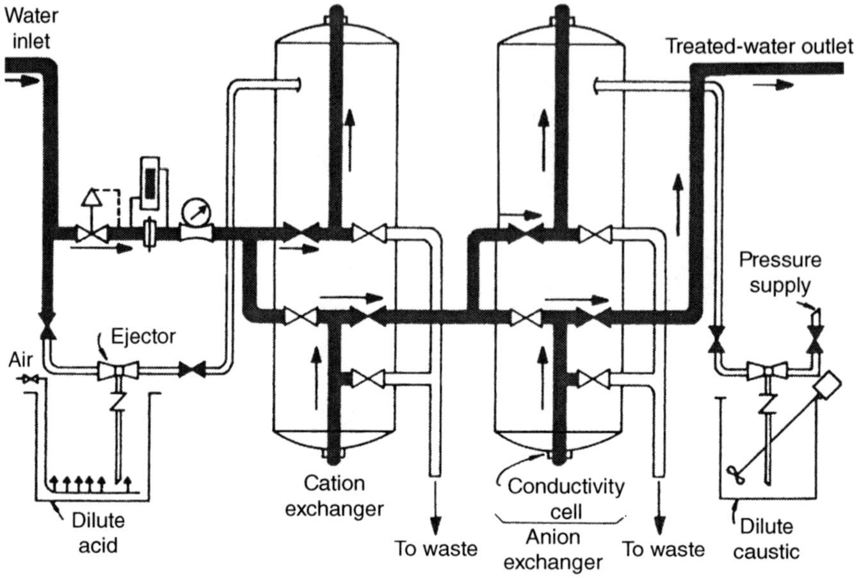

FIGURE 4.10 Typical dual bed-ion exchanger.

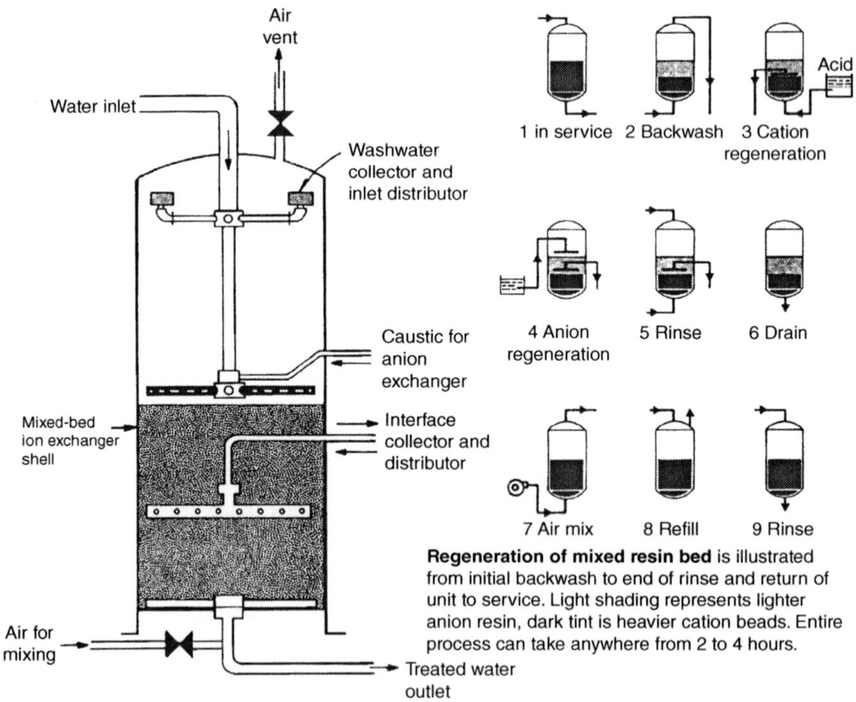

FIGURE 4.11 Typical mixed bed-ion exchanger.

The second is to regrade the resin beads so that new beads are on the top of the bed, which is where the heaviest duty from the beads is required. This is done by having the resin bed expanded by the reverse flow of water from its normal packed condition during use. Individual manufacturers have established the required flow rate of backwash that should be maintained. Too high a flow rate will blow resin out of the tank and into the drain. The flow rate of water should be high enough to scrub the beads together to increase the cleaning action. Higher than recommended flow rates will only waste water and provide no additional benefits.

The two chemicals used to regenerate cation resin beds are either a 93 percent solution of sulfuric acid or a 30 percent to 32 percent solution of hydrochloric acid, also called *muriatic acid.* As they flow through the columns, they replace the retained cations with hydrogen ions from the acid. Hydrochloric acid is used most often because it has the greatest efficiency; the amount of hydrochloric acid used is one-fourth the amount of sulfuric acid. Sulfuric acid is much lower in cost and is used when there is a large quantity of resin to be regenerated, making its use practical. The chemical used most often for regenerating anion resins is a 40 percent mixture of sodium hydroxide, which replaces the retained anion ions with hydroxyl ions. For mixed-bed units, the resins must be separated prior to regeneration.

The quality of the chemicals used for regeneration has an important effect on the maintenance of exchange capacity. Although chemically pure ingredients are not required, some contaminants found in these chemicals collect on the resins and eventually will cause difficulty in operation.

Technical grade acids, free of oils and other organic materials, are acceptable for regeneration of cation resins. They should be 66 degrees Baume (°Bé), free of suspended matter, and light in color. They should mix freely with water and not form any precipitate. Acid-containing inhibitors should not be used. Sulfuric acid is usually the most economical for large-scale use. Hydrochloric acid should be technical grade, a minimum of 30 percent HCl by weight (18°Bé) and shall not contain excessive amounts of iron and organic materials. HCl obtained by the salt-aid or the hydrogen-chlorine process has been found satisfactory. HCl obtained by the hydrolysis of chlorinated organic chemicals should be avoided, particularly if used to treat potable water.

Anion exchange resins are regenerated with 76 percent sodium hydroxide, which must be low in iron, chlorides, and silica to avoid fouling the strong base anion exchangers. Weak base anion exchangers are most economically regenerated with technical grade flake sodium hydroxide. Strong base exchangers are best regenerated using nylon or rayon grade sodium hydroxide, also 76 percent. If purchased in a 40 percent solution, use the grade previously indicated. All caustics must have a maximum of 2 ppm chlorates.

The flush cycle is the shortest. It is a concurrent process whose purpose is to flush away any remaining residue of the regeneration liquids and to repack the bed in preparation for the new run.

The entire regeneration cycle typically takes about 1 h. If the process requires continuous operation, a duplicate set of equipment is installed so that one is in use while the other is being regenerated.

To estimate the frequency of regeneration, first find from the manufacturer the exchange capacity in grains of the selected resin bed. Next, from the analysis of the raw water find the average total dissolved solids (TDS) level and convert this figure to grains per gallon. Dividing the flow rate in gallons into the grain capacity of the resin bed will give the time it takes to saturate the resin bed before regeneration is required.

Service Deionization. Service deionization (DI) is not another form of deionization, but rather a different type of equipment arrangement. In the regenerable type, the DI equipment is permanent and the regeneration is done on site by operations or maintenance personnel, who must handle and store the chemicals used for regeneration.

The service DI system uses individual cartridges or tanks for the anion, cation, and mixed beds. When the individual cartridges are exhausted, the supplier replaces them with recharged units on site, and the exhausted cartridges are removed to be regenerated at the supplier's premises. This arrangement considerably reduces the initial cost of the equipment, eliminates the need to store chemicals, and frees the operations or maintenance personnel from the job of regenerating the units. In addition, it saves water that does not have to be used for backwash. However, the operating costs are higher than for the permanent bed type.

Continuous Deionization. Continuous deionization (CDI), also known as electrodeionization (EDI), is a continuous water purification process that uses direct current, an alternating arrangement of cation- and anion-permeable ion exchange membranes that form parallel flow compartments (concentrating compartments) on either side of an additional flow compartment containing a thin layer of mixed-bed ion exchange resin (diluting compartment). The components, called a *cell pair*, are installed in a plate-and-frame device where the flow compartments form various flow paths for the product and wastewater. This arrangement is illustrated schematically in Fig. 4.12.

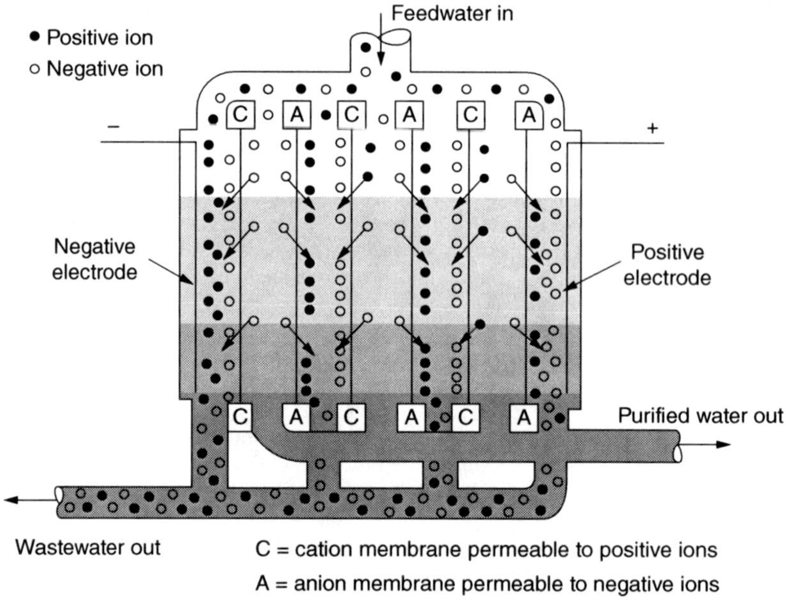

FIGURE 4.12 Schematic operation of a continuous deionization unit. In a continuous deionization cell, electric field pulls positive ions in feedwater through cation-permeable membranes (C), and negative ions through anion-permeable membranes (A). Blockage by next membrane confines all ions to one set of channels, purified water to the other set. The result is one purified stream, one waste stream.

The feedwater flows through the diluting compartment. When a direct-current (dc) field is applied across the pair of membranes, ions move from the diluting stream, through the ion beads and the membrane into the concentrate stream, thereby producing separation. With an alternating arrangement of cell pairs, the cations and anions are trapped in the concentrate stream where they are routed to drain. The resin bed serves as a highly conductive medium through which the ions flow. The various flow streams are hydraulically independent, allowing a high volume of high-purity water (product) and a low volume of concentrate (waste).

The resin-filled diluting compartment (cell) creates a low-level resistance path for ions. At the outlet of the diluting cell, under the proper combination of flow, temperature, water conductivity, and voltage, the ion exchange resins will regenerate automatically without the use of added chemicals. This process is continuous and results in a steady supply of high-purity water from the diluting compartment.

Continuous deionization (CDI) is very sensitive to feedwater impurities. Experience has shown that very few natural or potable feedwater supplies can meet the required feedwater specifications without softening. Typical feedwater system requirements are the same as those for potable water. Individual manufacturers may have other specific requirements. Because of these requirements, most processes that purify water for pharmaceutical applications use CDI for polishing purposes after reverse osmosis (RO). Another disadvantage of this process is that the membrane used is incompatible with standard chemical sanitizing agents and sensitive to the frequency of sanitizing.

Lime–Soda Ash Addition. The lime–soda ash method uses either hydrated lime or quicklime along with soda ash. When added to water, these chemicals react with the dissolved calcium and magnesium carbonate to form insoluble compounds. These compounds precipitate out of solution and are passed through a filter to remove them. They are then discharged to drain. This process is usually carried out during the clarification process rather than as a separate process and is reserved for large volumes of water.

Ion Exchange. The ion exchange process is a cation exchange used to remove insoluble and scale-forming iron, calcium, magnesium carbonate, and other multivalent cations, which are the primary causes of hardness, and replace them with sodium ions, which do not contribute to hardness. The removal of these impurities prevents the buildup of insoluble scale precipates on the piping and reverse-osmosis membrane. Ion exchange is accomplished by passing the water through a bed of granular sodium cation exchange resin. This process is commonly called *sodium cycle ion exchange.*

The resin bed typically occupies about two-thirds of the tank. The other one-third is needed for expansion of the resin bed during backwash. A generally accepted range of 0.4 to 3 gpm/ft^3 (0.34 to 2.51 L/m^2) of resin is used to determine the volume of resin and cycle time of the unit.

Microbial growth inside the unit is a concern in softening systems used for pharmaceutical and some laboratory purposes. The water softener is regenerated with a brine solution that does not destroy bacteria. The liquid brine solution storage and regeneration equipment also allows microbial growth in storage tanks that are exposed to the atmosphere. An alternative is to use a dry storage system, which generates a salt solution from water mixed with salt pellets only when necessary for regeneration. The dry storage system controls microbial growth better than wet systems, but constant maintenance is required to monitor the brine tank. The quality of the salt in all systems should be checked periodically to ensure that there are no added substances present.

The regeneration cycle is similar to that previously discussed. The difference is that salt is used to regenerate the resin bed. Industrial water softeners use rock salt for economy. Rock salt, because of its high mineral content, requires a special tank called a dissolver to dissolve it in water prior to use. The water softener is similar to the schematic single-bed ion exchanger illustrated in Fig. 4.9.

Ion Exchange System Design Considerations. One of the major decisions that must be made when selecting an ion exchange system concerns how much leakage to allow. Leakage is the presence of unwanted ions in the final treated water. The amount of leakage is a function of the completeness of the resin regeneration. For water softening, generally accepted leakage amounts range from 0.1 to 1 ppm. Since total regeneration of the resin bed is impractical, most water softeners operate at one-half to two-thirds of their ultimate capacity.

There is sodium leakage from cation exchangers and silica leakage from anion exchangers. Normally, mixed-bed units have negligible leakage.

In general, adequate purification of the water stream for high-purity applications is not possible in one pass through the system, so a polisher is necessary. A mixed-bed ion exchange system has an initial cost about 74 percent less than a dual-bed system when used as a polisher. A single-pass RO system is about equal to a dual-bed ion exchange system.

Usually, if the water demand for a facility is less than about 40 gpm, the greatest benefit will be derived from simpler and less costly equipment at the expense of higher operating costs. For this water demand, it is usual to have a mixed-bed unit without a degasifier, which is not required. For systems with a requirement of 200 gpm or more, most of the systems installed have multiple-bed units and a degasifier. Manufacturers must be consulted for the specific system, resin selection, and required equipment.

Where applicable, the use of weakly acidic and weakly basic resins minimizes chemical costs and reduces losses to waste because of high regenerative capacity.

There are some problems, such as microbial growth, associated with water softeners. Sanitization is usually accomplished during regeneration. Iron buildup in the unit could pass through to downstream purification equipment unless operating personnel constantly monitor the water quality.

Membrane Filtration and Separation

Membrane filtration and separation is a general term for a water purification process that removes contaminants from feedwater by means of a thin, porous barrier called a *membrane*. When used as a filter, a membrane is capable of removing impurities of a much smaller size than other types of filters. Filters of this nature are often called *ultrafilters* and *nanofilters*.

A semipermeable membrane limits the passage of selected atoms and/or molecules in a specific manner. Membrane filtration and separation, when used to produce pure water, is characterized by having the feedwater flow parallel to the membrane (often called *tangential flow*). Not all the feedwater is recovered. Many of the membranes used are also available as both depth filters and, in single thickness, disk filters. There are two general categories of membrane filtration: reverse osmosis using a semipermeable membrane and filtration using ultrafiltration and nanofiltration membranes.

Reverse Osmosis. In the natural osmosis process, when two solutions of different concentrations are separated by a semipermeable membrane, water molecules from the less concentrated solution will spontaneously pass through the membrane to dilute the more concentrated solution. This continues until a rough equilibrium is achieved. The driving force is a difference of pressure, called the *osmotic pressure* or *concentration gradient*, that exists across the membrane and is based on the degree of concentration of contaminants. The pressure in the stronger solution is lower than that in the weaker solution. This pressure is what drives the flow of solvent. The flow or flux will continue until the osmotic pressure is equalized, which then results in a higher pressure on the concentrated solution side that is equal to the osmotic pressure.

Reverse osmosis is the flow of solvent in the direction opposite to that of natural osmosis. If enough pressure is applied to the more concentrated solution, which in these discussions is water, pure water is diffused through the membrane, leaving behind the bulk of the contaminants. These contaminants are continuously flushed to drain, thereby removing them from the system. The purified water is called *permeate*, and the contaminant-containing water is called *reject* or the *reject stream*. In some cases, the reject stream is referred to as *salt*. The performance characteristics of the selected membrane determine how large a system is required. The flow rate is measured in membrane flux, which is a measurement of the flow rate of permeate that will pass through a given area of the membrane at a specific temperature and pressure. The ratio of purified water flow to the feedwater flow is called *recovery*. Most applications require a minimum 40 percent recovery rate to be considered practical. Rejection characteristics are expressed as a percent of the specific impurities retained and depend on ionic charge and size.

Membrane Module Configurations. There are four types of membrane module configurations used for RO applications: hollow fiber, spiral wound (SWRO), tubular (TRO), and plate and frame. Hollow-fiber and spiral-wound are the most common configurations. In each design, maximum turbulence is necessary to avoid concentration polarization.

HOLLOW-FIBER REVERSE OSMOSIS. The hollow-fiber configuration, illustrated in Fig. 4.13, consists of a perforated tube manufactured from ceramic, carbon, or porous plastic with inside diameters ranging from 1/8 to 1 in (8 to 25 mm). They require rigid support when mounted inside the pressure vessel. Feedwater can be introduced either into the center or outside depending on the manufacturer of the RO module. Fouling resistance is low.

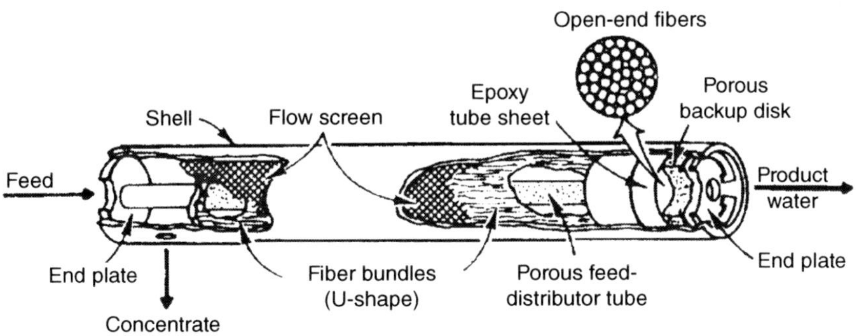

FIGURE 4.13 Hollow-fiber reverse-osmosis configuration.

SPIRAL-WOUND REVERSE-OSMOSIS (SWRO). This configuration, illustrated in Fig. 4.14, typically achieves a large surface area per unit volume. In this design, a flat membrane is formed around a fabric spacer closed on three sides with the open side terminating in a perforated produce water tube. The unit is then placed in a pressure vessel. Feedwater permeates the membrane and flows radially inside the enclosure toward the product tube.

TUBULAR REVERSE OSMOSIS (TRO). This configuration, illustrated in Fig. 4.15, consists of a perforated tube manufactured from ceramic, carbon, or porous plastic with larger inside diameters than the hollow-fiber configuration. The membrane is installed on the inside of the tube. A number of tubes are installed inside a pressure vessel. Feedwater enters the tube and permeates the membrane to be collected on the outside. The feedwater channels are much more open than the SWRO and less subject to fouling.

PLATE AND FRAME. This configuration, illustrated in Fig. 4.16, consists of a membrane that is fixed to a grooved plastic or metal plate with several plates stacked together in a frame that includes feedwater and drain ports. As the feedwater flows across the membrane surfaces, the purified water penetrates the membrane and flows along the frame for collection. The retentate continues to flow and could be recirculated or directed to drain.

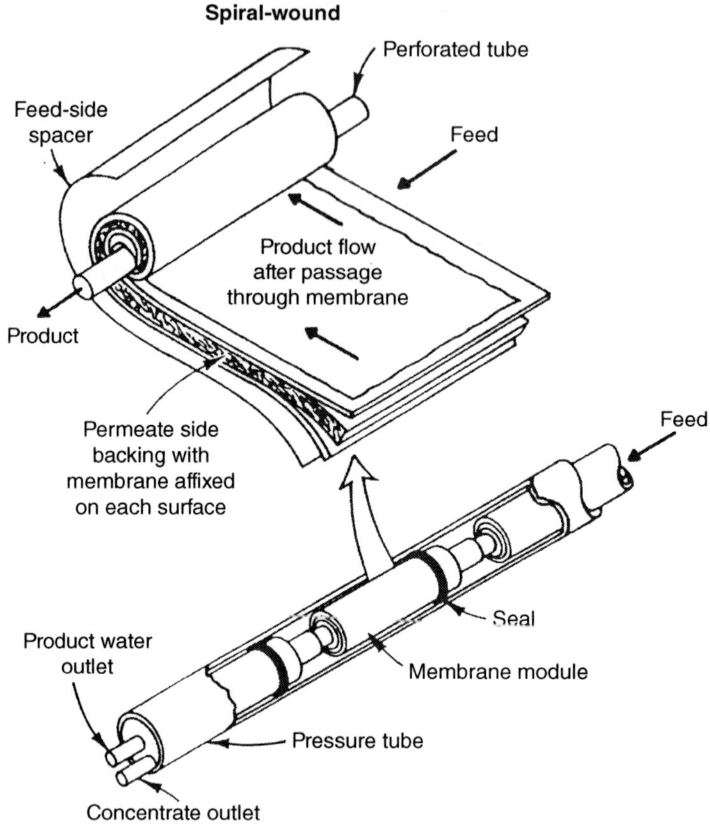

FIGURE 4.14 Spiral-wound reverse-osmosis configuration.

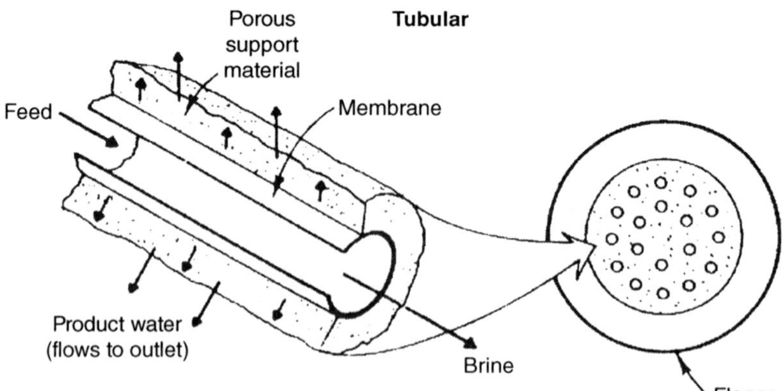

FIGURE 4.15 Tubular reverse-osmosis configuration.

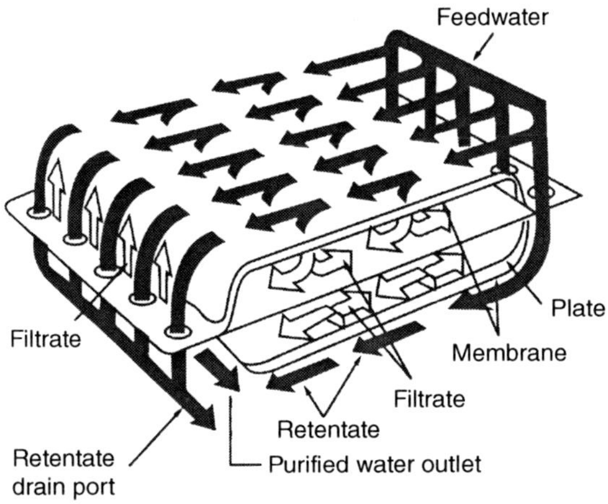

FIGURE 4.16 Plate and frame reverse-osmosis configuration.

This configuration is used mostly for filtration and rarely for RO systems. Packing density is low and resistance to fouling is very high. It is used for small to medium volumes, generally less than 20 gpm.

Membrane Selection. System performance is determined by considering the following factors that influence the capacity of the individual membranes selected:

1. Operating pH
2. Chlorine tolerance
3. Temperature of the feedwater
4. Feedwater quality, usually measured as SDI
5. Types of impurities and previous feedwater treatment
6. Membrane flux
7. Number of operating hours
8. Resistance to biodegradation and ability to be sanitized
9. Rejection characteristics. Typical RO systems remove the following contaminants to the following levels:
   - ***a.*** Inorganic ions 93% to 99%
   - ***b.*** Dissolved organics > 99%
   - ***c.*** Particulates > 99%
   - ***d.*** Microorganisms > 99%

The selection of a system configuration must consider the following:

1. Maximum recovery
2. Fouling properties and resistance
3. Production rate per unit volume

There are only a small number of polymers that have the necessary characteristics to function as semipermeable membranes. Typical characteristics and a comparison of these membranes are given in Table 4.9.

1. Thin-film composite of various polymer materials
2. Polyamide
3. Cellulose acetate
4. Cellulose triacetate
5. Polysulfone

Membrane Filtration. Ultrafiltration and nanofiltration describe membranes that are categorized by their pore size. Ultrafiltration membrane pore sizes range from 0.001 to 0.02 μm. Nanofiltration membranes have pore sizes that will allow the passage of solids to 10,000 daltons. The 10,000-dalton cutoff is recommended for the complete removal of pyrogens. Typical recovery rates for ultrafilters range between 95 and 98 percent, with the remainder flushed to drain.

A membrane is manufactured by bonding the membrane onto a porous, supporting substrate and then configuring it into elements. These filters are generally used for pretreatment in the removal of colloids, bacteria, pyrogens, particulates, and high-molecular-weight organics. Spiral wound and hollow fiber are the two configurations used most often.

Microbial Control

Chemicals. The most common disinfection method is the addition of oxidizing or nonoxidizing chemicals. Chemicals could be either *biocides,* which are substances that kill microbes, or *biostats,* which prevent the further growth of microbes. Commonly used chemicals are chlorine, chlorine compounds, hydrogen peroxide, and acid compounds. In order to be effective, the chemical must have a minimum contact time in the water. In addition, a residual amount of the chemical must be present to maintain disinfection.

Chemicals add impurities to the water and are not generally suitable for a pure water environment. They disinfect potable and process water and equipment by injection directly into the fluid stream by means of a metering pump. They must be removed from feedwater used for purification. Chlorine may produce trihalomethanes.

Ultraviolet Radiation. Ultraviolet (uv) radiation is an in-line process. UV light is generated using mercury vapor lamps. The uv spectrum is divided into three wavelengths: uvA (315–400 nm), uvB (280–315 nm), and uvC (less than 280 nm). Only uvB and uvC wavelengths can economically produce the intensity and energy output necessary for the intended germicidal treatment. Federal standard 209E and aseptic guidelines issued by the FDA provide some guidance for the use and application of uv irradiation.

The 254-nm wavelength operates in the germicidal region, sterilizing by destroying bacteria, mold, viruses, and other microorganisms. This wavelength is preferred for pure water systems and will significantly reduce the multiplication of organisms.

The 185-nm wavelength operates in the ozone-forming region, where it has the ability to break down organic molecules to carbon dioxide by the photooxidation process. It slowly breaks the bonds in organic molecules by direct radiation, and also oxidizes organisms by the formation of hydroxyl radicals. A flow rate of approximately 2 ft/s is an industry standard for effective sanitization of purified water. Flow rate through the uv device should be reduced compared to that of the circulation loop to extend the necessary contact time. The recommended location is ahead of deionization equipment.

Problems include generation of ions that lower the resistivity of water and the possible leaching of silica from the quartz sleeve of the uv device. Glass, plastic, rubber, and similar materials exposed to uv radiation will eventually crack, etch, discolor, and flake. Tests have shown that only one-half the energy used by a new bulb is actually transmitted to the water and that in time an additional 25 percent of the output will be lost compared to a new bulb.

Filtering. A filter will remove organisms from the fluid stream. Generally accepted practice is to use a 0.2-μm absolute filter for the removal of bacteria. However, some authorities question the effectiveness of this practice. Recommended current practice is to use a membrane filter with an absolute rating (cutoff) of 10,000 daltons. Cartridge filtration is the most common method.

Heat. Heating to 175°F (80°C) will effectively sanitize water under pressure. The heating can be accomplished by using steam, electricity, or other heat exchangers. It is common practice to circulate purified water at this temperature and to use heat exchangers to lower the water temperature at each point of use if necessary.

Ozone. Ozone (O_3) is an oxidizing gas generated from gaseous oxygen or catalytically from water. The most popular method to produce ozone is by a corona-discharge generator converting the oxygen in air to ozone. The air is passed between two electrodes where an electrostatic discharge across the gap converts oxygen to ozone.

The ozone system consists of a feed gas treatment unit, an ozone generator, a water–ozone contact mechanism, and a destruction unit to eliminate any residual ozone.

FEED GAS TREATMENT. The gas reaching the generator must have all particles larger than 0.3 μm and 95 percent of particles larger than 0.1 μm removed. In addition, aerosols, moisture, and hydrocarbons must be removed, as required by the manufacturer. This purification is usually supplied as part of a package.

OZONE GENERATOR. The three basic types of generators are the Lowther plate unit, Otto plate unit, and tube units. They differ only in the manner they are cooled. Ozone generators use large amounts of electric power, generally between 15 and 26 kWh per kilogram of O_3.

WATER–OZONE CONTACT MECHANISM. Ozone and water at a pH of 7.5 or less are mixed in direct contact with each other by the use of static or mechanical mixers, injectors, or columns that optimize the dissolution of the gas.

DESTRUCTION UNIT. Depending on the generator, ozone concentrations can vary from 100 to 3000 ppm. Because high concentrations are harmful to humans and metals, the ozone should not be allowed to escape to the atmosphere without being treated to a level below 0.1 ppm. Destruction can be accomplished by catalytic, thermal, or activated carbon. Thermal units operate at a temperature of 300°C and generally require a 3- to 5-min contact time to be effective.

The mechanism for ozone oxidation is through the generation of hydroxyl radicals. The gas is directly injected into the water stream. Problems with this method include incomplete oxidation of organic compounds and slow operation. Action is enhanced with the application of uv radiation at 185 nm.

UTILITY WATER TREATMENT

Water from wells, rivers, lakes, and streams is commonly used for cooling and washing purposes. Clarifying and treating this water in order to meet the purity requirements of the proposed end use might be necessary. This requires good monitoring and quick reactions to raw and treated water fluctuations. If the water is to be recirculated, the treatment methods are more stringent. If the water is to be used once and then discharged into the environment, local codes must be followed regarding allowable chemicals present in the discharged wastewater in order to avoid the need for waste treatment.

Potable Water Treatment

Water used for human consumption or intended to be part of food products must be treated to comply with the Safe Drinking Water Act (SDWA), local regulations, and also the Surface Water Treatment Rule, which is a part of the SDWA. In addition, compliance with regulation 10 CFR 141 is required. Substances that affect the quality of potable water are classified in four major categories:

Physical. Physical characteristics related to the appearance of the water include color, turbidity, taste, and odor. Physical quality is corrected mostly through the use of various types of filters.

Chemical. The chemical characteristics of water are related to dissolved minerals (mostly hardness), gases, and organics. The chemical quality of water is adjusted by the use of water softeners, ion exchange, RO units, and activated charcoal units to remove organic impurities.

Biological. This characteristic is concerned with microorganisms that will affect the health of the consumer. Biological treatment requires the use of biocides and biostats to eliminate and reduce the number of microbes present in the water and to create a residual amount of the chemical to maintain the required level of action required by code.

Radiological. This is concerned primarily with radon in areas where the water may have come in contact with radioactive substances. Retention and aeration will lower the radon count to acceptable limits in approximately 8 h.

A synopsis of general treatment methods for small potable water systems that shows general advantages and disadvantages is given in Table 4.10.

Domestic Water Treatment for Private Residences

The domestic water treatment most often used in home water softeners to reduce the amount of "hardness" found in the water is water softening. This generally involves the use of an ion exchange water softener, sometimes called a *cation exchange unit*. Water softening should be considered to reduce hardness when obtained from a surface source or potable water system when considered unacceptable or used for recirculated or cooling purposes.

Water hardness is expressed in units of grains per gallon (gpg) or parts per million (ppm). The conversion factor between the two is 1 grain per gallon (gpg) equals 17.1 mg/L (as $CaCO_3$). The term *grain* comes from the fact that it required 7000 average grains of dry wheat to equal 1 lb.

The term *ion exchange* is a descriptive phrase that describes a variety of processes. This is where a specific ion or group of ions of calcium is exchanged for another. Water softening by ion exchange involves the removal of soluble calcium ions by exchanging them for sodium ions before they can become scale. Water is considered soft when all the calcium and magnesium ions have been replaced with sodium ions.

Water enters the top of the softener through an common multiport valve that distributes the water through the resin bed where the ion exchange occurs. It passes through a series of strainers and is collected in the bottom of the softener, where it is returned to the multiport valve. The softened water passes through the valve and is diverted to the facility.

TABLE 4.10 Water Treatment Technology for Small Potable Water Systems

Technology	Advantages	Disadvantages
	Filtration	
Slow sand	Operational simplicity and reliability, low cost, ability to achieve greater than 99.9% *Giardia* cyst removal	Not suitable for water with high turbidity, requires large land areas
Diatomaceous earth (septum filter)	Compact size, simplicity of operation, excellent cyst and turbidity removal	Most suitable for raw water with low bacterial counts and low turbidity (< 10 NTU), requires coagulant and filter aids for effective virus removal, potential difficulty in maintaining complete and uniform thickness of diatomaceous earth on filter septum
Reverse-osmosis membranes	Extremely compact, automated	Little information available to establish design criteria or operating parameters, most suitable for raw water with turbidity < 1 NTU, usually must be preceded by high levels of pretreatment, easily clogged with colloids and algae, short filter runs, concerns about membrane failure, complex repairs of automated controls, high percent of water lost in backflushing
Rapid sand/direct filtration package plants	Compact, treats a wide range of water quality parameters and variable levels	Chemical pretreatment complex, time consuming, cost
Chlorine	Very effective, has a proven history of protection against waterborne disease, widely used, variety of possible application points, inexpensive, appropriate as both primary and secondary disinfectant	Potential for harmful halogenated by-products under certain conditions
Ozone	Very effective, no Triholomethanes (THMs) formed	Relatively high cost, more complex operation because it must be generated on site, requires a secondary disinfectant, other by-products
Ultraviolet radiation	Very effective for viruses and bacteria, readily available, no known harmful residuals, simple operation and maintenance for high-quality waters	Inappropriate for surface water, requires a secondary disinfectant

(Continued)

TABLE 4.10 Water Treatment Technology for Small Potable Water Systems (*Continued*)

Technology	Advantages	Disadvantages
	Organic contaminant removal	
Granular activated carbon	Effective for a broad spectrum of organics	Spent carbon disposal
Packed-tower aeration	Effective for volatile compounds	Potential for air emission issues
Diffused aeration	Effective for volatile compounds and radionuclides	Clogging, air emissions, variable removal efficiencies
Advanced oxidation	Very effective	By-products
Reverse osmosis	Broad spectrum removal	Variable removal efficiencies, wastewater disposal
	Inorganic contaminant removal	
Reverse osmosis	Highly effective	Expensive waste removal
Ion exchange	Highly effective	Expensive waste removal
Activated alumina	Highly effective	Expensive waste removal
GAC	Highly effective	Expensive waste removal

All softeners must be taken out of service when the exchange capacity is exhausted. The regeneration, or backwash, step is performed by first diverting water backward through the multiport valve, forcing it up through the top of the unit. This expands and redistributes the resin bed; removes dirt, debris, and broken beads from the unit; and reduces resin fouling. This is discharged to drain. Then the softener is regenerated using a brine (salt) solution. The brine strength, flow rate, and distribution relate directly to the softener's performance. The brine regeneration restores the exchange capacity of the resin beads by displacing the calcium and magnesium ions that have collected on the beads with sodium ions. The brine is discharged to drain.

The final step is a rinse to remove the excess salt from the unit. This usually takes 30 min to 1 h. There is a fast rinse and slow rinse possible; the selection is made by the user or required by the manufacturer.

Water softening should be considered to reduce hardness when obtained from a surface source or potable water system when considered unacceptable or used for recirculated or cooling purposes.

The size of a water softener is based on the flow through the unit and the ionic removal of the resin bed. A simplified method would use the following steps:

1. Determine the hardness of the raw water by sending it to be analyzed.
2. Divide the total hardness by 17.1 to find the total grains per gallon to be removed.
3. Determine the amount of water (in gallons) used per day, and calculate the amount of total grains to be removed per day.
4. Find the number of days desired between regeneration. Multiply the number of days by the total grains per day to find the total number of grains per regeneration.
5. Determine the flow through the softener. The unit is selected based on the flow rate and the amount of total grains removed per generation.
6. Trace amounts of radiation, such as that from uranium and radium, can be removed by water softening.

BIBLIOGRAPHY

Blake, R. T., *Water Treatment for HVAC and Potable Water Systems*, McGraw-Hill, New York, 2009.

Brown, J., N. Jayawardena, and Y. Zelmanovich, "Water Systems for Pharmaceutical Facilities," *Pharmaceutical Engineering Magazine*.

Cartwright, Peter S., "Reverse Osmosis and Nanofiltration System Design," *Plumbing Engineer Magazine*.

Collentro, W. V., "Pharmaceutical Water," Parts 1 and 2, *Ultrapure Water Magazine*.

Dow Chemical Corporation, *Water Conditioner Manual*.

Gorry, M., P. Amin, and D. W. Richardson, Sr., "Take the Guesswork out of Demineralizer Design," *Chemical Engineering Magazine*.

Janoschek, R., and G. C. du Moulin, "Ultraviolet Disinfection in Biotechnology: Myth vs Practice," *BioPharm Magazine*.

Meyers, P., "Trace Contaminant Removal," *Water Technology Magazine*.

Meyrick, C. E., "Practical Design of a High Purity Water System," *Pharmaceutical Engineering Magazine*.

Nalco Chemical Company, *The Nalco Water Handbook*, 2d ed., McGraw-Hill, New York, 1988.

Parekh, B. S., "Get Your Process Water to Come Clean," *Chemical Engineering Magazine*.

Stenzel, M. H., "Remove Organics by Activated Carbon Adsorption," *Chemical Engineering Progress Magazine*.

Yeh, K. L., and S. H. Lin, "Looking to Treat Wastewater-Try Ozone," *Chemical Engineering Magazine*.

CHAPTER 5
DESALINATION

GENERAL

Desalination, sometimes called *desalinization*, refers to any of several processes that remove excess salt and other minerals from seawater and brackish water for the purpose of making freshwater. Because only one percent of earth's water is fresh (potable), water can be in short supply in some parts of the world. As the population continues to grow, shortages of freshwater will occur more often, if only in certain locations. The oceans can be used as a means of supplementing the freshwater supply, where needed, to provide people with potable water. Water desalination is regarded as a cost-effective, environmentally safe, sustainable, and drought-proof water supply source. This process has been practised since ancient times.

The scarcity of freshwater resources and the need for additional water supplies are already critical in many arid regions of the world and will be increasingly important in the future. It is very likely that the water resources will be considered, similar to fossil fuel energy resources, to be one of the determinants of world stability. Many arid areas simply do not have freshwater resources in the form of surface water, such as rivers and lakes, and have only limited underground water resources that are becoming more brackish as water from the aquifers continues to be used and seawater infiltrates the aquifer.

Desalination or distillation is one of humans' earliest forms of water treatment, and it is still a popular treatment solution throughout the world today. In ancient times, many civilizations used this process on ships to convert seawater to drinking water. Today, desalination plants are used to convert seawater to drinking water on ships and in many arid regions of the world, and to treat water in other areas that is fouled by natural and humanmade contaminants. However, along with recycled water, this is one of the only non-rainfall-dependent water sources particularly relevant to countries that traditionally have relied on rainfall in dams to provide their drinking water supplies.

The worldwide availability of renewable energies and the availability of mature technologies in this field make it possible to consider the coupling of desalination plants with renewable energy production processes to ensure the production of water in a sustainable and environmentally friendly scheme for the regions concerned. Solar desalination is used by nature to produce rain, which is the main source of freshwater on earth. All available humanmade purification systems are mostly a duplication on a small scale of this natural process. Recently, considerable attention has been given to the use of renewable energy as sources for desalination, especially in remote areas, because of the high costs of fossil fuels, difficulties in obtaining them, attempts to conserve fossil fuels, interest in reducing air pollution, and the lack of electric power in remote areas.

The purpose of this chapter is to acquaint the reader with some of the various methods and technologies involved in the selection of desalinization methods, but not to present detailed design parameters. It is strongly recommended that an engineer select an experienced entity in the design of these systems.

ECONOMICS

A number of factors determine the capital and operating costs for desalination: capacity and type of facility, location, feedwater, labor, energy, financing, and concentrate disposal. Desalination stills now control pressure, temperature, and brine concentrations to optimize the water extraction efficiency. Nuclear-powered desalination might be economical on a large scale.

In California, the towns of Santa Barbara and Avalon have begun using desalination methods to remove impurities from seawater and make it suitable for drinking. A promising method to desalinate seawater is the *reverse-osmosis* method. Another more familiar method, *distillation*, involves heating seawater and capturing and condensing the resulting steam. In previous years, the high cost of desalination has kept it from being used more often, as it can cost more than $1000 to $2200 per acre-foot (1992 cost basis) to desalinate seawater compared to about $200 per acre-foot for water from normal supply sources. Desalination technology is improving and costs are falling, however, and Tampa Bay, Florida, is currently desalinizing water at a cost of only $650 per acre-foot. As the demand for both freshwater and technology increases, you can expect to see more desalination occurring, especially in areas such as California, the east coast, and the Middle East.

Large-scale desalination typically uses extremely large amounts of energy as well as specialized, expensive infrastructure, making it very costly compared to the use of treated freshwater from rivers or groundwater.

While noting that costs are falling, and generally positive about the technology for affluent areas that are proximate to oceans, one study argues desalinated water may be a solution for some water-stress regions, but not for places that are poor, deep in the interior of a continent, or at high elevation. Unfortunately, that includes some of the places with the biggest water problems, where one needs to lift the water 2000 m (6600 ft) or transport it over more than 1600 km (990 mi) to get transport costs equal to the desalination costs at a coast. Thus, it may be more economical to transport freshwater from somewhere else than to desalinate it. For cities on the coast, desalination is being increasingly viewed as an untapped and unlimited water source.

Nevertheless, desalination does not take into account recycling water and broken infrastructure. Water is reused (graywater) in Fountain Valley, California; Fairfax, Virginia; El Paso, Texas; and Scottsdale, Arizona. This process is an alternative to desalination that requires 50 percent less energy due to the significantly lower salt content and produces new water at 30 percent less cost to the consumer than desalinated seawater without the damage to marine life and ecosystems common to desalination plants.

To offset the energy use, the Sydney, Australia, plant cites mitigating concerns about harmful greenhouse gas emissions, a common argument used against seawater desalination due to the energy requirements of the technology. The purchase or production of renewable energy to power desalination plants naturally adds to the capital and/or operating costs of desalination.

Desalinating 1000 US gal (3800 L) of water can cost $3.

SALINE WATER

Saline water, the subject of this chapter is seawater obtained from the ocean. This could also be interpreted as "brackish water." Ocean water from different parts of the world is not uniform. Ocean water that is saline (referred to as "concentrations") is defined as containing 1000 parts per million (ppm) of dissolved salts. Water can be categorized as follows:

Freshwater—less than 1000 ppm
Slightly saline water—from 1000 to 3000 ppm
Moderately saline water—from 3000 to 10,000 ppm
Highly saline water—from 10,000 to 35,000 ppm

Ocean water generally averages about 35,000 ppm of salt.

Some Desalination Facts

1. It is estimated that some 30 percent of the world's irrigated areas suffer from salinity problems, and remediation is seen to be very costly.
2. In 2002, there were about 14,500 desalination plants around the world in 120 countries. They produce some 14 million m^3/day of freshwater, which is less than one percent of total world consumption.
3. The most important users of desalinated water are in the Middle East (mainly Saudi Arabia, Kuwait, the United Arab Emirates, Qatar, and Bahrain), which uses about 70 percent of worldwide capacity, and in North Africa (mainly Libya and Algeria), which uses about 6 percent of worldwide capacity.
4. Among industrialized countries, the United States is one of the most important users of desalinated water (6.5 percent), especially in California and parts of Florida.

COMPONENT SELECTION CONSIDERATION

Intake and Intake Structures

Access to a reliable supply of consistent quality seawater with minimum economic impact is one of the most fundamental issues to be addressed when evaluating potential sites for a coastal desalination plant. Intake designs are highly site-specific. Physical characteristics such as meteorological and oceanographic data, marine biology, and the possible effects of fouling, pollution, and navigation must be taken into consideration. The siting of new plants may require a current assessment of impingement and entrainment impacts as part of the environmental review and permitting.

The distance from the intake to the desalination plant affects the plant economics, and the available feedwater quality will directly affect the pretreatment process and the final determination of downstream quality. The ability to permit a new desalination facility may hinge on the methods used to mitigate environmental concerns and the marine environment. Consideration must be given to the potential loss of recreational facilities and to the visual impact that will result from certain intake arrangements.

There are typically three types of intake structures used: surface intakes, submerged inlets, and subsurface inlets.

1. *Surface intakes* float below the surface of the water. The intake water is usually prescreened using traveling water filter screens, mechanically screened bar screens, or well screens. In many cases the screening chamber is located onshore, and the actual intake pipe will extend into the sea. These screens must be cleaned by special screen cleaning pumps.

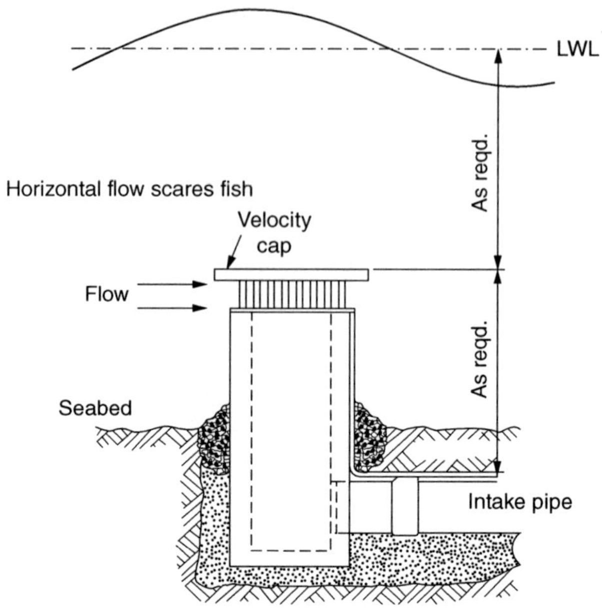

FIGURE 5.1 Typical submerged inlet.

2. *Submerged inlets* are placed on the seafloor. Figure 5.1 illustrates a typical intake structure, and Fig. 5.2 illustrates a typical schematic arrangement of piping to the plant.

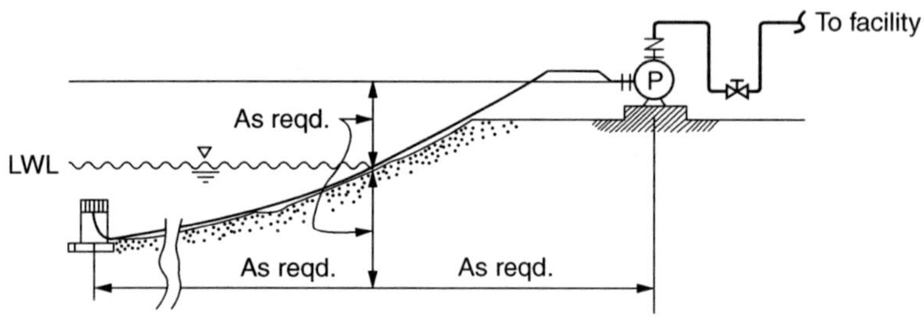

FIGURE 5.2 Schematic of typical piping arrangement.

3. *Subsurface inlets* are where the intake is collected via horizontal or vertical beach wells, infiltration galleries, or seabed filtration galleries. Refer to Fig. 5.3 for a typical beach well inlet. These inlets provide natural filtration and have the advantage of separating out many marine organisms from the intake.

There is great variation in the types of structures for specific plants. Good judgment and consultation with knowledgeable individuals familiar with this subject are strongly recommended.

A good intake design will not only protect the downstream equipment and reduce environmental impact on marine life but also reduce the operating costs of the pretreatment equipment.

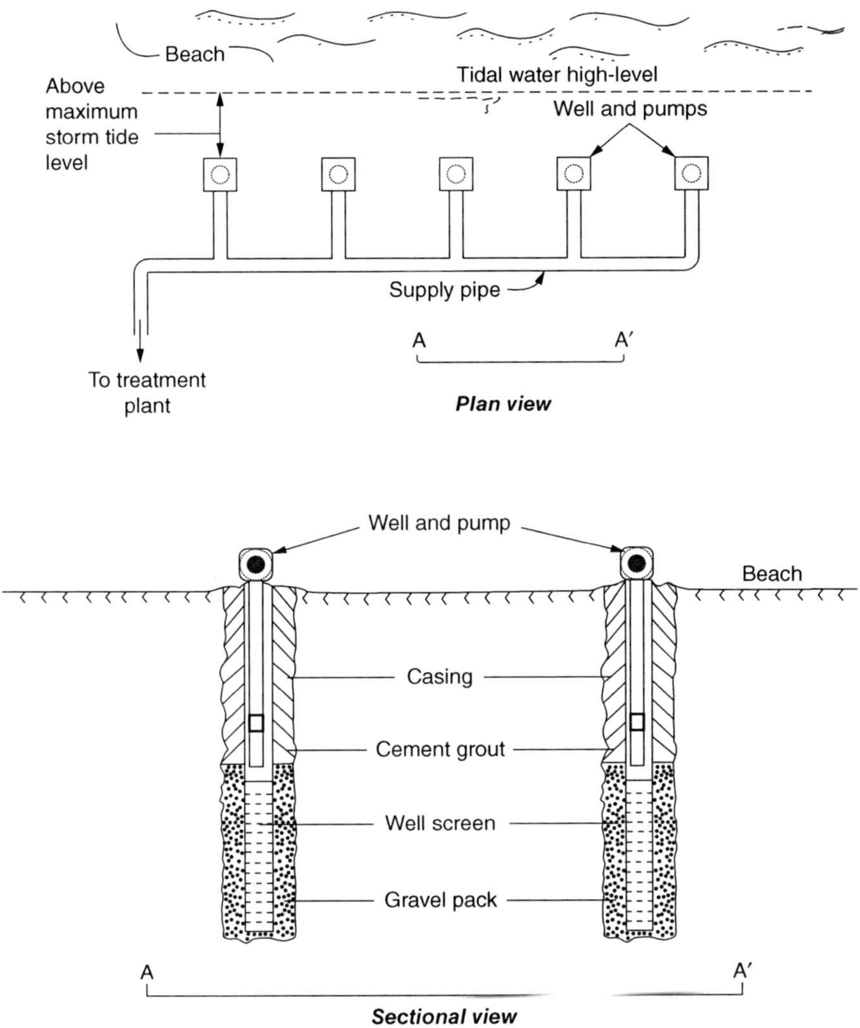

FIGURE 5.3 Typical beach inlet well.

Pretreatment and Posttreatment

The purpose of pretreatment is to remove particulate and biological organisms from the intake water. Chemical filtration agents are typically added to the inflow such as ferrochloride and acetylene. Common pretreatment facilities are sand media and ultrafiltration membranes. It is common practice to direct all pretreated water into a break tank for separation from the main plant pumping system. The pretreatment media must be periodically backwashed.

Posttreatment for potable water includes some form of chlorine added to the water for disinfection. An increase in hardness (for taste) may be desired, for which calcium hydroxide may be used. To balance the pH if required, carbolic acid may be used.

Pumps

In the past, seawater intake was generally done using immersion-type pumps such as a submerged or vertical pump. The main reason for this was the poor seawater pumping efficiency of land-based pumps. Difficult pumping conditions were commonly due to such factors as poor self-priming efficiency from the extrication of air and bubbles, which occurs easily in seawater, too great a distance from the water suction intake, and pipes crossing over embankments. With no strong self-priming pumps that could meet these demands, the solution closest at hand was forced pumping operation with submersible pumps.

However, for submerged pumps, because the pump motor is submerged together with the pump, there are concerns about the power supply, corrosion, and the danger of damage caused by rough seas. Added to this is the troublesome work of pulling up a submerged pump to perform inspection and maintenance.

As for vertical pumps, because the main part of a vertical pump is submerged in water, it is prone to the danger of flood tides and typhoons. In addition, the large size of the equipment means higher construction costs and the added trouble of performing inspection on the main part of the pump.

Disinfection

Disinfection is the process used to destroy disease-causing organisms or pathogens in drinking water and, to a lesser degree, in water used in commercial processes. They can be partially physically eliminated by filtration. After filtration, to ensure pathogen-free potable water, chlorine, in a process called *chlorination,* is the most often used chemical for disinfection of drinking water. This less expensive and powerful disinfection agent provides more benefits than shortcomings because of disinfection of by-products that have to be controlled.

Chlorination serves not only for disinfection but also as an oxidant for other substances, such as iron, manganese, and cyanide, among others, and for taste and odor control. Other chemical disinfectants generally used for large applications include ozone and chlorine dioxide.

ENVIRONMENTAL CONSIDERATIONS

Discharge

Environmental impacts associated with concentrate discharge have historically been considered to have the greatest single ecological impact when siting a saltwater desalination facility. Concentrated seawater has the potential to harm ecosystems, especially marine environments with low turbidity and high evaporation that have already elevated salinity, such as the Persian Gulf and coral lagoons and atolls. Concentrated seawater is denser than the surrounding sea and sinks to the bottom. Studies have concluded that with proper design and location of outfalls, concentrate disposal may not be a major impediment to desalination. Methods used for the discharge of the desalination concentrate have included deep well injection, dilution with ancillary water, and returning this to the sea.

Intake

The Clean Water Act mandated that the best available technology be used at the water intake to minimize environmental impact. Marine life impingement and entrainment resulting from intake operation was a greater, harder-to-quantify concern. *Impingement* occurs when marine organisms are trapped against the intake screens by the velocity and force of the water flowing through them. The fate of impinged organisms differs among intake designs and according to marine life species, age, and water conditions. Some hardy species may be able to survive impingement and be returned to the sea, but the survival rate of less robust species and/or juveniles may be less than 15 percent.

Entrainment occurs when small organisms pass through an intake screen and into process equipment. The mortality rate is 100 percent.

AVAILABLE DESALINIZATION TECHNOLOGY

Reverse Osmosis

Successful commercialization of reverse osmosis (RO) occurred in the early 1970s. In RO, seawater is forced through a semipermeable membrane using high pressure to separate salts from the water. As a portion of the water passes through the membrane, the remaining "feedwater" increases in salt concentration and is discharged to the sewer. Pretreatment is important in RO because the membranes are fine; suspended solids must be removed and the water pretreated so that salt precipitation or microorganism growth does not occur on the membranes. Usually the pretreatment consists of fine filtration and the addition of acid or other chemicals to inhibit precipitation. During the 1990s, the development of membranes that can operate efficiently with lower pressures and energy recovery devices has greatly reduced operating costs.

Right now, the high cost of desalinization has kept it from being used more often, as it can cost more than $1000 per acre-foot to desalinate seawater compared to about $200 per acre-foot for water from normal supply sources. Desalinization technology is improving and costs are falling, however, and Tampa Bay, Florida, is currently desalinizing water at a cost of only $650 per acre-foot. As the demand for both freshwater and technology increases, you can expect to see more desalinization occurring, especially in areas such as California and the Middle East.

Vacuum Distillation

Vacuum distillation is a method using the principle that boiling occurs when the pressure above the liquid mixture to be distilled is reduced to less than its vapor pressure (usually less than atmospheric pressure), causing evaporation. The water vapor is in turn condensed to form freshwater. Vacuum distillation is used with or without heating the solution. Thus, because of the reduced temperature, energy is saved. A leading distillation method is multistage flash distillation accounting for 85 percent of production worldwide in 2004.

Heat Distillation

A large percentage of the world's desalinated water is produced using heat to distill freshwater from seawater. The distillation process mimics the natural hydrologic cycle in that seawater is heated, producing water vapor, which is in turn condensed to form freshwater.

In a desalination plant, seawater is heated to its boiling point to allow maximum vaporization. For this to be done economically, the boiling point is lowered by reducing the atmospheric pressure. The reduction of the boiling point is important for two reasons: It allows multiple boiling that results in lower energy requirements, and it controls the buildup of carbonate and sulfate scale production on the apparatus. The distillation process is used successfully in many locations around the world.

Ion Extraction

Another method of desalinization is ion extraction, in which the ionized salts found in seawater are extracted through chemical or electrical means.

The chemical method is called *ion exchange*. In this method, granules of commercially prepared resin remove the positive ions from the seawater and replace them with ions that are loosely bound in the molecular structure of the resin. Other beds of resin are able to exchange negative ions. However, this process is too expensive to be used to desalinate large quantities of seawater.

Electrodialysis

In contrast to the chemical method, the electrical mechanism of ion removal, commercially introduced in the early 1960s, is much cheaper. It is called *electrodialysis*, since electric current pulls the ions through membranes that are permeable to only the positive or negative ions. Alternating positive or negative membranes, which number in the hundreds, are bound by a frame and form narrow compartments to trap the ions. When a direct electric current is applied, the positively charged ions tend to migrate through the membranes permeable to positive ions, and the negatively charged ions tend to migrate through the membranes permeable to negative ions. By this process, ions move between the compartments and become more concentrated.

The distance the liquid flows in the compartments, the intensity of the current flow, the permeability of the membranes used, and the distance between the membranes govern the efficiency of electrodialysis. The cost depends on the concentration of salts in the seawater, since the electric power used varies directly with the number of ions to be removed and their electrochemical characteristics. Usually, the electric power required to separate the ions from the water would be cheaper than the resin and chemicals used in ion exchange. Even then, it is usually so high that only brackish water, less salty than seawater, can be desalinated economically by electrodialysis for large-scale use.

In both ion extraction procedures, sediments and other impurities in the water can greatly reduce the success rate. Careful pretreatment of the water to remove undesirable materials is necessary.

Freezing Desalinization

Extensive work was done in the 1950s and 1960s to develop freezing desalinization. During the process of freezing, dissolved salts are naturally excluded from the lattice structure of ice crystals. Cooling the water to form ice under controlled conditions can desalinate seawater. Before the entire mass of water has been frozen, the ice is removed and rinsed to remove any salts adhering to the ice surface. It is then melted to produce freshwater.

Theoretically, freezing has some advantages over distillation, including a lower energy requirement, little scaling or precipitation, and minimal potential for corrosion. The disadvantage is that it involves handling ice-and-water mixtures that are difficult to move and process.

Solar Humidification

The use of direct solar energy for desalinating seawater has been investigated and used for some time. During World War II (from 1939 to 1945), small solar stills were developed for use on life rafts. These devices imitate the natural hydrologic cycle in that the sun's rays heat the seawater so that the production of water vapor (humidification) increases. The water vapor is then condensed on a cool surface, and the condensate is collected as freshwater. An example of this type of process is the solar greenhouse in Porto Santo, Portugal, in which the seawater is heated in basins, resulting in the condensation of water vapor on the sloping glass roof that covers the basins which spill to collection points.

Although the thermal energy may be free, the stills are expensive to construct. Additional energy is needed to pump the water to and from the facility, vapor can leak from the stills, and careful operation and maintenance is needed to prevent scale formation. Generally, these types of solar humidification units have been used for desalinating seawater on a small scale, where solar energy is abundant but electricity is not.

Cogeneration

Cogeneration is the process of using excess heat from power production to accomplish another task. In the sense of desalination, cogeneration is the production of potable water from seawater or brackish groundwater in an integrated, or "dual-purpose," facility in which a power plant is used as the source of energy for the desalination process. The facility's energy production may be dedicated entirely to the production of potable water (a stand-alone facility), or excess energy may be produced and incorporated into the energy grid (a true cogeneration facility). There are various forms of cogeneration, and theoretically any form of energy production could be used. However, the majority of current and planned cogeneration desalination plants use either fossil fuels or nuclear power as their source of energy. Most plants are located in the Middle East or North Africa, because of their petroleum resources and subsidies. The advantage of dual-purpose facilities is that they can be more efficient in energy consumption, thus making desalination a more viable option for drinking water in areas of scarce water resources.

Nuclear Reactors

Nuclear reactors can be used to produce large amounts of potable water. The process is already in use in a number of places around the world, from India to Japan and Russia. Eight nuclear reactors coupled to desalination plants are operating in Japan alone. Nuclear desalination plants could be a source of large amounts of potable water transported by pipelines hundreds of miles inland.

ENVIRONMENTAL IMPACTS

One of the main environmental considerations of ocean water desalination plants is the impact of the open ocean water intakes, especially when colocated with power plants. Many proposed ocean desalination plants' initial plans relied on these intakes despite perpetuating ongoing impacts on marine life.

To limit the environmental impact of returning the brine to the ocean, the stream of brine entering the ocean can be diluted with another source, such as the outfall of a wastewater treatment plant or power plant. While seawater power plant cooling water outfalls are not freshwater, as wastewater treatment plant outfalls are, the salinity of the brine will still be reduced. If the power plant is medium to large size and the desalination plant is not enormous, the flow of the power plant's cooling water is likely to be at least several times larger than that of the desalination plant and much less concentrated. Another method to reduce the increase in salinity is to spread the brine via a diffuser to mix with surrounding water in a mixing zone so that there is only a slight increase in salinity. For example, once the pipeline containing the brine reaches the sea floor, it can split off into many branches, each one releasing the brine gradually along its length through small holes. This method can be used in combination with the joining of the brine with power plant or wastewater plant outfalls.

The concentrated seawater has the potential to harm ecosystems, especially marine environments in regions with low turbidity and high evaporation that already have elevated salinity. Examples of such locations are the Persian Gulf, the Red Sea, and, in particular, coral lagoons of atolls and other tropical islands around the world.

Because the brine is denser than the surrounding seawater due to the higher solute concentration, discharge into water bodies means that the ecosystems on the bed of the water body are most at risk because the brine sinks and remains there long enough to damage the ecosystems. Careful reintroduction can minimize this problem. For example, for the desalination plant and ocean outlet structures to be built in Sydney, Australia, from late 2007, the water authority states that the ocean outlets will be placed in locations at the seabed that will maximize the dispersal of the concentrated seawater, such that it will be indistinguishable from normal seawater between 50 and 75 m (160 and 246 ft) from the outlet points. Sydney is fortunate to have typical oceanographic conditions off the coast that allow for such rapid dilution of the concentrated by-product, thereby minimizing harm to the environment.

DESALINATION COMPARED TO OTHER WATER SUPPLY OPTIONS

Increased water conservation and increased water use efficiency remain the most cost-effective priorities in areas of the world where there is a large potential to improve the efficiency of water use practices. While comparing ocean water desalination to wastewater reclamation for drinking water shows desalination as the first option, using reclamation for irrigation and industrial use provides multiple benefits. Urban runoff and stormwater capture also provide benefits in treating, restoring, and recharging groundwater. A proposed alternative to desalinization in the state of California and other areas in the American southwest is the commercial importation of bulk water either by very large crude carriers converted to water carriers or via pipelines.

EXPERIMENTAL TECHNIQUES AND OTHER DEVELOPMENTS

Iceberg Towing

Using icebergs as a source of freshwater is not a new idea. Captain James Cook used icebergs to replenish freshwater supplies aboard his ship *The Resolution* in 1773. However, it was not until the 1950s that serious consideration was given to towing icebergs from Antarctica to arid regions of the world.

Today, satellites could be used to find suitably sized icebergs that optimize the tradeoff between handling costs and ice volume. Once it is located, a bow could be cut in the iceberg and a Kevlar sheet wrapped around it to prevent melting. Powerful tugboats could tow the iceberg along favorable ocean currents to its destination. A trip to southern California is estimated to take one year and result in a 20 percent loss due to melting.

At present, the cost of transporting icebergs is prohibitive, and many technological barriers remain. Although towed icebergs would avoid major shipping lanes, towing large icebergs for long distances is not yet possible, and most icebergs are too thick to be towed into shallow seas and ports. Instead, other options for drinking water (such as desalinization) are more practical.

Passarell Process

A newer theoretical approach for desalination, focusing specifically on maximizing energy efficiency and cost effectiveness, is the *Passarell process*. This process depends on the fact that water will vaporize at a lower temperature when the atmospheric pressure is reduced, such as in a evaporation chamber. This improves the efficiency of the distillation process. When vacuum conditions are created at sea level, water vapor can be produced at lower temperatures. The Passarell process takes advantage of this principle in the distillation phase. The pure water generated by evaporation is drawn into a compression process where it is condensed and compressed using a specially designed compressor. The compression of the vapor causes the temperature to rise, and the water going into the tubes vaporizes. The water vapor on the outside of the tubes condenses as product water.

Geothermal Desalination

Using water obtained heated from deep underground entails the use of geothermal energy. From an environmental and economic point of view, geothermal water can be preferable, in many regions of the world, to the use of available surface and groundwater resources that already have long been under severe stress.

Nanotube Membrane

A *nanotube* is an impermeable membrane constructed of either a single, open-ended nanotube or a film composed of open-ended nanotubes that are oriented perpendicular to the surface of an impermeable film matrix much like the cells of a honeycomb. Impermeable is essential here to distinguish a nanotube membrane from other well-known semiporous membranes. Fluids and gas molecules may pass through the membrane at will. The operation consists of either a single, open-ended tube or a film composed of open-ended nanotubes that are oriented perpendicular to the flow of water. It was shown that water can pass through the graphitic nanotube cores of a membrane at speeds several magnitudes greater than classical fluid dynamics would predict, for both multiwall tubes and double-wall tubes. The flow of water through carbon nanotube membranes can be controlled through the application of electric current.

Recent research in the United States indicates that nanotube membranes may prove to be extremely effective for water filtration and may produce a viable water desalination process that would require substantially less energy than reverse osmosis.

Electric Field

On June 23, 2008, it was reported that Siemens Water Technologies had developed a new technology, based on applying an electric field to seawater, that desalinates 1 m^3 of water while using only 1.5 kWh of energy, which, according to the report, is one-half the energy that other processes use. The use of this process has been discontinued.

Low-Temperature Thermal Desalination

Low-temperature thermal desalination (LTTD) takes advantage of the fact that water boils at low pressures, even as low as ambient temperature. The system uses vacuum pumps to create a low-pressure, low-temperature environment in which water boils at a temperature gradient of 8 to 10°C between two volumes of water. Cooling water is supplied from sea depths of as much as 600 m (2000 ft). This cold water is pumped through coils to condense the evaporated water vapor. The resulting condensate is purified water. The LTTD process may also take advantage of the temperature gradient available at power plants, where large quantities of warm wastewater are discharged from the plant, reducing the energy input needed to create a temperature gradient.

BIBLIOGRAPHY

Levine, S. N., *Selected Papers on Desalination and Ocean Technology*, Peter Smith Publisher, Gloucester, U.K., June 2005.

Pankratz, T., "An Overview of Intake Facilities," CH2M Hill, January 2010.

Speigler, K. S., and Y. M. El-Sayed, *A Desalination Primer*, Balaban Publishers Water Desalination International, L'Aquila, Italy, April 2009.

CHAPTER 6
SITE DOMESTIC WATER SERVICE

This chapter discusses the source of the potable water supply for a facility. This could be water obtained from a public utility, wells, or other approved water sources on the site of a facility. The distribution piping extended from the source to a point just outside the building wall is included. Very often, the source will be a combined public water supply that provides both potable water and water for fire protection to the facility. Site fire protection water is discussed in Chap. 7, Site Fire Protection. Water from wells is discussed in Chap. 1, Water Wells.

WATER SUPPLIED FROM A PUBLIC UTILITY

General

Public utilities are the most reliable sources of water supply. The utility company will provide water, if available, of sufficient purity to the facility. This relieves the facility of the requirements for monitoring and treating the water for impurities, thereby eliminating ongoing maintenance and technical problems. This public service has a price, which must be balanced against the cost of electric power, treatment, chemicals, and maintenance of supplying water from alternate sources.

When a single service is used to provide potable and fire protection water, the supply is considered a combined service and requires conformance to additional code requirements. It is recommended that separate supplies be provided for fire protection and potable water if possible and cost effective. Some reasons for the separation are as follows:

1. High pressures used for firefighting may be detrimental to domestic piping and equipment. A pressure-reducing valve may be required on the domestic service.
2. Process use during an emergency may reduce the flow available for firefighting purposes.
3. System pressure may vary widely and be difficult to control.
4. Constant flow of water will increase corrosion of the combined main, in contrast to the relatively stagnant water in a dedicated fire main where much of the oxygen-causing corrosion is dissipated.
5. Separate supplies will be more reliable.

CODES AND STANDARDS

1. NFPA 24, for private water mains for fire service (if required)
2. AWWA codes for pipe materials, installation, and testing requirements
3. ASTM codes for pipe materials, installation, and testing requirements
4. Local plumbing and fire department codes and requirements
5. Fire insurance carrier requirements
6. Authorities responsible for backflow prevention
7. Model codes such as the Uniform Plumbing Code, International Plumbing Code, and National Plumbing Code

WATER SOURCE

Water source requirements for municipal water supplies are to provide sufficient pressure and quantity to meet the design demands of the fire protection and domestic water system. The points to be considered for the water supply include the following:

1. Quantity, static pressure at no flow and residual pressure at design flow, and availability of water
2. The overall fire demand, including duration of flow
3. The makeup and reliability of the source
4. The size, material of construction, and age of mains
5. Obstacles to be avoided, such as underground piping, trees, underground structures, and rock, that should be located

Quantity and Availability

The amount of water available from a network of underground site water mains, whether they be private yard mains or the public water supply, can be determined by the design engineer in several ways. The simplest method is to seek the history of fire flow tests in the area through the local water and fire departments, or plant personnel for private systems. If test results are not current or you suspect that the water supply has changed since the most recent hydrant flow test, a new hydrant flow test should be done by the engineer or requested through the local authorities such as the water department, fire department, or plant management.

Size and Age of Mains

The size and the age of the fire mains play an important part in the ability of the water supply to produce adequate flow. What may not be obvious to the casual observer, however, is how time and the corrosiveness of the water can affect the inside diameter of the supply mains. For example, a 30-year-old cast iron pipe with moderately corrosive water will have its friction factors increased over those of new cast iron pipe.

Hydrant Flow Test

To conduct a fire flow test, the following equipment and information are necessary:

1. Hydrant butt cap, with attached pressure gauge
2. Pitot tube and blade with an attached pressure gauge
3. Hydrant wrench
4. Nozzle pressure flow tables
5. Knowledge of the water main's sizing and piping layout

Once all the necessary equipment is assembled, a minimum of two adjacent operable fire hydrants to the site should be selected. It is recommended practice that the residual pressure hydrant (test hydrant) be as close as possible to the structure under design and downstream from the flow hydrant. After selecting the hydrants for the test, the engineer should remove the hydrant butt cap on the test hydrant and replace it with a pressure gauge. The butt cap on the flow hydrant should be opened to measure the actual inside diameter of the butt to confirm its diameter. The engineer should have the water department or maintenance personnel operate the hydrants if possible to limit the liability for damage. Special provisions may be required to accommodate the large volume of water that will discharge onto a street during a test.

The coefficient of discharge should be found, if necessary, by determining the construction and roughness of the inside lip of the test hydrant butt. (This information is available from standard engineering texts.) The butt cap should be replaced with one that has a bleed valve. Once in place, the hydrant should again be opened slightly to allow the air in the hydrant barrel to bleed off past the open bleed cock on the flow hydrant butt. After the air is bled off, the hydrant and the bleed cock can be closed. The pressure that registers on the gauge at this time is the static pressure (pressure with no flow). The second hydrant can now be approached.

To start the test, the test hydrant butt should be opened, The test hydrant with a pressure gauge shall have the pressure recorded. After these data have been recorded, the flow hydrant can be opened fully. Some caution should be exercised when opening the hydrant. It should never be opened rapidly, and the path of discharge should be investigated to ensure that personnel will not be injured and that property will not be damaged by the stream or the residual standing water.

After the test hydrant has been opened, it should be allowed to flow for 2 to 5 minutes to allow debris to clear the hydrant barrel and to stabilize the water flow before the pitot tube is inserted into the stream. An illustration of a pitot tube is shown in Fig. 6.1. When the pitot tube is inserted into the water stream, it should be placed in the centerline of the stream, at approximately one-half of the diameter of the butt opening. This is illustrated in Fig. 6.2. The value on the pressure gauge attached to the pitot tube then can be read. Simultaneously, the residual pressure (flowing pressure) must be read on the test hydrant gauge. Using the pitot tube pressure and the correct size butt [usually 2½ in (65 mm) is the most common], read the gallons per minute flow on the left side of Fig. 6.3.

The pitot reading and residual pressure now should be recorded. For best results, the pitot reading should not be less than 10 psi (67 kPa) or greater than 30 psi (200 kPa); similarly, the static pressure should drop 25 to 50 percent under the test conditions. It is a good idea to flow a quantity of water at least equal to the fire demand. Refer to Chap. 7, Site Fire Protection, for this information.

Now the data gathered during this test can be used to determine the quantity of water flow and the amount of water available for firefighting operations from the hydrant selected.

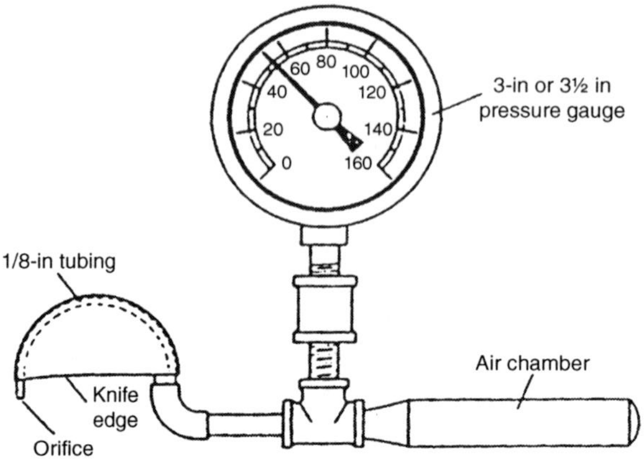

FIGURE 6.1 Pitot tube with gauge and air chamber.

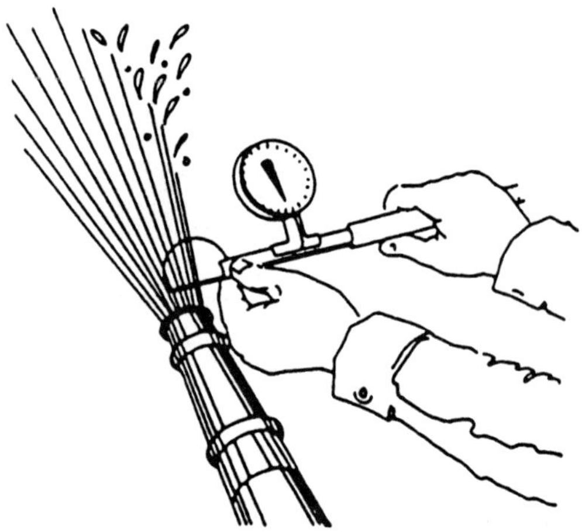

FIGURE 6.2 Taking nozzle pressure with a pitot tube.

Connection to Utility Company Mains

Connection to a public main can be made either with the water shut off (preferred) in the main or under pressure. Connections made with the water shut off are usually reserved for large building water services connected to large utility mains, but are also used if the main is too small to accept a new service line. A corporation cock that screws directly into a tapped hole made in the main is used for smaller connections. Very often, the utility company provides branch connections in the main when the main is constructed. It should be used if it is of adequate size and location. Their presence and location can be verified with the utility company.

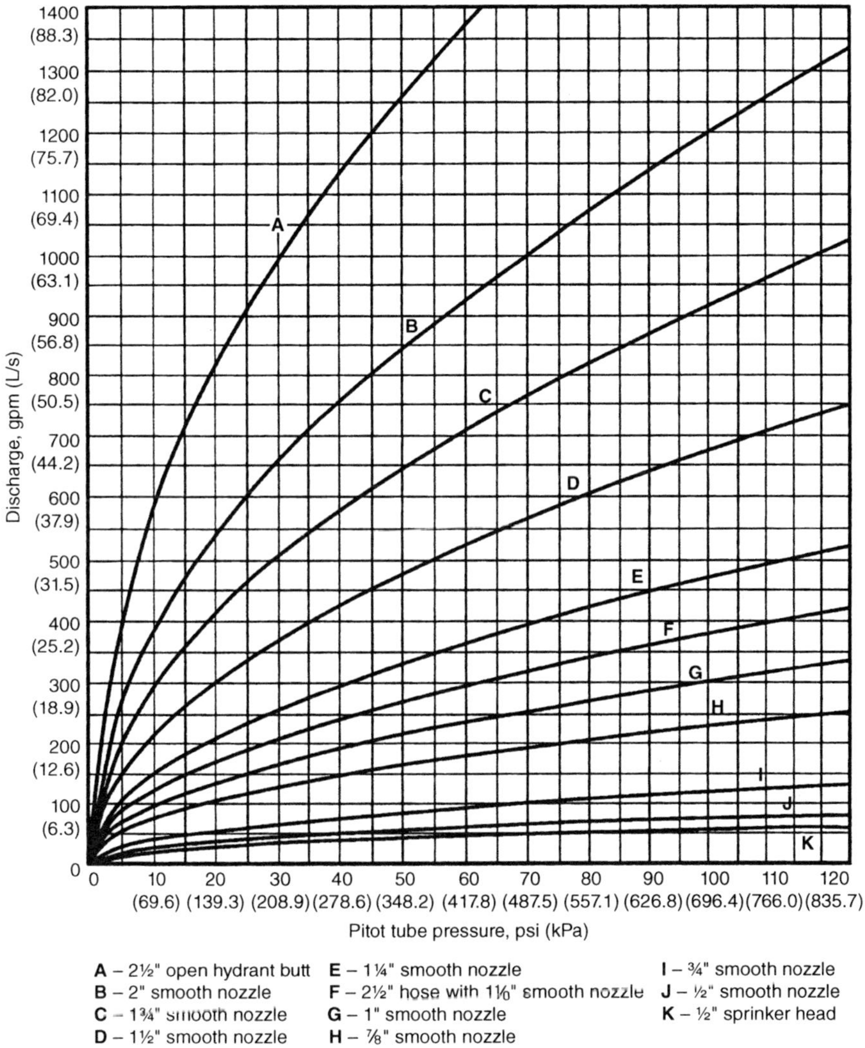

FIGURE 6.3 Relative discharge curves. (*Source: Chemical Engineers' Handbook,* McGraw-Hill.)

Connection to a pressurized metallic main is done with proprietary methods using a tapping machine. For larger mains, the hole is made by the tapping machine through a tapping valve and sleeve that mates with the main and retains water after the drilling bit and pipe coupon are removed. The valve has an outlet flange for a future connection. For recommended minimum excavation dimensions, refer to Fig. 6.4.

For branch connections 2 in (50 mm) or smaller, a service clamp with a threaded outlet for connection to a corporation cock is used. The corporation cock is a fitting that has a male thread on one end which screws into the outlet on the clamp assembly, an integral shutoff, and on the other end a connection to match the future building service pipe material and connection size.

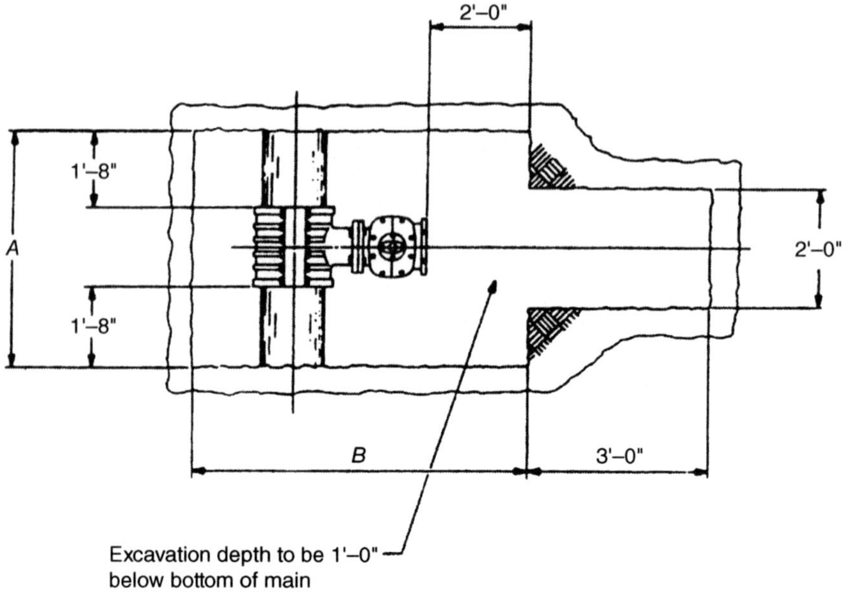

Pressure water main tapping. Excavation dimensions

Header size	A	B
3" Thru 8"	3' – 0"	3' – 6"
10" Thru 12"	5' – 6"	6' – 0"
14" Thru 24"	6' – 0"	7' – 0"

FIGURE 6.4 Minimum excavation dimensions.

A water supply piping system that consists of one large supply line feeding a series of dead-ended branch mains is far less efficient than a system of pipes that are looped and gridded together. For example, an 8-in (200-mm) main flowing at 2000 gpm (7500 L/min) that is fed from one direction will have 4.62 psi/1000 ft (30 kPa/333 m) of pipe friction loss. If that same main were fed from two directions, the flow could be balanced to 1000 gpm (3800 L/min) from each side (assuming equivalent pipe characteristics and lengths). The friction loss would then be reduced to 1.128 psi/1000 ft (4.26 kPa/333 m), or by a factor of four.

System Components

In addition to the piping and valves, the site system as a whole consists of water tanks (if required), valves, backflow preventers, water meters, pipe joint restraints (if required), fire hydrants, and air and vacuum release valves.

Elevated Water Storage Tanks. The elevated reservoir is reliable as it does not depend on electric power to provide the water pressure. The same reasoning holds true for private supplies. The elevated tank will usually share storage for both the domestic and fire water, and it is discussed in Chap. 7.

Valves. Valves in common use are the gate and butterfly valves recommended by the manufacturer and approved for underground service by the local authorities. Valves can be directly buried or installed in a valve pit. Installation of a valve in a pit permits easy access for repair and inspection, but is far more costly than direct burial.

Valves directly buried must have nonrising stems. The operation of the underground valves can be accomplished by two methods. The first provides a post indicator with an attached handle. This type of valve has an extension above grade on a post called a *post indicator valve* (PIV), which indicates whether the valve is open or closed. The PIV valve operator extends above grade on a post that makes its location obvious and is easily accessible. If the PIV is located near a road, it is common practice to protect the valve with guard posts. A detail of a typical guard post and a location diagram are illustrated in Fig. 6.5. A typical PIV is illustrated in Fig. 7.1.

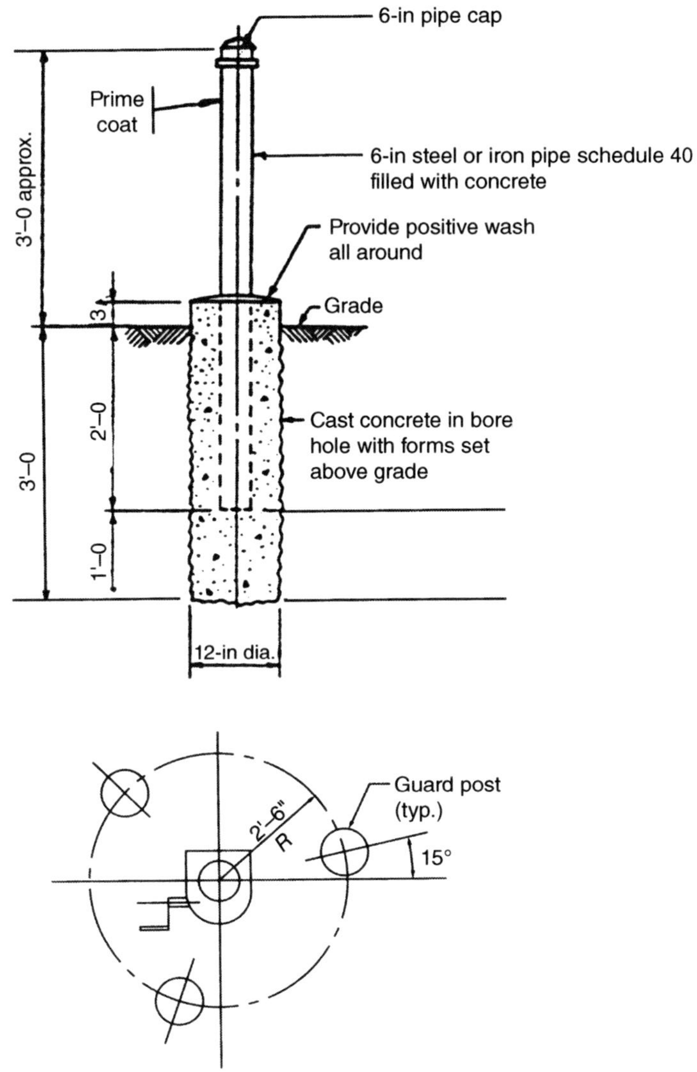

FIGURE 6.5 Typical guard post location.

The second is, the use of curb boxes instead of PIVs on a combined building service. This is discouraged for the following reasons:

1. Curb boxes are difficult to locate and supervise, since they are often covered by dirt, leaves, snow, ice, or paving materials.
2. Constant care is required to prevent dirt and stones from entering the hole and preventing valve operation.
3. Delays may be encountered if the operating T wrench is misplaced or stored at a remote location.

If the directly buried underground valve, called a curb valve in a street, is desired, it shall be provided with a nut instead of a handle. The operating nut is made accessible through a valve box extending up to the surface. Operation is accomplished by use of a T handle wrench. A typical underground curb valve, if used, is illustrated in Fig. 6.6.

If the connection is for service to a site consisting of several buildings, the utility company usually requires that an isolation valve be provided at the point of connection to the utility main. Based on utility company requirements, an exposed valve in a pit may be necessary, a curb valve could be directly buried, or a PIV could be furnished.

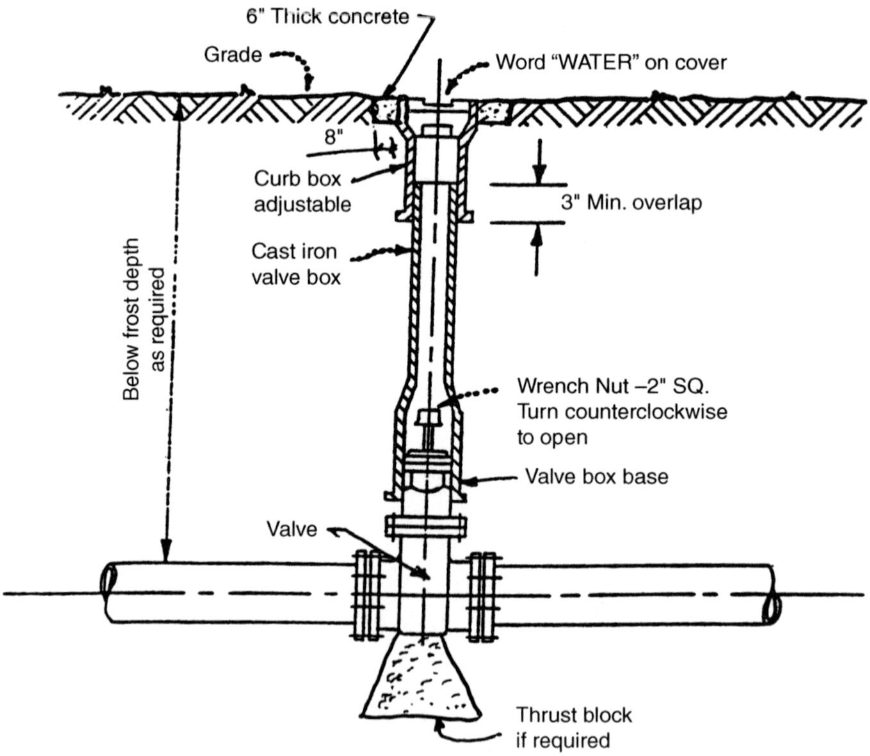

FIGURE 6.6 Typical curb valve.

On the site where water mains divide, it is good practice to provide an isolation valve, or section valves on all branch connections. For combined services, a post indicator valve is required.

Building Service Shutoff. All building water service branches must have a means of shutting off the water to the building from the site without personnel having to gain access to the inside of a building. When the building service main is a combined service, a PIV located 50 ft (15.5 m) from the building wall is recommended. If this distance is not possible, the PIV shall be located as far away as practical. In urban locations, a shutoff valve is usually installed in a public street or adjacent sidewalk. This shutoff valve is commonly called a *curb valve* and is installed using a curb box, as shown in Fig. 6.6.

In addition to the shutoff valve outside the building, code requires that another shutoff valve inside the building be provided as close to the water service entrance as space permits.

Backflow Prevention. Because of concerns about contamination, the installation of backflow preventers on the facility service line is generally required by local authorities if they determine that a potential for contamination is present. External backflow prevention installed on a building or site service is different in concept from the internal backflow prevention required on or at equipment or devices inside buildings. The reasoning is that any possible contamination of public water mains that could originate from the building or facility must be prevented. This is achieved by isolating the facility service connection from the public main and containing any contamination within the facility service piping or discharging the contaminated water to a sewer. Isolation and containment are accomplished by the installation of an approved device on the facility service pipe.

The only acceptable backflow preventers (BFPs) for service connections are the reduced pressure zone (RPZ), the double check valve (DCV), and the air gap. The type of device used is based on the potential hazard. The types and categories of hazards vary widely and are known by various names in different localities and in different locations within a building.

There are two primary causes of contamination. The first is the fire department. If the service main is combined, the fire department may use a nearby source of water instead of the public water supply. A tender may also have its own water supply for use immediately upon arrival to fight a fire until other apparatus arrives to make a connection to the public water supply. When the apparatus pumps contaminate water into the facility service, the water under high pressure must be stopped from entering the public mains. The other cause of contamination is the building itself, because if the public main breaks, it creates a vacuum that draws potentially contaminated water from the facility into the public main.

For industrial facilities using a storage tank, an air gap must be provided in the water supply into the tank. This will provide the best possible protection of the public water supply.

Determining the Degree of Hazard. The choice of protective device is based on an evaluation of the degree of potential hazards. The following should be considered:

1. Use, toxicity, nature, and availability of contaminants
2. Availability of a supplemental supply of water
3. Firefighting system evaluation

Hazard Classifications. The facility rating or classification should be made only after consulting with the proper authorities. The categories mentioned cannot list every circumstance or facility type. Judgment must be used in the final selection. The following ratings are often used based on the use, toxicity, nature, and availability of contaminants.

HIGH (SEVERE) HAZARD. Any facility that uses chemicals considered toxic or has the potential for discharge of toxic waste is considered a high hazard. Typical facilities are hospitals, chemical processors and manufacturers, pharmaceutical processors and manufacturers, laboratories, food processors and manufacturers, industrial manufacturers and processors, and water and sewage treatment plants.

MEDIUM (MODERATE) HAZARD. Commercial buildings and establishments, fire protection storage tanks and mains with no additives, and facilities that discharge water at higher than normal temperatures are medium hazards. The fire protection system will have only stagnant water present in the pipe.

NONHAZARDOUS (MINOR HAZARD). Private homes and commercial establishments without complex plumbing or fire protection systems are minor hazards.

The availability of a supplemental or auxiliary supply of water for firefighting purposes could create a situation that affects the hazard classification. If the facility is located close to a river or lake, the fire department could draw from this source to provide additional water during a fire. This water of indeterminate toxicity could be pumped back through the combined facility service into the public main, if not protected. Also, well water of unsatisfactory purity that is used for any purpose could find its way back into the public water supply.

An evaluation of the fire protection system shall be made separately because of the number of variables involved. Factors such as whether the system is wet or dry, the presence of storage tanks and pumps, whether there is a direct connection to the public mains, and interconnection with other water supplies must be studied to determine the degree of hazard. It is accepted practice to provide any water storage tank with an air gap between the pipe supplying the source of water and the highest possible level of water inside the tank. A gravity overflow two sizes larger than the supply pipe will provide sufficient protection from tank flooding.

Selection of the Appropriate BFP Device. Any facility classified as hazardous must be protected with either a reduced pressure zone (RPZ) BFP assembly or an approved air gap on the facility supply prior to connection to the public water main. For medium hazard, such as a fire department connection to the main, the selection of a DCV assembly is usually acceptable. For nonhazardous facilities, internal building BFPs are usually sufficient, with no requirement for external BFP in the facility service line.

The above selections are generalities and must be approved and accepted by all responsible code officials and facility authorities.

Location and Installation of a BFP Device. For a site service main, the external BFP must be located at or directly inside the property line of the facility. The installation requirements are obtained from the responsible authorities, which may be the local or state health department, fire department, the water utility, or the local building department. BFPs inside buildings must be located downstream of the meter or as close to the incoming service entrance as possible.

The installation of the BFP device depends on its type. Reduced pressure zone BFPs must be exposed, installed above grade or above the level of the water main inside the building. This requirement is to ensure that the water released as a result of backflow will be freely discharged from the device with no restriction. The wastewater must flow away from the device into a drain and not form a pool where the device is located. The drain must discharge by gravity and be large enough to accept the full flow expected from the device. The drain cannot be flooded and must be screened to prevent the entrance of dirt and vermin. For most installations, this requires that the device be installed in an aboveground structure. The enclosure must be protected from freezing. A schematic illustration of an exterior RPZ BFP in a pit is shown in Fig. 6.7. A schematic illustration of an exterior RPZ BFP above grade is shown in Fig. 6.8. The enclosure can be eliminated in a warm climate. A schematic illustration of an interior RPZ BFP is shown in Fig. 6.9. For the actual discharge from a RPZ BFP, refer to Fig. 6.10. Drains or scuppers must be adequately sized to accommodate the expected flow.

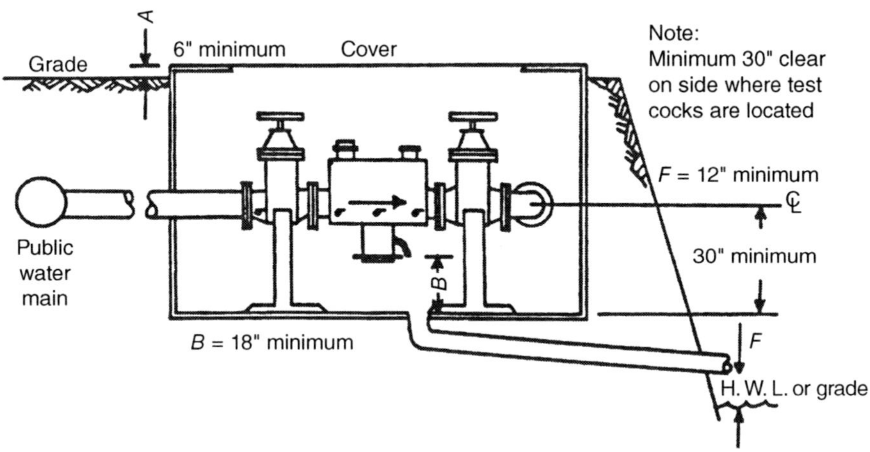

FIGURE 6.7 Detail of RPZ BFP in a pit.

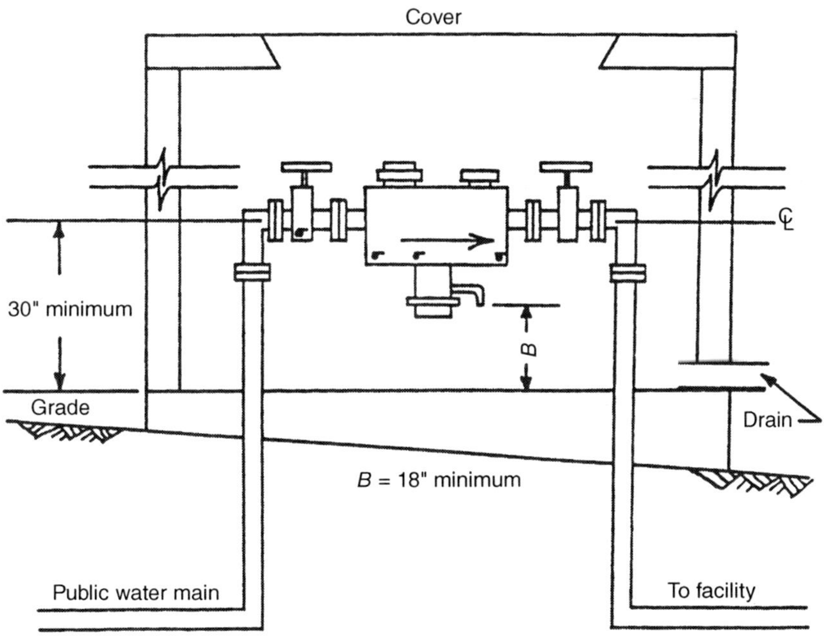

FIGURE 6.8 Detail of RPZ BFP above grade.

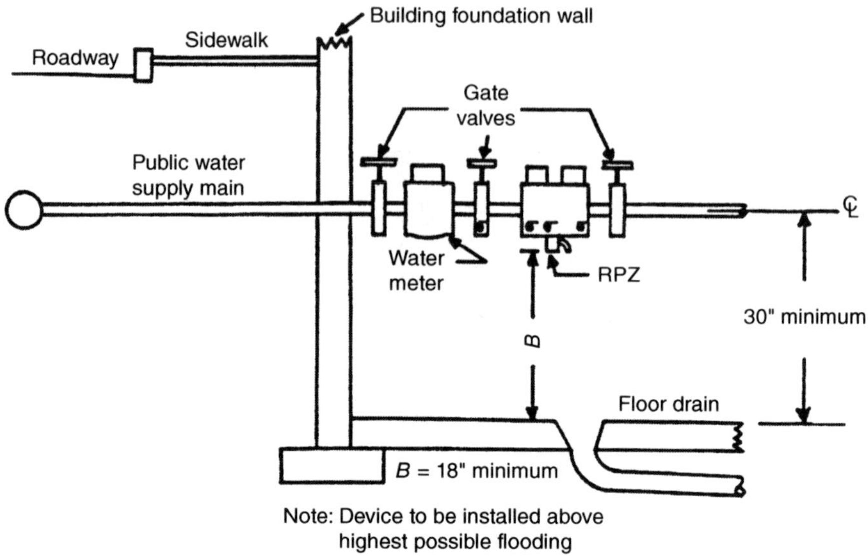

FIGURE 6.9 Interior detail of RPZ BFP.

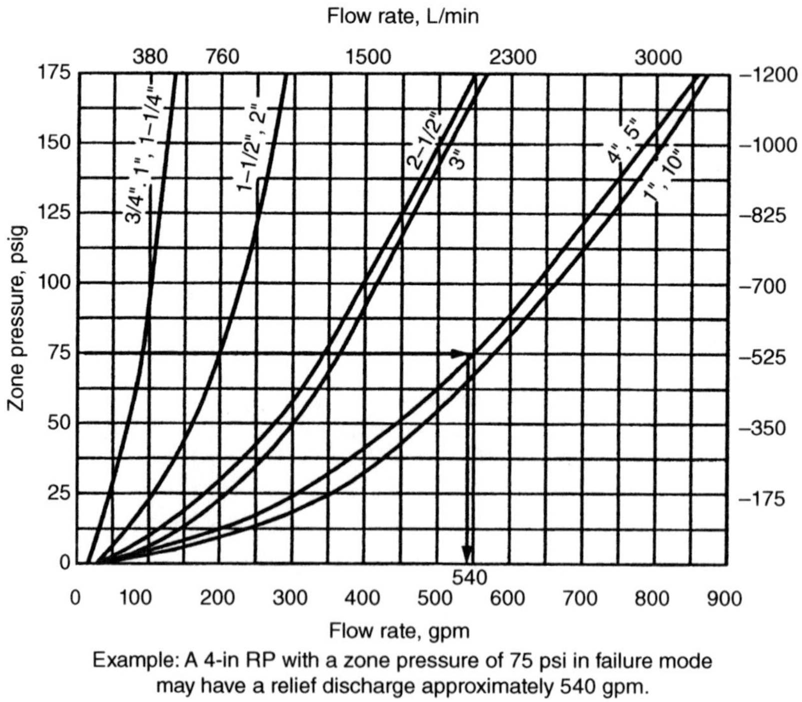

FIGURE 6.10 Discharge from a typical RPZ backflow preventer.

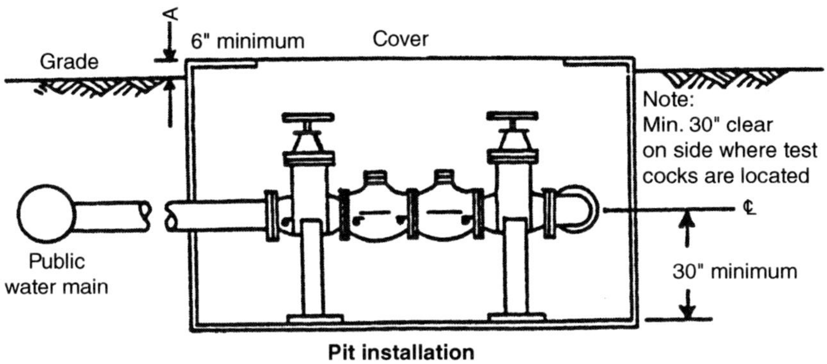

FIGURE 6.11 Detail of double check valve BFP in pit.

The DCV does not discharge water and can be installed in a pit below grade. It must be protected against freezing, and access must be provided for operation of the test cocks. If the service main is installed below the frost line, this is considered acceptable freeze protection. A schematic illustration of an exterior DCV BFP in a pit is shown in Fig. 6.11.

The BFP must be installed in such a manner as to allow easy access for testing and maintenance.

Water Meters. Since water obtained from a public utility must usually be paid for, a water meter is required to record the amount used. When there is no cost, meters are often installed by many facilities simply to record the amount of water used. Meters are selected by their function, accuracy, and pressure loss through the meter assembly. Registers on the meter are available to record usage in cubic feet or gallons (cubic meters or liters); remote reading and strip recording are usually available as options. Separate water meters are designed for fire protection or domestic water uses.

Meters can be installed at the property line, adjacent to a building, or inside a building. They can be installed aboveground or underground in a pit. The meter installation, called a *meter assembly,* usually includes valves, test tees, and a bypass around the meter to enable maintenance or replacement with no disruption to facility operation. The assembly arrangement is regulated by local authorities, the utility company, and the fire department. Consult them for installation requirements.

The meters for domestic use should be very accurate and selected based on accuracy at designed maximum flow and pressure loss through the meter. Where wide extremes of flow are expected, a compound-type meter is used. This meter consists of two separate meters: one to accurately register low flow and another to register higher flow.

Where reasonably steady, low to high flows are expected, turbine or disk meters are likely choices. Propeller-type meters are usually selected for large pipe sizes and very high, steady flows, or where low flows do not occur.

Smaller meters can be obtained to protect against frost with a breakaway bottom. If frozen, this is the only component that might have to be replaced. Special meters are required for hot water above 140°F (60°C).

Friction loss figures through meters at various flows are available from meter manufacturers.

Pipe Restraints. Mechanical joints and push-on joints seal when the elastomeric gasket becomes compressed in the space between the spigot end of the pipe and the bell. These joints are leaktight when installed on a straight line. When a fitting introduces a change of

TABLE 6.1 Pressure Exerted at Joints by Flowing Water
Total pounds resultant thrust at fittings at 100 psi water pressure

Nom. pipe diameter (in)	DN	Dead end	90° Bend	45° Bend	22½°	11½° Bend
4	100	1,810	2,559	1,385	706	355
6	150	3,739	5,288	2,862	1,459	733
8	200	6,433	9,097	4,923	2,510	1,261
10	250	9,677	13,685	7,406	3,776	1,897
12	300	13,685	19,353	10,474	5,340	2,683
14	350	18,385	26,001	14,072	7,174	3,604
16	400	23,779	33,628	18,199	9,278	4,661
18	450	29,865	42,235	22,858	11,653	5,855
20	500	36,644	51,822	28,046	14,298	7,183
24	600	52,279	73,934	40,013	20,398	10,249
30	750	80,425	113,738	61,554	31,380	15,766
36	900	115,209	162,931	88,177	44,952	22,585
42	1,000	155,528	219,950	119,036	60,684	30,489
48	1,200	202,683	286,637	155,127	79,083	39,733
54	1,350	256,072	362,140	195,989	99,914	50,199

Note: To determine thrust at pressures other than 100 psi, multiply the thrust obtained in the table by the ratio of the pressure to 100. For example, the thrust on a 12-in pipe with a 90° bend at 125 psi is 19,353 × 125/100 = 24,191 lb.

direction, the flowing water exerts a pressure on that fitting. Locations where thrust forces commonly occur are bends, wyes, reducers, offsets, tees, dead ends, valves, and hydrants.

Those pressures are listed in Table 6.1. The joint may not be capable of resisting this pressure; therefore a suitable means must be provided to restrain the joint from coming apart. The most common methods are integral joint restraint, clamps and tie rods, and thrust blocks.

Plastic piping with butt-fused joints has the same resistance as piping and does not require additional restraint. Split couplings with grooved pipe are also capable of resisting this pressure without restraint, if couplings designed for this purpose are used. This jointing method is generally recommended for new underground pipe installation.

Clamps and Tie Rods. Resistance is provided by clamps attached to the pipe at each side of the joint and rods connecting the clamps across each joint. Standard charts give the required restraint for different size pipes, depth of burial, number of joints to be restrained, and different fittings.

Integral Joint Restraint. This type of joint uses integrally cast glands for mechanical joint pipe and internally locked, grooved, and keyed joints for push-on joints.

Thrust Blocks. Thrust blocks consist of cast-in-place concrete blocks, the size and weight of which depend on the pipe size, water pressure (including surge loads), and safe soil-bearing load. These blocks must rest on undisturbed soil in order to resist the force exerted by the water. In general, thrust blocks are not recommended for the following reasons:

1. The blocks are comparatively large.
2. Often, undisturbed soil is not present.
3. The cost is high.
4. Space for the blocks may not be available due to the proximity of adjacent site utilities.

Air and Vacuum Relief Valve. When pressurized water is run in pipes at different elevations, air pockets often develop at high points in the system. Air pockets reduce the area of the pipe, thus lowering the flow rate and increasing the pressure loss. The device used to eliminate the air is called an *air release valve.* The following locations may require the installation of air release valves, but rarely at every location:

1. High points of the pipeline, in particular the hot water system
2. At aboveground locations, where the pipe rises up and then returns underground
3. Upstream from orifices, reducers, or other similar obstructions
4. On pumped systems

A vacuum can be created when a pipeline is emptied. If a service line is intended to be emptied regularly, a vacuum relief valve should be installed. Also, for long runs of thin-walled pipe, a vacuum relief valve should be installed to avoid pipe collapse.

Combination air pressure and vacuum release valves are available from manufacturers. These devices are selected based on main pipe size, main pressure rating, main pipe material, and wall thickness. An illustration of an exposed air and vacuum release is shown in Fig. 6.12.

Pipe Material Selection

For potable water piping 2 in (50 mm) and smaller, the most commonly used metallic piping is soft temper K copper tubing and copper fittings. Plastic piping includes PP and HDPE plastic pipe and fittings. For sizes 2½ in and larger, the most commonly used metallic piping is cement-lined ductile iron pipe and fittings.

Polyvinyl Chloride Pipe (PVC). This pipe is made by the extrusion of PVC. It is available from 10 mm to 1.2 m and with a variety of connection ends. It is subject to attack by certain organic chemicals. It is the pipe of choice for gravity, low-pressure, and vacuum sewer systems. It is pressure-rated in its class at 23°C.

High-Density Polyethylene (HDPE). This pipe is available from 100 to 1200 mm. It is pressure-rated in its class at 23°C and transients not exceeding two times the nominal pressure class. The pipe is normally joined with thermal butt fusion, but it takes a higher skill lever to accomplish. Ductile iron fittings are commonly used. This pipe is corrosion-resistant, and no coating or lining is provided.

Ductile Iron Pipe (DIP). Its primary application is for force mains. Because wastewater is corrosive, it is common practice to line the pipe with cement and use an asphaltic outer coating. Manufacturers recommend that the pipe be encased in a loose-fitting flexible polyethylene tube when installed in corrosive soils. It is available from 400 mm and smaller. It is also used for water distribution.

For combined systems, the two piping materials used most often are ductile iron and PVC, approved for use as fire protection mains.

SYSTEM DESIGN PROCEDURE

Preliminary Information Required

1. The location, size, material, and depth of the public water main available to the project under design must be known.
2. Determine the authority responsible for the need for, installation of, and approval of any BFP devices for the specific facility.

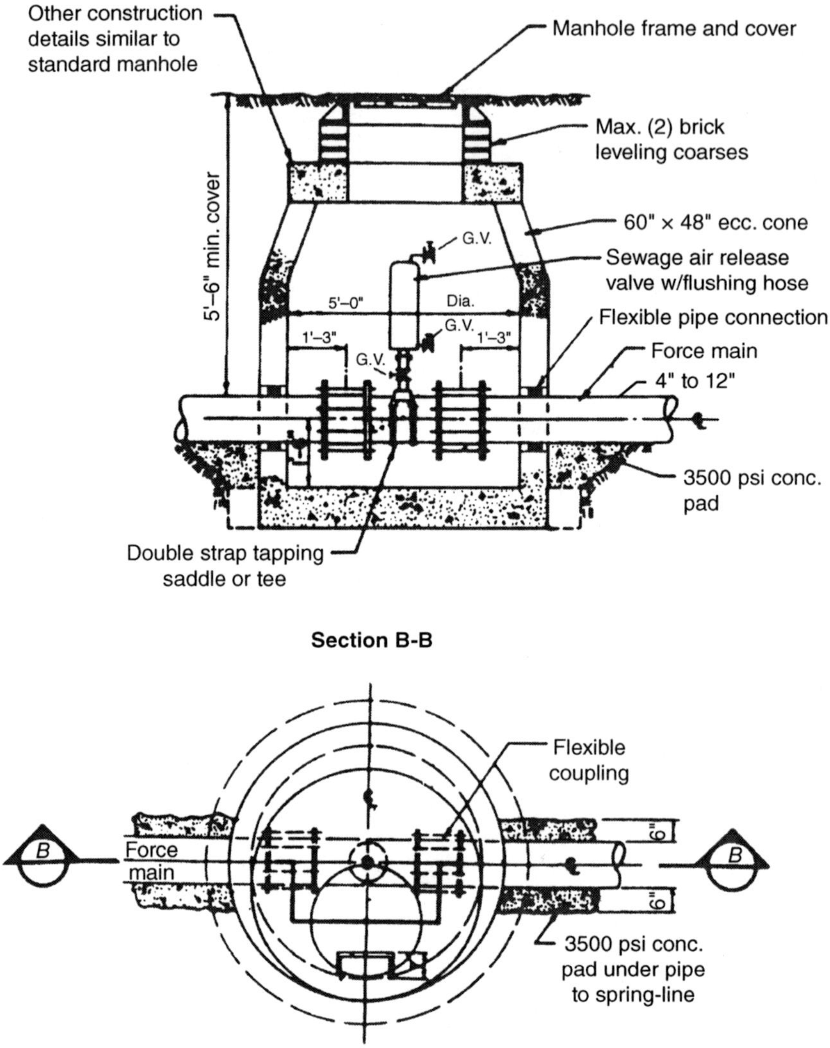

FIGURE 6.12 Air and vacuum release in manhole.

3. Obtain from the utility company or local authorities, the type of water meter required, the location of the meter, and the extent of work provided by the utility company versus the amount of work required to be done by a contractor. Usually, the utility company has standard details for meter assembly installations and the names of specific devices required to be part of the meter assembly and their arrangement.
4. The frost depth must be found for the area as well as the minimum cover over the proposed waterline.

5. Other special items or criteria that will affect the building service installation or add to the friction loss of the water service must be determined.

6. The maximum flow rate for the project must be calculated. This information will be obtained from the interior facility plumbing calculations.

7. Any obstructions, such as rocks, trees, and underground utilities, that would affect the run of pipe must be determined.

8. Very often, it is necessary to estimate the preliminary size water main into the building without adequate information. The best way is to use a preliminary fixture count if available. When this is not possible, the following information could be used.

   a. The number of proposed occupants could be used.

      (1) For a speculative building, use 100 ft^2 (9.3 m^2) per occupant.

      (2) For competitive building, use 150 ft^2 (13.9 m^2) per occupant.

      (3) For single-occupancy building, use 200 ft^2 (18.6 m^2) per occupant.

   b. Use a 10-h day and 10 gal per occupant.

   c. Use 0.125 gal/ft^2 of building area.

Calculation of Pipe Size

1. Obtain the static pressure and elevation of the main.

2. Obtain the maximum flow rate for the domestic water service.

3. Locate and lay out the service main from the point of connection with the utility service or site main to the building wall. Locate all devices (curb valve, BFP if required, meter, strainer, building valves, etc.). Take into account any site obstructions.

4. Select the pipe material.

5. Calculate the residual water pressure, starting with information obtained from the hydrant flow test, using the following guide. Record the time of day when the test is conducted. Add all the following losses in the service main:

   a. Losses from valves, meters, and BFP devices

   b. Pressure loss (or gain) from the difference in elevation (from step 4)

   c. Friction loss in service main to building entrance

   d. Subtract between 5 and 10 psig loss to the residual pressure to allow for future pressure drop in the public main. This adjusted pressure used shall reflect the possibility of other projects being built in the area.

6. Starting with the residual pressure, size the pipe based on maximum flow rate, using a friction loss that will result in a reasonable pressure loss figure along with the velocity of water in the pipe size selected. Select the most economical pipe size, using the equivalent run of pipe to establish the actual friction loss in the service. The equivalent length is found by adding a 50 percent allowance for fittings to the actual measured length.

7. Determine the available pressure for the building water distribution system.

8. A convenient calculation sheet indicating all the steps to be taken is illustrated in Fig. 6.13. If the pressure is too low, a method of increasing pressure inside the facility will be required.

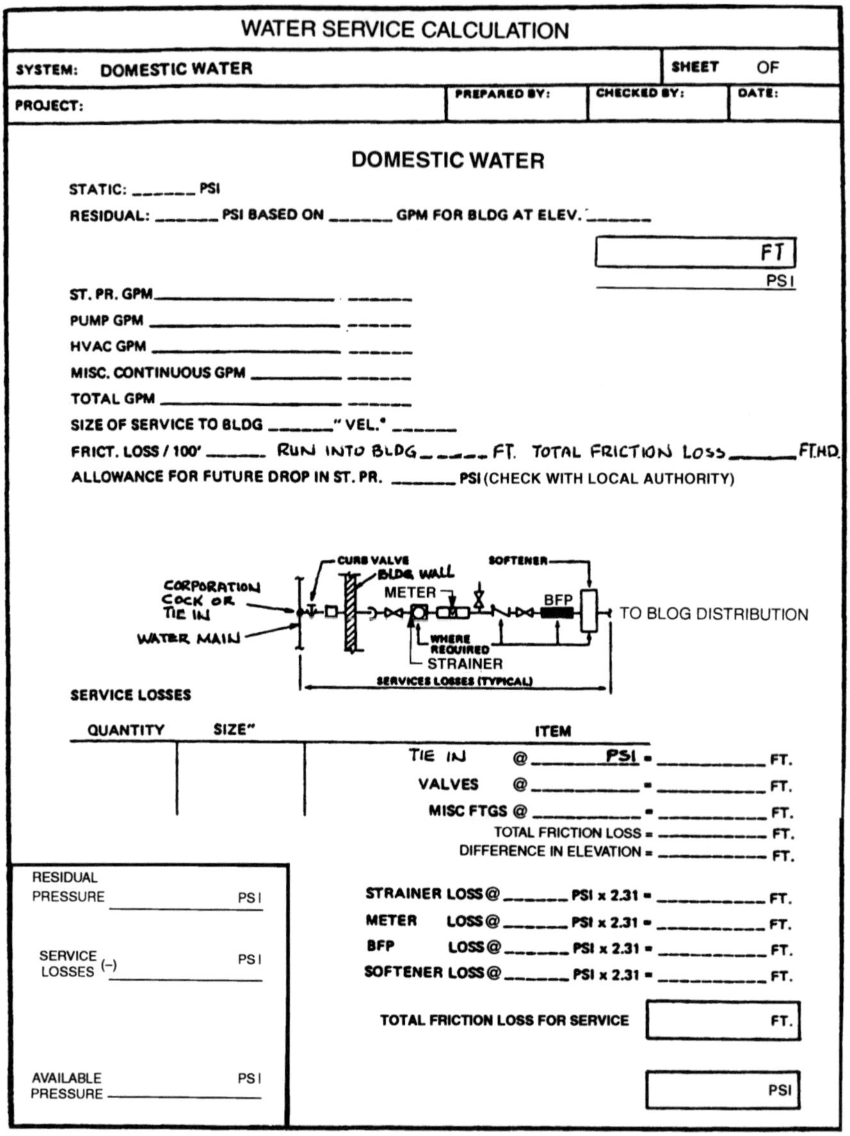

WATER SERVICE CALCULATION

SYSTEM: DOMESTIC WATER SHEET OF

PROJECT: | PREPARED BY: | CHECKED BY: | DATE:

DOMESTIC WATER

STATIC: ______ PSI

RESIDUAL: ______ PSI BASED ON ______ GPM FOR BLDG AT ELEV. ______

FT

PSI

ST. PR. GPM ____________ ______

PUMP GPM ____________ ______

HVAC GPM ____________ ______

MISC. CONTINUOUS GPM ____________ ______

TOTAL GPM ____________ ______

SIZE OF SERVICE TO BLDG ______" VEL.* ______

FRICT. LOSS / 100' ______ RUN INTO BLDG ______ FT. TOTAL FRICTION LOSS ______ FT.HD.

ALLOWANCE FOR FUTURE DROP IN ST. PR. ______ PSI (CHECK WITH LOCAL AUTHORITY)

SERVICE LOSSES

QUANTITY	SIZE"	ITEM
		TIE IN @ ______ PSI = ______ FT.
		VALVES @ ______ = ______ FT.
		MISC FTGS @ ______ = ______ FT.

TOTAL FRICTION LOSS = ______ FT.

DIFFERENCE IN ELEVATION = ______ FT.

RESIDUAL PRESSURE ______ PSI

SERVICE LOSSES (–) ______ PSI

AVAILABLE PRESSURE ______ PSI

STRAINER LOSS @ ______ PSI x 2.31 = ______ FT.

METER LOSS @ ______ PSI x 2.31 = ______ FT.

BFP LOSS @ ______ PSI x 2.31 = ______ FT.

SOFTENER LOSS @ ______ PSI x 2.31 = ______ FT.

TOTAL FRICTION LOSS FOR SERVICE ______ FT.

______ PSI

FIGURE 6.13 Water service calculations.

BIBLIOGRAPHY

American Society of Plumbing Engineers, *Book of Tables*, ASPE.

Church, J., *Plumbing Engineering Calculations.*

Davis, M. L., *Water and Wastewater Engineering*, McGraw-Hill, New York, 2010.

Frankel, M., *Facility Piping Systems Handbook*, 3d ed., McGraw-Hill, New York, 2010.

Mays, L., *Water Distribution Systems Handbook*, McGraw-Hill, New York, 2000.

National Fire Protection Association, *Fire Protection Handbook,* 1998.

NFPA 291, Recommended Practice for Fire Flow Testing and Marking of Hydrants, National Fire Protection Association, Quincy, Mass.

Smith, R., "Estimating Water Demand," *Plumbing Engineer Magazine,* November 2005.

Stutsman, C., "Hydraulic Shock," *Valve Magazine,* January 2009.

Tao, W., and K. Normann, "Plumbing System Design," *HVAC Magazine,* May 1995.

Westway, C. R., and A. W. Loomis, *Cameron Hydraulic Data*, Ingersol Rand.

CHAPTER 7
SITE FIRE PROTECTION

This chapter discusses the site fire protection water supply to a facility and describes the system components and criteria necessary to obtain the origination, storage, size, and flow rate for the water supply. This could be water obtained from a public utility, wells, or other approved surface water source. It is not uncommon for a fire department to use a nearby surface water source, such as a lake or stream, to supplement public water for a large fire. The distribution pipe extended from the public source to a point just outside the building wall is included. Very often, the public source may supply both the facility potable water and fire protection water, which is referred to as a *combined main*. For potable water site supply refer to Chap. 6.

FACILITY WATER SUPPLY

If the water supply to a facility is a combined service, the main shall be considered a fire service until the domestic connection to the building service is made. If the source is from a well, an aquifer performance analysis and investigation of the history of adjacent wells shall be made.

For many facilities, multiple water supplies from two separate public mains are a very desirable feature. This allows water to be supplied to the facility from either of two directions. If these connections are made to a single main, a sectionalizing valve installed in the source main somewhere between the connections will be necessary. Another desirable feature would be a fire hydrant installed on the building supply adjacent to a hydrant on the public main. This would allow a fire department pumper to increase the volume and pressure to the building if necessary.

Many of the items discussed in this chapter are part of a combined service that applies to both domestic and fire protection. If there is a combined site service, this service is discussed in Chap. 6. This chapter discusses only the fire protection water supply with certain exceptions.

GENERAL

A combined service requires conformance to more code requirements than those applied to only potable water. It is recommended that separate supplies be provided for fire protection and potable water, if possible and cost effective. These are some reasons for the separation:

1. High pressures used for firefighting may be detrimental to domestic piping and equipment. If the service is a combined one, a pressure-reducing valve may be required on the domestic service.

2. Process use during an emergency may reduce the flow available for firefighting purposes.
3. System pressure may vary widely and be difficult to control.
4. Constant flow of water will increase corrosion of the combined main, in contrast to the relatively stagnant water in a dedicated fire main where much of the oxygen-causing corrosion is dissipated.

CODES AND STANDARDS

1. NFPA 20, Standard for the Installation of Stationary Pumps for Fire Protection
2. NFPA 22, Standard for Water Tanks for Private Fire Protection
3. NFPA 24, Standard for the Installation of Private Fire Service Mains and Their Appurtenances
4. NFPA 291, Recommended Practice for Fire Flow Testing and Marking of Hydrants
5. AWWA codes for pipe materials, installation, and testing requirements
6. Local plumbing and fire department requirements
7. Fire insurance carrier requirements
8. Standards from authorities responsible for backflow prevention

CONNECTION TO UTILITY COMPANY MAINS

The primary agent for most fire extinguishing systems is water. The primary concern for the water source is the availability of municipal water supplies of sufficient pressure and quantity to meet the design demands of fire protection. Sprinkler systems must also be addressed. The points to be considered for the water supply include the following:

1. Quantity, static pressure at no flow, residual pressure at design flow and availability of water
2. The overall fire demand, including an estimate of the duration of flow
3. The makeup and reliability of the source
4. The size, material of construction, and age of mains
5. Obstacles to be avoided, such as underground piping, trees, underground structures, and rock, and their locations

Quantity and Availability

The amount of water available from a network of underground water mains, whether they are private yard mains or the public water supply, can be determined by the design engineer in several ways. The simplest method is to seek the history of fire flow tests in the area through the local water and fire departments, or for private systems, from plant personnel. If test results are not current or you suspect that the water supply has changed since the most recent hydrant flow test, a new hydrant flow test should be done by the engineer or requested through the local authorities (i.e., water department, fire department, or plant management). Discussion of a hydrant flow test is discussed in Chap. 6.

The best water main arrangement on a site is in the form of a loop around the entire site with section valves. This will permit the flow of water in two directions, cutting friction losses, and allowing service to most of the site if there is a break in any portion of the site main. Refer to further discussions in Chap. 6. A typical connection to a main is illustrated in Fig. 6.1.

SYSTEM COMPONENTS

Valves

If the water service connection is for the fire department only to a site consisting of several buildings, the requirement is that a shutoff or isolation valve be provided at the point of connection to each building, if a combined system is used. Based on utility company requirements, a valve pit may be necessary or the post indicator valve (PIV) could be directly buried. All valves shall be fire-rated.

All building fire protection water service branches must have a means of shutting off the water to the building from the site without personnel having to gain access to the inside of a building. When the building service main is a combined service, a post indicator valve (PIV) on the fire branch, indicating whether the valve is open or closed, located 50 ft (15.5 m) from the building wall is recommended. If this distance is not possible, the PIV shall be located as far away as practical. In urban locations, a shutoff valve is usually installed in a public street or adjacent sidewalk. This shutoff valve is commonly called an *underground curb valve* and is installed using a curb box. A PIV is illustrated in Fig. 7.1. A curb valve is illustrated in Fig. 6.6.

In addition to the shutoff valve outside the building, code requires that another shutoff valve inside the building be provided as close to the water service entrance to the building as space permits.

Location and Type of BFP Devices

Contamination of the public water supply is a prime concern of the water company that supplies water to the project. The determination of the acceptable device for backflow prevention of the fire protection water supply will be made by the health department, the water purveyor, or the plumbing subcode official.

There are two primary causes of contamination. The first is the fire department. If the service main is combined, the fire department may use a nearby source of water instead of the public water supply. A pumper may also have its own water supply for use immediately upon arrival to increase pressure in the standpipe system to fight a fire until other apparatus arrives to make a connection to the public water supply. When the apparatus pumps contaminated water into the facility service, the water under high pressure must be stopped from entering the public mains and the facility piping. The second primary cause of contamination is within the facility itself, because if the public main breaks, it creates a vacuum that draws contaminated water from the fire protection of the facility into the public water main.

The availability of a supplemental or auxiliary supply of water for firefighting purposes could create a situation that affects the hazard classification. If the facility is located close to a river or lake, the fire department could draw from this source to provide additional water during a fire. Then this water of indeterminate toxicity could be pumped back through

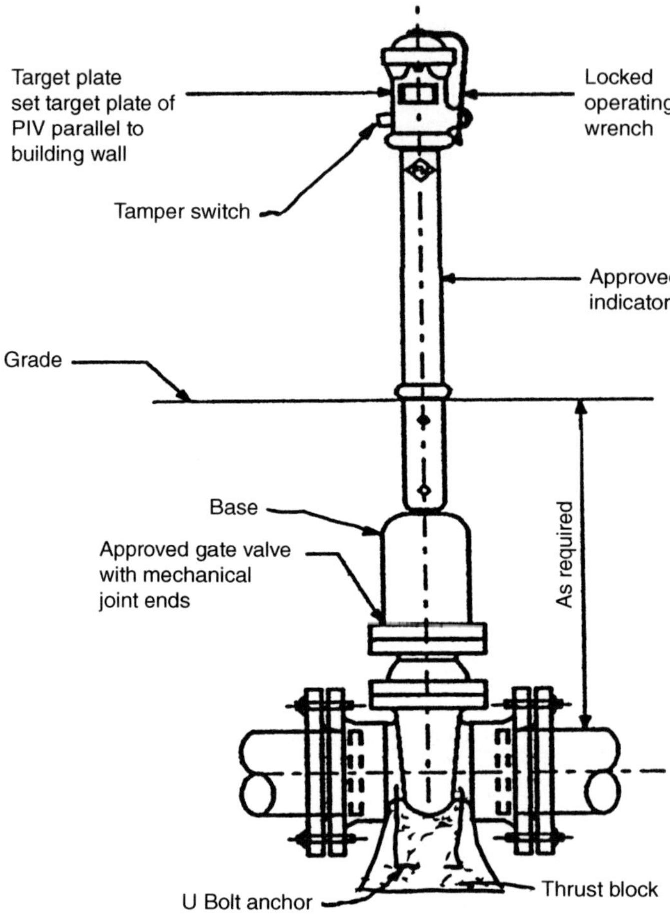

FIGURE 7.1 Typical post indicator valve.

the combined facility service into the public main that was not protected. Also, well water of unsatisfactory purity that is used for any purpose could find its way back into the public water supply.

An evaluation of the fire protection system shall be made separately because of the number of variables involved. Factors such as whether the system is wet or dry, the presence of storage tanks and pumps, whether there is a direct connection to the public mains, and interconnection with other water supplies must be studied to determine the degree of hazard. It is accepted practice to provide any water storage tank with an air gap between the pipe supplying the source of water and the highest possible level of water inside the tank. A gravity overflow two sizes larger than the supply pipe will provide sufficient protection from tank flooding.

The type of device used is based on the potential hazard. The types and categories of hazards vary widely and are known by various names in different localities. The authority having jurisdiction (AHJ) shall be consulted regarding the type and placement. The most common Backflow Preventer (BFP) used on a fire protection service is the double check valve (DCV) and the air gap for the water tank supply. The various types of reduced pressure zone and DCV backflow preventers are discussed in Chap. 6 and illustrated in Figs. 6.7, 6.8, 6.9, and 6.11.

If the installation is for fire protection only, a double check valve is acceptable for installation because of the low potential hazard to prevent water from entering the public water supply.

Meters

Since water obtained from a public utility must be paid for, a water meter is required to record the amount used. When there is no cost for fire protection system water, meters are often eliminated or installed by many facilities simply to record the amount of water used. Meters are selected by their function, accuracy, and pressure loss through the meter assembly. Registers on the meter are available to record usage in cubic feet or gallons (cubic meters or liters); remote reading and strip recording are usually available as options.

Meters can be installed at the property line, adjacent to a building, or inside a building. The mains shall be buried below the frost line in areas where freezing is a possibility. The meters can be installed in protected locations aboveground or underground in a pit. The meter installation, called a meter assembly, usually includes valves, test tees, and a bypass around the meter to allow maintenance or replacement with no disruption to facility operation. The assembly arrangement is regulated by local authorities and the utility company. Separate water meters are designed for fire protection or domestic water uses.

Meters used for fire protection purposes are suitable only for large flows and minimal pressure loss. They are not very accurate. A "detector" function can be added to signal flow in a dedicated fire main and connected to an alarm. A check valve can also be an integral part of a fire water meter. The most popular meter is the detector-check type. An illustration of a typical fire protection meter in a pit is shown in Fig. 7.2. Construction of the meter

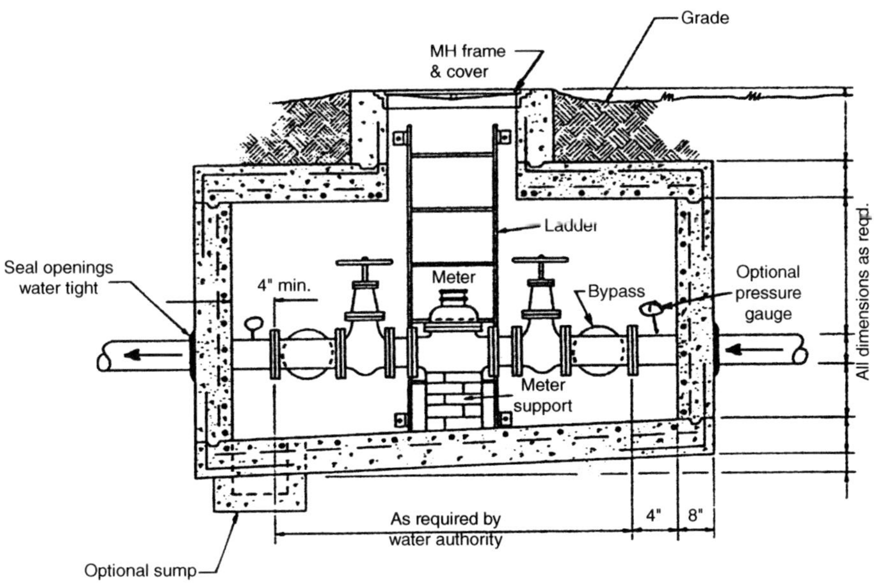

FIGURE 7.2 Typical meter assembly in pit.

pit, normally built by the contractor, shall be as required by the AHJ. Friction loss figures through meters at various flows are available from meter manufacturers.

Fire Hydrants

There are two categories of fire hydrants: dry barrel to prevent freezing and wet barrel. The dry barrel does not have any water from the connection to the main to the valves on the hydrant. The wet barrel is continuously full of water. The operating nut shall be uniform throughout all areas, installed and in conformance with local authorities. All hydrants shall be approved by Factory Mutual or Underwriters Laboratories.

The dry barrel has a device that allows the water in the barrel to drain down to below the frost line after the valve that controls water, has been shut off. The dry type is recommended for all areas that have any remote possibility of freezing. Experience has shown that freezing only once in 25 years is sufficient reason to install dry-type hydrants. For a detail of a typical dry-type hydrant, refer to Fig. 7.3. The drain is omitted for a wet hydrant installation. Shutoff valves for the hydrant are useful, but optional. Check local code and client preference for installation requirements.

The sizes of connections available on hydrant barrels vary. All connections have a threaded end onto which fire hose is connected. These threads must be compatible with the local fire department and the facility on whose property the hydrant has been installed. The threaded ends are protected by a cap (butt) attached to the hydrant by a chain. A single internal valve, terminating in an operating nut on top of the hydrant, controls all the connections. A special wrench is used to turn the nut which by convention is turned in a counterclockwise direction to open. Special hydrants can be obtained with individual outlet valves on each outlet. The most common outlet sizes are nominal 2½ in (60 mm), 4 in (100 mm), and 6 in (150 mm). The two larger sizes are called *pumper connections*. The most common arrangement consists of two 2½-in (60-mm) connections, mounted at a 90° angle, facing the road. When a larger pumper connection is added, the pumper connection faces the road and the other, smaller connections are arranged at 90° angles on either side. Generally accepted practice is not to provide a pumper connection unless specifically required by local code or client preference. Pumper connections are used when a cross-connection is needed between two water mains or to increase pressure in the fire main itself. Fire hydrants shall be rated for a minimum of 300 psig (2000 kPa).

Hydrants are typically located 250 to 300 ft (78 to 90 m) apart. For large sites, it is recommended that hydrants be installed 200 ft (65 m) from any point where a significant hazard exists. They should be located about 50 ft (15 m) from a building wall under normal conditions, with a longer distance required if the building stores or uses explosives, flammable liquids, chemical products, petroleum products, or high-piled storage. Excessive radiant heat and the danger from falling building walls are factors in considering longer distances. Hydrants are usually placed 2 to 10 ft (0.6 to 3 m) from the center of the hydrant to the curb of a road. Hydrant outlets are usually located from 1 ft 3 in to 2 ft (37 to 60 cm) above grade.

Hydrants are generally located adjacent to roads that will allow fire department apparatus to approach near enough to the hydrant to allow easy hookup. If this is not possible, a clear path over a lawn, e.g., or from a road shall be available. If reasonable, they shall be located on all sides of the building being protected.

An optional shutoff valve is often located between the water main and the hydrant to allow servicing of the hydrant without shutting down the water main. Opinion is divided about the need for this valve, since hydrant servicing is rarely necessary. The installation of the valve is a client or AHJ preference.

Since the hydrant is located adjacent to roads where trucks travel, it is a generally accepted precaution to provide guard posts to protect the hydrant. Guard post locations are illustrated in Fig. 6.5. Fire hydrants are often painted different colors or marked to

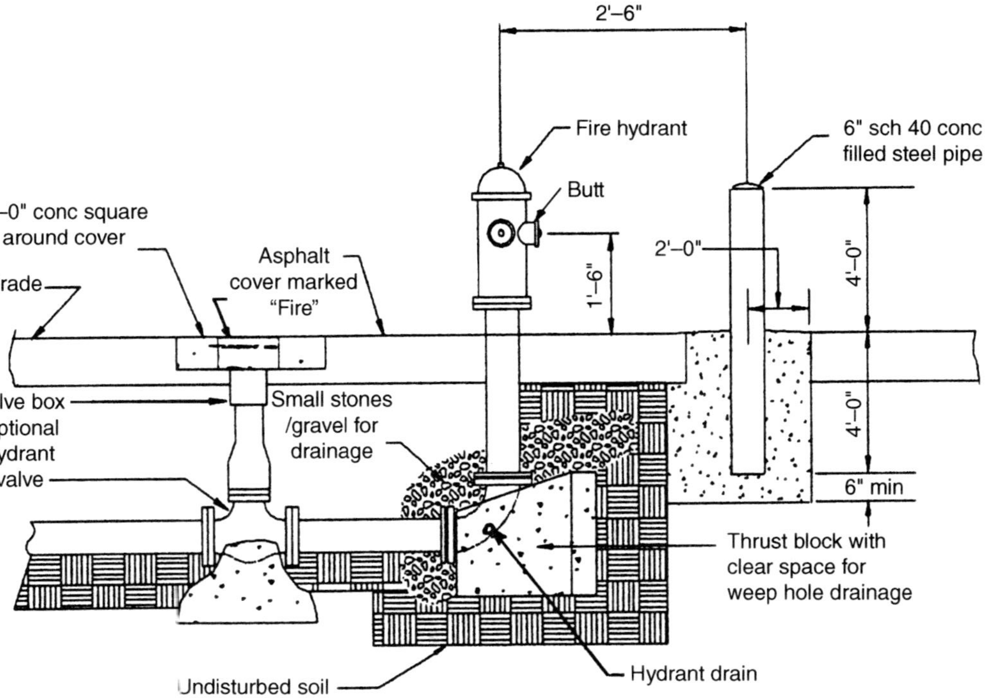

FIGURE 7.3 Fire hydrant and valve installation detail.

identify the flow capacity and/or pressure of the individual hydrant. This is regarded as a great help to fire department personnel and should be considered if different pressures are available.

Hose Houses

In some cases, at the option of the owner or the suggestion of the fire department or plant safety officials, it is necessary to provide lengths of hose and a hydrant in a protective house located at strategic places around a large campus or site. These hose houses also contain other miscellaneous firefighting accessories such as axes, nozzles, spanners, and other items that would be recommended by the owner. This hose would be used by trained facility personnel for fighting a large fire. It can be put in operation in a shorter time than it takes to run up a hose cart, connect the hose, and run out the hose. The house provides protection from the elements and would be used by trained facility personnel.

At least 100 to 150 ft (35 to 50 m) of 2½-in (60-mm) hose would be directly connected to one of the hydrant outlets with the other left for additional use. Additional hose shall be provided as required to reach the top floor or around the farthest side of a building, allowing the hose to be used at a safe and effective distance.

Monitors

A *monitor* is a permanently mounted, high-volume fire protection nozzle assembly, connected to a water main. It is intended to provide a large amount of water at sufficient pressure. The monitor is capable of being rotated and elevated. It would be installed where large amounts of combustible materials are located in yards or protection of other hazards is required. The nozzle can be a stand-alone unit or attached to a fire hydrant. A monitor installed on a hydrant is illustrated in Fig. 7.4. Monitors are usually installed as safeguards for significant hazards such as chemical and petroleum product storage or combustible material such as piled wood. When mounted on towers, monitors can be used for vapor suppression from specific hazards. The advantage of a monitor is that it can be placed into service very quickly, can be aimed in the proper direction, and, once set, can be left unattended.

Monitors are available in 500 to 2000 gpm (3150 to 12,600 L/min) models. They shall be located about 150 to 200 ft (45 to 62 m) apart, with all protected areas capable of being reached with two streams. A general requirement of 80 psig (550 kPa) minimum pressure is necessary for proper operation, which usually requires a fire pump. Consideration shall be given to not place the monitors too close to a hazard because of the potential heat generated by fire.

Fixed Water Spray Systems

A water spray system would be used on the site for exposure protection (cooling) at building walls and tanks of combustible liquids. Other protected equipment would be transformers, cable racks, openings in firewalls, and rotating electrical machinery.

Fixed water spray systems are specifically designed to provide an automatic discharge and volume over the surface or area to be protected. They shall be connected to a reliable supply of water.

The actual selection of spray nozzles will take into consideration the nature of the hazard to be protected, the purpose of the system, and other factors that can only be analyzed individually for each installation.

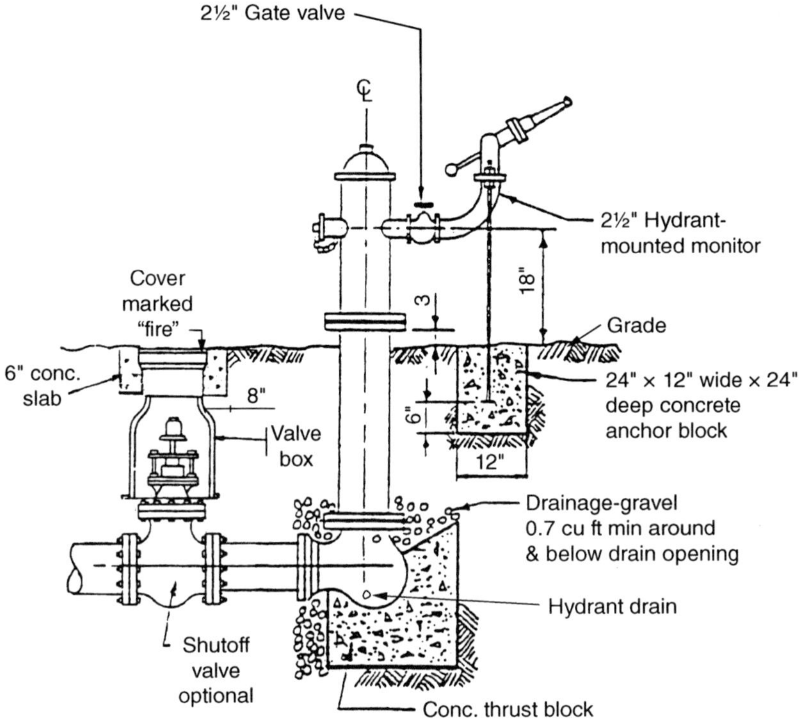

FIGURE 7.4 Detail of combination hydrant and monitor nozzle.

Fixed Water Spray Demand and Pipe Sizing. The water supply must be adequate to supply all the spray nozzles at adequate pressure. The pressure required would depend on the selection of nozzles and the density of water required. Another consideration would be any additional water for fire hose. The pipe size shall be found by hydraulic calculations to ensure adequate pressure and discharge.

Fixed Spray Drainage. Fixed pipe water spray systems discharge, by nature, large amounts of water. To limit the spread of flammable liquids, special drainage and disposal facilities are important. Trenches and dykes are often used to contain flammable liquid floating on water which, if on fire, is often left to burn itself out.

Water Storage Tanks

In many cases, it is not possible to supply the required water supply for firefighting purposes from the public supply or other available sources. For these situations, the use of a water storage tank is called for. These tanks can be either elevated or installed on the ground. The design of these tanks is outside the scope of this handbook. Connections from the water supply to a storage tank shall terminate 1 ft (0.9 m) above the overflow level. The overflow shall be two pipe sizes larger than the water supply pipe size. Where necessary, the tanks shall be protected against freezing.

A water supply's reliability can be determined by evaluating the method by which the piping network is fed. Water supplies on site consist of three types: elevated reservoirs (water towers), direct pump, or combined. The distribution of the water supply can be a dead end, grid, or loop pipe system or a combination of all three.

The elevated reservoir is reliable as it does not depend on electric power to provide the water pressure. With reliability in mind, a designer should first make an attempt to design an automatic fire protection system that can function properly without the introduction of a booster fire pump. If this cannot be accomplished, then a fire pump shall be added to provide the water supply. (See the discussion under "Fire Pumps" later in this chapter.)

A water supply piping system that consists of one large supply line feeding a series of dead-ended branch mains is far less efficient than a system of pipes that are looped and gridded together. For example, an 8-in (200-DN) main flowing 2000 gpm (7500 L/min) that is fed from one direction will have 4.62 psi/1000 ft (31 kPa/330 m) of pipe friction loss. If that same main were fed from two directions, the flow could be balanced to 1000 gpm (3800 L/min) from each side (assuming equivalent pipe characteristics and lengths). The friction loss would then be reduced to 1.128 psi/1000 ft (8.6 kPa/330 m), or reduced by a factor of four.

A typical elevated fire tank is illustrated in Fig. 7.5.

Size and Age of Mains

The size and the age of the fire mains play an important part in the ability of the water supply to produce adequate fire flows for fire protection systems. What may not be obvious to the casual observer, however, is how time and the corrosiveness of the water can affect the inside diameter of the supply mains. For example, a 30-year-old cast iron pipe, with moderately corrosive water, will have its friction factors increased over new cast iron pipe. A minimum pipe size of 6 in (DN 150) is recommended for any fire protection main.

Joint Restraint

When run in a straight line, properly specified joints will provide a leakproof installation. When there is any change in direction of the pipe that is made by fittings, pressure is exerted by the flowing water on the fitting. Joints not designed for this purpose may not provide resistance to the force exerted by the pressure of the water against the fitting. To prevent joint failure due to the pressure exerted, pipe joints shall be restrained in some manner.

The preferred method of restraining joints is by means of the split-ring couplings for grooved pipe. The use of properly selected joints will not require any additional method of restraint. If steel piping is used, welded joints do not require restraints.

Mechanical joints and push-on joints seal by a rubber-type gasket compressed in a space between the spigot end of the pipe and the bell. These joints will not adequately resist the pressure of the water on the joint. The method of restraint with external clamps and rods should be considered. Friction clamps are bolted around the pipe on both sides of the joint. These clamps engage steel tension rods, sometimes called tie rods, across the joint, preventing it from separating. This will combine the resistance of these joints to prevent separation.

Another method of restraining joints is the use of integrally cast glands for mechanical joint pipe with internally laced, grooved, and keyed push-on joints. This type of restraint is usually recommended only for repair of existing systems.

The last method, which is not generally recommended, is the use of thrust blocks.

For further discussion of joint restraint, refer to Chap. 6.

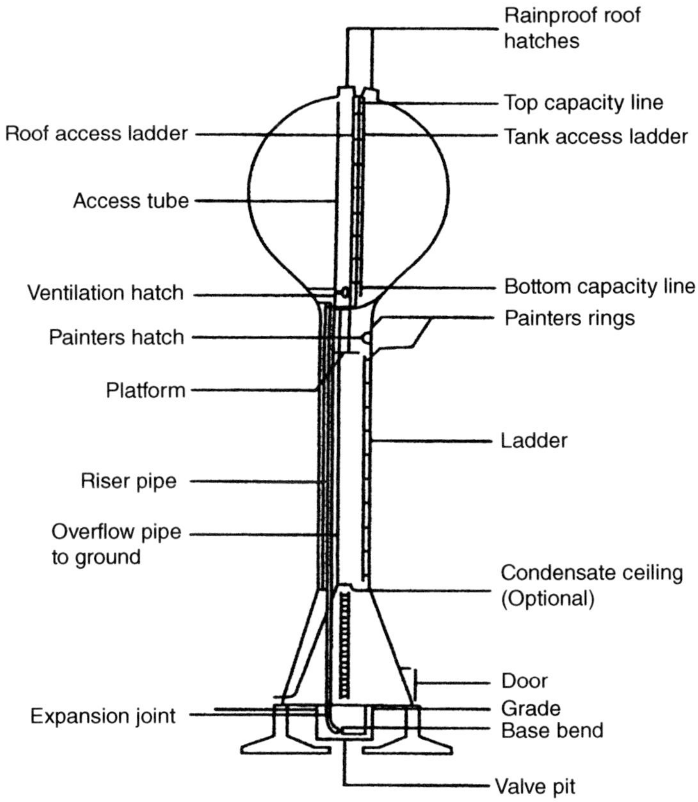

Capacity		Diameter of sphere	
m^3	gal	m	ft
95	6,555	5.80	15
114	7,866	6.15	16
151	10,419	6.74	17.5
189	13,041	7.21	18.7
284	19,600	8.24	21.7
379	26,150	9.02	23.5
473	32,640	9.71	25.20
568	39,200	10.32	27

FIGURE 7.5 Typical elevated storage tank.

Fire Pumps

With reliability in mind, a designer should make an attempt to design an automatic fire protection system that can function properly without the introduction of a booster fire pump. If this cannot be accomplished, then a fire pump shall be added to provide adequate water supply volume and pressure.

Fire pumps are required when the local municipal water system cannot provide sufficient pressure to meet the pressure requirements of the fire sprinkler or standpipe system. This usually occurs when the building is a high-rise building or in systems that require a relatively high pressure to flow a large volume of water from the sprinklers, such as early supression fast response (ESFR)–sprinklered storage warehouses. Fire pumps also are needed if the fire protection water supply is provided from a ground-level water storage tank. A jockey pump is used to keep a minimum pressure in the main.

The fire pump becomes active when the pressure in the fire main drops below a set threshold pressure provided by the jockey pump. Once operating, the fire pump provides additional water and pressure to the sprinkler system.

Types of pumps used for fire service include horizontal split case, vertical split case, vertical inline, vertical turbine, and end suction.

Labeled fire pumps are made in specific sizes ranging from 25 to 5000 gpm (95 to 19,000 L/min). Pressure selections may range from 40 to 475 psi (270 to 3200 kPa) for fire pumps.

Per NFPA 24, the fire pump must meet the following requirements:

1. The net pump shutoff (churn) plus the maximum static suction pressure shall not exceed the pressure for which the system components are rated.
2. At 150 percent of the rated capacity, it shall develop at least 65 percent of its rated head and shall not exceed 140 percent of the rated head at zero capacity.
3. Suction size and test components shall be per Table 7.1.
4. The maximum pump brake horsepower (bhp) must not exceed the rating of the particular driver.
5. Each fire pump must have listed pressure gauges and be fitted with a suitable air relief valve. With certain exceptions, a ¾-in (18-mm) casing relief valve is required to prevent overheating of the pump when it operates against a closed valve. Where the pump pressure may exceed the safe working pressure of the system, and always when a diesel driver or variable-speed driver is used, a listed main relief valve must be furnished.

Fire Pump Drivers. Fire pumps are driven by either electric motors or diesel engines.

If the fire pump is electric, power must be uninterruptible with properly protected power cabling and emergency power. Alarms should sound if normal power is interrupted (i.e., loss of phase or phase reversal). All fire pump motor drivers are required to be rated for continuous duty and must not be used at voltages in excess of 110 percent of the rated voltage. At rated voltage and frequency, the full-load ampere rating must not be exceeded under any pumping conditions.

If the fire pump is diesel, provisions must be made for a day tank that will supply fuel immediately before drawing from the main storage tank. Other items of concern are for engine cooling, exhaust pipe discharge location, and sufficient airflow for cooling, combustion, and ventilation.

The performance requirements and accessories for pumps, whether electric motor or diesel engine driven, are basically the same. The related components of the specific drivers, however, vary in installation, operation, and maintenance.

The motor control equipment must be factory-assembled, -wired, and -tested as well as specifically approved for fire service and complying with NFPA 24.

TABLE 7.1 Summary of Typical Centrifugal Pump Data

Pump rating		Minimum pipe sizes (nominal)							
L/min	gpm	Suction, in	Discharge, in	Relief valve, in	Relief valve discharge, in	Meter device, in	Number of hose valves	Size of hose valves, in	Hose header supply, in
946	250	3½	3	2	2½	3½	1	2½	3
1136	300	4	4	2½	3½	3½	1	2½	3
1514	400	4	4	3	5	4	2	2½	4
1703	450	5	5	3	5	4	2	2½	4
1892	500	5	5	3	5	5	2	2½	4
2839	750	6	6	4	6	5	3	2½	6
3785	1000	8	6	4	8	6	4	2½	6
4730	1250	8	8	6	8	6	6	2½	8
5677	1500	8	8	6	8	8	6	2½	8
7570	2000	10	10	6	10	8	6	2½	8

Jockey Pump. The jockey pump is normally required in all pressurized systems. This automatic electric pump has a capacity of 5 to 10 gpm (20 to 40 L/min) or less. The intent is to maintain pressure in the piping when pressure is lost to minor leaks, not to keep up with sprinkler discharge. Its controller is set to start at about 5 psi above the start signal for the fire pump and to stop at full pressure.

Fire Pump Sizing Estimate

$$\text{bhp} = \frac{\text{head} \times \text{gpm} \times \text{specific gravity}}{3960 \times \text{efficiency}} \tag{7.1}$$

$$\text{bhp} = \frac{\text{psi} \times 2.13 \times \text{specific gravity}}{3960 \times \text{efficiency}} \tag{7.2}$$

Example. Calculate brake horsepower given the following conditions:

- Standpipes = 3
- Fully sprinklered = yes
- Flow demand = 1000 gpm
- Building height = 50 ft
- Efficiency = 65%
- Specific gravity = 1.0

$$\begin{aligned} \text{bhp} &= (50 \times 1000 \times 1.0)/(3960 \times 0.65) \\ &= 50{,}000/2574 \\ &= 19.42\text{, say, } 20\text{ hp} \end{aligned}$$

Additional losses to consider are internal pipe loss, pump loss, backflow prevention equipment, and safety factors.

Fire pumps are used to add to the supply of water from public water mains, tanks, reservoirs, and other sources. They are installed to increase the pressure of water supplied to the fire protection system from the above sources. The pumps are usually classical designs that include ancillary items for the complete installation such as pumps, starters, and control devices. This discussion covers only those pumps installed on site. Pumps located inside buildings are outside the scope of this handbook.

Fire Pump System Components. The following components are necessary for a complete installation; all must be in accordance with local AHJ, NFPA 24, and manufacturers' requirements.

1. *Main firefighting pump.* Usually this is diesel driven and supplied from a source that is from a large tank or public water supply and is considered practically inexhaustible.
2. *Jockey pump.* This is a smaller pump used to maintain pressure in the fire protection mains.
3. *Driver.* The purpose of the driver is to provide the fire pump with enough motive force under all conditions.
4. *Controllers.* These are used to start the engine and report various possible malfunctions.
5. *Commonly used accessories*
   - ***a.*** Casing relief valve, used to prevent no-flow condition when pump is running at shutoff
   - ***b.*** Automatic air relieve valve, to vent trapped air

c. Hose valve heads, to allow fire pump tests in lieu of flowmeter

d. Main relief valve

e. Increasers and/or reducers when the size of the suction does not match the main connection

f. Other accessories necessary as determined by code, AHJ, facility personnel, or manufacturer

A detail of a typical fire pump arrangement is illustrated in Fig. 7.6. Not all the accessories may be pictured. A summary of typical fire pump sizing data is given in Table 7.1.

Water Supply

Based on NFPA 20 and most fire protection agencies, the adequate water supply required for the proper performance of standpipe systems is as follows:

1. For class I and class III systems, the minimum flow rate for the hydraulically most remote standpipe shall be 500 gpm (475 L/min). The minimum flow rate for additional standpipes shall be 250 gpm (950 L/min) per standpipe, with the total not exceeding 1250 gpm (4750 L/min). For buildings that are fully sprinklered throughout, the total calculated standpipe flow rate may be reduced to 1000 gpm (4000 L/min).
2. Hydraulically calculated standpipe systems shall be designed to provide the water flow at a minimum residual pressure of 100 psi (700 kPa) at the outlet of the hydraulically most remote 2½-in (DN 65) hose connection and 65 psi (435 kPa) at the outlet of the hydraulically most remote 1½-in (DN 40) outlet. (Some building codes, fire marshals, and authorities having jurisdiction may allow for a variance for a lower standpipe pressure.)
3. The pressure for a 1½-in hose connection shall not exceed 100 psi (700 kPa) at the top of the most remote standpipe. The pressure for a 2½-in (DN 65) hose connection shall not exceed 175 psi (1225 kPa) at the top of the most remote standpipe. Where the available residual pressure exceeds the outlet rating, an approved pressure-regulating device shall be provided to limit the residual pressure at the flow required.

If the water supply to the standpipes cannot provide the required volume and pressure, the system must be upgraded to ensure an adequate supply, or a fire pump must be added to the system.

Sizing the Fire Protection Water Service

All the information required for design has previously been obtained for the domestic water system. The fire marshal, the fire protection subcode official, and the insurance company shall be contacted for their installation requirements. This includes static and residual pressures.

What constitutes an adequate water supply for fire protection has generated much discussion over time. There are many factors that, when added together, will provide an adequate water supply. The water supply should be capable of supplying the largest demand, which is usually the largest sprinkler system and expected hose flow under reasonably adverse conditions. Other factors, such as building occupancy, yard storage, external structures that must be protected, exposure protection, and a catastrophic hose demand, will all add to the possible maximum flow rate. Generally accepted practice is to have a minimum size of 6 in (DN 250) for any fire protection main.

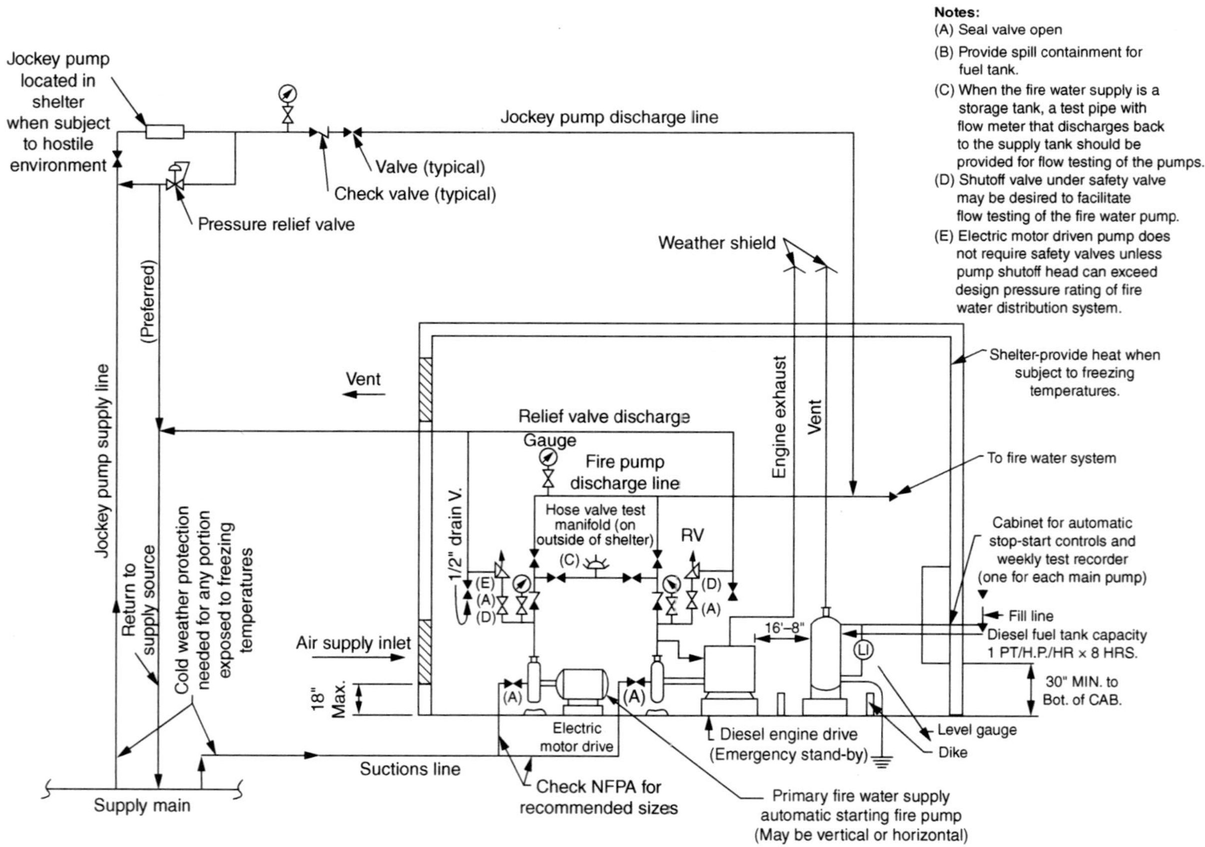

FIGURE 7.6 Typical automatic fire pump installation.

Demand Flow Rate

The following figures are presented for discussion purposes only and should be considered preliminary estimates. They are not to be used for actual design, which should only be made on a specific project basis and only with consultation with authorities.

The flow requirement for the sprinkler system is the hydraulically calculated flow rate. The base hose streams depend on the occupancy hazard:

Light and ordinary hazard, 500 gpm (1900 L/min)

Extra hazard, group 1, 750 gpm (2900 L/min)

Extra hazard, group 2, 1000 gpm (3800 L/min)

High-piled storage, as required by insurance company

The hose demand shall be increased by 25 percent for the following conditions:

1. Combustible construction
2. Possible delay in response by public fire department
3. Minimum protection less than recommended by insurance company requirements
4. Limited access to remote interior sections
5. Additional requirements for monitors: allow 500 gpm (1900 L/min) for each monitor

STORAGE TANK CAPACITY

If the gravity tank is the sole source of water, the tank shall be capable of being filled in 8 h. Evaluation of the total capacity should take into consideration the following storage capacity based on the fire pump discharge:

Light and ordinary hazard occupancies, group 1, 2 h

Ordinary hazard occupancies, group 2, 3 h

Extra hazard occupancies, 4 h

The design of the elevated water tank is dependent on the local officials including the fire marshal, the plumbing department, and the NFPA standard, all of which should be consulted for the actual design of the tank. The combined water tank will have a domestic supply at the top of the tank and a fire reserve at the lower portion of he tank. The actual number of gallons in each reserve will be selected after consultation with the local plumbing official and the fire department. A detail of a typical elevated tank is shown in Fig. 7.5.

PIPE MATERIAL SELECTION

For combined systems, the two piping materials used most often are ductile iron and PVC, approved for use as fire protection mains. They shall have a pressure rating of 150 psi (1035 kPa). Other less used pipe materials include steel pipe. A generally accepted leakage rate is 0.005 gph for each joint.

Preliminary Information Required

1. The location, size, material, and elevation of the public water main available to the project under design must be known.
2. The water pressure must be determined for the specific elevation closest to the point where the proposed service will be connected. This requires that a flow test be conducted to be accurate. This test will give the static pressure (with no water flowing), the residual pressure (the water pressure observed with a flow rate), and the time of day when the test is conducted. Several flow rates and residual pressures should be recorded to allow plotting on a flowchart. A hydrant flow test is discussed in Chap. 6.
3. Determine the authority responsible for installation and approval of (and the need for) any BFP devices for the specific facility.
4. Obtain from the utility company or local authorities the type of water meter required (if any), the proposed location of the meter, and the extent of work provided by the utility company versus the amount of work required to be done by a contractor. Usually, the utility company has standard details for meter assembly installations and the specific devices required to be part of the meter assembly and their arrangement.
5. The frost depth must be found for the area as well as the minimum cover over the proposed water line. For guidance, refer to Fig. 22.3.
6. Other special items or criteria that will affect the building service installation or add to the friction loss of the water service must be determined.
7. The maximum flow rate for the project must be determined.

Calculation of the Pipe Size

1. Find the static and residual water pressure of the main and the elevation at which the pressure was taken.
2. Determine the maximum flow rate for the fire protection water service.
3. Locate and lay out the service main from the building wall to the point of connection with the utility service or site main. Locate all devices (curb valve, BFP if required, meter, strainer, building valves, etc.).
4. Calculate the static pressure loss based on the elevation of the water main and the elevation of the service entrance into the building.
5. Select the pipe material.
6. Add all the losses in the service main including losses from valves, meters, BFP devices, etc.
7. Calculate the pressure loss (or gain) from the difference in elevation (from step 4).
8. Find the friction loss in the service main to the building entrance.
9. Add an allowance of 5 to 10 psig loss for future pressure drop in the public main. The pressure used shall reflect the possibility of other projects being built in the area. Size the pipe based on maximum flow rate and the above criteria, using the friction loss as a variable along with the velocity of the pipe size selected.
10. Determine whether the available pressure is high enough for the selected building water distribution system. If it is determined that additional pressure is required, a fire pump will be necessary.
11. Select the pipe size based on the available pressure required less pressure losses.

BIBLIOGRAPHY

American Water Works Association, *Steel Water Storage Tanks*, McGraw-Hill, New York, 2010.

Fire Protection Handbook, National Fire Protection Association, Quincy, Mass.

Frankel, M., *Facility Piping Systems Handbook*, 3d ed., McGraw-Hill, New York, 2010.

Mays, L., *Water Distribution Systems Handbook*, McGraw-Hill, New York, 2000.

NFPA 291, Recommended Practice for Fire Flow Testing and Marking of Hydrants, National Fire Protection Association, Quincy, Mass., April 2011.

CHAPTER 8
SITE FUEL GAS SYSTEMS

GENERAL

The natural fuel gas system delivers fuel gas to provide heat energy in sufficient volume and pressure required for the satisfactory operation of all connected devices and for all purposes. Propane systems are used as a backup for natural gas, and where natural gas is not available or cost effective, it is also used as a heat source. This chapter discusses only natural gas and propane that are installed on the site.

FUEL GAS DESCRIPTION

Many gases are used as a fuel gas. Where easily and cheaply available, two major fuel gases, gaseous natural gas (NG) and liquefied petroleum gas (LPG), are preferred. Refer to Table 8.1 for the physical and combustion properties of natural gas and propane.

NG and LPG are hydrocarbon compounds obtained from the separation of gas mixtures occurring naturally at the wellhead of crude oil–producing or gas-producing wells or as a by-product of the crude oil refining process. NG is primarily composed of methane with minor percentages of other gases that result in variations in heating content. LPG is a derivative of NG, and it is a term applied to a group of hydrocarbons such as propane, butane, isobutene, and pentane or to various mixtures of each. Propane and butane are the principal constituents of LPG, with propane the most common. Since NG and LPG are odorless, an odorant, such as ethyl mercaptan, is added to make detection of the gas possible in the event of a leak. It has been found that the odorant added to natural gas has been reduced or completely dissipated in the distribution pipe within a system over time. This is called *odor fade* and occurs where the odorant is lost and a gas leak cannot be detected by smell alone. It could be caused by cutting oils, dirt or debris, moisture, or stagnation.

There are variations in the hydrocarbon mixtures that produce variations in the specific gravities and British thermal unit (Btu) contents of NG and LPG obtained from various suppliers. Natural gas from various locations will have a different Btu content. The values presented here are average values and sufficiently accurate for the design of these systems. It is recommended that an analysis of the products actually furnished by the supplier for each specific project be obtained if extreme precision is required.

TABLE 8.1 Average Physical Properties of Natural Gas and Propane

	Propane	Natural gas (methane)
Formula	C_3H_8	CH_4
Molecular weight	44.097	16.402
Melting (or freezing) point, °F	–305.84	–300.54
Boiling point, °F	–44	–258.70
Specific gravity of gas (air = 1.00)	1.52	0.60
Specific gravity of liquid 60°F/60°F (water = 1.00)	0.588	0.30
Latent heat of vaporization at normal boiling point, Btu/lb	183	245
Vapor pressure, lb/in^2, gauge at 60°F	92	
Pounds per gallon of liquid at 60°F	4.24	2.51
Gallons per pound of liquid at 60°F	0.237	
Btu per pound of gas (gross)	21,591	23,000
Btu per ft^3 gas at 60°F and 30 in Hg	2516	1050
Btu per gallon of gas at 60°F	91,547	
Cubic feet of gas (60°F, 30 inHg)/gal of liquid	36.39	59.0
Cubic feet of gas (60°F, 30 inHg)/lb of liquid	8.58	23.6
Cubic feet of air required to burn 1 ft^3 gas	23.87	9.53
Flame temperature, °F	3595	3416
Octane number (isooctane = 100)	125	
Flammability limit in air, upper	9.50	15.0
Flammability limit in air, lower	2.87	5.0

1 inHg = 3.37 kPa   60°F = 15.6°C   1 lb = 2.2 kg   1 ft^3 = 0.03 m^3

CODES AND STANDARDS

There are a number of codes and standards governing the manufacture, design, installation, and testing of LPG and NG systems, piping, and components. The principal codes, standards, and regulations are as follows:

1. ANSI/NFPA 30, Flammable and Combustible Liquids Code
2. ANSI/NFPA 54, National Fuel Gas Code
3. NFPA 55, Combustible Gas and Cryogenic Fluids Code
4. ANSI/NFPA 58, Standard for the Storage and Handling of Liquefied Petroleum Gases
5. ANSI Z83.3, Standard for Gas Utilization Equipment in Large Boilers
6. ANSI/UL 144, Pressure Regulating Valves for LP Gas
7. American Gas Association (AGA) codes and standards
8. Local utility company rules and regulations
9. Model national codes

Some insurance carriers, such as Factory Mutual and Industrial Risk Insurers, have standards that in many aspects may be more stringent than those listed above.

NATURAL GAS

General

Natural gas is obtained from a franchised public utility obligated to provide gas to every customer that requests this service. As part of this service, the utility company usually supplies and installs the service line free of charge from the utility main to the facility meter, in addition to providing a regulator-meter assembly in or adjacent to the building.

If the utility company has an existing main in the vicinity of the project, which is most often the case, then the initial start-up and utility testing of the branch to the facility is borne by the utility. Sometimes, however, an installation fee is charged if the utility company regulations concerning the installation cost compared to the time of payback from expected revenue does not justify the cost of installing the service. The same might be true if the use of gas for a commercial or industrial facility is considered too little. It would then be the responsibility of the owner to pay for the design, installation, and testing of the complete site service and meter assembly in conformity with utility company regulations.

There are several different types of service that a utility company may provide, each with a different cost (or rate). Specific types of service may be unavailable or known by different names in various localities.

1. *Firm service.* This type of service provides a constant supply of gas under all conditions without exception. This service has the highest rate.
2. *Interruptible service.* This type of service allows the utility company to stop the gas supply to the facility under predetermined conditions and then to start it again when these conditions no longer exist. The most common condition usually occurs when the outside temperature falls below a certain point. The rate is lower than for firm gas, and users with this service will require a backup source of fuel, such as fuel oil or LPG.
3. *Light or heavy process service.* This type of service is provided for industrial or process use and is reserved for quantities of gas that the utility companies define for this class of service.
4. *Commercial or industrial service.* This type of service is provided for heating and cooling system loads for this class of building usage.
5. *Transportation gas service.* This service uses NG purchased directly from a company at the wellhead, not the public utility, with the gas actually carried to the site by the utility company mains. In addition to the cost of the gas, the utility company will charge an additional fee for the use of its main. This type of service is always available.

The following criteria and information should be obtained in writing from the public utility company at the start of a project:

1. The British thermal unit content of the gas provided
2. The minimum pressure of the gas at the property line where the engineer's connection will be made
3. The minimum pressure to be provided at the outlet of the meter
4. The extent of the installation work done by the utility company and the point of connection by the facility construction contractor
5. The location of the utility supply main and the proposed run of pipe on the site by the utility company

6. The acceptable location of the meter and/or regulator assembly and any work required by the owner to allow the meter assembly to be installed (such as having a meter pit dug or a slab installed on grade)
7. The types of service available and the cost of each
8. Any auxiliary service, such as a necessary telephone or electrical connection.

For the utility company to provide these data, it requires that the following information be provided to the utility company:

1. The total connected load. The utility will use its own diversity factor to calculate the size of the service line. The design engineer will use another diversity factor since she or he is responsible for the interior facility installation.
2. The minimum and maximum pressure requirements for the most demanding device.
3. The site plan indicating the location of the proposed building on the site and the specific area of the building where the proposed NG service will enter the building.
4. Preferred location of the meter-regulator assembly.
5. Expected date for the start of construction.

Major Natural Gas System Site Components

Drip Pots. Drip pots are rarely required, but are necessary to keep water that is in the mains (if any) from reaching the building. Modern technology has reduced, but not eliminated, the presence of moisture in the utility company mains. Each utility company has standard methods of requiring and installing drip pots, depending on the pressure in the main. Drip pots are also installed if "wet" gas is known to be present, such as after large pressure-reducing valves.

Gas Line Filters. Filters for natural gas are installed on the site service to protect the regulator and meter from injury due to particulate clogging, which may damage the equipment inside the building. Filters should be considered when

1. Line scale, dirt, or rust is known to be present.
2. Dirty gas is obtained from a transmission company.

The filter consists of a housing and a cartridge filter element. Selection of the housing is based on the highest flow rate and pressure expected and the size of the proposed building service. The housing should be capable of having interchangeable elements to allow replacement of the original filter, if desired. Having a filter oversized is considered good practice since this allows a longer service life if the cost of the larger size is not excessive.

The filter element is usually cellulose or synthetic fiber. Generally, the cellulose type is used for sizes of 3 in (DN 75) and less and fiber is used for 4 in (DN 100) and larger. Fiber is stronger and should be considered where the pressure differential may be excessive when recommended by the manufacturer. A filter rating of 10 μm is suggested as a starting point, with actual operating experience by the utility being the final criterion.

The size of the filter is based on the *actual cubic feet per minute* (acfm [L/min]) and velocity across the filter. The pressure drop across the filter when dirty shall be considered in the actual pressure loss allowed for the service assembly. Since the filter should be changed when the pressure difference approaches 10 percent of the operating pressure, it is good practice to provide pressure gauges on both sides of the filter. If no filter is installed on the service, a spool piece should be provided if there is a possibility that one may be required. Most filters are mounted horizontally.

Gas Meters. Gas meters are part of a service assembly that may also consist of filters, valves, regulators, and relief valves. The complete assembly is usually supplied and installed by the utility company. For the rare instances when this is not the case, gas meters are selected using the local utility company standards with the size and arrangement of the entire meter assembly based on the flow rate and pressure. Because steady pressure is necessary for accurate metering, a regulator is installed upstream of the meter. The utility company will provide sketches and details, with space requirements, of typical meter assemblies for various capacities and pressures. This will permit accurate determination of the dimensional and space requirements for the area where the complete assembly will be installed. Obtain from the utility company all requirements for the specific installation under design. Usually duplex meters will be required.

A contractor other than the utility company will be responsible for providing a pit for the assembly or a concrete slab under the complete meter assembly when installed outside the building. This will include meter slab size for outdoor meter installations and pit sizes and access openings for meters installed in pits. Additional requirements will usually include a weatherproof telephone outlet and a 120-V electric outlet at the meter site. For facilities with large demands for boilers and other equipment, a duplex meter assembly with dimensions of 6 ft (2 m) wide by 25 ft (8 m) long is not uncommon.

Generally, if the contractor is installing the site service, the utility company will either supply the meter or install the entire assembly at no charge based on the utility's requirements. This must be confirmed. Requirements of various utility companies differ regarding the placement of the meter assembly. It could be installed in an underground, exterior meter pit, at an aboveground exterior location exposed to the weather, or inside the building in a well-ventilated area or mechanical equipment room.

Pressure Regulators. The line pressure regulator is usually pilot-operated and provided with an internal or external relief valve. The regulator is pressure-rated to withstand the highest pressure expected. The relief valve is installed downstream from the regulator and is set to trip at a pressure of about 10 percent higher than the highest set pressure. The line regulator is placed upstream from the meter in order to provide the meter with a constant pressure, allowing accurate measurement. This line pressure regulator is most often selected and installed by the utility company as part of the gas meter assembly.

An intermediate regulator is used within a facility where high pressure used for distribution purposes must be reduced to a lower pressure required by the terminal appliances. There are two types of intermediate regulators used, and the choice is determined by the accuracy desired and the ability to install a relief valve and associated gas vent discharge. One type of regulator has an internal relief valve set to discharge when the pressure rises above the set point. This has the least initial cost and least accuracy. Another type of intermediate regulation, called a *monitor type*, consists of two pilot-operated regulators and does not require a relief valve. They can be installed in two configurations, both of which use regulators in series. The first uses the upstream regulator to initially reduce the inlet pressure to some intermediate value and the downstream regulator to further reduce the pressure to the final set point. This arrangement puts less stress on each of the regulators. The second configuration uses the upstream regulator wide open and the downstream regulator to do all of the pressure reducing. If the gas pressure goes above the set point, the upstream regulator closes to partially lower the pressure. This installation has a high initial cost but must be used where a relief valve cannot be installed. The intermediate-type regulator arrangement is the most often used.

The appliance regulator is used to control the pressure of gas directly connected to the terminal appliance or equipment inside the facility. The appliance regulator is most often provided by the manufacturer as part of the equipment gas train. The *gas train* is an assembly of piping, valves, regulators, relief vents, etc., used to directly connect the gas supply to the terminal equipment. For larger pieces of equipment such as boilers, the gas train arrangement is dictated by insurance carrier requirements. An additional requirement of most gas trains is small relief vents from various devices. These vents must be piped outside the building to a point where they can be diluted by the outside air and will not pose a threat to the public or create a fire hazard. This is usually above a roof or the highest point of the structure.

A piped gas vent must be provided from regulators to a point several feet above the roof of any adjacent structure or of a safe place that satisfies utility company requirements. It must be properly sized to carry the amount of gas that will be discharged. This is to protect the system from overpressure in the event of a malfunction or failure to fully lock up. Each individual regulator vent must be separately carried to a nonhazardous location away from any potential source of ignition. Common vent lines are not permitted.

Regulator Selection Considerations. It is common practice to oversize the capacity of a regulator by about 15 percent to provide a margin for accurate regulation.

1. For large loads, regulators in parallel are often used to keep the pressure drop to a minimum. The adjustment range of a regulator should be approximately 50 percent over the desired regulated pressure to 50 percent under that pressure.
2. The utility company may require that regulators be of the "lockout" type. This feature will stop regulator operation when the pressure falls below a predetermined set point.

Site Distribution

The site distribution portion of the system starts at the property line of the customer and usually ends above grade (or the concrete pad) at the valve after the meter. The connection to the utility company mains is done by the utility company. Piping to be used inside the building is connected by the plumbing contractor outside the building at a connection point agreed to by the utility company. Obstacles to be avoided, such as underground piping, trees, underground structures, and rock, should be located.

The most often used piping material is high-density polyethylene (HDPE) and polypropylene type (PP-R). The rating for pipe 4 in (DN 100) and smaller should be standard design ratio (SDR) 11, and for 5 in (DN 125) and larger SDR 13.5. The pipe shall be buried a minimum of 3 ft (1 m) belowground, and a 14 AWG corrosion-resistant tracer wire placed in the pipe trench in over the pipe. Another detection method is to put a warning tape containing metallic material with the words "Natural Gas" on it. This is to allow location of plastic piping by a metal detector and to warn of the gas line immediately below the tape if digging has taken place without trying to locate the pipe beforehand.

The pipe aboveground shall be metal; therefore, a transition fitting is required if HDPE is used for the supply service. The transition fitting shall comply with ASTM D-2513. Refer to Fig. 8.1.

Testing and Purging

After installation of the pipe is complete, the site system shall be tested, purged of air, and then filled with natural gas. The testing phase shall be done with compressed air at a gauge

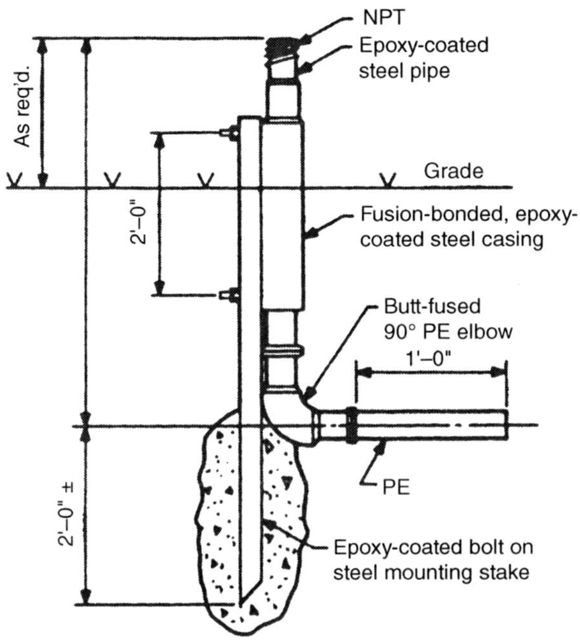

FIGURE 8.1 Transition fitting from underground plastic to aboveground steel pipe.

pressure 50 percent higher than the highest pressure expected in the main. The larger the total volume of the pipe, the longer the test shall last. No loss of pressure shall be allowed. The test period is based on the size of the line and the length of run. Table 8.2 gives the recommended test periods based on actual length of run.

After successful testing, the line shall be purged of air with dry nitrogen. The reason is to prevent a flammable mixture of gas and air when the pipe is filled with natural gas for the first time. Calculate the volume of the piping and introduce an equal volume of nitrogen into the pipe. After the nitrogen purge, natural gas is introduced until the nitrogen is displaced and then left under pressure.

Site Service Sizing Procedure

The following information must be determined or calculated by the engineer to size the site service main:

1. The pressure available at the point of connection with the utility companies main
2. The pressure required at the meter to ensure adequate pressure inside the facility
3. The friction loss for the piping
4. The total connected load and the calculation of the maximum probable demand
5. The equivalent run of site piping
6. The pipe sizing method acceptable to local code

TABLE 8.2 Maximum Length of Pipe Being Tested, ft, Mains to Operate at Less than 125 psig (840 kPa)

Main size		Test duration, h												
DN	in	2	3	4	6	8	10	12	14	16	18	20	22	24
25	1 or less	6,590	11,000	15,400	24,200									
50	2													
75	3	1,700	2,830	3,960	6,220	8,490								
100	4	743	1,240	1,730	2,730	3,720	4,710	5,700						
150	6	439	732	1,030	1,610	2,200	2,780	3,370	3,960	4,540	5,130			
200	8	189	315	442	694	947	1,200	1,450	1,710	1,960	2,210	2,460	2,730	2,970
300	12	108	180	252	396	541	685	829	973	1,120	1,260	1,410	1,550	1,700
350	14	48	80	112	177	241	306	370	435	499	563	628	692	757
400	16	41	68	96	151	206	261	316	371	426	482	537	592	647
500	20	30	50	70	110	150	191	231	271	311	352	392	432	472
600	24	19	31	44	69	95	120	146	171	197	222	247	273	298
		13	21	30	48	65	83	100	118	135	153	170	188	205

General. There are only two possibilities for the routing of the site work: the consulting engineer would be responsible for the entire site main from the property line to the meter location, or the utility company would be responsible for the entire installation of the piping from the property line up to and including the meter. The selection for the routing of the pipeline shall be made first.

Pressure Available from the Utility Company. If the engineer is designing the piping from the tie-in point with the utility main to the meter at the facility, the minimum pressure that the utility company will guarantee at the tie-in shall be obtained from the utility company, in writing. The utility company is usually responsible for providing the piping from its main to the designated point at the property line and any valving necessary. The meter(s) themselves would be selected by the utility company and the location chosen by either the contractor or the utility company. If there is a specific requirement for equipment that needs a higher pressure, such as that for a boiler, the utility company must be advised of this requirement and guarantee this pressure in writing. If the higher pressure is not possible, a gas pressure booster will be necessary. This is a rare occurrence. The minimum pressure necessary for the meter to register accurately shall also be provided. A duplex meter is highly recommended, and it may be a requirement of the utility company. The utility company shall provide a diagram of the proposed piping, the valving arrangement, the meter pad size, and the connection point for the contractor. The complete meter layout and ancillary apparatus (such as telephone and electric outlets) shall be approved by the utility company as well as the selection of the pressure-reducing valve (PRV), if necessary. If the meter train is to be installed by the utility, connection points shall be clearly indicated. A higher site pressure will result in smaller pipe sizes.

If the utility is providing the entire site main, up to and including the meter, the important information for the utility to have is the proposed location of the main on the site, the highest pressure required for any equipment of the facility to operate properly, and the proposed location of the meter. The utility will indicate the tie-in point from the meter into the building for the contractor. This tie-in point is a valve on the outlet of the meter. From this point, the facility is responsible for the pipe.

Required Pressure Inside the Facility. The pressure selected shall be high enough to support the operation of the equipment with the highest pressure requirements based on the pressure and the interior friction loss selected by the engineer for the piping. If the high pressure is needed for only one piece of equipment, it would be more practical to provide a pressure-reducing valve for the low-pressure line.

Friction Loss in the Site Piping System. The pressure in the utility company's main is usually very high. It generally ranges up to 50 psig. It is usual to keep the original pressure on site up to the meter to minimize the pipe size. When the appropriate sizing chart that matches the pressure in the main is selected, the friction loss is not much of a consideration as long as the pressure delivered to the PRV is accessible. Any changes in elevation can be ignored, as the gas gains pressure at a higher elevation, offsetting the friction loss.

Maximum Probable Demand. The maximum probable demand for the building is the actual expected use of gas, under the assumption that not all the connected devices will be used at the same time. The total connected load is not used to size the piping system. For design purposes, it is necessary to apply a diversity factor to the total connected load. The diversity factor for residential facilities will be as follows.

This calculation first requires the listing of every device using gas in the building or facility and the demand in British thermal units per hour for each. Use the most demanding figures. The manufacturer of each device should be consulted to find the actual input gas consumption of that device. Average gas demand values for typical devices are listed in Table 8.3.

TABLE 8.3 Average Gas Demand Values for Typical Equipment
(British thermal units per hour)

Appliance	Input, Btu/h (MJ/h)
Commercial kitchen equipment:	
Small broiler	30,000 (31.7)
Large broiler	60,000 (63.3)
Combination broiler and roaster	66,000 (69.6)
Coffee maker, 3 burner	18,000 (19)
Coffee maker, 4 burner	24,000 (25.3)
Deep fat fryer, 45 lb (20.4 kg) of fat	50,000 (52.8)
Deep fat fryer, 75 lb (34.1 kg) of fat	75,000 (79.1)
Doughnut fryer, 200 lb (90.8 kg) of fat	72,000 (76)
2 deck baking and roasting oven	100,000 (105.5)
3 deck baking oven	96,000 (101.3)
Revolving oven, 4 or 5 trays	210,000 (221.6)
Range with hot top and oven	90,000 (95)
Range with hot top	45,000 (47.5)
Range with fry top and oven	100,000 (105.5)
Range with fry top	50,000 (52.8)
Coffee urn, single, 5 gal (18.9 L)	28,000 (29.5)
Coffee urn, twin, 10 gal (37.9 L)	56,000 (59.1)
Coffee urn twin, 15 gal (56.8 L)	84,000 (88.6)
Stackable convection oven, per section of oven	60,000 (63.3)
Residential equipment:	
Clothes dryer	35,000 (36.9)
Range	65,000 (68.6)
Stovetop burners	40,000 (42.2)
Oven	25,000 (26.4)
30-gal (113.6-L) water heater	30,000 (31.7)
40- to 50-gal (151.4- to 189.3-L) water heater	50,000 (52.8)
Log lighter	25,000 (26.4)
Barbecue	50,000 (52.8)
Miscellaneous equipment:	50,000 (52.8)
Commercial log lighter	50,000 (52.8)
Bunsen burner	10,000 (10.6)
Gas engine, per horsepower (745.7 W)	10,000 (10.6)
Steam boiler, per horsepower (745.7 W)	50,000 (52.8)

Note: The efficiency of the individual units may require additional input to that shown.

For multiple-unit dwellings, a direct reading of the quantity of gas based on the number of apartments, with a diversity factor built in, is presented in Fig. 8.2 for buildings of up to 50 apartments and in Fig. 8.3 for buildings with more than 50 apartments. For laboratories,

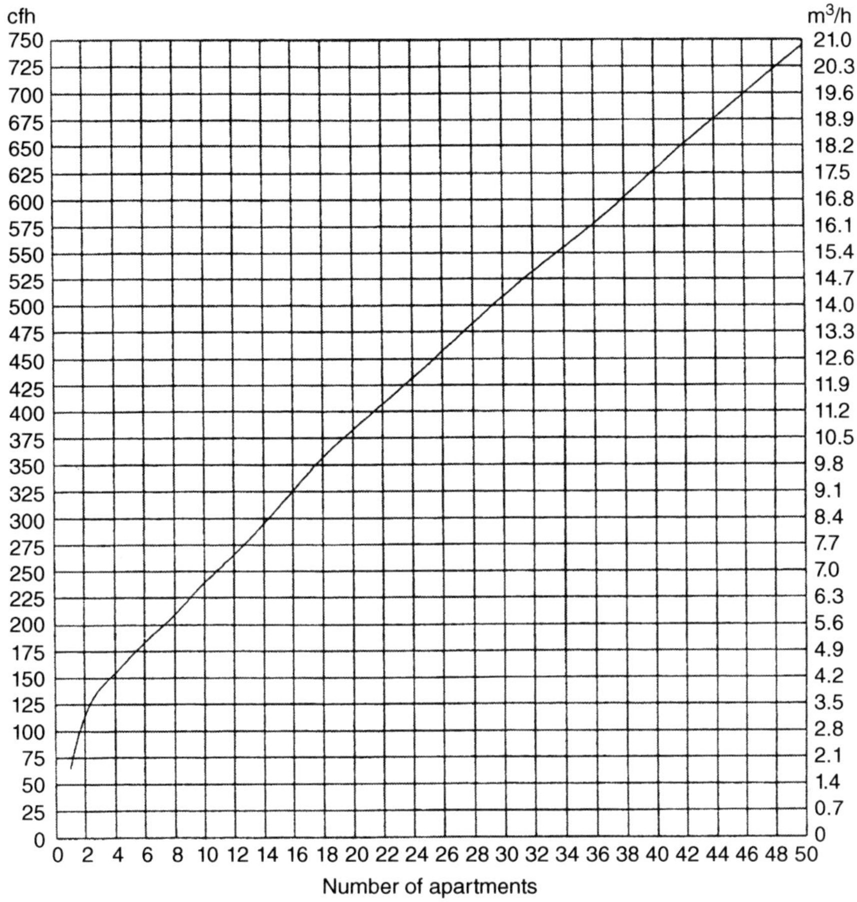

Gas demand for multiple-unit dwellings with less than 50 apartments.

FIGURE 8.2 CFH Demand in 1000s.

refer to Table 8.4 for diversity factors. Where laboratories are part of a school, use no diversity for entire rooms, and consult with the school authorities to find the total number of rooms with Bunsen burners that might be in use at once. If there is no conclusive answer, use no diversity.

The maximum demand for trailer parks is given in Table 8.5. For industrial or process installations and for major gas-using equipment such as boilers and water heaters in all building types, a diversity factor is generally not used because it is possible that all connected equipment could be in use at the same time.

Equivalent Length of Piping. The *equivalent length of piping* is the actual measured length of the site pipe plus an allowance made to account for the loss through valves and fittings. It is one of the items needed to use the sizing chart. It is common practice

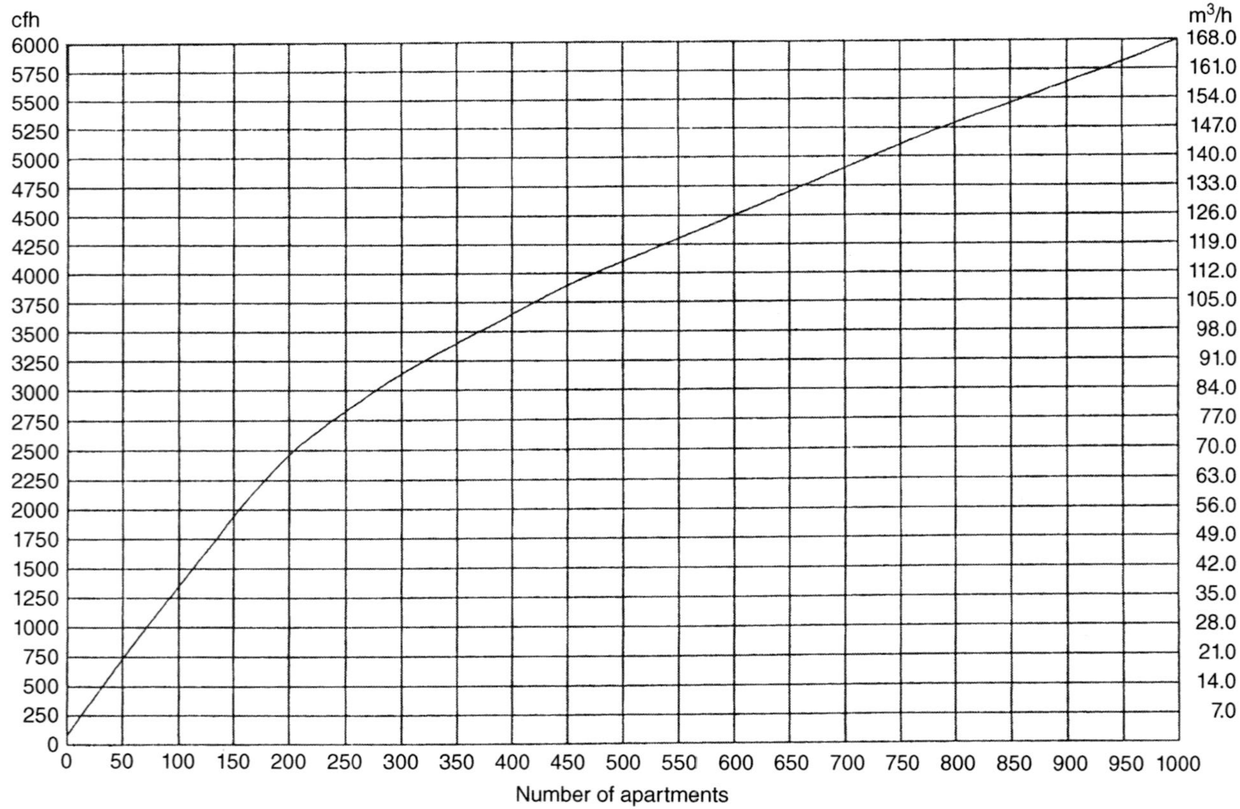

Gas demand for multiple-unit dwellings with more than 50 apartments.

FIGURE 8.3 CFH Demand in 1000s.

TABLE 8.4 Laboratory Diversity Factors

Number of outlets	Average percent of use	Maximum percent of use
1–5	100	
6–10	75	90
11–20	60	75
21–70	40	60
71–150	30	50
Over	20	40

TABLE 8.5 Maximum Demands for Trailer Parks

Number of trailer sites	Btu/h per trailer site	MJ/h
1	125,000	(132)
2	117,000	(123)
3	104,000	(109)
4	96,000	(101)
5	92,000	(97)
6	87,000	(92)
7	83,000	(88)
8	81,000	(86)
9	79,000	(85)
10	77,000	(81)
11–20	66,000	(69)
21–30	62,000	(65)
31–40	58,000	(61)
41–60	55,000	(58)
Over 60	50,000	(53)

to calculate the equivalent length of piping by first finding the measured length of the proposed piping run and then adding one-half of that length to the measured length to calculate the equivalent length. If a more precise equivalent run is desired, refer to Table 8.6 for equivalent lengths of fittings and valves for the flow of gas in a pipeline. This will be added to the measured run. The equivalent length will be the length used in the sizing chart.

Site Service Pipe Sizing Methods

The most conservative method for sizing NG piping systems is by the use of tables, such as those prepared by the American Society of Plumbing Engineers (ASPE) and in NFPA 54, National Gas Code. Calculators are also available. The calculators are considered more accurate than the prepared tables.

Tables are provided that will allow sizing for a majority of systems. Table 8.7 is for a pressure of 10 psig (67 kPa), Table 8.8 is for a pressure of 20 psig (125 kPa), and Table 8.9 is for a pressure of 50 psig (335 kPa). If higher pressure is encountered, compressed air

TABLE 8.6 Resistance of Valves and Fittings to the Flow of Gas
(Expressed in equivalent feet of pipe)

		Screwed fittings[a]				90° Welding elbows and smooth bends[b]					
		45° ell	90° ell	180° close return bends	Tee	$R/d = 1$	$R/d = 1\frac{1}{2}$	$R/d = 2$	$R/d = 4$	$R/d = 6$	$R/d = 8$
k factor		0.42	0.90	2.00	1.80	0.48	0.36	0.27	0.21	0.27	0.36
L/d' ratio[d] *n*		14	30	67	60	16	12	9	7	9	12

Nominal pipe size, in	Inside diam. *d*, in, Sched. 40[f]	L = equivalent length in feet of Schedule 40 (standard weight) straight pipe[f]									
½	0.622	0.73	1.55	3.47	3.10	0.83	0.62	0.47	0.36	0.47	0.62
¾	0.824	0.96	2.06	4.60	4.12	1.10	0.82	0.62	0.48	0.62	0.82
1	1.049	1.22	2.62	5.82	5.24	1.40	1.05	0.79	0.61	0.79	1.05
1¼	1.380	1.61	3.45	7.66	6.90	1.84	1.38	1.03	0.81	1.03	1.38
1½	1.610	1.88	4.02	8.95	8.04	2.14	1.61	1.21	0.94	1.21	1.61
2	2.067	2.41	5.17	11.5	10.3	2.76	2.07	1.55	1.21	1.55	2.07
2½	2.469	2.88	6.16	13.7	12.3	3.29	2.47	1.85	1.44	1.85	2.47
3	3.068	3.58	7.67	17.1	15.3	4.09	3.07	2.30	1.79	2.30	3.07
4	4.026	4.70	10.1	22.4	20.2	5.37	4.03	3.02	2.35	3.02	4.03
5	5.047	5.88	12.6	28.0	25.2	6.72	5.05	3.78	2.94	3.78	5.05
6	6.065	7.07	15.2	33.8	30.4	8.09	6.07	4.55	3.54	4.55	6.07
8	7.981	9.31	20.0	44.6	40.0	10.6	7.98	5.98	4.65	5.98	7.98
10	10.02	11.7	25.0	55.7	50.0	13.3	10.0	7.51	5.85	7.51	10.0
12	11.94	13.9	29.8	66.3	59.6	15.9	11.9	8.95	6.96	8.95	11.9
14	13.13	15.3	32.8	73.0	65.6	17.5	13.1	9.85	7.65	9.85	13.1
16	15.00	17.5	37.5	83.5	75.0	20.0	15.0	11.2	8.75	11.2	15.0
18	16.88	19.7	42.1	93.8	84.2	22.5	16.9	12.7	9.85	12.7	16.9
20	18.81	22.0	47.0	105	94.0	25.1	18.8	14.1	11.0	14.1	18.8
24	22.63	26.4	56.6	126	113	30.2	22.6	17.0	13.2	17.0	22.6

Miter elbows[b] (no of miters)					Welding tees		Valves (screwed, flanged, or welded)			
1–45 in	1–60 in	1–90 in	2–90 in	3–90 in	Forged	Miter[c]	Gate	Globe	Angle	Swing check
0.45	0.90	1.80	0.60	0.45	1.35	1.80	0.21	10	5.0	2.5
15	30	60	20	15	45	60	7	333	167	83
			[e]	[e]						
L = equivalent length in feet of Schedule 40 (standard weight) straight pipe[f]										
0.78	1.55	3.10	1.04	0.78	2.33	3.10	0.36	17.3	8.65	4.32
1.03	2.06	4.12	1.37	1.03	3.09	4.12	0.48	22.9	11.4	5.72
1.31	2.62	5.24	1.75	1.31	3.93	5.24	0.61	29.1	14.6	7.27
1.72	3.45	6.90	2.30	1.72	5.17	6.90	0.81	38.3	19.1	9.58
2.01	4.02	8.04	2.68	2.01	6.04	8.04	0.94	44.7	22.4	11.2
2.58	5.17	10.3	3.45	2.58	7.75	10.3	1.21	57.4	28.7	14.4
3.08	6.16	12.3	4.11	3.08	9.25	12.3	1.44	68.5	34.3	17.1
3.84	7.67	15.3	5.11	3.84	11.5	15.3	1.79	85.2	42.6	21.3
5.04	10.1	20.2	6.71	5.04	15.1	20.2	2.35	112	56.0	28.0
6.30	12.6	25.2	8.40	6.30	18.9	25.2	2.94	140	70.0	35.0
7.58	15.2	30.4	10.1	7.58	22.8	30.4	3.54	168	84.1	42.1
9.97	20.0	40.0	13.3	9.97	29.9	40.0	4.65	222	111	55.5
12.5	25.0	50.0	16.7	12.5	37.6	50.0	5.85	278	139	69.5
14.9	29.8	59.6	19.9	14.9	44.8	59.6	6.96	332	166	83.0
16.4	32.8	65.6	21.9	16.4	49.2	65.6	7.65	364	182	91.0
18.8	37.5	75.0	25.0	18.8	56.2	75.0	8.75	417	208	104
21.1	42.1	84.2	28.1	21.1	63.2	84.2	9.85	469	234	117
23.5	47.0	94.0	31.4	23.5	70.6	94.0	11.0	522	261	131
28.3	56.6	113	37.8	28.3	85.0	113	13.2	629	314	157[f]

Note: For SI units, 1 ft = 0.305 m.

Values for welded fittings are for conditions where bore is not obstructed by weld spatter or backing rings. If appreciably obstructed, use values for "Screwed Fittings."

[a] Flanged fittings have three-fourths the resistance of screwed elbows and tees.

[b] Tabular figures give the extra resistance due to curvature alone to which should be added the full length of travel.

[c] Small-size socket-welding fittings are equivalent to miter elbows and miter tees.

[d] Equivalent resistance in number of diameters of straight pipe computed for a value of $f – 0.0075$ from the relation $n – k/4f$.

[e] For condition of minimum resistance where the centerline length of each miter is between d and 2½.

[f] For pipe having other inside diameters, the equivalent resistance may be computed from the above n values.

TABLE 8.7 Gas Pipe Sizing, Initial Gas Pressure 10 psig, 1 psig Pressure Loss Capacity of Horizontal Gas Piping, cfh

Length, ft	Nominal: Actual:	0.50 0.622	0.75 0.824	1 1.049	1.25 1.380	1.5 1.610	2 2.067	2.5 2.469	3 3.068	3.5 3.548	4 4.026	5 5.047	6 6.065
10		1,023	2,164	4,120	8,561	12,914	25,256	40,388	72,079	106,206	148,773	271,820	443,686
20		723	1,531	2,913	6,047	9,131	17,780	28,559	50,967	75,164	105,198	191,580	313,733
25		647	1,369	2,606	5,415	8,168	15,902	25,543	45,586	67,171	94,093	171,915	280,605
30		590	1,293	2,379	4,943	7,456	14,517	23,318	41,615	61,318	85,894	156,936	256,162
35		546	1,157	2,202	4,576	6,903	13,440	21,588	38,528	56,770	79,523	145,294	237,411
40		511	1,082	2,060	4,281	6,457	12,572	20,194	36,040	53,103	74,386	135,910	221,843
45		482	1,020	1,942	4,036	6,088	11,853	19,039	33,978	50,066	70,132	128,137	209,156
50		457	968	1,843	3,829	5,776	11,245	18,062	32,260	47,497	66,533	121,562	198,422
60		417	883	1,682	3,475	5,272	10,265	16,488	29,426	43,358	60,736	110,970	181,133
70		387	818	1,558	3,236	4,881	9,503	15,265	27,243	40,142	56,231	102,739	167,698
80		362	765	1,457	3,027	4,566	8,889	14,279	25,483	37,550	52,599	96,103	156,868
90		341	721	1,373	2,854	4,304	8,381	13,462	24,026	35,401	49,590	90,607	147,895
100		323	684	1,303	2,707	4,084	7,952	12,772	22,793	33,585	47,542	85,957	140,306
125		289	612	1,165	2,422	3,653	7,112	11,423	20,387	30,040	42,080	76,882	125,493
150		264	559	1,063	2,210	3,334	6,492	10,428	18,611	27,422	38,413	70,184	114,559
200		228	484	921	1,914	2,888	5,622	9,031	16,117	23,746	33,266	60,780	99,211
300		187	395	752	1,563	2,358	4,591	7,374	13,160	19,390	27,162	49,628	81,006
400		162	342	652	1,353	2,042	3,975	6,348	11,397	16,793	23,523	42,979	70,153
500		145	306	583	1,210	1,826	3,556	5,712	10,193	15,020	21,039	38,441	62,747
1,000		102	216	412	856	1,291	2,514	4,039	7,208	10,620	14,877	27,182	44,368
1,500		82	177	336	699	1,054	2,053	3,298	5,885	8,671	12,148	22,194	36,227
2,000		72	153	291	605	913	1,778	2,856	5,160	7,510	10,519	19,221	31,373

TABLE 8.8 Gas Pipe Sizing, Initial Gas Fressure 20 psig, 2 psig Pressure Loss Capacity of Horizontal Gas Piping, cfh

		Pipe diameter, in										
	Nominal:	0.50	0.75	1	1.25	1.5	2	2.5	3	3.5	4	5	6
Length, ft	Actual:	0.622	0.824	1.049	1.380	1.610	2.067	2.469	3.068	3.548	4.026	5.047	6.065
10		2,040	4,329	8,240	17,121	25,820	50,285	80,771	144,079	212,400	247,753	543,611	887,323
20		1,446	3,060	5,826	12,332	18,262	35,556	57,113	101,929	150,190	210,385	384,391	629,684
25		1,293	2,738	5,211	10,828	16,334	31,803	51,084	91,168	134,334	188,174	343,810	561,192
30		1,180	2,499	4,757	9,885	14,911	29,032	46,633	83,225	122,630	171,779	313,854	512,296
35		1,093	2,314	4,404	9,150	13,791	26,878	43,174	77,051	113,532	159,035	290,572	474,294
40		1,022	2,165	4,120	8,561	12,913	25,143	40,385	72,674	106,200	148,764	271,806	443,662
45		964	2,040	3,885	8,071	12,175	23,705	38,075	67,953	100,127	140,257	256,260	418,288
50		915	1,936	3,685	7,657	11,550	22,488	36,122	64,466	94,988	133,059	243,110	396,823
60		835	1,767	3,340	6,990	10,544	20,529	32,974	58,848	86,712	121,466	221,928	362,248
70		773	1,636	3,114	6,471	9,761	19,006	30,528	54,483	80,276	112,455	205,466	335,377
80		723	1,530	2,913	6,053	9,131	17,779	28,556	50,964	75,095	105,193	192,196	313,716
90		682	1,443	2,746	5,707	8,609	16,761	26,924	48,050	70,794	99,176	182,750	294,774
100		646	1,369	2,606	5,414	8,167	15,902	25,670	45,584	67,167	94,087	171,905	280,596
125		578	1,224	2,331	4,843	7,305	14,223	22,845	40,770	60,076	84,153	153,756	250,973
150		528	1,118	2,128	4,421	6,668	12,110	20,855	37,219	54,842	76,822	140,360	229,106
200		457	968	1,843	3,828	5,775	11,244	18,061	32,232	47,494	66,529	121,555	198,411
300		374	790	1,504	3,126	4,715	9,181	14,747	26,318	38,779	54,322	99,299	162,002
400		324	685	1,303	2,707	4,083	7,950	12,771	22,792	33,583	47,043	85,952	140,299
500		289	612	1,141	2,421	3,653	7,111	11,423	20,386	30,038	42,077	76,878	125,487
1,000		204	433	824	1,712	2,583	5,028	8,076	14,415	21,240	29,753	54,361	88,732
1,500		166	354	673	1,398	2,109	4,106	6,595	11,770	17,342	24,293	44,385	72,449
2,000		145	306	583	1,210	1,826	3,557	5,712	10,193	15,018	21,039	38,439	62,743

TABLE 8.9 Sizing, NG Pipe with Initial Pressure of 50 and 5 psig Pressure Drop

Length, ft		Capacity of horizontal gas piping, cfh											
		Pipe diameter, in											
	Nominal:	0.5	0.75	1	1.25	1.5	2	2.5	3	3.5	4	5	6
	Actual:	0.622	0.824	1.049	1.380	1.610	2.067	2.469	3.068	3.548	4.026	5.047	6.065
10		5,850	12,384	23,575	48,984	73,889	143,864	231,083	412,407	607,670	851,220	1,555,221	2,538,598
20		4,137	8,757	16,670	34,637	52,248	101,727	163,400	291,616	429,688	601,903	1,099,729	1,795,060
25		3,700	7,832	14,910	30,980	46,732	90,988	146,150	260,829	384,324	538,359	983,627	1,465,660
30		3,377	7,150	13,611	28,281	42,660	83,060	133,416	238,103	350,839	491,452	897,925	1,356,938
35		3,127	6,619	12,601	26,183	39,495	76,899	123,519	220,441	324,813	454,996	831,317	1,356,938
40		2,925	6,192	11,787	24,492	36,945	71,932	115,541	206,203	303,835	425,610	777,626	1,269,299
45		2,758	5,838	11,113	23,091	34,832	67,818	108,934	194,411	286,459	401,269	733,152	1,196,706
50		2,616	5,538	10,543	21,906	33,044	64,338	103,343	184,434	271,758	380,677	695,529	1,135,295
60		2,388	5,056	9,624	19,998	30,165	58,732	94,339	168,364	248,080	347,509	634,929	1,036,378
70		2,211	4,681	8,910	18,514	27,927	54,376	87,341	155,875	229,678	321,731	587,830	959,500
80		2,068	4,378	8,335	17,319	26,124	50,864	81,700	145,808	214,844	300,952	549,864	897,530
90		1,950	4,128	7,858	16,328	24,630	47,955	77,028	137,469	202,557	283,740	518,417	846,199
100		1,850	3,916	7,456	15,490	23,366	45,494	73,075	130,415	192,162	269,179	491,814	802,775
125		1,655	3,503	6,668	13,855	20,899	40,691	65,360	116,646	171,875	240,761	439,891	718,024
150		1,510	3,197	6,087	12,648	19,078	37,146	59,665	106,483	156,900	219,784	401,564	655,463
200		1,308	2,769	5,271	10,953	16,522	32,169	51,672	92,217	135,879	190,339	347,765	567,648
300		1,068	2,261	4,304	8,943	13,490	26,266	42,190	75,295	110,945	155,411	283,949	463,482
400		925	1,958	3,727	7,745	11,683	22,747	36,537	65,207	96,081	134,590	245,907	401,388
500		827	1,751	3,334	6,927	10,450	20,345	32,680	58,323	85,938	120,381	219,946	359,012
1,000		585	1,238	2,357	41,898	7,389	14,386	23,108	41,241	60,767	85,122	155,525	253,860
1,500		478	11,011	1,925	4,000	6,033	11,746	18,868	33,673	49,616	69,502	126,986	207,276
2,000		414	876	1,667	3,464	5,225	10,173	16,340	29,162	42,969	60,190	109,973	179,506

TABLE 8.10 Multipliers to Be Used for Specific Gravities Different from 0.60

Specific gravity	Multiplier	Specific gravity	Multiplier
0.35	1.31	1.00	0.78
0.40	1.23	1.10	0.74
0.45	1.16	1.20	0.71
0.50	1.10	1.30	0.68
0.55	1.04	1.40	0.66
0.60	1.00	1.50	0.63
0.65	0.96	1.60	0.61
0.70	0.93	1.70	0.59
0.75	0.90	1.80	0.58
0.80	0.87	1.90	0.56
0.85	0.84	2.00	0.55
0.90	0.82	2.10	0.54

tables can be used for sizing, reducing the flow rate based on a factor found in Table 8.10 for IP units using the specific gravity of air, which is 1.00. For lower pressure of 0.5 psig, refer to Table 8.16

Calculate the equivalent run of pipe from the tie-in point to the inlet of the meter or regulator. Using the chart with the correct system pressure, find the lowest cubic feet per hour (cfh) [liters per hour (L/h)] figure that equals or exceeds the calculated cfh (L/h) in conjunction with the equivalent run. Read the pipe size at the top of the chart. If the design engineer is responsible for sizing the meter and regulator, the following information must be calculated or established.

1. The adjusted maximum flow rate, in cfm (L/min), for all connected equipment in the facility. Typical Btu values for equipment are given in Table 8.4. The cfh demand for multiple dwellings is given in Fig. 8.2 (for 50 apartments and less) and in Fig. 8.3 for more than 50 apartments. All equipment in the facility must be identified.
2. The highest pressure required for the most demanding equipment inside the building.
3. The pressure selected for use in the building service main upstream of the regulator including the allowable pressure loss for the piping system inside the facility.

The regulator and meter can now be sized using literature and catalog information from the manufacturers.

Pipe and System Materials for Site Installation

All the piping materials must be listed by NFPA 54 and applicable local codes. In addition to the codes, the recommendation of the utility company, which has experience in the area of the project, should be considered. The most often used material for underground lines is high-density polyethylene (HDPE). Larger lines and piping for high pressure are steel pipe protected against corrosion by wrapping with a plastic coating.

Joints for HDPE should be butt-type, heat-fused joints. Socket-type joints have been found to introduce a stiffness in the joint area that is undesirable. Joints for steel pipe shall be screwed for pipe sizes 4 in (DN 100) and smaller. Sizes 5 in (DN 125) and larger should be welded. Where flanged connections are necessary to connect some devices, flanges are heat-welded to the HDPE pipe end.

LIQUEFIED PETROLEUM GAS

General

Liquefied Petroleum Gas (LPG) is supplied to facilities from a tank truck (or railroad tank car if the storage tank is large enough) and is stored on-site as a liquid. Where usage is large, permanent large tanks are installed on the facility property and refilled directly from a tanker truck. If the usage requirements are small, small propane tanks containing liquid may be installed and the tanks replaced after they are emptied.

The liquid must be vaporized to produce a gas. Depending on the actual installation and flow rate, the liquid can be vaporized in the storage tank (using heat gained through the storage tank wall) or in an auxiliary vaporizer that uses an outside source of heat. The gas may be used directly as vaporized, or if it is a substitute supply for natural gas, it may have to be mixed with air to provide a lower British thermal unit content suitable for the specific application and equipment.

Storage Tanks

Tanks used for the storage of propane are made of steel. They can be installed either above or below grade. Although underground tanks have an advantage of a more constant environment, most tanks are placed aboveground because of the lower initial cost and Environmental Protection Agency (EPA) requirements. Because of the relatively high pressure developed by the propane vapor, all tanks must be designed in conformance with the American Society of Mechanical Engineers (ASME) code for unfired pressure vessels. The vapor pressure developed in aboveground tanks is based on the ambient outside air and can be found from Fig. 8.4. For underground tanks, a figure of 50°F (10°C) is used. Typical capacities of large, standard tanks are shown in Fig. 8.5.

Although it is common practice to classify containers as either portable or stationary, there are too many exceptions to make these classifications practical. Containers and storage tanks are referred to as either *Department of Transportation* (DOT) *cylinders* (generally portable) or *ASME tanks* (generally stationary). In the following discussions, all capacities referring to gallons are IP gallons of water, and all weights are of liquid propane, unless indicated otherwise.

DOT cylinders range in capacity from 1 to 420 lb (0.5 to 190 kg). Cylinders built in 1966 or earlier may bear the Interstate Commerce Commission (ICC) designation. ASME tanks range in capacity from 500 to 60,000 gal (1900 to 227,000 L).

The advantages of an underground tank are as follows:

1. The tank is not visible if aesthetics are a consideration.
2. There is greater vaporization of liquid in the winter due to the constant temperature, which is about 50°F (10°C).

The disadvantages are as follows:

1. The tank may have to be anchored to prevent floating in areas of high water tables.
2. The initial cost is higher.
3. Inspection and maintenance are more difficult to perform.
4. If the underground tank is subject to traffic or potential damage, the top should be a minimum of 2′-0″ ft (0.6 m) below grade. In remote locations, 6 in (125 mm) of cover is considered adequate.

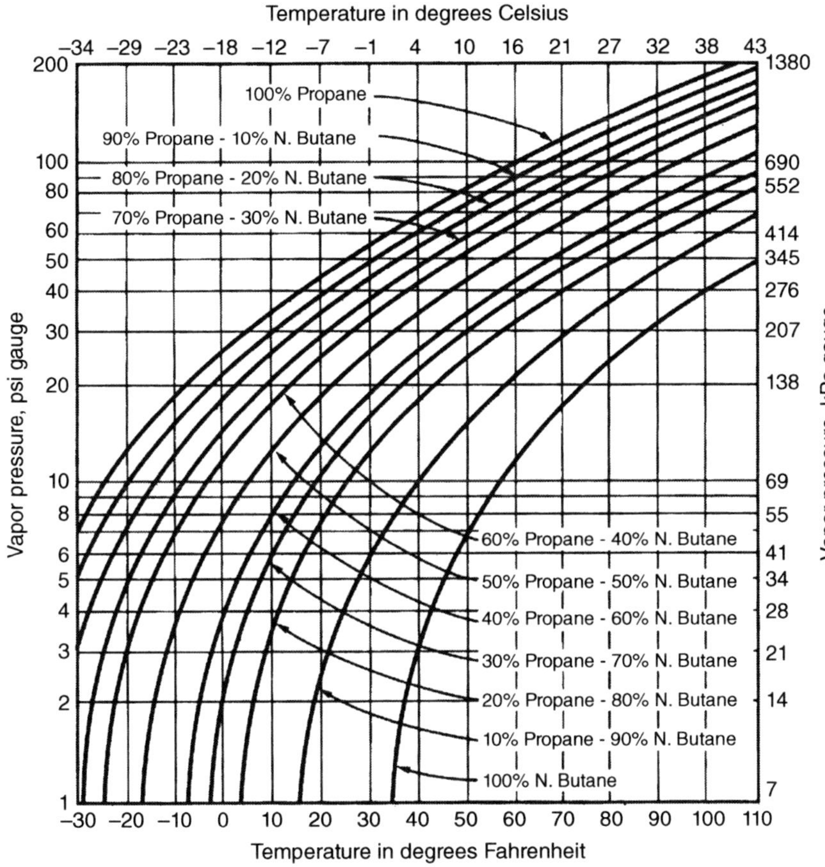

FIGURE 8.4 Vapor pressures of butane-propane mixture. (*Courtesy Fisher Regulator.*)

5. A manhole giving access to the valves, gauges, connections, and so on must be provided for maintenance and inspection.

New tanks must be purged of both air and water prior to being placed in service. A concentration of 6 percent air is the maximum limit acceptable. This is normally done by purging the tank multiple times. Water should be removed from the tank by the manufacturer, and the tank should be shipped to the site sealed.

Maximum Content of Liquid in Tanks. The maximum allowable quantity of LPG permitted in a container is limited by NFPA 58 to a filling density based on tables and formulas provided in that code. In future discussions of tank sizing, this maximum allowable quantity must be compared to that calculated by other methods to make sure that the maximum is not exceeded.

Tank Foundations and Support. Tank foundations must provide a stable means of support. Uneven settlement will lead to errors in fuel gauging and stresses, leading to broken

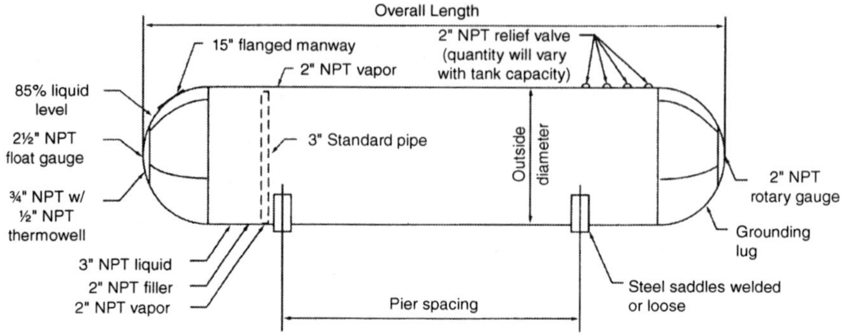

Dimensions and Capacities of Larger Tanks				
Capacity, gal	Dia, in	Length	Pier spacing	Estimated weight, lb
6,565	84	25'-10"	8'-0"	11,260
9,200	84	35'-4"	17'-6"	15,920
12,000	84	44'-10"	27'-0"	20,035
14,500	84	54'-4"	36'-6"	25,230
18,000	109	40'-11"	21'-0"	30,502
20,000	109	46'-11"	27'-0"	35,300
25,000	109	56'-5"	36'-6"	43,100
30,000	109	65'-11"	46'-0"	50,900
30,000	121	53'-3"	32'-4"	50,000
30,000	121	46'-9"	24'-11"	50,100

FIGURE 8.5 Dimensions and capacities of larger propane tanks. (*Courtesy Fisher Regulator.*)

lines. The weight of the tank should be calculated using water since there may be a requirement for hydrostatic testing during the life of the system. An approximation of the gross weight of a steel tank can be obtained by multiplying the capacity in gallons by 11 for water and by 6 for propane. Some larger tanks may require a temporary, intermediate center support during the time they are filled with water for hydrostatic testing.

Location of Equipment on the Site

These factors are to be considered in locating the equipment:

1. Code-required clearance from other buildings, property lines, and any other equipment. These clearances also depend on insurance carrier requirements and client preferences. Factory Mutual requirements, indicated in Fig. 13.8, are considered conservative. Distances specified in NFPA 54 and other insurance carrier codes vary but are not long.
2. Accessibility for fuel delivery.
3. Location of underground utilities.
4. Site elevations. Avoid placing the equipment at low points.
5. Client preferences.

Regulators. The regulator for LPG is different from that used for natural gas service. The purpose of an LPG regulator is to reduce a variable inlet pressure to a constant outlet pressure under variable flow conditions. The following information is required to size a regulator:

1. Minimum significant and maximum flow rate possible
2. Maximum and minimum container pressure
3. Required outlet pressure desired
4. Manufacturer's rating curves for the regulator

The pressure developed inside the storage tank is produced by the vaporization of product inside the storage tank. This pressure is called the *vapor pressure*. This figure varies, depending on the ambient temperature. Figure 8.4 is a direct reading chart to determine the vapor pressure for various mixtures of propane and butane that might be provided by suppliers. Outside temperatures can be found by requesting this information from the National Oceanographic and Atmospheric Administration (NOAA).

Regulators are manufactured in single- and double-stage models and are selected using the capacity and rating curves supplied by the manufacturer.

Experience has shown that when regulators are kept in service longer than 15 years, the probability of malfunction increases. This figure is only a generalization, with actual climate and service conditions playing an important factor in the service life of a regulator.

Pressure Relief Devices. The purpose of a pressure relief device is to automatically vent propane vapor to the atmosphere upon reaching a predetermined high pressure. It can be an integral part of a pressure regulator, separately installed on the storage tank, part of vaporizers and mixers, or in the piping system itself.

The relief valve flow capacity is calculated on the basis of a full tank and the resultant vaporization of that volume of propane. For an underground tank, the capacity will be 30 percent of the aboveground value. Refer to Table 8.11 to determine the required flow of propane based on the square foot area of an aboveground storage tank. All larger tanks must have a separate relief valve installed directly on the tank. If a single relief valve does not have the required flow capacity, it will be necessary to have multiple relief valves mounted on the top of the tank to provide the necessary flow rate. The final pressure setting of the relief valve is nominally 88 to 100 percent of the design pressure of the tank, with a permitted error of +10 percent. Refer to Table 8.12 for pressure settings.

Another pressure relief device within the piping system is the external bypass valve, which relieves excessive pressure in an LPG pumping system upon reaching a high-pressure set point and returns the discharge back to the storage tank instead of to the atmosphere, as is done by a relief valve. All LPG pumps require an external bypass valve. Different types of bypass valves are manufactured for different applications.

Excess Flow Valve. An excess flow valve permits the flow of vapor or liquid in both directions but shuts off the flow of liquid or vapor in only one direction when the flow exceeds a predetermined limit. It is recommended that an excess flow device be placed on all connections to a larger tank, except for the safety relief valve. The filler connection should have an integral valve. The capacity is calculated from the largest expected flow and its mounting position (horizontal or vertical). Valves are selected to open at between 150 and 200 percent of the expected maximum flow.

Service Line Valves. Service line valves control the flow of propane from a DOT cylinder. This is a multipurpose valve that could contain a shutoff valve, relief valve, gauge, and filler valve all in one body.

Filler Valves. A filler valve permits the flow of liquid in one direction only and is used for filling larger tanks with liquid. This valve contains an integral back-pressure check

TABLE 8.11 Relieving Capacity for Relief Valves on Tanks and Vaporizers

(Minimum required rate of discharge in cubic feet per minute of gas at 120 percent of the maximum permitted start-to-discharge pressure for relief valves to be used on containers other than those constructed in accordance with Department of Transportation specifications)

Surface area		Flow rate		Surface area		Flow rate		Surface area		Flow rate	
ft^2	m^2	cfm	m^2/min	ft^2	m^2	cfm	m^2/min	ft^2	m^2	cfm	m^2/min
20	17	626	1,064	170	141	3,620	6,154	600	498	10,170	17,289
25	21	751	1,277	175	145	3,700	6,290	650	540	10,860	18,462
30	25	872	1,282	180	149	3,790	6,443	700	581	11,550	19,635
35	29	990	1,623	185	152	3,880	6,596	750	613	12,220	20,774
40	33	1,100	1,870	190	158	3,960	6,641	800	664	12,880	21,827
45	37	1,220	2,014	195	162	4,050	6,885	850	705	13,540	22,010
50	42	1,330	2,271	200	166	4,130	7,021	900	747	14,190	24,123
55	46	1,430	2,480	210	171	4,300	7,310	950	788	14,830	26,147
60	50	1,540	2,672	220	175	4,470	7,600	1,000	830	15,470	26,183
65	54	1,640	2,885	230	179	4,630	7,890	1,050	877	16,100	26,211
70	58	1,750	2,975	240	184	4,800	8,015	1,100	913	16,720	28,424
75	62	1,850	3,145	250	188	4,960	8,432	1,150	954	17,350	28,632
80	67	1,950	3,310	260	192	5,130	8,721	1,200	996	17,960	28,683
85	71	2,050	3,516	270	196	5,290	8,993	1,250	1,038	18,570	31,027
90	75	2,150	3,721	280	232	5,450	9,265	1,300	1,079	19,180	32,606
95	80	2,240	3,925	290	240	5,610	9,537	1,350	1,120	19,780	33,122
100	86	2,340	3,910	300	249	5,760	9,810	1,400	1,158	20,380	33,711
105	90	2,440	4,101	310	253	5,920	10,034	1,450	1,191	20,980	35,666
110	94	2,530	3,995	320	268	6,080	10,336	1,500	1,245	21,570	36,021
115	98	2,630	4,470	330	273	6,230	10,591	1,550	1,281	22,160	36,414
120	102	2,720	4,624	340	282	6,390	10,780	1,600	1,319	22,740	38,658
125	106	2,810	4,777	350	290	6,540	10,921	1,650	1,352	23,320	39,005
130	112	2,900	4,930	360	299	6,690	11,370	1,700	1,382	23,900	39,428
135	116	2,990	5,110	370	307	6,840	11,630	1,750	1,452	24,470	39,862
140	120	3,080	5,191	380	315	7,000	11,900	1,800	1,494	25,050	42,585
145	122	3,170	5,465	390	323	7,150	12,201	1,850	1,530	25,620	43,554
150	126	3,260	5,542	400	331	7,300	12,410	1,900	1,571	26,180	44,506
155	130	3,350	5,731	450	340	8,040	12,701	1,950	1,610	26,750	45,475
160	134	3,440	5,920	500	348	8,760	12,911	2,000	1,660	27,310	46,427
165	138	3,530	6,004	550	352	9,470	16,000				

Note: Buried tanks require only 30 percent of the vent capacity shown.
Source: Factory Mutual.

TABLE 8.12 Final-Stage Relief Valve Settings

Regular delivery pressure	Relief valve start-to-discharge pressure setting (percent of regulator delivery pressure)	
	Maximum, percent	Maximum, percent
1 psig or less	200	300
Above 1 psig but not over 3 psig	140	200
Above 3 psig	125	200

valve that will prevent the loss of vapor or liquid from the tank if the fill hose or a fitting ruptures. A preferred location is at the filling connection used by the delivery truck hose.

Vapor-Equalizing Valves. When large tanks are filled, the liquid propane added to the storage tank compresses the vapor in the tank. A vapor equalizing line is connected between the vapor area of the receiving tank and the vapor area of the delivery truck during propane deliveries to equalize pressure in both containers and to lessen the head requirements of the delivery truck supply pump. The vapor equalizing valve is mounted in the equalizing line tank truck hose connection and permits the flow of vapor both ways.

Liquid-Level Gauges. When a larger tank is filled on the basis of volume, it is important that a liquid level gauge be installed on the tank to indicate the exact liquid level inside. There are several types available.

The slip tube gauge is mounted on top of a tank and consists of a vertical rod extending from the liquid level inside to a point outside the tank for measurement. There is a float on one end of the tube that rides on the liquid, automatically rising and falling with the level.

A magnetic float gauge consists of a gauge faceplate mounted on the side of a tank and a connecting rod with a float at the end that rides on the liquid level. As the float rises and falls with the liquid, it automatically moves a magnetic device on the faceplate indicating the level inside. Because there is no direct opening to the tank outside, this is the most widely used gauge. Since it is a mechanical device, it is recommended that this type be used with the slip tube as a backup.

A *rotary liquid-level gauge* is a manually operated gauge that uses an angled dip tube inside the tank connected to a bleeder valve on a calibrated faceplate on the outside of the tank. The bleeder valve is opened, and the dip tube is manually turned until the vapor escaping changes to a liquid. The level is then read from calibrated markings on the faceplate.

The fixed-level gauge is an external needle valve connected to a short length of tubing extending inside the tank. There may be a number of these gauges installed at various calibrated levels on the tank. As the filling proceeds, the valve is cracked open until the vapor changes to liquid. When the desired level is reached, the filling operation stops.

Miscellaneous Equipment. In addition to the devices indicated, it is often desirable to install a liquid temperature gauge and a pressure gauge on a tank to aid in diagnosing any potential problem.

Vaporizer. A *vaporizer* is a device that converts liquid propane into a vapor in larger quantities than would be possible from ambient heat alone. It is required when sufficient quantities of liquid propane inside the storage tank cannot be vaporized quickly enough to satisfy the maximum demand such as in locations with cold weather.

There are two basic types of vaporizers: direct fired and indirect fired. The direct-fired vaporizer uses the propane itself as fuel for the direct flame. It has the lowest initial cost and no ancillary power requirements. Disadvantages include the problem of pilot light extinguishment in high winds, production of sludge in the propane being vaporized, and a useful life expectancy of only 5 years. Optional accessories to be considered would be an automatic ignition, manual drain, liquid strainer, and propane/air mixer. An indirect type of vaporizer is available that uses steam, glycol-water bath, or the less-used electric resistance heaters as a heat source.

Propane Mixers. *A mixer,* also called a *blender* or *proportioner,* is a device used to combine pure propane and air together into a mixture of both gases. It is required when propane will be used as a direct substitute for natural gas because the much higher British thermal unit content of 100 percent propane will not allow the use of the same burner orifices and gas train as natural gas. Experience has shown that a mixed gas content of 1450 Btu/ft^3 (44 Btu/m^3) with a specific gravity of 1.30 provides the best burning characteristics. Refer to Table 8.13 for properties of various mixtures of propane and air.

Several methods can be used to accomplish mixing. The simplest is the use of a venturi. Intended for exterior installation, it consists of a nozzle inside of a larger body. Propane passing through the nozzle creates a partial vacuum that pulls air from around the opening, thereby creating a mixed gas. Because the venturi principle is accurate only within a very limited flow range, this type of unit is either on or off. If the demand pressure is higher than 5 psi (35 kPa), the use of an air compressor will be required to raise the mixed gas pressure. Generally, this system has the least initial cost. Four conditions have a strong bearing on the accuracy and composition of the mixture:

1. Venturi tube dimensions and throat diameter
2. Nozzle or orifice size and location of the venturi in the throat
3. Pressure of both gases at the venturi
4. Desired discharge pressure

Another more costly device is the modulating proportional blender. Suitable only for interior installation, these units are generally used for larger quantities and higher pressures than available from a venturi unit. They use regulators to stabilize both the air and propane in separate supply lines. A proportioning valve is then used to meter both supplies into a common discharge line. An air compressor is often used to supply air at the higher pressure required. The vaporizer and mixer are also available as a factory-assembled package on a single skid, ready for installation.

LPG System Design. Prior to sizing components, initially decide what equipment (such as pumps or vaporizers) may be required and if the tank is to be aboveground or belowground, and select the proposed location of the tank and other required equipment on the site. Determine if the propane system will be in constant use or used only periodically, such as for emergency operation or as an occasional substitute for interrupted natural gas. Calculate the maximum hourly and daily fuel gas demand, and determine if this demand is continuous, such as that for a process used all day, or intermittent, such as that used for heating purposes. If only a small part of the demand is continuous, the entire load should be considered continuous.

Storage Tank Sizing and Selection

The storage tank volume is based on one of two factors: a reasonable return schedule for the local supplier or the total amount of liquid propane that has to be vaporized by the ambient air in order to satisfy the maximum demand.

TABLE 8.13 Properties of Various Propane/Air Mixtures

	Propane/air mixtures			
British thermal unit per cubic foot of mixture	Percentage of propane by volume	Percentage of air by volume	Percentage of oxygen by volume	Specific gravity of the mixture
2,550	100.00	0.00	0.000	1.523
2,500	98.04	1.96	0.409	1.513
2,450	96.08	3.92	0.819	1.502
2,400	94.12	5.88	1.288	1.492
2,350	92.16	7.84	1.639	1.482
2,300	90.19	9.81	2.050	1.472
2,250	88.24	11.76	2.458	1.461
2,200	86.27	13.73	2.869	1.451
2,150	84.31	15.69	3.279	1.441
2,100	82.35	17.65	3.688	1.431
2,050	80.39	19.61	4.098	1.420
2,000	78.43	21.56	4.506	1.410
1,950	76.47	23.53	4.918	1.400
1,900	74.51	25.49	5.317	1.390
1,850	72.55	27.45	5.737	1.379
1,800	70.58	29.42	6.149	1.369
1,750	68.62	31.38	6.558	1.359
1,700	66.67	33.33	6.964	1.349
1,650	64.70	35.30	7.378	1.338
1,600	62.74	37.26	7.787	1.328
1,550	60.78	39.22	8.197	1.318
1,500	58.82	41.18	8.606	1.308
1,450	56.86	43.14	9.016	1.297
1,400	54.90	45.10	9.246	1.287
1,350	52.94	47.06	9.835	1.277
1,300	50.98	49.02	10.245	1.267
1,250	49.02	50.98	10.654	1.256
1,200	47.06	52.94	11.064	1.246
1,150	45.09	54.91	11.476	1.236
1,100	43.13	56.87	11.886	1.226
1,050	41.17	58.83	12.295	1.215
1,000	39.21	60.79	12.705	1.205
950	37.25	62.75	13.115	1.195
900	35.29	64.71	13.524	1.185
850	33.33	66.67	13.934	1.174
800	31.37	68.63	14.344	1.164
750	29.41	70.59	14.753	1.154
700	27.45	72.55	15.163	1.144
650	25.49	74.51	15.573	1.133
600	23.53	76.47	15.982	1.123
550	21.56	78.44	16.394	1.113
500	19.61	80.39	16.892	1.103
400	17.65	82.35	17.221	1.092
200	7.84	92.16	19.261	1.041

Vaporization directly from a tank, if economically feasible, is preferred and used to avoid having to add another mechanical device (a vaporizer) that requires constant maintenance. If a single tank sized to optimum criteria will not provide the required vaporization rate, a single oversized tank or two smaller, separate tanks are often installed if space conditions and initial cost are acceptable.

Tank Size Based on Return Schedule. Using the return schedule, a preliminary starting point for determining actual capacity is a 10-day usable supply for continuous demand and between 3 and 5 days of usable supply for intermittent or standby purposes. With the maximum propane level based on the maximum quantity for the allowed filling density and the minimum level between 10 and 15 percent of the capacity, the usable tank capacity is about 75 percent of the total tank capacity.

Tank Size Based on Vaporization Rate of Propane. When the vaporization rate is used as the critical factor in selecting the size of the storage tank, the tank must be large enough to vaporize the maximum flow rate of propane required by the facility when the outside air temperature and liquid level are at their lowest. For aboveground tanks, the rate of vaporization is calculated using the wetted area of liquid propane in the tank. For underground tanks, the entire area of the tank is used even if the tank is partially full.

Two methods will be discussed. The first, and most accurate, will use various formulas to calculate the wetted area of a tank that will vaporize the required amount of propane. The second, although slightly less accurate, is the direct reading of the vaporization rate from prepared charts. This is a more conservative method and by far the easier.

Formula Method. Using this method, the wetted area of the proposed aboveground tank necessary to vaporize the required amount of propane will be calculated. If an underground tank is proposed, the area calculated shall be that of the entire tank. The basic formulas are as follows:

Total heat of vaporization from an aboveground tank

$$Q = UA \times \mathrm{TD} \tag{8.1}$$

Total heat of vaporization from a belowground tank

$$Q = UA \times 15 \tag{8.2}$$

Equation (8.1) transposed to find the wetted area is

$$A = \frac{Q}{U(\mathrm{TD} \times \mathrm{TD} \text{ or } 15)} \tag{8.3}$$

Actual amount of propane vaporized

$$V = \frac{AQ}{L} \tag{8.4}$$

where Q = total amount of heat required to vaporize a quantity of liquid propane, Btu

U = rate of heat transfer through walls of steel tank, per square foot of wetted area/temperature difference °F (°C). Generally accepted figures are 2 Btu/(ft^2·°F) for aboveground tanks based on severe conditions and 0.5 Btu/(ft^2·°F) for underground tanks.

A = area of wetted surface of aboveground tank containing liquid propane, or entire surface area of an underground tank, ft^2 (m^2)

TD = temperature difference between ambient outside air reached during coldest part of day and temperature of propane in tank reached during warmest part of the day, °F (°C)

TABLE 8.14 Difference between Air Temperature of Frost Formation

Lowest air temp		Relative humidity, percent							
°C	°F	20	30	40	50	60	70	80	90
−34.4	−30	—	—	—	—	8.0*	5.0*	2.5*	1.0*
−28.9	−20	—	20.0*	15.0*	11.5*	8.5	5.0	3.0	1.5
−23.3	−10	27.5*	20.5	16.0	12.0	9.0	6.0	3.0	1.5
−17.8	0	29.0	21.5	16.5	12.5	9.0	6.0	4.0	2.0
−12.2	10	30.0	22.5	17.0	13.0	9.5	6.5	4.0	2.0
−6.7	20	31.5	24.0	18.0	14.0	10.0	7.0	4.0	2.0
−1.1	30	33.0	25.0	19.5	15.0	11.0	8.0	5.0	3.0
4.1	40	35.0	27.0	21.0	16.5	12.0	9.0	8.0	8.0

*If the full temperature difference is used in these cases, the minimum tank pressure may be too low for satisfactory performance.
1°F = 0.55°C.

One factor that must be taken into consideration is the formation of frost on the outside of the storage tank. Since frost acts as an insulation, its formation must be avoided. Table 8.14 gives the minimum temperature to avoid frost formation. When predicted conditions are outside the figures required to enter the table, use the true temperature difference.

15 = 15°F (−11°C), which is the generally accepted temperature difference for underground tanks

V = volume of propane vaporized under design conditions, gal

L = latent heat of vaporization of propane, Btu/gal. This is the amount of heat required to change liquid propane to a vapor (Table 8.15). Interpolate to find intermediate values when necessary.

TABLE 8.15 Latent Heat of Vaporization for Propane

Ambient air temp		Propane		
°C	°F	Btu/lb	Btu/gal	Btu/kg
−40	−40	180.8	765	85
−34.4	−30	178.7	755	80
−28.9	−20	176.2	745	75
−23.3	−10	173.9	735	78
−17.8	0	171.5	725	77
−12.2	10	169.0	715	76
−6.7	20	166.3	704	75
−1.1	30	163.4	691	74
4.4	40	160.3	678	73
10.0	50	156.5	662	70
15.6	60	152.6	645	69

The following methodology is used to select a tank with propane vaporization rate as criteria, by solving for the wetted area that would give the required rate of vaporization [Eq. (8.4)]:

1. Establish the lowest predicted ambient air temperature, highest predicted relative humidity, and the required propane gas demand in Btu/(h·gal).
2. Convert the British thermal unit per hour (Btu/gal) demand into gallons per hour. Refer to Table 8.1.

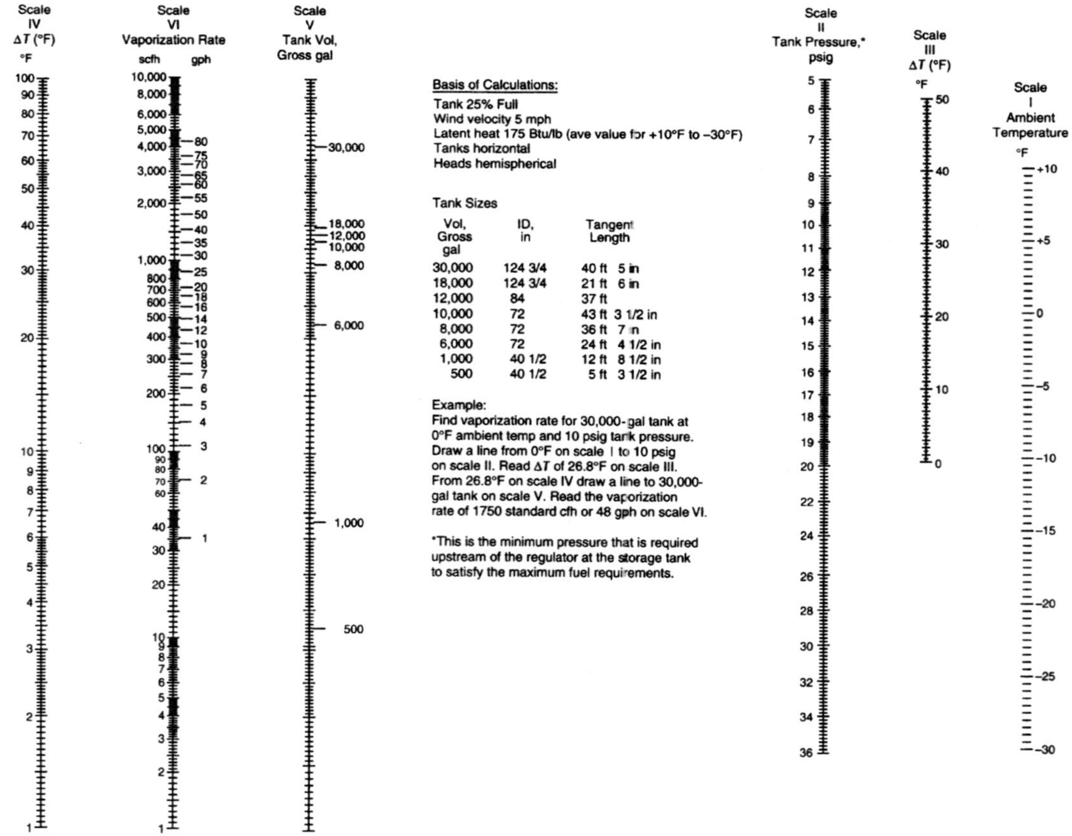

FIGURE 8.6 Propane vaporization of large propane tanks. (*Courtesy Phillips Petroleum Co.*)

3. Transpose Eq. (8.2) to find Q. Refer to Table 8.15 to find L.
4. Substitute all figures into Eq. (8.3), and solve for the wetted area. After calculating the required minimum area, consult manufacturers' catalogs for a size tank providing a figure with the minimum level of propane in the storage tank.

The result of the previous calculation is a tank area that will contain the absolute lowest volume of propane that will adequately supply facility needs. A higher volume should be used as an actual "low level." The additional capacity will be required to allow enough time for a delivery before the volume in the tank falls below the level needed to supply the facility. This low level will be a percent of the total capacity of the tank that will be used to select the actual, total capacity of the storage tank in conjunction with the proposed schedule of the supplier. Select a "standard" tank with a slightly greater volume than that calculated. If a single tank selected for adequate vaporization results in a size that is impractical, two tanks can be used. If the tanks are still too large or have too high an initial cost, the only other solution is to provide a vaporizer.

Direct-Reading Method. Figure 8.6 provides a chart for a direct reading of propane vaporization from larger tanks 25 percent full. Reduce the figures by 10 percent to approximate tanks 15 percent full.

LPG Gas Pipe Sizing Methods. A widely used pipe sizing method for low-pressure, 100 percent propane gas systems appears as tables used in NFPA 54. Another reliable table is given in Table 8.16. To use Table 8.16, start with the equivalent length of piping on the left side and the CFH being used for the facility. Find the CFH figure equal to or larger than the demand, and read the size pipe at the top.

TABLE 8.16 Low-Pressure Propane Gas Pipe Sizing Schedule, cfh
Maximum capacity of pipe for gas pressure of 0.5 Psig or less and a pressure drop of 0.3 in of water column, specific gravity of 1.52

Length of pipe		Diameter of pipe, in (mm)										
		in	¾	1	1¼	1½	2	3	4	5	6	8
m	ft	DN	20	25	32	40	50	75	100	125	150	200
4.5	15		170	315	617	970	2,016	5,418	11,214	19,530	30,870	63,000
9	30		120	246	428	680	1,386	3,843	7,875	13,860	21,735	44,730
13	45		98	202	353	561	1,134	3,150	6,489	11,340	17,640	36,540
18	60		85	176	308	485	1,008	2,709	5,607	9,765	15,435	31,500
23	75		76	158	271	428	882	2,394	5,040	8,820	13,800	28,350
28	90		69	145	252	391	819	2,205	4,599	7,875	12,600	25,830
31	105		64	132	233	365	756	2,079	4,284	7,434	11,655	23,940
36	120		60	126	217	343	693	1,953	3,969	6,930	10,962	22,680
38	150		55	113	195	309	630	1,701	3,528	6,174	9,765	20,160
54	180		50	101	176	280	567	1,575	3,213	5,607	8,821	18,270
63	210		46	95	164	258	529	1,449	3,024	5,229	8,190	17,010
72	240		43	88	154	243	493	1,355	2,772	4,851	7,749	15,750
81	270		41	82	145	230	466	1,292	2,646	4,599	7,245	14,805
90	300		38	79	139	217	447	1,229	2,520	4,410	6,930	14,175
135	450		32	63	113	176	359	1,008	2,016	3,528	5,607	11,340
180	600		27	55	95	151	308	851	1,764	3,024	4,788	9,765

Courtesy Fisher.

Proprietary charts are available that are less conservative and more accurate. For propane use with NG tables, a conversion factor of 0.63 to reduce the indicated flow rate must be used.

Propane is often mixed with air to provide a direct substitute for natural gas. Field experience has found that if mixed gas with the same British thermal unit value (1050 Btu/ft^3) as natural gas is used, the mixture will not burn properly. As a result, it is common practice to provide a mixed gas with a value of 1450 Btu/ft^3. This mixture has a specific gravity of 1.30.

Since gas flow varies inversely as the square root of the specific gravity, a conversion factor can be used with readily available natural gas charts to size the mixed gas–natural gas piping system. The conversion factor for 1450 Btu/ft^3 mixed gas is 0.69, which is used to multiply the capacity found in NG tables to convert the chart figure to the actual mixed gas capacity value. The method of using the charts was discussed previously for NG sizing.

Piping Materials. The piping materials are the same as for the NG system.

BIBLIOGRAPHY

Clifford, E. A., "A Practical Guide to LP-Gas Utilization," *LP-Gas Magazine*.

Corken Guide to Liquefied Gas Transfer Equipment, Corken International.

Denny, L. C., et al. (eds.), *Handbook of Butane-Propane Gases*, 4th ed., Chilton Company, Los Angeles.

Frankel, M., "Designing Propane Systems," *Plumbing Engineer Magazine*.

"Gas Systems," *American Society of Plumbing Engineers Design Handbook*, vol. 2, Chap. 7.

CHAPTER 9

FATS, OILS, AND GREASE INTERCEPTORS, AND MOTOR OIL SEPARATION UNITS

When fats, oils, and grease (FOG) are a normal part of the wastewater from a facility, it is important that they be separated from the sanitary waste discharge prior to being routed into the public sewer. This chapter discusses the separation, interception, retention, and removal of FOG, which leaves the treated water suitable for routing onto the site and discharge into public sewers. Included are design configurations of the various interceptor devices and accepted industry sizing methods that are the primary concern of the design engineer.

This chapter includes facilities such as commercial, educational, and food service establishments. Industrial facilities, where personnel are required to test for industrial effluent standards and must perform numerous types of analysis, are outside the scope of this handbook.

Volatile oil from other than food establishments is discussed in Chap. 15, Volatile Liquids: Treatment and Disposal. Oil separators are necessary to prevent wastewater contaminated with oil from discharging into site or public sanitary sewers.

It must be pointed out that the final decision for the interceptor size and installation rests with the authority having jurisdiction (AHJ), who must be consulted prior to the design and installation of an interceptor. The installation requirements discussed are generic and should not be used as the basis for the specific design and installation of any interceptor.

GENERAL

Municipally owned wastewater treatment facilities throughout the United States are detrimentally impacted by FOG, commonly referred erroneously as grease. FOG is usually introduced primarily from food service establishments that prepare and serve large volumes of food, and smaller amounts of FOG can also originate from other individual facilities. Discharged grease, if not intercepted, floats on the surface of the water as it flows through the drainage system. As the initial water temperature cools, a portion of the grease clings to the pipe walls. Once in the sewer system, the remainder of the FOG can clog sewer lines as it coats the pipes, and this coating can become as hard as concrete. Over time sewer pipe flows can be restricted, and the lines can even seal completely. The restricted flow can cause the sewer system to back up, resulting in a sanitary sewer overflow (SSO). These overflows can then contaminate the ground, local water bodies, and any property that the sewage comes into contact with, and eventually they find their way into wetlands, rivers, and lakes. This is a violation of the Federal Clean Water Act and will result in a fine issued to the individual or sanitary sewer district. Wastewater treatment plants have the authority, through the adoption of local ordinances, of requiring businesses to practice these methods of FOG discharge control.

While the laws of physics dictate that FOG separates from water at a slower rate as the temperature decreases, recent studies indicate that the separation rates at room temperature and at elevated temperatures [testing has been done up to 200°F (93°C)] are so close that the other benefits of room-temperature water outweigh the slight improvement in separation rate. The only difference is the rate of separation in colder water, not the separation itself.

The improper management of FOG not only is a hazard to the environment and potentially to the public's health, but also results in added financial burdens to both food service establishments and wastewater treatment plants. Lines clogged with FOG deposits require cleaning by the food establishment when it involves business-owned pipes and by the wastewater treatment plant operator for sewer system pipes. Operations at treatment plants can also be negatively impacted.

Types of Grease and Oil

Restaurants and similar establishments create three types of grease; white, brown and yellow.

White grease is collected from olive oil, cooking pans, and similar equipment. It has been found that this type of grease is a high-value product that can be recycled for uses such as cosmetics, glycerin products, and some animal feed supplements. Because of its value, this type of oil is usually of better quality than the other types of FOG and often is collected in a separate container or tank.

Brown grease has come in contact with water, such as that from washdown of hoods. The brown grease is not permitted to be recycled into any use that involves contact or consumption by animals or humans. Acceptable uses include paints, polymers, biodiesel, and composting.

Yellow grease is distinct from brown grease and is typically used as frying oil from deep fryers.

Oil, for this handbook, generally consists of motor oil, cutting oil, and any discharge from commercial establishments such as cutting oil, and metal shops are discussed later in the chapter.

Fatty acids, particularly from brown grease, have free fatty acids that are corrosive to sewer lines in the same manner as hydrogen sulfide.

CODES AND STANDARDS

Codes and standards are now a part of all models and most local codes. Many define the required size of interceptors on liquid retention time based upon the maximum anticipated rate of flow. Rules and regulations cover effective methods that can be employed at food service establishments and other facilities to prevent significant amounts of FOG from flowing through and clogging the disposal pipes. These methods involve the capture of the FOG at the source and disposal of the FOG for collection or recycling. This requirement varies from city to city or region to region, so it must be left to the appropriate administrative authority to establish this requirement through testing, mostly from third parties. These requirements currently vary from simple statements of minimum basin capacity size to achieve sufficient retention times based upon flow rates, to formulas that make assumptions about the amount of water used per meal served.

1. PDI-G 101, Testing and Rating Procedures for Grease Interceptors, Plumbing and Drainage Institute
2. ASME A112.14.3, Grease Interceptors
3. ASME A112.14.4, Grease Removal Devices (various types)
4. CSA B481, Grease Interceptors
5. All major model codes contain methods or sizing instructions.

Since remotely located outdoor interceptors must deal with conditions different from interior point-of-use interceptors, the design requirements will vary. In certain areas of the country, existing codes will require an interceptor whose sole specification is size. The local authorities must be consulted regarding the approved size.

Principle of Operation

Most grease interceptors operate on the principle that there is a difference in specific gravity between the FOG and wastewater. Interceptors can be categorized as *gravity interceptors* where the separation of FOG from the wastewater stream is free of mechanical devices, *hydromechanical interceptors* where gravity flotation is free of mechanical devices with the addition of air introduced into the wastewater stream, and *biological* separation systems where FOG is converted into separate molecules that permit discharge into the sanitary system. Finally, FOG disposal systems are those that collect and take away FOG with no human intervention.

DEFINITIONS

The following definitions are for this chapter only.

1. *Fats, oils, and greases (FOG):* A nonpetroleum polar compound of fat, oil, or grease present in a wastewater discharge derived from animal and/or plant sources that contain multiple carbon-chain triglyceride molecules. These substances are detectable and measurable using analytical test procedures established in 40 CFR 136, as may be amended from time to time. All are referred to as *grease* in this handbook.
2. *Cooking oils:* Nonpetroleum light compounds that are not fats or grease, used by commercial, institutional, or industrial food service, or separate compounds from food processing establishments consisting of vegetable oil, canola oil, olive oil, or similar compounds that are edible by humans. Cooking oil is generally saved in a separate container away from FOG due to its value as salvage.
3. *Generator:* Any person or entity who owns or operates an establishment where a grease interceptor is installed, or whose action or process generates or discharges FOG.
4. *Grease interceptor:* A device installed in a drainage system to separate, store, or eliminate nonpetroleum fats, oils, and greases from a wastewater discharge at its source, preventing it from entering the facility sanitary system or a publicly owned sewer.
5. *Gravity grease interceptor (GGI):* A device installed in a sanitary drainage system to intercept and retain nonpetroleum FOG from wastewater discharge. It separates the

FOG by gravity, using internal baffles that direct the flow of wastewater in such a manner as to increase retention time within the device and help separate the FOG from the wastewater.

6. *Hydromechanical grease interceptor (HGI):* A device installed in a sanitary drainage system to intercept, separate, and retain nonpetroleum FOG from wastewater discharge by gravity. The design incorporates baffles for air entrainment to the wastewater stream and promotes efficient separation of the FOG by means of passive interior baffling and/or barriers, in combination or separately.
7. *Grease removal device (GRD):* Any mechanical device that automatically disposes of nonpetroleum FOG into a separate container after separation from the interceptor. The design incorporates a mechanism that transfers FOG by means of either a moving or a fixed film, the control of which could be either automatic or manually initiated.
8. *Biological separation device:* A device installed in a sanitary drainage system to intercept, separate, and dispose off nonpetroleum FOG from wastewater discharge. The design incorporates a method of introducing a bacterium that disposes of FOG by decoupling the molecular structure, leading to mass and volume reduction.
9. *Grease waste:* Material present in and collected from a grease interceptor in the sanitary sewer service line of a commercial, institutional, or industrial food service or processing establishment, including the solids resulting from dewatering processes.
10. *Publicly owned treatment works (POTW):* A treatment works which is owned by a state or municipality as defined by Section 502(4) of the Clean Water Act. This definition includes any devices and systems used in the storage, treatment, recycling, and reclamation of municipal sewage or industrial wastes of a liquid nature. It also includes all sewers, pipes, and other conveyances that convey wastewater to a POTW treatment plant. The term also means the municipality, as defined in Section 502(4) of the act, that has jurisdiction over the indirect discharges to and the discharges from such a treatment works. For the purposes of this ordinance, the terms *public sanitary sewer system* and *POTW* may be used interchangeably.
11. *Transporter:* A person who is registered with and authorized to transport grease interceptor waste in accordance with current regulations.
12. *Removal:* Taking away from the interceptor, the accumulated FOG collected from a grease interceptor by a licensed removal service.
13. *Disposal:* Elimination of FOG by means of a dedicated container external to an interceptor with no discharge allowed to the facility sanitary system or a publicly owned sewer.
14. *Oil:* For the purposes of this handbook, oil is a nonedible petroleum-based liquid discharged from motor vehicles, garages, vehicle storage and industrial facilities that shall be prevented from entering a sewer system by regulations of federal and state codes.
15. *Immiscible:* Oil that will disperse into water, but will not dissolve.

Types of Interceptors

Previously, all designs of interceptors were referred to by the now obsolete term *grease traps*. Based on recent regulations and code language interpretations, AHJs did not permit a double trap on a fixture; therefore a grease trap and a fixture trap were not allowed to be used on the same branch. Since there has been previous confusion because both designs were referred to as grease traps, the term *grease trap* has been revised in recent years to refer to devices as *gravity grease interceptors* (GGIs), *hydromechanical grease interceptors* (HGIs), *fixed film bioreactor separation systems*, *moving-bed bioreactor separation systems* and *bioremediation systems*. It is necessary to include a fixture trap on the fixture inlet to an interceptor prior to discharge into the sanitary drainage system. This has been done with the leadership of the International Association of Plumbing and Mechanical Officials (IAPMO) and a consensus of the industry.

Separation can occur by gravity alone or by biological remediation. Grease removal devices discharge into a separate container for removal. For any grease interceptor to function well, three things must be present:

1. The unit must be properly designed, sized, and tested for efficient operation.
2. The installation must be correct in order to function according to rated parameters.
3. It must be properly maintained.

Gravity Grease Interceptor. The GGI incorporates the Bernoulli principle, allowing the time required for the buoyancy of grease in water to cause separation by gravity without any mechanical assistance. The unit is identified by volume. The time required has been established as 30 min for the separation to occur. The FOG initially floats on the surface of the discharged water from a fixture as they flow into the interceptor. The GGI design incorporates a larger volume to slow the velocity of water entering the interceptor. Internal baffles are added only to increase the path of wastewater to allow additional time for separation to occur. The entire operation is passive without mechanical assistance. Some codes do not permit a gravity interceptor small enough to be installed indoors. Where it is permitted, an illustration of a typical interior gravity interceptor is shown in Fig. 9.1. The gravity interceptor is generally much larger than a hydromechanical interceptor and is mostly located outdoors. The entrapped FOG must typically be removed from the interceptor manually. A typical larger unit is schematically illustrated in Fig. 9.2.

The top of externally installed concrete units shall meet AASHTO A-20 guidelines for exterior units where there is a possibility that trucks or other vehicles could pass over the unit. If the unit is in a nontraffic area, the unit shall meet minimum top-loading standards.

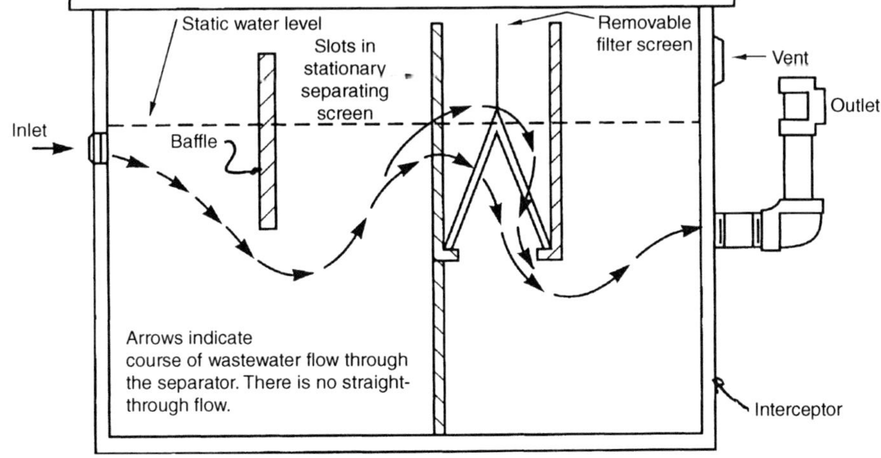

FIGURE 9.1 Detail of small gravity interior grease interceptor.

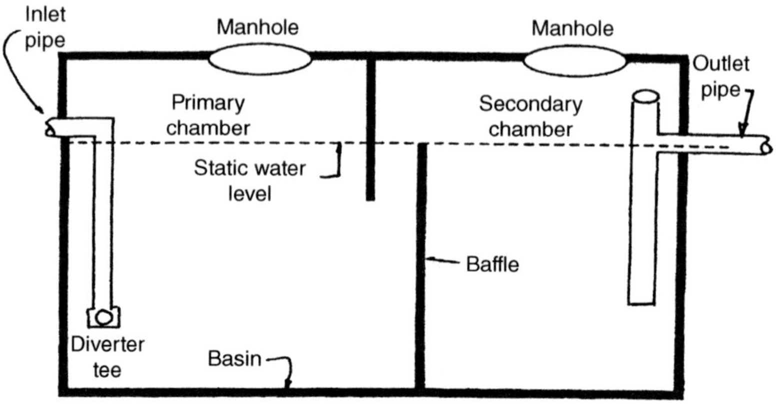

FIGURE 9.2 Detail of large gravity exterior grease interceptor.

Hydromechanical Interceptor. The HGI is a misnomer because it has no mechanical assistance. It continuously separates the FOG by gravity as it enters the interceptor, similar to a GGI. The unit is identified by its flow rate. Internal baffles are added to increase the flow path and allow additional time for separation to occur without any mechanical assistance. These baffles also have an additional function that provides air entrapment into the wastewater to assist in separation. A flow control device on the inlet is necessary. This type of interceptor is covered by PDI-G101, which requires performance testing.

The passive baffles serve to lengthen the flow path of the effluent to increase the time of separation while providing a nonturbulent environment for separation to take place. They allow air to be entrained into the wastewater stream without any mechanical assistance. The entrained air will separate FOG from the effluent quickly. As it does so, it accomplishes two things: First, the escaping air accelerates the separation of FOG as it rises rapidly to the surface of the water in the separation chamber. The rising air bubbles literally pull the FOG globules to the top of the water. Second, the air released then provides a small amount of positive pressure above the contents of the separation chamber to regulate the internal running water level of the grease interceptor. The entrapped FOG must be removed from the interceptor manually. An illustration of a typical hydromechanical separator is shown in Fig. 9.3.

The major manufacturers of grease interceptors do not label their units as hydromechanical.

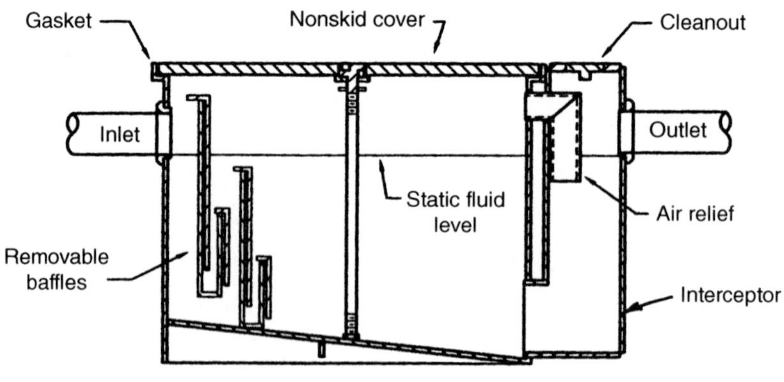

FIGURE 9.3 Typical hydromechanical separator.

Grease Removal Device. Similar to a HGI, the Grease Removal Device (GRD) is used to provide an automatic means by which grease is removed and emptied into a dedicated separate container. It is believed that an automatic removal system produces a more efficient interceptor by removing FOG as it accumulates. This can be accomplished by a level device that automatically empties the FOG from the separation chamber of the interceptor, an electric timer that empties the separation chamber at predetermined intervals, or the use of hot water introduced into the interceptor that forces the FOG out into a separate container.

Other GRD types use a mechanical device to remove accumulated FOG to a separate container. The first is a fixed film separation system, and the second uses a moving skimming wheel separation system and bioremediation system. The adjacent containers are emptied manually when full. A detail of a skimming wheel installed inside an interceptor is shown in Fig. 9.4.

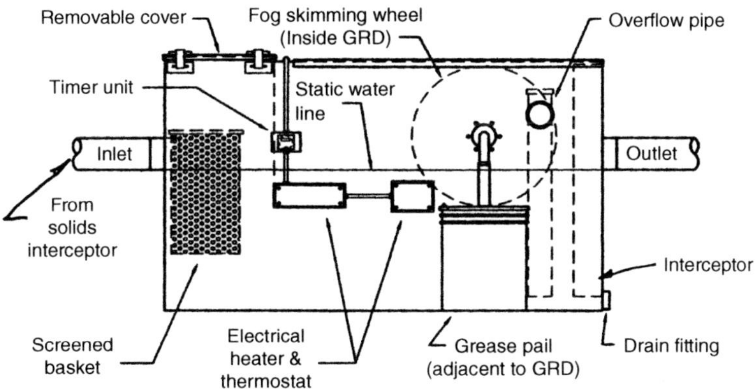

FIGURE 9.4 Skimming wheel grease removal device. (*Courtesy of ASPE.*)

Typically, devices for larger discharges are specifically engineered with either mechanical fixed-film or moving-bed types. Either type may have such features as dissolved air flotation, automatic sludge removal, mechanical aerators, or sequential batch processing.

These other systems are proprietary in nature, and the manufacturer should be contacted to learn the advantages and disadvantages for the system. However, all systems shall conform to ASME A-112-14.4, Grease Removal Devices.

Biological Separation System (BSS). The device traps and separates FOG from the waste stream, and converts the FOG within the interceptor into water-soluble, biodegradable products which can be safely discharged into the sanitary system. It accomplishes this by means of external mechanical methods that introduce bacteria into the device. It does not have to be emptied every day. Biological separation does not eliminate the need for monitoring the effluent quality, routine maintenance to remove undigested materials, or inspections to ensure all components are clean and functioning properly.

For an illustration of a typical biological interceptor, refer to Fig. 9.5.

Auxiliary Devices

Flow Control Devices. Flow control devices designed for use with grease interceptors have a number of important purposes. This is an essential part of the installation. The understanding of its function and how the device performs those functions is critical. The flow

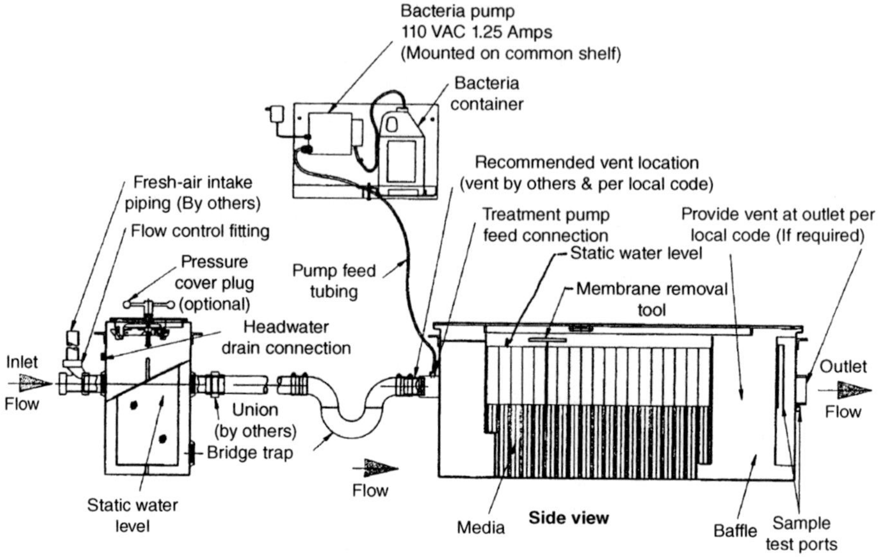

FIGURE 9.5 Biological separation system.

control is manufactured in a variety of configurations depending upon the manufacturer. The device works with the interceptor to contribute the following important functions.

1. It limits the flow of influent into the interceptor for desired efficiency.
2. It passively introduces air into the influent to accelerate separation.
3. It provides air into the interceptor that regulates internal control of the running water level without mechanical assistance.
4. The flow control device in reality is nothing more than an orifice. Inflow into the interceptor is restricted by means of an orifice sized to limit the inflow to the optimal flow required by the specific unit and head of inflow. Downstream of the flow control will be an area of low pressure that is connected to a source of air needed for the device to perform its intended function. The amount of air drawn in is proportional to the flow. Although the interceptor is required to have a source of air to perform as intended, this connection is often incorrectly assumed to be a vent, but its name is inappropriate. For an illustration of typical flow control devices, refer to Fig. 9.6. For the recommended size orifice based on flow, refer to Table 9.1.

The flow control fitting furnished with Plumbing and Draining Institute (PDI) certified interceptors must be installed in the waste line ahead of the interceptor. It should be located beyond the last connection from the fixture and as close as possible to the underside of the lowest fixture to minimize the effects of head pressure. When the wastes of two or more sinks or fixtures are combined to be served by one interceptor, a single flow control fitting may be used. Any flow control fitting installation not in conformance with these recommendations requires manufacturer consultation. The gravity interceptor would not generally require a flow control device to be installed.

Most codes contain language requiring a means of preventing the interceptor from becoming air bound. Typically, that language will state: "Venting. Interceptors and separators

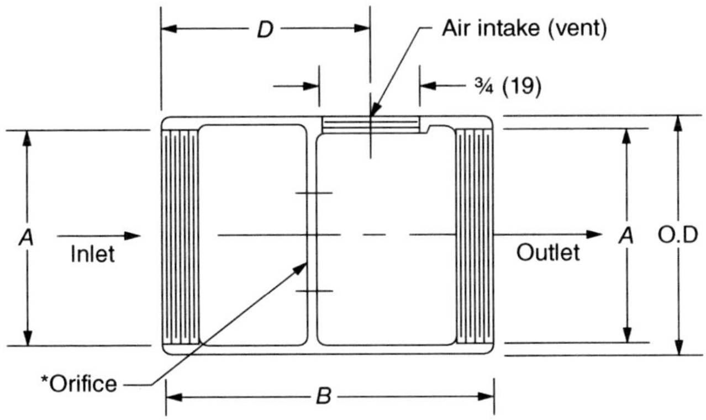

*Orifice size varies according to size of interceptor.

FIGURE 9.6 Typical flow control device.

TABLE 9.1 Grease Interceptor Recommended Orifice Size Based on Flow

Flow rate, gpm	Standard size orifice, in
4	5/8
7	5/8
10	15/16
15	15/16
20	1¼
25	1¼
35	1½
50	1¾
75	2¼
100	2¼
125	2-5/8
150	2-5/8
200	3¼
250	3¼
300	4

1 gpm = 3.8 L/min
1 in = 25 mm

shall be so designed that they will not become air bound when airtight covers are used." The air intake for the flow control may terminate under the sink drain board as high as possible above the flood level of the sink to prevent overflow. It may also terminate in a return bend at the same height outside the building. When the fixture is individually trapped and back-vented, the air intake may intersect the vent stack. All installation recommendations are subject to the approval of the local plumbing code authority.

Solids Interceptor. A solids interceptor is essentially a strainer designed to remove solids such as lettuce leaves, spoons, soda straws, and most importantly rice grains and all other items that should not be in an interceptor. It is strongly recommended that one be installed at the discharge of each dishwashing machine and each large sink.

The solids and the FOG are likely to form a globule where the specific gravity is heavier than that of FOG alone and will most likely sink and then discharge from the interceptor. Keeping the FOG from attaching to the solids is important because the combined material will sink. Since there is hot water discharged from any kitchen fixture, the grease will be more viscous and most likely separate freely from the solids due to the higher temperatures. With the separated solids and FOG, the FOG is likely to be retained.

FOG-laden solids passing directly into the interceptor create two problems. The globules can become hard and create a problem in the drainage system. In addition, FOG reaching the treatment facility makes treatment more difficult because FOG decomposition is slow and costly to remove.

Many of the codes now in existence require the discharge of food grinders and dishwashers to bypass a FOG interceptor to prevent solids from entering. If there is a solids interceptor, there is no reason for this to occur. A typical solids interceptor is illustrated in Fig. 9.7.

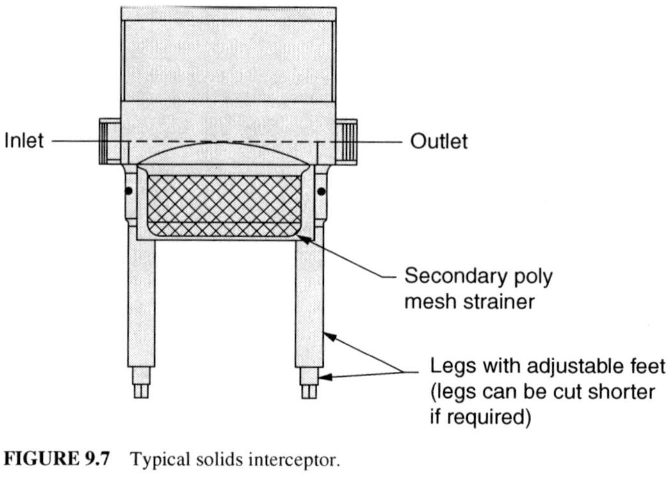

FIGURE 9.7 Typical solids interceptor.

Interceptor Sizing Procedures

General. It is critical that any sizing procedure selected by the design engineer be reviewed by the manufacturer upon selection and approved by the local authorities prior to installation. It is probable that local authorities have code and standard requirements that must be adhered to. All sizing for interceptors given in this handbook shall be considered preliminary in nature and must be confirmed by the local authorities. They shall be used for the purpose of selecting interceptors for such considerations as size and weight before the project has progressed to allow specific sizing.

Sizing a Gravity Interceptor. The gravity interceptor is little used inside a building because of the large size required. This will usually require an outdoor installation. The size is expressed in gallons of volume, based on the expected drainage fixture units.

The method for sizing a gravity interceptor is to use the drainage fixture units (DFUs) obtained either by calculating the DFU values of fixtures or from information obtained from the owner of the facility. If the DFUs are known, refer to Table 9.2 to select an interceptor. Using a retention time of 30 min and 25 gpm flow rate, this figure multiplied by 30 would equal a 750-gal capacity interceptor. There are also code sizing requirements using DFUs in many jurisdictions that must be followed if applicable.

TABLE 9.2 Drainage Fixture Unites Converted to Volume for Gravity Interceptor

DFUs	Pipe size, in	Gravity interceptor volume, gal
8		500
21		750
35	3	1,000
90		1,250
172	4	1,500
216		2,000
307		2,500
342	5	3,000
428		4,000
576		5,000
720	6	7,500
2,112		10,000
2,640		15,000

Based on 30-min retention time.

There is a difference of opinion from various authorities regarding the connection of fixtures and flow that the interceptor should use to determine the size. One school of thought is to calculate the total gpm flow discharged from all the known grease-producing fixtures that would be expected from all the volumes of fixtures that might discharge at one time. Another method is to use the DFU assigned to each fixture and the flow rate from the appropriate source. A third method is to use the fixture trap size based on the maximum flow entering the interceptor, calculated by standard engineering calculations when the pipe is flowing full at ¼-inch (2.0-cm/m) pitch. For information that would give a gpm flow from trap sizes not listed, refer to Table 9.3. The last method would assume

TABLE 9.3 Selection of Interceptor When Flow Rate Is Known

PDI size symbol	4	7	10	15	20	25	35	50	75	100
Flow rate										
gpm	4	7	10	15	20	25	35	50	75	100
L/min	15	26	38	57	77	95	132	191	230	378
Grease capacity										
lb	8	14	20	30	40	50	70	100	150	200
kg	3.6	6.4	9.1	13.8	18.2	22.7	31.8	45.4	68	90.8

Note: When testing interceptors in excess of 100 gpm, the PDI size shall be the rated flow and the grease capacity shall be twice the flow rate in pounds of grease.

Courtesy PDI

that for a properly designed system, the pipe would flow at 50 percent full and therefore, use one-half of the flow of the third method. This last method would result in a slower drain-down time.

There have been recent revisions to the code requirements and conformance standards for gravity interceptors in many areas, but not all. The authorities at the location where it is to be installed have performance standards for sizing that must be complied with. This standard for discharge is usually that 90 percent or better of the FOG must be removed. Without a code to follow, it seems like the most accurate approach would be the first option. This would be a best management practice and give the best result.

Sizing a Hydromechanical Interceptor. There is no specific label for a "hydromechanical" grease interceptor appearing in manufacturers' catalogs. The hydromechanical grease interceptor is generally selected because of its smaller size. This sometimes limits the size made by various manufacturers. The capacity is based on the gpm flow rate entering the interceptor and on guidelines published by PDI and ASME codes. The capacity varies between jurisdictions. The total load is based on all possible fixtures being served that may discharge simultaneously. Each manufacturer has a schedule of interceptor sizes to choose from based on the total gpm value.

The hydromechanical grease interceptor is sized in conformance to PDI Standard PDI-G101 based on the gpm flow entering the unit. This standard is tested and certified to operate efficiently at its maximum flow rate in gpm while retaining a minimum capacity, in pounds of grease, twice its rated gpm flow. Based on PDI standards, Table 9.3 has been prepared for sizing based on gpm flow. If flow is rated in DFU, use Table 9.4 to obtain the actual gpm value to use. Consult a manufacturer's catalog for the correct sizes.

TABLE 9.4 Drainage Fixture Units Converted to Flow for Sizing Hydromechanical Interceptor

DFUs	Flow, gpm
8	20
10	25
13	35
20	50
35	75
172	100
216	150
342	200
428	250
576	350
720	500

*Based on intermittent full flow in drainage lines.

When sizing for fixtures not listed, the trap size for each fixture will determine the individual DFU or gpm count in accordance with the local code. The total number of fixtures would determine the total DFUs expected. Based on the total unknown DFU count, Table 9.5 shall be used to find the gpm value necessary to size the interceptor. The flow is based on the capacity of only the trap shown.

TABLE 9.5 Conversion of Pipe Size or Fixture Units to Flow Rate Pipe Flowing Half Full at ¼-in Pitch (20 cm/m) for Hydromechanical Interceptors Only

Drain outlet or fixture trap size		Drainage fixture	Peak flow	
in	DN	unit value	gpm	L/min
1½	40	3	22.5	85
2	50	4	30	120
2½	65	5	50	190
3	80	6	60	230
4	100	8	130	500
6	150	10	300	1140

Sizing the FOG Disposal System. The FOG disposal system shall be sized in consultation with the selected manufacturer to allow discharge of effluent with no more than 100 mg/L of FOG. It is general practice in codes to provide an approved listing of acceptable referenced standards.

Additional Interceptor Sizing Guidelines

One of the most controversial issues relating to installation concerns which fixtures or sources must be part of the FOG interceptor system. There is little controversy about connecting pot sinks. There is some controversy, and possible code restrictions, about connecting dishwashers. There are some questions relating to floor drains, but discharge from food grinders (or garbage disposals) is almost universally required to bypass the grease interceptor or to have the pulverized solids removed from the waste stream before it enters the interceptor. The applicable code would usually address this question.

The food grinder (and the associated prerinse station at the dishwasher) is one of the single greatest sources of FOG. Yet despite that fact, many codes forbid food grinder discharge from passing through a grease interceptor. Technologically, if the solids have been removed, there is no reason for the waste stream to bypass the grease interceptor.

The following considerations are recommended for the selection of a grease interceptor where there is no code to follow.

1. A rectangular interceptor is believed to be the optimum shape. An approximate ratio of depth to width shall be 1 to 1.5. There should be at least 4 in (102 mm) above the water level for venting.
2. For an outdoor gravity interceptor, an allowance of 6 to 12 in (152 to 300 mm) shall be provided on the bottom for sludge accumulation.
3. A sample port should be provided to allow ease of sample taking.
4. If the calculated flow falls between sizes, select the larger interceptor.

The most important consideration is conformance with all code requirements or standards of the authorities having jurisdiction.

When the size of the dishwasher is known and it is going to discharge into the interceptor, the flow rate can be increased using the following flow.

Dishwasher tank capacity		
Up to 30-gal (115-L) water tank capacity	15 gpm	57 L/min
Up to 50-gal (190-L) water tank capacity	25 gpm	95 L/min
Up to 100-gal (380-L) water tank capacity	40 gpm	152 L/min

For establishments other than restaurants, the maximum flow rate method was selected because it relates the flow rate for fixtures to the size of the interceptor.

1. Determine the number and size trap of fixtures and the size of a dishwasher (if any) in the establishment discharging into the interceptor.
2. Add the dishwasher and only the single largest sink together. If there is no dishwasher, add the two largest flows, in gpm (L/min). This is the maximum probable flow rate into the interceptor. This figure is based on the probability that no more than two fixtures could discharge at the exact same time.
3. For a gravity interceptor, multiply the flow rate by 30 to calculate the minimum pounds of grease required to be retained. (This figure has a proven history of success.) Pick a standard size interceptor with a capacity equal to or larger than the calculated flow rate size.
4. In some jurisdictions, a minimum size shall be chosen on a sliding scale based on the size of the establishment, as follows:
   - *a.* For small shops, such as a pizza parlor or other similar establishments, 200-gal (760-L) capacity
   - *b.* For medium-size shops and those with higher FOG potential, such as meat markets, 500-gal (1900-L) capacity
   - *c.* For large establishments, such as regular supermarkets, 1000-gal (3800-L) capacity
   - *d.* For very large supermarkets and other similar establishments, 1500-gal (5700-L) capacity

For restaurants, the size of the interceptor should be based on the number of seats and the number of meals served, as indicated in Eq. (9.1). This formula is based on information taken from the manual written by the EPA. This method was chosen because it appears to have the most realistic sizing criteria of all methods examined.

$$F = W \times \frac{H}{2} \times \mathrm{LF} \tag{9.1}$$

where F = discharge flow, gpm
W = 5 gal (20 L) of waste per meal served
H = number of hours restaurant is open
LF = loading factor

$$= \begin{cases} 1.25 & \text{where located on interstate freeway} \\ 1.00 & \text{for other freeways and recreational areas} \\ 0.8 & \text{for main highway} \\ 0.5 & \text{for other highways (most often used regardless of location)} \end{cases}$$

To select an interceptor based on the discharge of multiple-compartment sinks, use Table 9.6 to determine the flow from each compartment of sinks filled to 75 percent capacity. The drainage period will be selected by the design engineer.

For a detail of a typical installation of a grease interceptor, refer to Fig. 9.8.

TABLE 9.6 Drainage from Multiple-Compartment Sinks at 75 Percent Capacity

Selection Chart (Metric Equivalents Omitted for Simplicity)

Fixture compartment size for each compartment, in	Number of compartments	Drainage load, gal	Recommended PDI size grease interceptor: 1-min Drainage period	Recommended PDI size grease interceptor: 2-min Drainage period
18 × 12 × 6	1	4.2	7	4
16 × 14 × 8	1	5.8	7	4
20 × 18 × 8	1	9.4	10	7
18 × 16 × 8	2	15.0	15	10
20 × 18 × 8	2	18.7	20	10
30 × 20 × 8	1	15.6	20	10
24 × 20 × 12	1	18.7	20	10
22 × 20 × 8	2	22.9	25	15
22 × 20 × 12	2	34.3	35	20
24 × 24 × 12	2	44.9	50	25
22 × 20 × 12	4	68.6	75	35
24 × 24 × 12	4	89.8	100	50

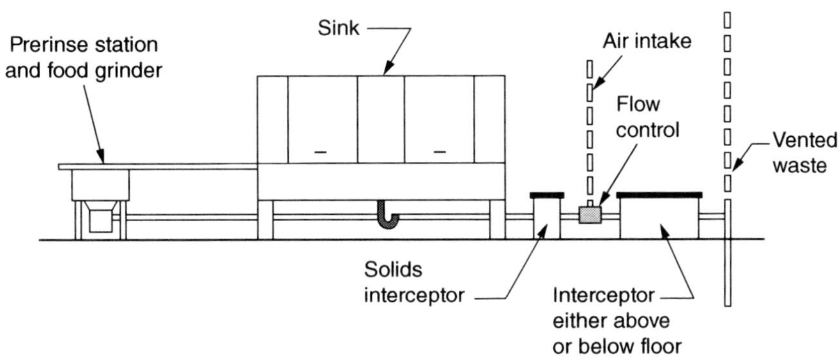

FIGURE 9.8 Typical interceptor installation.

System Design Considerations

Even the best designed interceptors, properly installed, will fail if they are not maintained. The precise requirements for maintenance are not possible to define since conditions at each installation vary. In terms of the typical code, maintenance must be performed before the grease in the wastewater downstream from the interceptor exceeds local limits. The importance of continuous maintenance cannot be overemphasized since it prevents any accumulated FOG from entering the house drain and eventually onto the site where it might cause stoppages.

A properly sized and designed grease interceptor may not work or may work less efficiently if it is installed incorrectly. As obvious as it seems, the interceptor must not be installed backward. Based on experience, far too many interceptors that upon inspection, are condemned for not working have merely been installed backward. The problems relating to installation, however, go beyond the obvious. Regardless of whether the interceptor is a hydromechanical interceptor or a large gravity interceptor, one of the most important installation practices to follow must be to locate the interceptor as near as possible to the source of the FOG-laden water. As stated previously, this is important because every foot of piping between the source of FOG-laden wastewater and the interceptor is unprotected and is a potential maintenance problem. A second reason for locating the interceptor near the fixture is that FOG separates best when the effluent is relatively hot.

Some interceptors have been installed under sinks without clearance for removal of the cover. Other interceptors have been placed in the floor and tiled over; still others have been located so that they are literally hidden from view; and some large outdoor interceptors have actually been paved over. The placement should allow the cover to be visible and easily removable for cleaning, and clearances should be such that the internal baffling can be serviced. With the cover removed, all wetted surfaces should be visible. This is necessary not only for access to clean the interceptor, but also to have the capability to easily inspect the interior for potential problems such as damaged baffles and blocked air relief bypasses.

While that is a simple statement to make, it is impossible for the user of a grease interceptor to determine when those limits have been exceeded. The method for determining when an interceptor's rated capacity has been reached is fairly simple if it is a PDI-certified interceptor. A PDI-certified interceptor has a rated retention capacity equal to twice its flow rate expressed in pounds. For example, a 35 gpm interceptor is rated to retain at least 70 lb. of grease. A user may determine a cleaning schedule by measuring how much grease has been trapped over a period of time.

Grease will weigh about 7 lb/gal, and if it is determined that a 35 gpm interceptor accumulates a total of 70 lb of grease (about 5 gal of grease every 4 days), it would be easily and correctly assumed that the interceptor must be cleaned no less than once a week. In fact, if the user must comply with a code that limits grease to 100 parts per million (ppm), the cleaning interval recommended would be every 2 or 3 days. When cleaning is discussed, it should be understood that cleaning an interceptor should always include the removal of grease from the top of the interceptor and should become a part of the daily routine. It usually will only require about 15 min, and there will be limited or no objectionable odors.

It has been determined that when food grinders are part of the waste system, and a properly sized solids interceptor, cleaned daily, is located ahead of the grease interceptor, the odors normally associated with the grease interceptor are not present because the food particles that decay and cause odors never reach the interceptor. Use of the solids interceptor improves the grease quality to a condition that the restaurant may be compensated for the recovered grease, since it can be recycled into a variety of products.

When regular maintenance is not performed, the obvious result is a grease interceptor that becomes unable to separate the FOG due to overloading, thus passing these materials downstream. Unless it is equipped with an electronic, sensor-controlled, positive inlet closure valve to prevent such overloading, no grease interceptor will automatically shut itself down to prevent overload discharge. Apart from violating codes or ruining any on-site wastewater treatment system, sewer blockages and the associated health risks are likely.

Some FOG producers would rather do almost anything but clean a grease interceptor. FOG producers have several options, some of which are acceptable alternatives, and others of which are possibly legal, but nevertheless unacceptable. One alternative is to engage the services of a company that specializes in cleaning interceptors. This is not an inexpensive approach, and in the case of large interceptors, this may be required. If the service is performed as often as necessary, it ensures the interceptor will function as intended.

MOTOR OIL SEPARATION UNITS

General

The oil discussed in this chapter is from hydrocarbons, similar to motor oil and similar substances. An oil separation unit is a device intended to separate oils and other light-density liquids present in a wastewater system from a facility by gravity and to prevent the oil from being discharged into public sewers. It also includes gravity disposal of the oil that is separated. Facilities that require a separator are repair garages, car washing facilities, and industrial facilities where oil-bearing or flammable waste may be discharged.

Oil in Water

Oil in water exists in several forms:

1. It exists as free oil. It is immiscible, which means that it cannot dissolve in water.
2. Mechanically dispersed oil is made up of fine droplets ranging in size from micrometers to fractions of a millimeter. It is stable due to electric charges and other forces, but not because of the presence of surface-active agents.
3. Chemically stabilized emulsions are made up of fine droplets that are stable because of surface-active agents.
4. Dissolved and dispersed oil is either actually dissolved or suspended in such a small size (typically 5 μ or less) that ordinary filtration is not possible.
5. Oil-wet solids are particulates in which oil adheres to their surface.

Methods of Separation and Treatment

Oil spills and leaks are best treated in their most concentrated state, which is at their source or as close as is reasonable. The primary methods used to separate and remove free oil and oil-wet solids are filtering, flotation, and centrifugation. This depends upon the difference in specific gravity. Secondary treatment, such as chemical treatment, coalescence, or filtration, is then used to break up oil–water emulsions and remove dispersed oil. This method is outside the scope of this handbook. Finally, tertiary treatment, such as ultrafiltration, biological treatment, or carbon adsorption, will remove the oil to required levels prior to discharge. This section will discuss the general principles of the separation methods and devices only.

The American Petroleum Institute (API) has established criteria for the large-scale removal of globules larger than 6 inches (150 mm). In abbreviated form, they are as follows:

1. The horizontal velocity through the separator may be up to 15 times the rise velocity of the slowest-rising globule, up to a maximum of 3 ft/s (1.0 m/s). The depth of flow in the separator shall be between 3 and 8 ft (0.9 and 2.4 m).
2. The width of the separator shall be between 6 and 20 ft (1.8 and 6.0 m).
3. The depth-to-width ratio shall be between 0.3 and 0.5.
4. An oil retention baffle should be located no less than 12 in (300 mm) downstream from a skimming device.

The separators could be furnished with sediment buckets, filter media that would remove small suspended droplets and other smaller materials from the wastewater. Baffles are an integral part inside the separator to increase the flow path, with the configuration determined by the manufacturer.

Gravity Separators. Gravity separation is the primary separation method. It is based on the specific-gravity difference between immiscible oil globules and water. Since all these liquids are lighter than an equal volume of water, gravity separators operate on the principle of flotation. As water and oil flow through the unit, the oil floats to the top and is trapped inside a series of internal baffles. Since the oil remains liquid, it is easily drawn off. Refer to Table 15.2 for the specific gravity of many common oils and other substances.

Flotation Devices. For service on a larger scale, the flotation of oil and oil-wet solids to the top of the flotation chamber can be increased by the attachment of small bubbles of air to the surface of the slow-rising oil globules. This is done by adding compressed air to the bottom of the flotation chamber in a special manner that will create small bubbles which will mix with, and attach themselves to, the oil globules.

Centrifugal Separators. For service on an even larger scale, the centrifugal separator is used. This device operates on the principle of inducing the combined oil–water mixture to flow around a circular separation chamber. The lighter oil globules will collect around a central vortex, which contains the oil removal mechanism, and the clear water will collect at the outer radial portion of the separation chamber. Methods have evolved that can produce effluent water with only 50 to 70 ppm of oil; proprietary devices exist that can lower oil content to 10 ppm.

Filtration. Using chemical methods first to break oil–water emulsions and then using filters to remove the destabilized mixture have proved effective in removal of oil globules in sizes between 1 and 50 mm. The velocity and flow rate of the mixture must be carefully controlled to allow optimum effectiveness of the system.

Smaller Systems. Oil separators for small flows usually take the form of a single unit consisting of a drain grating into which the effluent flows and inside which the oil remains to be drawn off manually. Another type of unit uses an overflow arrangement that sends the trapped oil to a remote oil storage tank.

Because there is the possibility that the vapor given off by the flammable liquid could also ignite, it is important to provide a separator vent that terminates in the open air at an approved location above the highest part of the structure. Some codes require that a flame arrester be installed on the vent.

The most common material used for an oil interceptor is cast iron, although steel can be used for less severe service. Gratings must have the strength for the type of vehicle expected.

Refer to Fig. 9.9 for an illustration of a typical small oil interceptor. Figure 9.10 shows the installation of a typical oil interceptor with gravity oil draw-off for garage floor drains.

Sizing for Typical Installations

The basic sizing procedure would involve determining the maximum anticipated flow, in gpm, that would be discharged into the separator it serves. To keep the separator as small as possible, a flow control fitting is recommended to be installed on the inlet. Floor or trench drain units are the preferred units for installations.

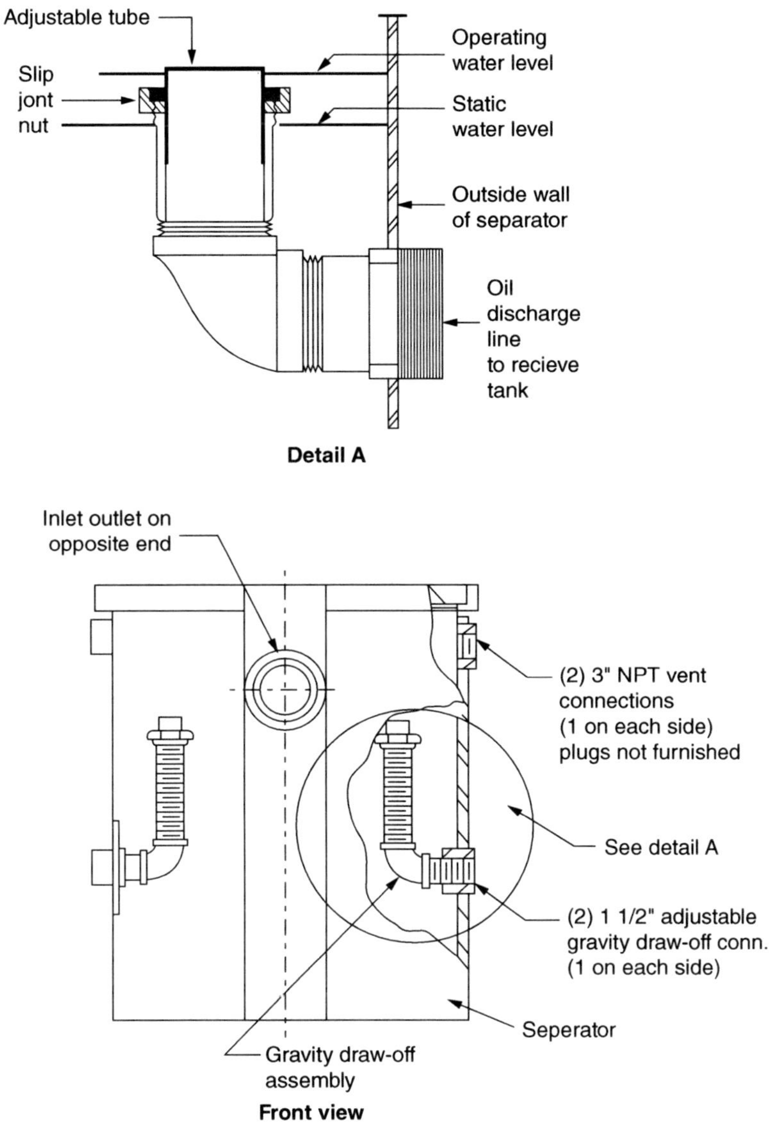

FIGURE 9.9 Typical old separator.

The oil draw-off contains oil that is retained inside the unit and would be removed either manually or automatically from an oil draw-off built into the separator. The automatic draw-off would be piped directly into a dedicated storage tank in order for the separated oil to be continuously and automatically drained.

Facilities such as garages, vehicle storage, servicing, and car washing should have a combination separator drain where the maximum flow rate can be determined by the owner.

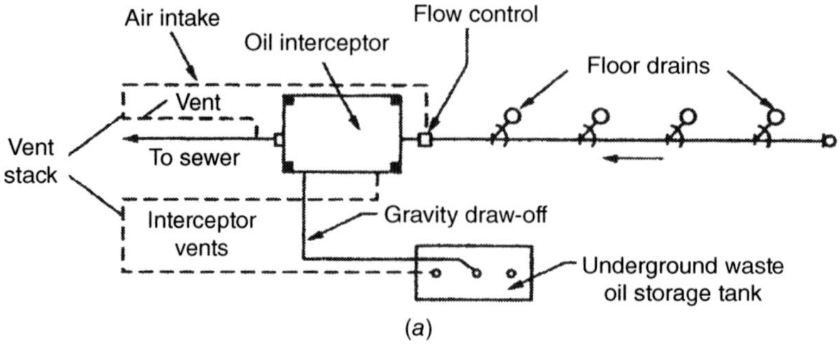

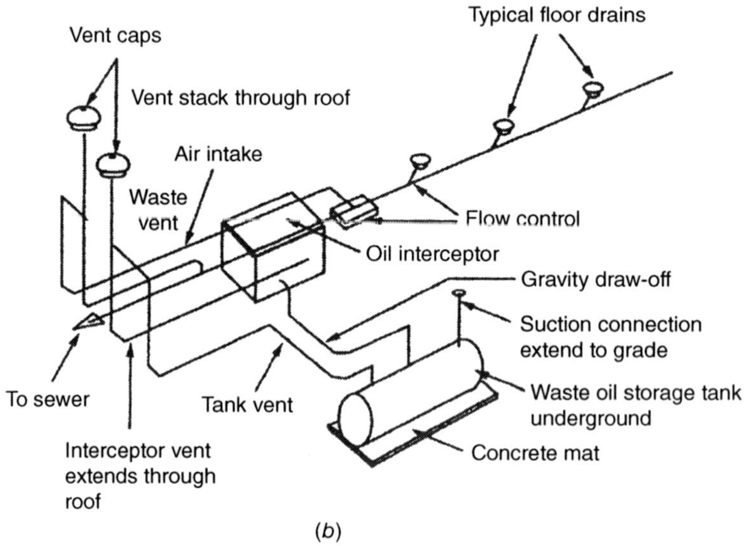

FIGURE 9.10 Typical gravity draw-off installation. (*a*) Plan, (*b*) isometric.

Owner actions such as floor washing, rainfall on outside vehicle storage, and maximum washing flow shall be considered. Care shall be exercised in maintaining a manufacturer-recommended static water level in the drain if applicable.

The size for the oil separators is often controlled by the local codes where the facility is located. The codes shall be consulted prior to having any separator designed.

BIBLIOGRAPHY

Ackroyd, R., "Basic Principals for Sizing Grease Interceptors," *Plumbing Standards Magazine*.

Aymong, G., "Automatic Grease Removal Systems," *Plumbing Systems & Design*.

Batten, B., "Isolation, Containment and Grease Separators," *PM Engineer*.

Frankel, M., *Facility Piping System Handbook*, 3d ed., McGraw-Hill, New York, 2010.

Hamilton, S., "Grease Collection—Nothing but a Nuisance," *PM Engineer.*

Nourmohammadi, S., "FOG and Restaurant Plumbing Design," *Plumbing Systems & Design.*

PDI, "Guide to Grease Interceptors," Plumbing and Drainage Institute.

Standard PDI-101, Testing and Rating Procedure for Hydro-mechanical Grease Interceptors, Plumbing and Drainage Institute.

Wehrenberg, J., "Design, Installation and Keys for Proper Interceptor Installation," *Plumbing Engineer.*

Weiss, M., "Grease Removal," *PM Engineer.*

CHAPTER 10
pH NEUTRALIZATION SYSTEMS

GENERAL

A pH waste drainage and neutralization system collects, neutralizes, and removes liquid wastes with a pH lower or higher than 7.0 from laboratory fixtures, equipment, and all areas of a facility. This discharges either into an appropriate local treatment facility or into the sanitary drainage system. A vent system equalizes flow in the neutralized drainage system in the same manner as the sanitary drainage system. For the purposes of this chapter, discussion is limited to small-scale facilities with little or no solids content, rather than large-scale production. The pH waste is divided into general categories such as laboratory, classrooms, and industrial waste which includes both acid and caustic liquids.

A laboratory could be any room or area within a building where investigation, teaching, testing, experiments, and research are conducted. Pharmaceutical facilities generally prepare, manufacture, and package drugs of all kinds. Manufacturing is generally considered any facility where a product is the end result of having material or components packaged or assembled from parts obtained elsewhere or made within the facility. These facility definitions cover an extremely wide range of possible drainage categories discharging from various sources, each with different effluent characteristics and design requirements.

Laboratory waste consists primarily of diluted and concentrated mixtures of liquid chemical substances of mineral and organic origin, and water. Acids of many types are usually present. Laboratory waste is discharged from sinks, cup sinks, fume hoods, other similar fixtures, and equipment. Ordinary waste from floor drains, autoclaves, glass washers, and condensed water from various sources are also included. All these ordinary fixtures would require a separate vent and should be considered as a completely separate pH drainage system. The pH values of common liquids are shown in Table 10.1.

Acid and caustic waste drainage from industrial facilities will consist of discharge from laboratories, accidental spills originating from tanks and piping, and anticipated waste from manufacturing equipment, often into drains. Very often the drainage piping would have to carry many of the liquids that are used as part of a process.

Where spills are directed into holding tanks, the drainage piping, tanks, and pumps, necessary to convey the effluent to treatment facilities will normally be part of the plumbing engineer's responsibility.

TABLE 10.1 pH Values of Common Liquids

Liquid	pH
Acidic acid	2.4
Apple juice	2.9–3.3
Baking soda	8.0
Beer	4.0–5.0
Detergent	9.5–10.0
Grapefruit juice	3.0–3.3
Hydrobromic acid	1.1
Lemon juice	2.2–2.4
Lime juice	1.8–2.0
Milk	6.2–6.6
Nitric acid	1.0
Perchloric acid	0.8
Phosphoric acid	0.9
Soda	2.0–4.0
Sulfuric acid	2.1
Vinegar	2.4–3.4
Wine	2.8–3.8

Definition

In chemistry, pH is a measure of the acidity or basicity of an aqueous solution. Pure water is said to be neutral, with a pH of 7.0 at 77°F (25°C). Solutions with a pH less than 7 are said to be acidic, and solutions with a pH greater than 7 are basic or alkaline. The pH values for specific common liquids are given in Table 10.1. The pH values of a general nature are given in Fig. 10.1.

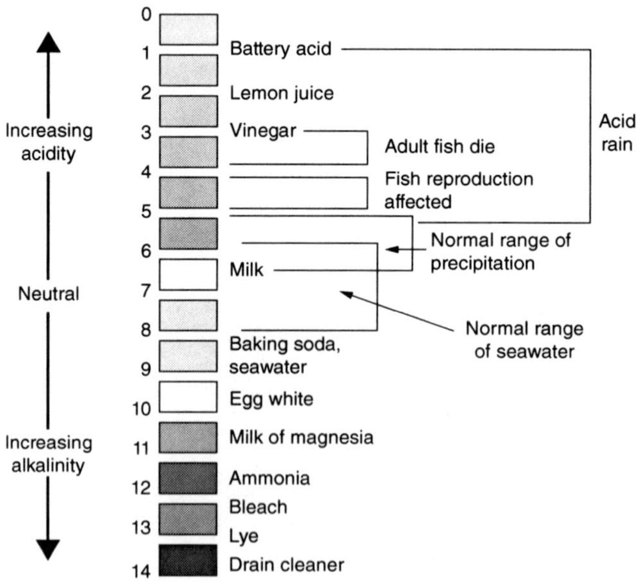

FIGURE 10.1 pH scale.

CODES AND STANDARDS

There is no consistent pH figure used by the authorities to allow direct discharge into a public sewer. It varies in various models, states, and city codes or by many authorities. Common accepted practice limits the direct discharge of an acid to a pH of 5.0, although the pH of lemon juice is around 2.2 and that is constantly being put into a sewer. A caustic limit is not yet defined by any national standard. The decision is up to the local authority having jurisdiction (AHJ). Although the EPA explains that no upper limit is set, many municipalities find it useful to establish one. An upper limit of a pH of 12.0. for caustic is a generally accepted figure.

There is a variance, depending upon the location of the project, in the exact pH figure. It must be pointed out that the final decision for the interceptor system approval rests with the AHJ, who must be consulted prior to the design and installation of any system. In many jurisdictions, additional or prior approval may be necessary from other agencies such as the health department or the EPA.

HEALTH AND SAFETY CONCERNS

All grades and concentrations of acids and caustic can cause severe damage to eyes and tissues of the body. Contact with the skin will cause irritation and burns. Contact with the eyes could cause blindness. Inhaling the mist or vapors could cause lung irritation or burns. Ingestion will destroy tissue of the mouth, throat, and stomach. Extreme care should be exercised in the handling and cleanup of all acids.

Where the generation of acid vapor is possible, manually operated fog nozzles using water to suppress the vapor and foam systems to prevent vapor from rising should be considered to mitigate the effects of an accidental spill.

Spills of concentrated acids from tanks onto floors and on equipment should be washed off and flooded with water, which can then be routed to the acid drainage system for neutralization. Tanks and piping that contain this spillage should be of a suitable material. Since water reacts rapidly with the acid and splatters, caution should be exercised. Heat and fumes are also given off. Breathing of the fumes will cause throat and lung injury. Where this situation is possible, suitable emergency breathing apparatus should be provided. An emergency shower should be installed in the immediate vicinity of acid storage and pipe routing.

COMMON TYPES OF ACIDS

Acids are widely used chemicals in the laboratory and chemical processing industry. The most commonly used acids are the following.

Sulfuric Acid. Sulfuric acid (H_2SO_4), among the most used of the acids, is commercially available in many concentrations and as various percentages of oleum. Oleum is sulfuric acid containing sulfur trioxide dissolved in the acid, which is called the "fuming" grade.

Phosphoric Acid. Phosphoric acid (H_3PO_4) is available in concentrations of 75 to 87 percent. OSHA has limits for human exposure to fumes from this acid.

Hydrochloric Acid. Hydrochloric acid (HCL), also known as muriatic acid, is available in four strengths designated as degrees Baumé (an equivalent notation of specific gravity).

Nitric Acid. Nitric acid (HNO_3) is available in three grades, designated by percentage of concentration by weight: 56 to 70, 70 to 84, and 97.5 to 100 percent.

Hydrobromic Acid. Hydrobromic acid (HBR) is commercially available in two concentrations: 70 and 99.95 percent.

Perchloric Acid. Perchloric acid ($HCLO_4$) is available in a concentration of 69 to 72 percent strength and is the strongest of all the inorganic acids.

SELECTION OF PIPING AND COMPONENTS

General

The pH is a designation given to show the range of acids and caustic, with 14.0 being the most caustic (or strongest) and 0.0 being the lowest, or most acidic. The pH of 7.0 is neutral. A pH scale is shown in Fig. 10.1. The most important considerations in the selection of piping, valves, and tanks for acid and caustic are type of discharge, concentration, and temperature.

The majority of the effluent from an "average" laboratory or research facility consists of primarily water and acid. Chemicals used for experiments are usually confined to fume hoods if they are toxic to the staff. Biological facilities discharge effluent from decontamination areas for disposal. For industrial discharges, obtain from the end user the extent and concentration of all the acids expected to be used in the laboratory and the facility. Accidents and spills must be allowed for by selecting pipe for the most concentrated acid being used in the facility.

The most cost-effective piping above the floor from laboratory fixtures is generally fire-retardant polypropylene, either with heat-fused socket or proprietary "screwed mechanical" type of joints. Other acceptable materials are glass with compression joints and high-silicon cast iron with caulked or compression gasket joints. Although PVC and CPVC have the least initial cost, they also have a limited range of chemical compatibility, with PVC having a low temperature rating. PTFE is the most resistant to the widest variety of chemicals and has the highest temperature rating and highest cost. The plastic piping should be continuously supported on angle irons every 4 ft (1.2 m) to prevent future sagging resulting from hot effluent.

Vent pipe shall be of the same material as the drain pipe. The vent shall be carried up to above the roof level. Vent piping penetrating the roof shall not be glass. Use an adapter and provide any other acceptable acid-resistant pipe material through the penetration.

Piping underground could also be polypropylene with heat-fused socket joints or high-silicon cast iron with compression gasket joints. Glass piping should be encased in a sleeve of polyethylene for protection. Following is a discussion of piping compatible with various acids.

Sulfuric Acid

Generally recommended piping materials for sulfuric acids at low temperatures of 140°F (60°C), and lower and up to 90 percent concentration could be PVC, CPVC, PP, PVDF, ETFE, HDPE, plastic, glass, Alloy 20, Duriron, and fiberglass reinforced plastic (FRP) piping with special resins. At 90 percent concentration and higher, carbon steel schedule

80 is often used. Stainless steel is generally unsuitable, except for oleum. Vent lines should be of the same material as the drain line.

Sulfuric acid is nonflammable, but it is highly reactive. Below a concentration of 75 percent it reacts with carbon steel and other metals to form hydrogen. It is particularly hazardous when in contact with carbides, chlorates, nitrates, fulminates, picrates, and powdered metals. In higher concentrations it will ignite combustible materials such as oily rags and sawdust. Dry chemicals or carbon dioxide is the fire suppression method of choice.

Oleum spills, because of the danger of fumes, should be contained by curbs, and the liquid diverted away from the area of a spill to a containment area where the liquid will be neutralized. The remaining liquid should be absorbed with diatomaceous earth, expanded clay, or other nonreactive material. This material will then be carted away for suitable disposal.

Valve types for sulfuric acid include ball, gate, and diaphragm, with gate valves being the most commonly used. For low pressure and temperatures suitable for the specific plastic pipe, plastic is often used. For higher temperature and pressures, stainless steel (SS) alloy 320 is preferred. In all cases, because of differences in manufacture, pipe vendors should be consulted as to the suitability of materials for specific acid piping service.

Centrifugal pumps constructed of SS alloy 320 with Teflon packing are in common use. Other manufacturers use FRP and plastic pumps. Also available are metallic pumps lined with plastic or glass. Temperature limits should be carefully checked for material suitability.

Phosphoric Acid

Drainage and vent piping, valves, and pumps are similar to those used for sulfuric acid. Recommended pressure piping is SS type 316 extra low carbon (ELC). Spills, safety, and health concerns are similar to those for sulfuric acid.

Hydrochloric Acid

Piping materials for hydrochloric acid drainage and vent piping, valves, and pumps are similar to those used for sulfuric acid. Spills, safety, and health concerns are similar to those for sulfuric acid, except that caustic soda should not be used because hydrochloric acid reacts with this chemical.

Nitric Acid

For nitric acid, recommended pressure piping material in concentrations up to 95 percent is 304L SS. Above this concentration, aluminum piping is recommended. Pumps for concentrations up to 95 percent should be constructed of 304L SS. Above this concentration, titanium, aluminum type 3003, or silicon iron is commonly in use. Recommended materials for gate, ball, plug, and globe valves are 347 SS or 304L SS. Drain lines should be glass.

Spills, safety, and health concerns are similar to those for sulfuric acid, except that temperature and humidity have an effect on the reaction of nitric acid on such metals as copper, brass, and zinc. Nitric acid reacts violently with organic substances, occasionally causing explosions. Self-contained breathing apparatus is required when approaching spills because of the emission of nitrogen oxides, commonly called nitrous fumes, which are extremely hazardous.

Hydrobromic Acid

Recommended pressure piping materials are glass- and rubber-lined steel pipe, PVC, PE, and PTFE. Glass pipe could be used for drainage in addition to the pressure piping. Valves are often ball or plug type with PVC, PE, and PTFE lining. Rubber-lined pinch valves are commonly used. Pumps are similar to those used for sulfuric acid, with the addition of Hasteloy B material.

Spills, safety, and health concerns are similar to those of phosphoric acid, but the vapors are much more hazardous. This acid reacts with metals and produces explosive hydrogen gas.

Perchloric Acid

Recommended pressure piping materials are glass and PTFE. Drain lines could be glass or Duriron. Valves are often ball and plug type manufactured from PTFE and Duriron. Pumps manufactured from PTFE are the most commonly used.

Spills, safety, and health concerns are similar to those of phosphoric acid, except that when heated to 150°F (64°C), perchloric acid can cause objects not normally considered combustible, such as rubber gloves and human skin, to burst into flames.

Caustic Liquids

PTFE and PP are the most recommended materials for any kind or temperature of caustic expected to be used in any facility.

pH WASTE TREATMENT AND COMPONENT SELECTION

The two primary considerations when choosing and sizing an acid waste treatment system are the pH properties of the waste stream and the flow rate. Regardless of the pH of the waste stream, the two primary approaches are a continuous flow system or a larger batch system. In general, the following considerations should be used in selecting a treatment system.

Continuous Flow

1. The influent flow is relatively constant and not subject to large surges.
2. The influent chemistry is relatively constant and does not have large pH swings.
3. The effluent chemistry is not critical, such as that required for a multistage system, and pH must be kept within specific parameters.

Batch Flow

1. Influent wastewater stream is discharged in large batches or when there are large fluctuations in influent flow.
2. Influent chemistry is characterized by large swings in pH or contains concentrated acid or bases.
3. Effluent makeup is not critical, and pH must be kept within specific parameters.

The most often used treatment to neutralize smaller acid flow is direct, continuous contact with limestone chips of 2 to 3 in (50 to 75 mm) in an acid-neutralizing basin. For caustic, a continuous treatment in an automated neutralization basin is recommended with the system utilizing automatic chemical feed neutralizing.

For general laboratory waste, several methods for treatment using limestone chips are available. For single isolated sinks, a small acid-neutralizing sump should be considered. It is too heavy to be placed in a cabinet. A typical basin, illustrated in Fig. 10.2, is also available in a size range of small to large that will treat the effluent from a facility to a larger number of laboratory sinks. If an acid-neutralizing sump is to be selected based primarily on the number of fixtures and the dimension of the basin, refer to Table 10.2. If the selection is to be made on the basis of the tank size, refer to Table 10.3. Table 10.3 also includes the weight of limestone required for neutralizing acid in basins. Figure 10.3, which is very conservative, gives the recommended contact time for neutralization based on information from one manufacturer.

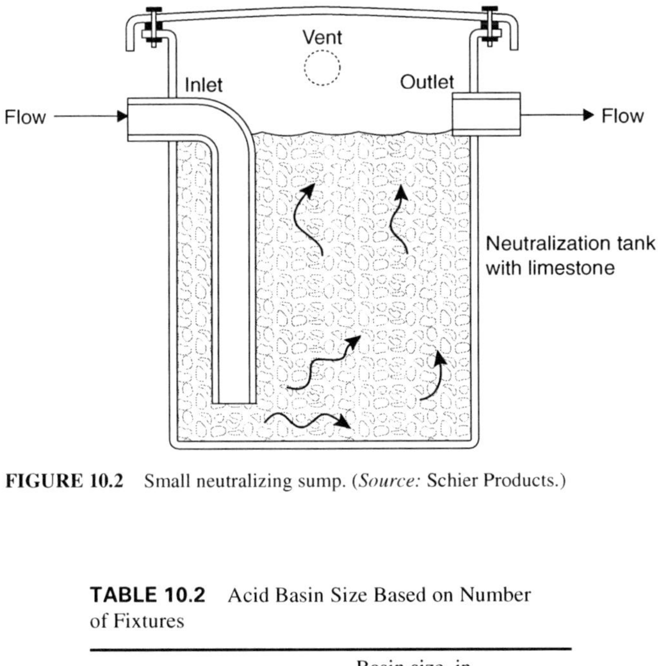

FIGURE 10.2 Small neutralizing sump. (*Source:* Schier Products.)

TABLE 10.2 Acid Basin Size Based on Number of Fixtures

Number of fixtures	Basin size, in	
	Diameter	Height
1	12	12
2	18	12
3–6	18	24
7–20	24	36
21–50	30	57
51–100	36	70
51–100	42	52

Source: Alfred Steele, PE, *Advance Plumbing Technology*.

TABLE 10.3 Acid Waste–Neutralizing Tank Sizing

Number of tab sinks	Tank size gal	Tank size L	Weight of Limestone lb	Weight of Limestone kg
2	5	20	50	25
4	15	60	150	75
8	30	120	200	100
16	55	210	500	250
22	75	285	750	140
27	90	340	900	400
30	108	410	1,000	450
40	150	570	1,500	740
50	175	665	1,750	870
60	200	800	2,500	1,250
75	275	1,045	3,000	1,400
110	360	1,370	4,000	1,910
150	500	2,000	6,000	3,000
175	550	2,100	6,500	3,200
200	650	2,470	8,000	3,600
300	1200	4,800	12,000	4,900
500	2000	7,600	15,000	5,500
600	3000	11,400	20,000	9,800

Note: For commercial and industrial laboratories, the number of lab sinks should be multiplied by a 0.5 use factor.

If there is expected discharge of oil or grease in the laboratory waste stream, the installation of an acid-resistant grease interceptor is recommended before the acid. Some objectionable contaminants can coat individual chips and prevent proper chemical action from taking place to neutralize the acid.

In cases where a facility uses caustic to kill microbial wastes, it is common to provide a single neutralizing tank, with necessary probes and a direct feed of neutralizing solution in the effluent. The use of probes throughout the treatment system is important to monitor the waste stream, to act as an alarm in the event of failure, and to indicate when the neutralizing media must be replaced.

A 30-min contact time between the acid and chips is generally accepted time for neutralization. Average figures have determined that 100 lb of limestone chips treats 97 lb of sulfuric acid and 75 lb of hydrochloric acid. Effluent consisting mostly of sulfuric acid should be treated with dolomite limestone chips.

An acid-neutralizing basin operates on the principle of a chemical reaction between the acid and the limestone chips. Each basin shall be designed by the manufacturer to allow sufficient contact time for the chemical reaction to accomplish neutralization based on the anticipated maximum flow rate. As the acidic wastewater reacts with the limestone chips, it forms water, carbon dioxide, and calcium salts. These salts precipitate into a sludge that sinks to the bottom of the basin and must be cleaned out periodically, ranging from 6 months to 2 years depending on the rate of flow into the tank. This precipitate coats the limestone and negatively affects its ability to neutralize the acid. All limestone that remains must be eventually removed and replaced.

Where a larger number of fixtures or equipment and treatment by limestone chips alone are not practical, a batch neutralizing system consisting of single or multiple basins and/or a mixing tank should be installed. If it is located at a low level, a pump will be required to discharge up to the level of the sewer. A sophisticated arrangement of probes, chemical feed

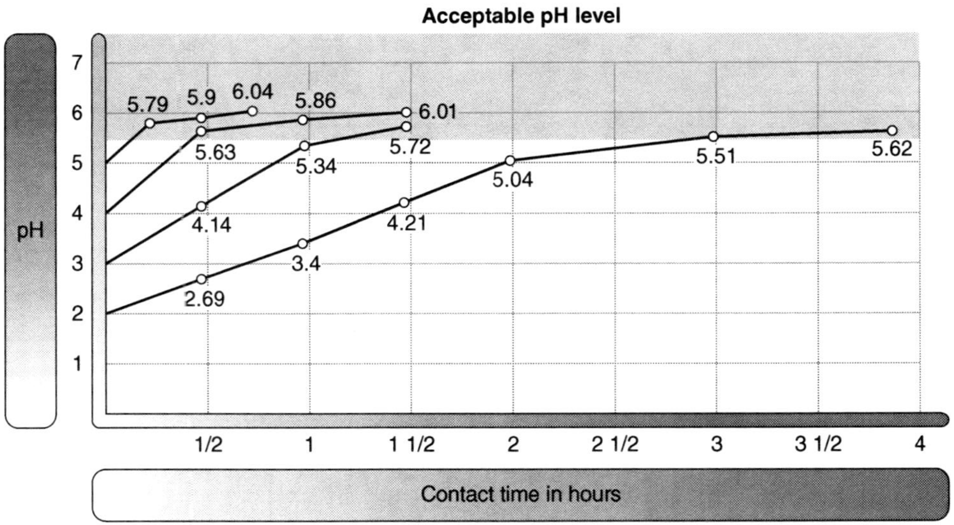

Schier products

Note: Data based on using clean limestone 2–3 inches in diameter with minimum 90 percent calcium carbonate content and hydrochloric acid diluted with an alkalinity of 120 parts per million and total dissolved solids of 170 parts per million.

FIGURE 10.3 Contact time for limestone-filled neutralizer.

pumps, level indicators, and alarms will be required. An agitator or mixer may be installed in the basin to mix the acid with the caustic. The addition of a recorder may be desired. The acid-neutralizing system operates on the principle of automatically adding proper amounts of caustic to the incoming acid waste, thereby neutralizing the acid. The probe is connected to an automatic caustic feed pump that introduces the proper amount of neutralizing liquid into the basin or mixing tank. The most commonly used neutralizing chemical is caustic soda. Continuous treatment may also require additional downstream sensing probes and chemical additive locations to ensure that the discharge is within acceptable limits. For an illustration of a large-scale batch neutralizing system employing two basins that could be used for acid and/or caustic, refer to Fig. 10.4 which shows a typical continuous waste treatment system. A neutralizing system using only one basin for caustic is shown in Fig. 10.5. Manufacturers have various proven methods of acid treatment. A typical large basin installation is shown in Fig. 10.6 and Fig. 13.2.

A batch neutralization is also used when both acids and bases are present in the waste stream. When strong acids and bases react, neutralization is slow when the combination is not close to being made neutral. However, when the reverse is true and the neutralization is almost complete, the reaction is rapid. When this occurs, small amounts of reagent go a long way when the mixture comes close to neutral. It is important that the correct process controller be used; otherwise, it will lead to continuous loops of overdosing, which means wasting money because of wasted reagent.

It is good engineering practice to have the discharge from the neutralizer separately routed into the sanitary house drain outside of a building for dilution with facility waste prior to ultimate discharge into the public sewer. This may also be necessary for monitoring of the waste stream by local authorities without having to enter a building.

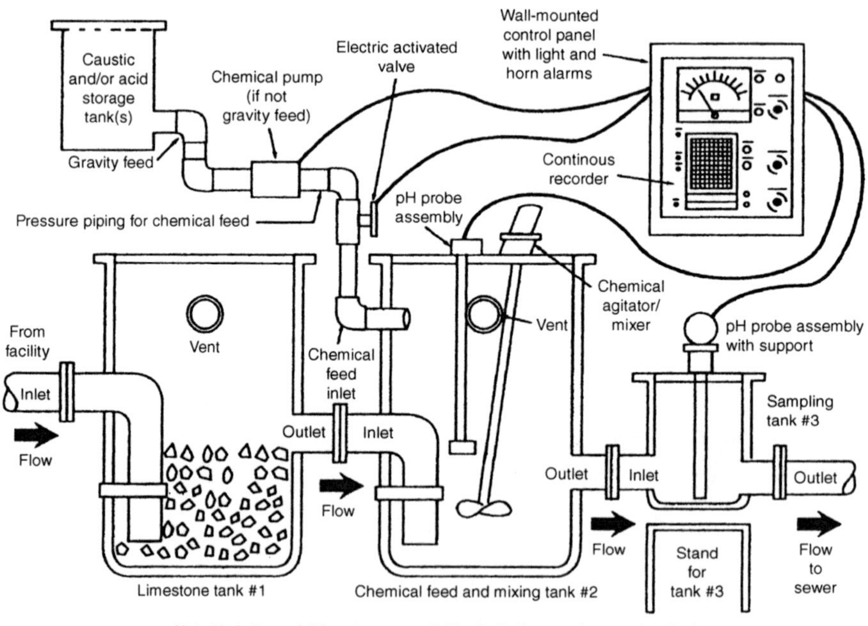

Note: Variations of this setup are available, including one large tank with three compartments instead of three separate tanks.

FIGURE 10.4 Typical bulk neutralization system, acid or caustic drainage.

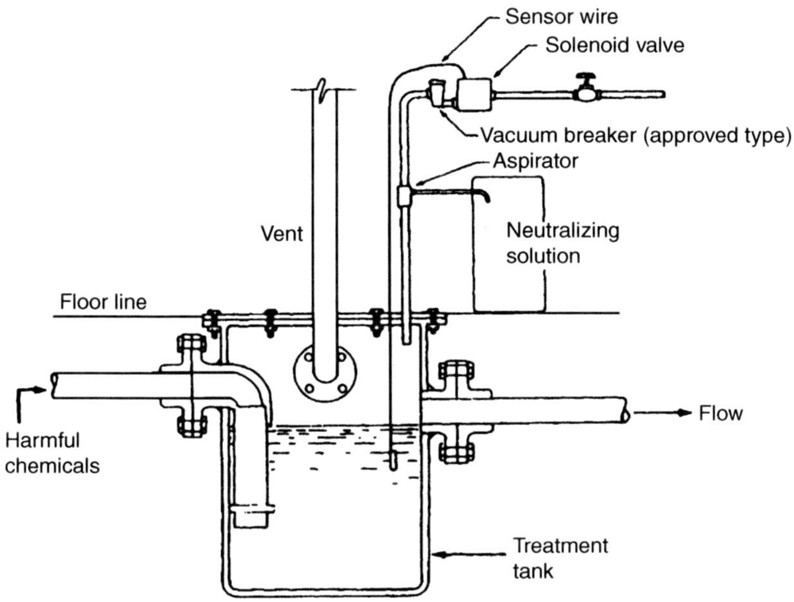

FIGURE 10.5 Small-scale caustic-only neutralization basin.

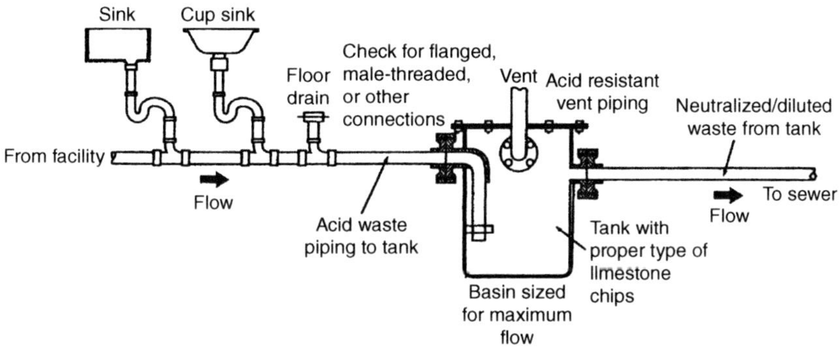

FIGURE 10.6 Larger acid-neutralizing basin arrangement.

For preliminary determination of a neutralization basin when the project first starts, calculate the number of sinks that will be required for average laboratories. With no fixtures shown, the method is to allow one laboratory sink for each 200 ft^2 of laboratory area. Each sink will discharge 0.5 gpm. Cup sinks will discharge 0.25 gpm. For a maximum flow rate, assume that 5 percent of the cup sinks could discharge simultaneously. Add the two types of sinks together for the maximum discharge and tank selection. Very often cup sinks are ignored.

Solids Interception

Solids handling in any chemical drain should be considered for every project. In teaching laboratories, items such as broken glass, latex gloves, etc. can go down the drain. Any number of hazardous solids could also be present. A typical solids interceptor of acid resistant material is illustrated in Fig. 10.7.

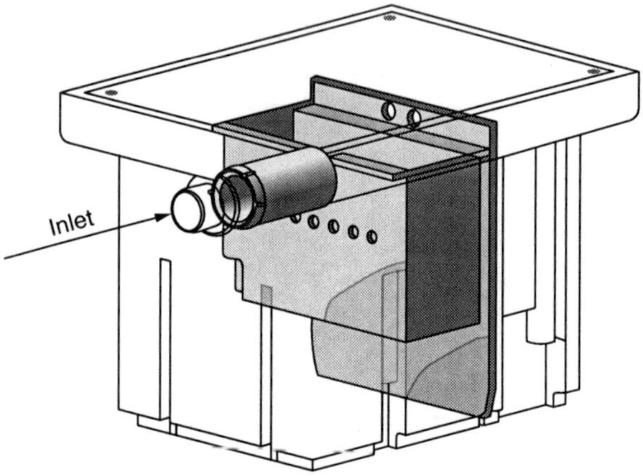

FIGURE 10.7 Laboratory solids interceptor.

SYSTEM DESIGN CONSIDERATIONS

The laboratory and facility drainage system shall have the same general piping system design considerations as sanitary drainage, including placement of cleanouts. Each fixture shall be individually trapped and vented. Clean water, such as that discharged from air compressors and other condensate drains, could also spill into the acid drainage system when convenient to dilute the waste. Because of possible stoppages that could back up and flood all the pipe, the adjacent facility waste system to the laboratory shall be of the same acid-resistant piping material.

Where the only waste discharge is from laboratory fixtures, the use of fixture unit schedules for pipe sizing is acceptable, except that simultaneous use should be factored into the sizing process. The pipe shall be sized using the required pitch and a three-fourths full pipe from standard engineering work. Cup sinks, except where the client suggests otherwise, should be ignored but provided with a pipe size of 1½ in (DN 40). The acid drainage and vent system shall be separate from all other systems until the waste is adequately treated and possibly diluted on the site with other waste lines. If a manhole is required in the acid waste line, it should be constructed of acid-resistant material.

BIBLIOGRAPHY

Frankel, M. *Facility Piping Systems Handbook*, 3d ed., McGraw-Hill, New York, 2010.

Ismart, L., "Laboratory Waste Design."

CHAPTER 11
INFECTIOUS AND BIOLOGICAL WASTE DRAINAGE SYSTEMS

GENERAL

This chapter is intended for design professionals who work for small-scale biological, commercial, and institutional facilities that could be exposed to any infectious or biological material. It is limited to an understanding of the basics of liquid infectious and biological waste decontamination and relates to the nature of the final effluent that is intended to eventually discharge onto the site and to a public sewer. Typical waste, in addition to laboratory waste, includes liquid animal waste from blood, urine, feces, and other necropsy services; insectaries that contain pathogens; and biological waste generated by laboratories and research and production facilities. The waste from all the processes will eventually discharge into a public sewer after neutralization. The actual design of the laboratories and sterilization equipment that discharges to a sanitary sewer along with ordinary plumbing fixtures that are part of the facility is outside the scope of this handbook.

Waste generated in biosafety facilities has the same basic characteristics as that of other laboratory and production facilities, but with the addition of biohazardous material. Biohazardous material consists of live organisms suspended in the waste stream that have the potential to cause infection, sickness, and other very serious, life-threatening diseases, if not contained. This waste will be completely treated and be discharged by gravity or under pressure from many sources including

1. Process centrifuges
2. Laboratory hand-washing and process sinks
3. Containment area floor drains
4. Necropsy table drains
5. Autoclave drains
6. Contaminated condensate drains
7. Kill tanks and tissue digesters
8. Fermentation tanks and equipment

Containment is the method used to isolate and confine biohazardous material. The facility equipment and design shall conform to acceptable and appropriate containment practices based on the hazard potential. A containment category is used to describe an assembly of both primary and secondary preventive measures that provide personnel, environmental, and experimental protection. Primary barriers are specific pieces of equipment such as the biological safety cabinet (which is the biologist's equivalent of the chemist's fume hood) and glove boxes. Secondary containments are features of facility design surrounding and supporting the primary containment. These features are described and classified in many publications such as those of the National Institutes of Health in Bethesda, Maryland.

The classifications for biological containment in laboratories consist of four biosafety levels, BL1 through BL4. Publications describe the work practices, equipment, and BL selection criteria based on the activity of that particular laboratory. This technology for treating the waste changes rapidly.

CODES AND STANDARDS

Mandated guidelines and regulations include the following:

1. OSHA Bloodborne Pathogen regulations
2. NIH guidelines for the use of recombinant microorganisms
3. FDA cGMP regulations
4. CDC/NIH Guidelines for Biosafety in Microbiological and Biomedical Laboratories

If the process wastewater is already identified as hazardous waste, the plumbing codes require the hazardous wastewater to be piped completely separate from the domestic wastewater. The two wastewaters cannot be combined before the pretreatment of the hazardous wastewater.

BIOLOGICAL SAFETY LEVELS

CDC/NIH Guidelines for Biosafety in Microbiological and Biomedical Laboratories in the United States are summarized below in an abbreviated form for the following laboratory containment levels. Other countries may be different.

Biosafety Level 1 (BL1) Containment

This is the typical biological research facility classification for work with low-hazard agents. Viable microorganisms not known to cause disease in healthy adults are used at this level. Work activity is done on an open bench, and any hazard present can be controlled by using standard laboratory practice. Standard features consist of easily cleaned, impervious bench surfaces and hand-washing sinks, and lab work is separated from general offices, animal rooms, and production areas. Contaminated liquid and solid waste shall be treated with a suitable disinfectant to remove biological hazards before disposal into laboratory sinks. Wastes containing DNA materials or potentially infectious microorganisms shall be decontaminated before disposal. Facilities to wash hands are required in each laboratory. The wastes shall be properly treated and shall be pH neutral when discharged. Emergency eyewash and showers are required to be installed outside the room in an easily convenient location.

Biosafety Level 2 (BL2) Containment

This facility designation is similar to BL1, except that the microorganisms may pose some risk and safety cabinets are often present. Equipment and work surfaces shall be wiped down with a suitable disinfectant. Sinks shall be scrubbed daily with a chlorine-containing abrasive and flushed to the sanitary system with a suitable disinfectant. All liquid waste generated shall be immediately decontaminated by mixing with a suitable disinfectant. Nearly all laboratories operate under level 1 or level 2 containment. At these levels, the facility is engaged in research, diagnostic, or production activities thought to pose little or no risk to workers. The wastes shall be properly treated and shall be pH neutral when discharged.

Biosafety Level 3 (BL3) Containment

Level 3 activity involves organisms that pose a significant risk or represent a potentially serious threat to health and safety. Biosafety cabinets are required, and all penetrations to outside the facility must be sealed to prevent leakage. These seals must be capable of being cleaned. Liquid waste is kept within the laboratory or facility and is steamed or chemically sterilized prior to discharge or disposal. Vacuum inlets must be protected by appropriate filters and/or disinfectant traps. Laboratory animals require special housing. Personnel must be appropriately protected with full suits and respirators. A hand-washing sink shall be located adjacent to the facility exit. All wastes hall be routed to a biowaste sterilization system for decontamination. Vents from plumbing fixtures must be filtered. All personnel shall shower with protective equipment on before leaving the area, and the discharge shall be routed to the decontamination system.

Biosafety Level 4 (BL4) Containment

This is a rarely used classification that applies to those facilities in which the activities require a very high level of containment. The organisms present life-threatening potential and may initiate a serious disease of epidemic proportion. All the BL3 requirements apply. In addition, showers shall be provided for personnel at the air lock where clothes are changed upon entry and exit. A biowaste treatment system shall be provided within the facility to sterilize liquid waste. This waste shall be pH neutral, properly decontaminated, and allowed to discharge into a public sewer.

LIQUID WASTE DECONTAMINATION SYSTEM

A liquid waste decontamination system (LWDS) collects and sterilizes (decontaminates) liquid waste. Effluent containing potentially hazardous biomatter is collected in a dedicated drainage system, generally discharging by gravity into a storage tank below the floor level within the facility. From the tank, effluent is pumped into a "kill tank" where the actual sterilization (decontamination) occurs. A kill tank is a pressure vessel into which steam or chemical disinfectant can be injected to kill, neutralize, and decontaminate any organism. The kill tank system should be qualified to the same biosafety level as the facility from which it receives the discharge. The kill tank system must be a batch process since time is needed to complete the sterilization and decontamination. The sterilization is based on the contact time of the process used. It is set up in such a manner that effluent is routed to one tank while the decontamination occurs in the other tank.

System Components

In addition to piping, the system consists of the sump or tank to receive contaminated discharge from the drains and equipment of the facility, a pump to remove the contaminated effluent from the collection pit and up into the kill tank(s), and the kill tanks that will decontaminate and sterilize the effluent to a point permitting disposal into the same system as the sanitary waste from the facility. This is generally into a public sanitary sewer that has the proper instrumentation to warn of incomplete decontamination in the branch serving the decontamination system.

Sump Pit. The sump pit into which the effluent drains shall have a gasketed, waterproof cover. The controls are similar to those provided on a plumbing sump pump and shall be capable of being sterilized chemically or steam-sterilized. The pit is sized in conjunction with the sizing of the pump so that the pump stays on for a minimum of 1 min to avoid too frequent starting. Other factors, such as having the pit contain one batch of product if necessary, may be considered.

Kill Tank Assembly. The kill tank consists of a duplex tank arrangement, allowing one batch to be decontaminated while the other tank is filling. The size of the tanks will vary based on the individual facility, but common practice is to have each tank capable of containing 1 day's effluent plus the chemicals used for decontamination. Another consideration is to have sufficient tank size to hold a catastrophic spill of a specific unit. There is usually an agitator to mix the effluent with the deactivation chemicals. In addition to the kill tanks, tanks containing disinfectant chemicals to be injected are required. A fully automatic control system must be provided to ensure the timely addition of the required chemicals in the correct amounts and for the required duration of deactivation of the biomatter. Alarms and status should be displayed in an appropriate panel located in a facility control room or other easily visible area.

Tissue Digester. The tissue digester is essentially a giant pressure cooker, which is intended to digest up to 10,000 lb of waste or carcasses.

It uses an alkaline hydrolysis process at an elevated temperature of 300°F (149°C). It converts the carcass bones into a solid and the remainder into a sterile aqueous solution. The liquid, although at an elevated pH, can be discharged into a sanitary sewer with pH adjustment. No smoke or fumes are given off because the digester is sealed airtight.

Drainage System and Components

The biowaste drainage system must be closed and airtight, which requires sealed floor drains and valved connections to equipment when not in use. Since the HVAC system maintains a negative pressure, it is important that the traps on all floor drains have a seal 2½ in (65 mm) deeper than the negative difference in air pressure. The traps of floor drains are filled with a disinfectant solution when not needed, to eliminate the possibility of spreading organisms between different areas served by the same connected sections of the piping system. Very often the floor drains are sealed. An adequately designed and protected vent shall be provided to allow air into the drainage piping to make drainage possible.

The choice of drainage piping material is based on the suspected composition of effluent chemicals and the sterilization method. If the local authorities determine that the biowaste is hazardous, a double-contained piping system with leak detection may be required. Stainless steel or PTFE pipe is usually chosen where higher-temperature effluent may be discharged or steam sterilization may be required. PVC, CPVC, PP, or lined FRP pipe could be used where effluent temperatures are lower and also where chemicals will provide the method of sterilization.

If waste from pressurized equipment is discharged into a gravity system, the system must be adequately sized to carry away the proposed flow rate with pipe flowing half full and adequate vents provided to equalize the internal pressure to ensure that the pipe is always at atmospheric pressure.

Valves shall be of the diaphragm type, capable of being sterilized using the same method as the pipe. After appropriate decontamination, the kill tank effluent shall be discharged to drain. This effluent must be treated prior to discharge into a public sewer system for disposal.

The disinfection shower and dunk tanks permit personnel and objects to exit the containment area and ensure that nothing is on the surface of the protective suits that could cause any contamination to escape out of the protected environment. In addition, dunk tanks, filled with disinfectant, are provided for objects to pass into and out of the suite. A breathing air system is provided. Large objects are required to pass through a sterilizer. The shower is usually a spray system with disinfectant added to the water spray that would ensure contact with all surfaces of the protective suit. A manually operated deluge shower can be activated if the spray system fails. All the waste from the shower is routed to the kill tank.

If the process wastewater is already identified as hazardous waste, the plumbing codes require the hazardous wastewater to be piped completely separate from the domestic wastewater. The two wastewaters cannot be combined before the pretreatment of the hazardous wastewater.

SYSTEM DESIGN

Vents

Vents from pipe, fixtures, sealed sump pits, and kill tanks must be filter-sterilized prior to leaving the system by using a HEPA or a 0.2-m filter. In the event of an accident, OSHA has rules to aid personnel responding to emergencies involving any hazardous material. The treated discharge from any containment treatment shall be separately routed to the sanitary system outside the building to allow for monitoring and sampling.

BIBLIOGRAPHY

Georgehegan, R. F., and H. W. Meslar, "Containment Control in Biotechnology Environments," *Pharmaceutical Magazine.*

U.S. Environmental Protection Agency, Code of Federal Regulations.

Yrjanainen, K., and P. Halamar, "High Containment Plumbing Design," *Plumbing Systems & Design.*

CHAPTER 12
NUCLEAR WASTE

GENERAL

This chapter is intended to acquaint the design professional with the nature of radiation and the thought process that will enable the responsible disposal to the site and discharge into a public sewer of very low amounts of nuclear waste.

Hospitals and laboratories are generally considered "institutional" facilities characterized by low quantities and levels of radioactive waste and are therefore subject to a less stringent set of regulatory requirements. These wastes are intended to be discharged into public sewers where they will not provide an unacceptable source of radiation. Those types of facilities having higher quantities of radioactive material and levels of radiation fall under a different, much stricter set of regulatory requirements. The principles of drainage system design are applicable to all kinds of projects, some of which may have significantly higher levels of radiation than most. The design philosophy is the same, but the submission of documentation for the protection of the public and workers in the event of any accident is considerably more complex. Because of the amount of radioactive material present at industrial facilities, larger storage and treatment systems are provided, as well as severe safety requirements that are not necessary at a site where lesser quantities of radioactive material are present.

With the exception of providing radiation shielding if necessary, the requirements for a radioisotope laboratory are essentially no different from the requirements of other laboratories handling toxic chemicals or pathogens. The ultimate aim is to keep the exposure of workers, staff, and the general public to nothing. Since this is not realistic, it is required not only to prevent overexposure but also to keep any exposure to radiation as low as reasonably achievable. The design shall implement criteria that will eliminate or reduce to allowable levels the radiation exposure of workers and maintenance personnel and prevent exposure of the general public to unacceptable amounts of radiation by waterborne radioactive waste (radwaste).

CODES AND STANDARDS

1. Nuclear regulatory commission
2. Code of federal regulations

THE NATURE OF RADIATION

Radioactivity is the spontaneous emission of "harmful" particles from the unstable nucleus of an atom. *Nuclear radiation* is the propagation of these harmful emissions through space. In an effort to become stable, neutrons are spun out of the orbit of the atom, where they collide with the nucleus of another atom, causing a *fission,* or splitting. This splitting forms a new element and at the same time releases heat, particles (including light), and new neutrons to fly out of orbit and split other atoms. There are many intermediate steps in the stabilization cycle that include the formation of other less complex radioactive by-products called *isotopes.* These by-products in turn decay to form other unstable isotopes as the cycle continues. The end result is an element that is highly stable. As an example, the end product of uranium is lead. One of the intermediate by-products of uranium is radon gas.

One of the most misunderstood concepts of radioactive materials is the potential for explosion. To clarify this, consider gunpowder. If the grains are spaced widely apart, one grain might ignite and possibly cause the ignition of the grain next to it. There could be not enough energy generated to remotely cause an explosion. But when the same amount of gunpowder is placed in a confined case and ignited, it burns so quickly so as to virtually explode. In the same way, unless the atoms of a highly unstable (radioactive) element are closely packed together, or enriched, there is no possibility of an explosion. For example, the fuel in the cores of many nuclear power reactors is enriched to only about 3 percent, which will allow only a self-sustaining reaction. The cores have to be packed to 95 percent to explode.

Radiation is a general term. The degree and nature of such ionized radiation depend on the energy of the individual particles, not on their number. It encompasses any of or all the following: alpha rays, beta rays, gamma rays, neutrons, x rays, and other atomic particles. There are three general classifications of radiation that are concerns to people, namely, alpha, beta, and gamma. *Alpha radiation* is actually a helium atom with a high velocity. *Beta radiation* is an electron with a high velocity. *Gamma radiation* is a particle similar to a photon, which is light. Alpha and beta radiation can generally be stopped by the skin or clothing, paper, or other similar light material. Alpha loses energy very quickly in air and is of no practical concern for distances greater than 12 in. High-energy beta radiation is commonly contained by only 1 in of solid, dense plastic. Beta is denser, carries more energy through greater distances than alpha, and will burn bare skin and, in particular, damage the eye, but will generally not penetrate the body to cause any internal damage. The greatest danger with beta radiation is to the skin and eyes, particularly when the eye is directly exposed close to the source.

Gamma radiation is electromagnetic in nature. It carries the most energy and therefore is the most dangerous to humans. Its wavelength is shorter than light waves. When generated, it is similar to x rays and behaves in a manner similar to light waves. When released from a source, gamma rays have a mass and velocity that have a measurable energy potential.

The best way to visualize the manner in which radiation harms cells in the body is to think of all the particles as bullets. These particles have velocity and energy. When stopped by the skin, these particles actually bounce off the skin, and the friction produces burns in the same way that sandblasting causes irritation. Sensitive tissue can also be killed by this friction. Gamma rays pass through the body and will change or damage any cell they touch on the way through. All three of the above radiation types are considered *ionizing radiation,* which means that the radioactive particles emitted from the source produce ionization. *Ionization* is the conversion of neutral particles to charged particles. Since cells are made mostly of water, the most common changes produce radicals that affect the bonds of cell molecules, causing breaks in DNA molecules and abnormalities in cells. Many cells are destroyed outright. Most affected cells die and are naturally replaced by the body. Some cells, particularly in the reproductive system, brain, and eyes, cannot be reproduced by the body and are gone forever. Other cells become mutant and reproduce so quickly that they multiply out of control. That is the start of a cancer.

Radiation Measurement

Radioactivity is a general term used for the total release of radiation of all types from a source. It is measured in *disintegrations per second* (dps). This measurement is possible for gamma radiation because in most radioactive materials, disintegrations also produce a known amount of gamma radiation. However, the best manner of measuring gamma radiation is from the energy it produces per kilogram of air. Because the instruments needed to measure radiation in this way are very expensive, they are not widely used outside of the laboratory. The so-called Geiger-Müeller counter is the most common. This instrument measures penetration of the particles that enter into a tube where the particles react with a gas in the tube, creating an electric charge that can be measured. If an amplification device is used, the charge can be heard in the form of static. The more modern instruments have a digital readout.

Units of Radiation Dose

Particulate radiation is measured by the number of disintegrations per unit of time. One curie (Ci) is equal to 3.7×10^{10} dps. One millicurie is equal to 0.001 Ci, or 3.7×10^{7} dps. One *rad* (*r*adiation *a*bsorbed *d*ose) is defined as the dose corresponding to the absorption of 100 ergs per gram (erg/g) of tissue. One roentgen (R) measures ions carrying a total of 2.58×10^{4} coulombs (C) of electric energy.

The term *radiation* is intended to mean ionizing radiation, which consists of particles that are energetic enough to detach electrons from atoms or molecules, thus ionizing them. Since the term *radiation* is a general one, a more specific term must be used to indicate the effect of radiation on humans. That measurement is called a *dose.* A *dose* is defined as the total quantity of radiation absorbed by the body or any portion of the body. Much of the time, the dose will be modified by reference to a unit of time. This differs from radioactivity because not all radiation is absorbed by the body. A *rad* is a measure of the dose to body tissue in terms of energy absorbed per unit mass. Gamma radiation is the most common of this type.

The most important measurement is the *radiation equivalent to man,* or rem. A rem is the measure of ionizing radiation passing through or absorbed by the body in terms of the biological effect relative to a dose of 1 R of x rays. The relation of the rem to other dose units depends on the actual biological effect to the particular part of the body being studied and the actual conditions and amount of time of the irradiation. One rem is the equivalent of 1 R due to x or gamma radiation, and also 1 rad due to x, gamma, or beta radiation. One rem of high-flux neutrons is roughly equivalent to the 14 million neutrons per square centimeter incident to the body.

SI Measurement Units

The becquerel (Bq) is the unit of radiation. One Bq is one disintegration per second.

The coulomb per kilogram (C/kg) is the basic unit of ionizing radiation.

One curie is 37 billion Bq.

One Sievert (Sv) equals 100 rem.

The roentgen is the older traditional unit that is obsolete and almost out of use.

Allowable Radiation Levels

There is no exact radiation level that is certain to cause permanent harm to the human body. On the other hand, many scientists believe there is no level below which it is certain that harm will *not* occur. A certain constant amount of naturally occurring radiation, called the *background level of radiation,* exists all over the world. The most common source is the sun, and its radiation is called *cosmic radiation.* In addition, there are many other sources of radiation such as fly ash from burning organic fuels (particularly coal), granite, and other natural substances that contain trace isotopes of elements. One of the most common of these trace elements is carbon 14, used by scientists to date many materials.

The Nuclear Regulatory Commission (NRC) is the government body that has the responsibility for establishing criteria for the field of radioactivity. These criteria appear in the federal government's Code of Federal Regulations. As an example, Title 10, Part 50 is commonly referred to as "10 CFR 50." The levels that have been established by the NRC are as follows (measured in rems for one calendar quarter for the exposure):

Individual in a restricted area:	
Whole body (head, trunk, gonads, eye lens, marrow)	1¼ rem
Hands, forearms, feet, ankles	18¼ rem
Skin of whole body	7½ rem
Individual in an unrestricted area:	
Whole-body dose for 1 h	2 mrem
Whole-body dose for 7 days	100 mrem
Whole-body dose for 1 year	0.5 rem

All personnel working at any site that has a possibility of exposure to radiation are required to wear some type of exposure detection device that will allow accurate determination of their actual exposure. The photographic badge is the most common and is used where sensitivity is required. A pen-shaped device called a *dosimeter* is commonly used where there is less need for accuracy. It is used where instant determination of dose is necessary. This can be increased to 3 rem to the whole body, provided that no prior dose was received in the previous calendar quarter.

An *unrestricted area* is any area within a facility that is not specifically controlled for the purpose of protecting any individual from radiation or radioactive materials. A *restricted area* is access-controlled. Another term, *environs,* may also be used to describe areas adjacent to a restricted or high-radiation area. A *high-radiation area* is defined as any accessible area within a facility that is capable of allowing the body to receive 100 mrem of radiation in a 1-h period.

Shielding

The purpose of shielding is to reduce or eliminate radiation emanating from any source within the facility. The most effective material has the greatest density, and so lead has been universally used for this purpose. Another material that is commonly used is concrete. In terms of shielding thickness, 4 mm of lead is the equivalent of 12 in of concrete. If it is known that a building will need shielding, concrete is the most often chosen building material as it provides ample structural support as well as shielding from radiation. It is up to the radiation safety officer (RSO) to determine the type of shielding and its placement to lower radiation in specific areas. Radiation travels in a straight line; therefore, if a tank or a length of pipe has to be shielded, the proper manner is to form a labyrinth so that the shine from the tank cannot escape in a straight line.

The most common materials used for shielding purposes are concrete and sheet lead. Other materials that have proved effective are (1) lead-lined concrete blocks, (2) lead-lined lath for plaster, and (3) lead-lined panels and gypsum boards. The barriers set up to reduce radiation levels are primary barriers, which are the first line of defense, and secondary barriers, which are used to eliminate leakage radiation and scattered radiation.

Radioactive Materials

Radioactive materials are used for the following general types of activity:

1. Imaging sciences
2. Electric power generation
3. Medical treatments and diagnostic tests
4. Manufacturing
5. Research

Almost all the materials used are isotopes. An *isotope* is a form of an element with a different (or excess) number of neutrons in its nucleus. Because of this difference, the atom is unstable. These isotopes are identified by their atomic weight, which is the weight of the number of neutrons and protons in the nucleus.

There are a great number of isotopes in use today. Some of the more common are

Iodine 131 (8-day half-life)
Phosphorus 32
Technetium 99 (6-h half-life)
Calcium 45
Carbon 14
Strontium 90
Radium 226

Since any given amount of radioactive materials remains active for different periods of time, it is not possible to predict when any material will become completely stable. The most often used figure is the time required for a specific material to lose one-half of its radioactivity, which is called its *half-life*.

SYSTEM DESIGN CRITERIA

The Approval Process and Application Requirements

Where there is a very low amount of radioactive material, such as in cardiac tracing, and the material has a very short half-life, no license is necessary.

The use of a larger amount of radioactive material requires the licensing of the site for a specific purpose and amount of radioactive material. Application for this license is made to either the NRC or a particular state. Those states that have elected to adopt the NRC regulations and provide their own staff for the purpose of issuing and approving licenses are called *agreement states*. In some cases these states make additional regulations of their own. Those states that rely on the NRC to review and issue licenses are *nonagreement states*. The application in those states is made to the appropriate department of the NRC.

For a hospital type of facility, the granting of the license mandates the presence of some form of a committee assembled for the express purpose of ensuring compliance with the terms and conditions of the site license as well as reviewing all proposals for the actual use of the radioactive materials present at the site. This team may consist of a physician specialist in internal medicine, a physician specialist in nuclear medicine, and a radiation safety officer. A health physicist is also commonly included. For a research type of facility, the members of the team will include a member of the research team in lieu of a physician.

The administrative duties of the RSO include monitoring personnel exposure limits and controlling any release of radionuclides into the sewer system. In addition, the RSO works with engineers in the design phase of the facility to ensure that the piping runs and all other mechanical work will result in a low exposure to people within the facility. For the most part, this work is directed to ensuring that neither facility personnel nor nonstaff members are subjected to more than the maximum permissible radiation dose allowed under the applicable codes for any particular type of radioactive material present. The RSO is also responsible for the following:

1. Teaching facility staff about the potential dangers
2. Keeping the necessary records for the facility
3. Keeping an inventory of material and records disposal
4. Concentrating materials at the facility
5. Assisting the engineer in design of mechanical systems
6. Designating areas within the facility to be restricted

General Design Criteria

The prime consideration in the design of any facility is a concept controlling the exposure of personnel to radiation. This concept is called ALARA, an acronym for "as low as reasonably achievable." This requires that the design of the facility consider every reasonable method to limit the possible exposure of personnel inside the facility and to keep the presence of radioactivity in any unrestricted area to a figure as low as reasonably achievable. This must take into account the current state of technology, the economics of further improvements in relation to the benefits to public health and safety, and other aspects of socioeconomic considerations in relation to the utilization of radioactive material in the general public interest. The facility must also make a reasonable effort to eliminate residual radiation. One of the overriding concepts is the "worst-case" possibility, by which contingency plans are made for the worst possible combination of circumstances to determine the highest possible level and amount of time of radiation exposure. This concept should not be overdone, and a general rule is to have only one "accident" at a time. As an example, a serious spill and a fire would not be considered as occurring simultaneously.

Pipe Material Selection

The pipe selected for the radioactive drainage system depends on the type of radiation and level of radioactivity, which in turn depend on the amount and type of radioactive material at the facility. In general, an ideal radwaste piping should have the following properties:

1. It must be nonporous.
2. It must be easy to clean and decontaminate.
3. It should be acid-resistant.
4. It should be nonoxidizing.
5. The joints should not form a crud trap.
6. Joint materials must not be affected by radiation exposure.

It is possible in very high radiation areas to have a pipe affected by the radiation present. The oxides of the pipe can become radioactive, or the pipe itself can be weakened. Another factor is the weakening of elastomeric seals or gaskets because of high levels of radiation. For this reason, Teflon is never used where anything more than a very low level of radiation is present. Other materials should be investigated as to their suitability for use at the levels anticipated.

All the commonly used materials (cast iron, ductile iron, copper, steel, and glass) and the joints normally used to put the pipes together fall far short of the ideal. However, all are suitable for low-level waste and radioactive source materials found in conventional health institutions. Plastic piping is not acceptable for radwaste systems under any circumstances, because of the possibility that the plastic may be affected by the radiation. It is only when the radiation levels of the waste materials get into the high-radiation-level category that they fail one or more of these conditions. As a result, stainless steel with welded joints has emerged as the material of choice for all the industrial types of waste products. Type 316L is the most common.

Welded joints are the only type of joint that meets the criteria for not allowing a crud trap.

General Design Considerations

Human waste, even that contaminated with radioactivity, is exempt from all NRC regulations, requiring only compliance with local plumbing codes as far as disposal, sizing, and all other criteria applicable to standard drainage systems are concerned. Also many low-level isotopes are exempt from regulations regarding disposal into the public sewer because of the extremely low level of radioactivity.

Another requirement is that the liquid radwaste to be discharged shall be diluted with the ordinary sanitary system effluent from the rest of the facility before being discharged to the public sanitary sewer system. This usually requires that the radwaste piping be first kept separate from the rest of the facility's piping, then joined to the main system before leaving the building for discharge into a public sewer. There are no restrictions regarding the combining of any radwastes together that are permitted to be discharged separately. A method should be provided for the RSO to take a grab sample of the radwaste stream if desired, such as a valved outlet from both the radwaste line and the combined discharge. Keep the pitch of the piping as steep as possible, in order to empty the pipe quickly and to provide a scouring action that will keep the radioactive solids in suspension.

It is common practice in laboratories to have radiation confined to glove boxes or protected fume hoods. The small amount of liquid waste produced from this equipment would be stored in shielded containers below the equipment that is removed periodically. If storage of larger quantities of low-level radwaste is required, it is piped to a holding tank. A common holding time is 10 half-lives of the effluent. Usually, radwaste is stored for disposal on the site, outside of a building where easy transfer of the liquid is possible. The removal must be done by licensed waste disposal contractors who remove the waste from the holding tank into a special truck that transports the liquid waste to a designated site suitable for disposal of low-level waste. The solid wastes, such as gloves and wipes, are stored in special containers that are removed to the disposal area with the liquid radwaste.

Floor drains are normally not desired in laboratories. If there is a spill of radioactive material, it is wiped up by hand, using absorbent material, and the solid containing the spill is put in a special radwaste holding container within the lab. If a floor drain is called for in the design, it should be made of stainless steel, which is available from all the major manufacturers. For testing purposes and to close off a drain when it is not expected to be used, each drain should be supplied with a closure plug. If there are areas where there may be a spill, the floor must be pitched to a floor drain. A generally accepted value for pitch of the floor is 1 in 20 ft. The thickness of the slab must be closely coordinated because the slab is thinnest at the drain and made thicker at the ends of the area served to make up the pitch. It is not practical to cast the slab evenly and add a topping because there is a tendency to chip the topping and possibly have a radioactive spill get under the top coating. Since the longer the run to the drain, the deeper the slab, it is necessary to indicate the top of drain elevation at each drain. This also makes it easier for the shop fabricator to make up accurate pipe spools.

Drains also require special treatment in a high-radiation facility. Like the drains in a low-radiation facility, they should be manufactured of stainless steel. There will be different types of drains in different areas, and they may be installed at different elevations. Because of this and the probability that the piping will be made in spools, it is a good idea to number all the individual drains on the design drawings. A box next to each drain can be used to provide information regarding type, number, and elevation.

Since fittings are a natural crud trap, avoid running piping in, under, over, or adjacent to unrestricted areas in a facility. If this is not possible, place the line where additional shielding can be added either at the time of construction or after the start of actual use, when the RSO may determine by survey that additional shielding is necessary. Much of the time, the ability to take apart the joint and flush out any crud is an advantage. Any of the popular joints for no-hub or grooved pipe are acceptable, as well as glass pipe if used in a laboratory for chemical resistance.

Be generous with cleanouts. They may be needed to flush out the line to reduce spot high radiation rather than to rod it out.

Decontamination

Decontamination of both personnel and equipment may also be required. Often, valves, small lengths of piping, or instruments must be taken out of service for repair or replacement. Personnel decontamination areas must be provided in the event that some radioactive contamination may accidentally spill on somebody. Three methods are generally used to decontaminate equipment:

1. Ultrasonic decontamination
2. Electropolishing decontamination
3. Washing with soap, water, and/or brushes

The most commonly used decontamination method is washing equipment with a detergent solution and a stiff wire or bristle brush. This method can be used only if the part is lightly contaminated and there is a lack of close proximity that any person must have to the part being cleaned. If the cleaning is on a system that is unrestricted, the crud relieved can simply be washed down any drain. The RSO must determine that the part is not radioactive enough to cause exposure problems. If the contamination is more than the allowed minimum, the part is placed in a container. Next a high-pressure stream of water is used to remove the loose crud, and then a brush is used to finish the job. Finally the container is removed to a permanent storage facility for eventual disposal.

If the part is determined to have a significant amount of radiation and the crud cannot be removed by washing, ultrasonic cleaning is the next choice. Except for very large systems that are built in place, most of the ultrasonic cleaning systems are factory-packaged. The cleaning system consists of a tank that contains a detergent cleaning solution with a wetting agent, water heater, ultrasonic generators, filters, and a water circulating system to remove the soil from the detergent solution in the tank. The part to be cleaned is put in the tank, whereupon the filters remove the loosened crud suspended in the water. The filters are in a cartridge that is removed periodically and disposed of along with the other low-level radwaste.

If ultrasonic cleaning is not effective, the final step is to use an electropolishing system. This cleaning system is similar to electroplating, except that the part to be cleaned is the sacrificial anode, and a layer of metal is deposited on the cathode of the system. This system also uses a liquid bath, and the removed crud is filtered out and contained in a filter cartridge for proper disposal.

All these areas contain floor drains, discharging to either a public sewer if the radiation is from exempt material or, if a higher level, to a holding tank for disposal or to a radwaste treatment system.

Personnel decontamination (decon) consists of regular showers, with one caution: hot water cannot be used. Hot water opens the pores of the skin and may allow contaminated particles to enter the body. People go from a dirty area where they shed their protective clothing, next they get scanned, and then they go either to the clean area for a regular shower or to the decon area for a cold shower, during which they use a stiff brush with detergent for scrubbing the contamination off the skin. A rescan determines if it is safe for them to take a regular hot shower. Again, if the material used in the facility is exempt and the worst case determined by the RSO is acceptable, the effluent from the showers can all discharge into the public sewer system. If not, then it shall be routed to a holding tank or radwaste treatment equipment for disposal. It is highly unlikely that a decontamination facility will be necessary in any facility using only exempt products.

BIBLIOGRAPHY

U.S. Environmental Protection Agency, Code of Federal Regulations.

CHAPTER 13
INDUSTRIAL WASTE

GENERAL

This chapter describes and discusses fundamental information for the design engineer in identifying the pollutants and local treatment for waste produced by small commercial facilities that are ultimately intended for discharge into a municipal sewer system or storage until removed. Since these facilities produce a varied number of waste products, guidance is given only for treatment.

Local treatment shall reduce or eliminate pollutants to a level required by the AHJ and federal regulations before discharge is permitted. If such local treatment is beyond the scope of work that an engineering firm contract might entail, the waste shall be stored on site in order to be identified and removed by an appropriate vendor. If any waste is intended to be discharged directly into the municipal sewer system, approval from the AHJ is mandatory. Except for the neutralization of acid effluent, the treatment methods of volatile liquids and of fats, oils, and grease are outside the scope of this handbook.

Unless specifically noted otherwise, all the waste streams discharging into public sewers will be assumed to have the approximate flow characteristics of water. Pipe sizing criteria are based on this assumption.

Much of the time, engineering firms engaged in projects that may release any pollutant have specialists who are on staff or contracted to supervise all work of the firm and are familiar with the permitting process for waste streams.

System Approval Requirements

Industrial wastewater is a generic term used to describe nonsanitary (plumbing) effluent, such as that typically found in chemical, pharmaceutical, industrial, and other manufacturing facilities. It may also be applied to wastewater from commercial facilities, such as self-service laundries or large restaurants. The definition also includes storm water runoff containing anything considered harmful by the U.S. Environmental Protection Agency (EPA) discharged from sites originating from any industrial or construction activity other than sanitary and common stormwater sources.

Whether discharged, surface waters, pumped into deep wells, or from the environment (land and groundwater), industrial wastewater and some storm water runoff are subject to government permitting requirements. In most cases, these wastes must be treated for pollution abatement before discharge to municipal sewers. Permits specify the maximum allowable concentrations of pollutants in the discharge and the frequency and type of monitoring required to show compliance. Pollution abatement by dilution is no longer permitted. The segregation of incidental water streams, such as from process wastewater, is almost universally required. Even clean, incidental wastewater streams from industrial sources, such as from cooling water or storm water runoff, often require a permit.

Most nonaqueous liquids (including solvents, oils, and sludge) and some solids and gases are regulated as hazardous substances or hazardous wastes during their generation, use, collection, storage, transportation, treatment, and disposal. Some aqueous wastes that are not regulated under a wastewater permit are regulated as either hazardous substances or hazardous wastes.

As a result of these regulations, engineers must either consult with an experienced environmental engineer or become familiar with the various environmental requirements to ensure an acceptable installation. New facilities must meet both environmental and plumbing code requirements. For novel manufacturing processes, the designer, environmental engineer, and owner may be required to work with regulatory authorities during the design stage to ensure compliance with the intent of the various codes and regulations. The designer is responsible for producing an installation with a low probability of failure. For plumbing design, this means minimizing the possibility of leaks and providing a means to limit the impact of spills on the public's safety and the environment.

Where production and manufacturing facilities discharge waste, it is a general practice to engage the services of professionals experienced in wastewater treatment and environmental issues to ensure compliance with all the latest applicable regulations and maintenance of an acceptable treatment system.

Obstacles to be avoided, such as underground piping, trees, underground structures, and rock, should be located.

Definitions

Hazardous Substances. Under Section 311 of the Clean Water Act, the EPA has compiled a list of hazardous substances (40 CFR 116). If a substance on this list is spilled or discharged, it must be reported to the EPA and a permit is required. The list is too large to be reproduced here.

Priority Pollutants. The Natural Resources Defense Council and the EPA determine priority toxic pollutants. These pollutants have been incorporated into several regulatory programs, including the National Pollution Discharge Elimination System (NPDES) permits (40 CFR 261), pretreatment standards in the Code of Federal Regulations (40 CFR 403), hazardous wastes (40 CFR 261), and CERCLA (40 CFR 541).

Hazardous Wastes. The EPA has adopted regulations to control hazardous wastes under the Resource Conservation and Recovery Act (RCRA). These regulations (40 CFR 2651) list hazardous wastes, including specific chemicals and mixtures defined by their characteristics. Note that controls under RCRA apply to waste only and not to hazardous substances that are being stored prior to use in product manufacturing or that are to be reclaimed, recycled, or reused. RCRA regulates the generation, transportation, storage, treatment, and disposal of hazardous wastes.

Hazardous Materials. This term means substances or materials that have been determined by the Secretary of Transportation, under 49 CFR 172, to be capable of posing an unreasonable risk to health, safety, and property when transported in commerce. Chemicals included in this definition are hazardous substances, hazardous wastes, most of the priority pollutants, and many other chemicals in commerce that are too numerous to mention here. In this chapter, *hazardous materials* is a generic term used to describe all the previously defined materials and substances.

For detailed listings of these and other regulated chemicals and wastes, refer to the regulations cited in the above definitions.

CODES AND STANDARDS

The most important federal environmental legislation affecting the design of plumbing systems for hazardous material and waste facilities serving industrial plants are the Clean Water Act (CWA), the Resource Conservation and Recovery Act (RCRA), and the Comprehensive Environmental Response, Compensation, and Liability Act (CERCLA or Superfund). These regulations, promulgated by the EPA and enforced by either the EPA or its counterpart state agencies, provide a comprehensive framework of pollution control. The National Pollutant Discharge Elimination System (NPDES) permitting program has recently been revised to include the discharge of storm water from selected industrial sites. Current EPA regulations can be found in the Code of Federal Regulations (CFR), which is an annually updated compendium of all federal regulations. CWA regulations are contained in the sections beginning with 40 CFR, Part 100, and 40 CFR, Part 400; RCRA regulations begin at 40 CFR, Part 300. Any updates or regulation changes after the last publication date of the CFR can be found in the *Federal Register,* a daily government newspaper in which all agencies and departments publish their notices, proposals, and final regulations. Most states with counterpart regulations have similar codifications and newspapers.

There are two general local jurisdictional bodies that regulate different aspects of special waste systems. The first is the local and regional authorities who create and enforce plumbing and health codes. Included are the local authorities charged with review and approval of plumbing systems design. They are concerned primarily with regulating the sizing and design of the piping systems within buildings. Their regulations generally mention special system piping material, treatment, and system configuration. They do not mandate specific use of double-wall piping or leak detection. These requirements are mandated by federal, state, and local agencies. Recommendations for piping material usually require that any pipe material be approved by the local authority and be capable of resisting degradation and the effects of the nature and results of concentration of any special waste. System design is concerned only with adequacy of the pipe to carry away the design flow. System configuration usually requires that traps be provided on fixtures and floor drains, and that the venting system conform to good plumbing design practice. The effect that the very general regulations have in terms of conformance with local and model plumbing codes is that all elements of the system design are left to the experience of the design engineer and also must comply with the requirements of other regulating agencies. The design must be approved by the AHJ.

The second jurisdictional body, with far more stringent regulations, comprises the various agencies concerned with protecting workers, the public, and the environment from the discharge of toxic substances. Included are federal, state, and local authorities responsible for preventing discharge into the general environment, public sewers, and public treatment systems of any substances considered harmful. Discharges can occur either lawfully or unlawfully, or as a result of spills and accidents. To prevent this discharge, it is common practice for these agencies to inspect facilities, mandate treatment systems, and require the use of specific piping systems such as double-wall piping and leak detection to advise of such leakage. The effect of some very specific regulations in terms of conformance with agency requirements is to rely on the experience of the design engineer to provide pipe materials and systems capable of complying with regulations and receiving approval from those agencies. Other regulations, such as current Good Manufacturing Practice (cGMP) and validation, are constantly being revised because of technological and design changes.

The great body of codes and regulations can easily be a labyrinth, even for those persons with a good understanding of the overall outline and purpose of the statutes. Anyone with questions should not hesitate to seek advice from trade groups, regulatory officials, environmental engineers, lawyers, and other specialists, as needed. Numerous current event reporting services are available that digest federal and state regulatory actions and publish readable evaluations as well as reprints of important regulations, policies, and case notes.

One disturbing aspect of some model codes is that they allow dilution of some waste until a level is reached that is acceptable. This past, obsolete convention that the "solution to pollution is dilution," if written, is no longer acceptable.

The Clean Water Act

The Clean Water Act (CWA) establishes a mechanism for setting uniform national standards for discharge to surface waters and sewers. The EPA has established categorical effluent standards, usually prorated to production volume, for 28 industrial classifications under the Effluent Guidelines Program Plan. The affected industries are listed in Table 13.1. The EPA has also promulgated general standards for discharge to public sewer systems. These standards restrict pollutants that interfere with sewage treatment, pass through the system untreated, damage sewer lines or treatment facilities, or overload treatment processes.

TABLE 13.1 Categorical Standard Summary Status

1.	Aluminum forming
2.	Battery manufacturing
3.	Coal mining
4.	Coil coating
5.	Copper forming
6.	Electrical and electronic components, I
7.	Electrical and electronic components, II
8.	Electroplating
9.	Inorganic chemicals manufacturing, I
10.	Inorganic chemicals manufacturing, II
11.	Iron and steel
12.	Leather tanning and finishing
13.	Metal finishing
14.	Metal molding and casting
15.	Nonferrous metals forming
16.	Nonferrous metals manufacturing, I
17.	Nonferrous metals manufacturing, II
18.	Ore mining and dressing
19.	Organic chemical, plastic, and synthetic fibers
20.	Pesticide chemicals
21.	Petroleum refining
22.	Pharmaceuticals
23.	Plastics molding and forming
24.	Porcelain enameling
25.	Pulp, paper, and paperboard
26.	Steam electric
27.	Textile mills
28.	Timber products

Source: EPA

State laws and regulations follow the federal format, with a few important differences. All states are allowed to make their regulations more stringent than the federal standards. Additionally, some states regulate discharges to the land that might produce pollution of the groundwater. States also set goals for water quality levels in streams, lakes, and coastal waters. They then determine the allowable loading of each pollutant and allocate portions of that loading based on low-flow conditions where dilution is minimal. Water quality–based discharge permit limits are almost always more stringent than industrywide limits of the categorical effluent standards.

Two types of water discharge permits may require special treatment processes. Permits for a direct discharge into a surface water (stream, lake, ocean) are called NPDES permits. These permits may be issued by the EPA or a state or jointly, depending on the location. Application and approval are required well in advance of initiating a discharge.

The other type of industrial discharge permit is obtained from a Publicly Owned Treatment Works (POTW), the modern name for the local sewage and wastewater treatment plant. Industrial discharges to a POTW are called *indirect discharges* and are regulated by pretreatment ordinances to ensure that the POTW meets the conditions of its NPDES permit. The ordinances are usually administered by the POTW, except when the industrial discharge is large, when the POTW fails to meet its permit requirements, or when the discharge is from an industry regulated under the Effluent Guidelines Program Plan.

RCRA and CERCLA

These two laws, together with the federal and state regulations derived from them, have had a major impact on the industrial management of hazardous substances and hazardous wastes. Almost all nonaqueous liquids, many aqueous mixtures, and many solids and gases are regulated when they become wastes or are spilled.

The purpose of CERCLA is to limit the uncontrolled release or threat of release of hazardous substances into the environment and to provide for a coordinated and effective response to mitigate accidental releases. CERCLA requires industrial risk evaluations, in the form of contingency plans, and establishes a mechanism for government response to environmental and health hazards. CERCLA does not require any permits and does not force changes in current hazardous substance handling methods. Nevertheless, a great deal of publicity can result when the mishandling of hazardous substances leads to an environmental incident. There also can be an enormous cost to dispose of the hazardous wastes generated during a spill. These are strong inducements to chemical manufacturers and users to install process systems that minimize risk.

RCRA's purpose is similar to CERCLA's, except that RCRA regulates hazardous wastes from ongoing manufacturing activities. The goals of RCRA are accomplished through strict licensing and operational standards for every aspect of hazardous waste management.

RCRA requires identification numbers and/or permits for every hazardous waste activity. The permit application requirements are lengthy and technical for some of the regulated activities. In some cases, the regulations give performance standards that the designer must meet to obtain a permit for the facility. In other cases, the compliance method is almost completely specified. Obtaining RCRA facility may take 6 months and may require public hearings and disclosure of detailed process and waste information, including chemical material safety data (MSD).

When evaluating waste streams to determine if they are regulated as hazardous wastes, it is important to check both federal and state regulations. This is so because states are allowed to be more stringent than the EPA. States may define certain wastes as hazardous when the EPA does not consider them so. For example, waste oil is considered hazardous by many states, but not by the EPA.

The relationship between RCRA and CERCLA is shown by the following example. A tank containing a hazardous substance begins to leak onto the ground. Under CERCLA, the owner must notify the EPA and the state, stop the leak, and clean up the spilled material. If the owner fails to take action, the regulatory agency may act under CERCLA and then seek reimbursement up to triple the cost of its expenses. Under RCRA, the spilled hazardous substances and any contaminated soil become hazardous waste. The hazardous waste must be removed, stored, transported, and disposed of in accordance with RCRA requirements.

DESIGN CONSIDERATIONS

For engineers, the challenge is to design systems that minimize the chance of leaks, contain any spills that might occur, and segregate hazardous substances from both nonhazardous process streams and incompatible hazardous process streams. The design of a system that anticipates the potential for leaks and spills must include suitable materials, reliable jointing, good fabrication, and provision for the secondary containment of liquids in areas and systems that pose a high spill risk. The guidance of specialists in this field is highly recommended.

Many leaks occur as a result of material incompatibility between the equipment and either the hazardous substances handled or the atmosphere in which the equipment is utilized. The incompatibility can be physical, such as polyvinyl chloride (PVC) pipes melting at high temperatures or plastic pipes dissolving in solvents they were not designed to contain. Table 13.2 lists the general properties of the most common tank and pipe materials. Specific applications should be checked with the appropriate chemical compatibility references from the manufacturer.

Even the best-designed liquid handling systems are subject to occasional failure, particularly during liquid transfer operations. Secondary containment is an important aspect of any hazardous material system design to protect employees and the environment. Secondary containment may include a dike around a tank or tank farm or pipes within pipes for systems handling extremely hazardous liquids. Common secondary containment systems typically have the following features:

1. Containment floors, pads, ponds, and dikes constructed of materials impervious to the substance stored
2. Perimeter diking and storage reservoirs sized to contain 110 percent of the largest tank plus the maximum rainfall predicted to occur over 24 h once in 10 years in exterior areas (or 20 min of sprinkler water flow for interior areas)
3. Pumps, drain valves, or siphons to empty the secondary containment area to either a storage tank for disposal or treatment facility which are themselves secondarily contained
4. Controls and procedures to prevent the accidental release of contained spills and an alarm system to notify operations, should a spill occur

Minimal equipment and practices for preventing transfer spills should include overfill prevention, including level sensors; gauges and a high-level alarm; automatic valve and pump shutdown; established transfer procedures, including an operator on duty; proper curbing and containment; redundant valves and controls; vapor recovery; alarms; regular inspections; and a maintenance program. The equipment used in water treatment is operated to achieve the following: mixing and flocculation, sedimentation, clarification, filtering, removal of turbidity, removal of metals, and disinfection.

Historically, underground tanks up to 10,000-gal (37,854-L) capacity, and occasionally larger, have been preferred for hazardous materials, primarily to minimize fire risks.

TABLE 13.2 General Properties of Materials Used for Storage Tanks and Piping

Containment materials	Advantages	Disadvantages
Carbon steel	Compatible with petroleum products and dry organics and incompatible with many aqueous solutions	Subject to attack by corrosive soils and chemicals
Stainless steel	Better corrosion resistance than carbon steel and higher structural strength. There are more than 70 standard types of stainless steel and many special alloys	Corroded by chloride and exposure to reducing environments
Fiberglass-reinforced plastic (FRP)	Compatible with a wide range of petroleum and chemical products, if proper resin is selected	Lacks the structural strength and impact resistance of steel tanks
Polyvinyl chloride	Excellent chemical resistance to acids, alkalies, and gasoline	Lower structural strength than steels, generally not suited for the storage or handling of organic solvents such as benzene, carbon tetrachloride, and acetone, or use at temperatures above 140°F (60°C)
Concrete	Generally good resistance to alkaline chemicals, epoxy coatings often applied to concrete to increase chemical resistance	Subject to cracking and spalling with changes in temperature such as during freeze/thaw cycles; uncoated concrete absorbs solvents
Polypropylene	Resistance to all aqueous solutions except strong oxidizers	Low structural strength
Lined steel	Chemical resistance of plastic and structural strength of steel	Relatively high cost for material and installation

Unfortunately, many of these tanks were kept in service too long and, because of corrosion, have leaked and contaminated drinking water supplies. As a result, designers now must compare the environmental and fire risks between belowground and an aboveground tank. In either case, greater attention must be paid in the design to ensure an installation resistant to leaks and capable of containing spills. Soil conditions and groundwater levels, which affect the design, must be evaluated for each installation. Secondary containment of these tanks is the primary method of ensuring against leakage.

ELEMENTS OF AN INDUSTRIAL WASTEWATER HANDLING SYSTEM

pH Control

Neutralization for pH control involves the chemical reaction of an acid with an alkaline substance (a base), resulting in the formation of a salt and water. In an aqueous solution, the acid or base molecules dissociate and form ions. Sulfuric acid is presented in the solution as H^+ (hydrogen) and SO_4 (sulfate) ions and caustic soda as Na^+ (sodium) and OH^- (hydroxide) ions. The H^+ ions of the sulfuric acid and the OH ions of the caustic soda have a strong attraction for one another and combine to form H_2O (water). For example, in the neutralization reaction of H_2SO_4 (sulfuric acid) and NaOH (sodium hydroxide), the following process occurs:

$H_2^{+}SO_4 + 2(Na^{+}OH)$

$Na_2^{+}SO_4 + 2H_2O$

If excess hydrogen ions remain, the liquid will be acidic, and with a surplus of hydroxide ions, the liquid will be alkaline.

The acidity or alkalinity of a solution is expressed on the pH scale, with neutral water at a pH of 7, in the middle of the range between extremely acid (pH = 0) and extremely alkaline (pH = 14). The scale is logarithmic, so a pH of 3 is 10 times more acidic than a pH of 4. There are also buffers, such as bicarbonate/carbonate, which undergo a chemical change when strong acids or bases are added to a solution and thereby act as capacitors that must be filled before the pH will change.

A typical two-stage, continuous-flow pH waste neutralization process and oil separator for larger installations is shown in Fig. 13.1. Similar details are Figs. 10.6 and 13.2. For flows less than 10 gpm (37.85 L/min), it is sometimes preferable to neutralize in a batch basis, with two tanks alternating between collection and treatment. Small discharges with variations of pH, such as those from schools and laboratories, are best treated locally. Refer to Chap. 10 for further discussion.

The most critical feature of the pH adjustment system is the controller that activates the chemical feed pumps. The controller must have the ability to prevent overfeeding of either acid or base, which would cause wide pH swings and subsequent repetitive chemical additions. Overfeeding is most probable in wastewater with no buffering capacity near the pH set point (e.g., deionized and soft water). Controllers with multirate response adjustment should be specified for each application.

Sulfuric acid is the most costly method for pH adjustment purposes, although in some cases the commercial 93 percent acid must be diluted prior to use. The manufacturer's recommendations for materials of construction, control of the heat of dilution, and safety precautions should be carefully followed. Carbon steel pipe and tanks are commonly used for 93 percent sulfuric acid, with stainless alloy 20 (a high-nickel alloy) valves, as well as PVC and chlorinated polyvinyl chloride (CPVC) plastics. For sulfuric acid at concentrations below 93 percent, polypropylene (PP), fiberglass-reinforced plastic (FRP), PVC, CPVC, and lined steel are preferred.

Sodium hydroxide (caustic or caustic soda) in 50 percent solution is the most convenient commercial alkaline material for pH adjustment. Unfortunately, 50 percent caustic freezes at 54°F (12.2°C), so indoor storage or heated tanks are necessary. Note that 20 percent caustic freezes at 18°F (−27.7°C). Caustic solutions may be handled in carbon steel, stainless steel, and PVC tanks and pipe. Lime and hydrated lime are considerably less expensive than caustic but must be fed with dry feeders and/or slurry tanks, which require considerable maintenance. Consequently, lime is favored in applications where the cost outweighs the convenience of caustic.

If the only required method of treatment is pH adjustment prior to discharging into the building sanitary drainage or public sewer system, most authorities have requirements for acid drainage systems that must be followed. A pH value of 4.0 is the generally acceptable level for direct discharge into a public sewer system. Refer to Chap. 10 for further discussion of pH neutralization systems.

Removal of Dissolved and Suspended Metals

Another common industrial wastewater treatment requirement is the removal of dissolved and suspended metals. The most popular method for removing dissolved chemicals is to separate cyanide and chromium VI wastewater sources from one another and all other

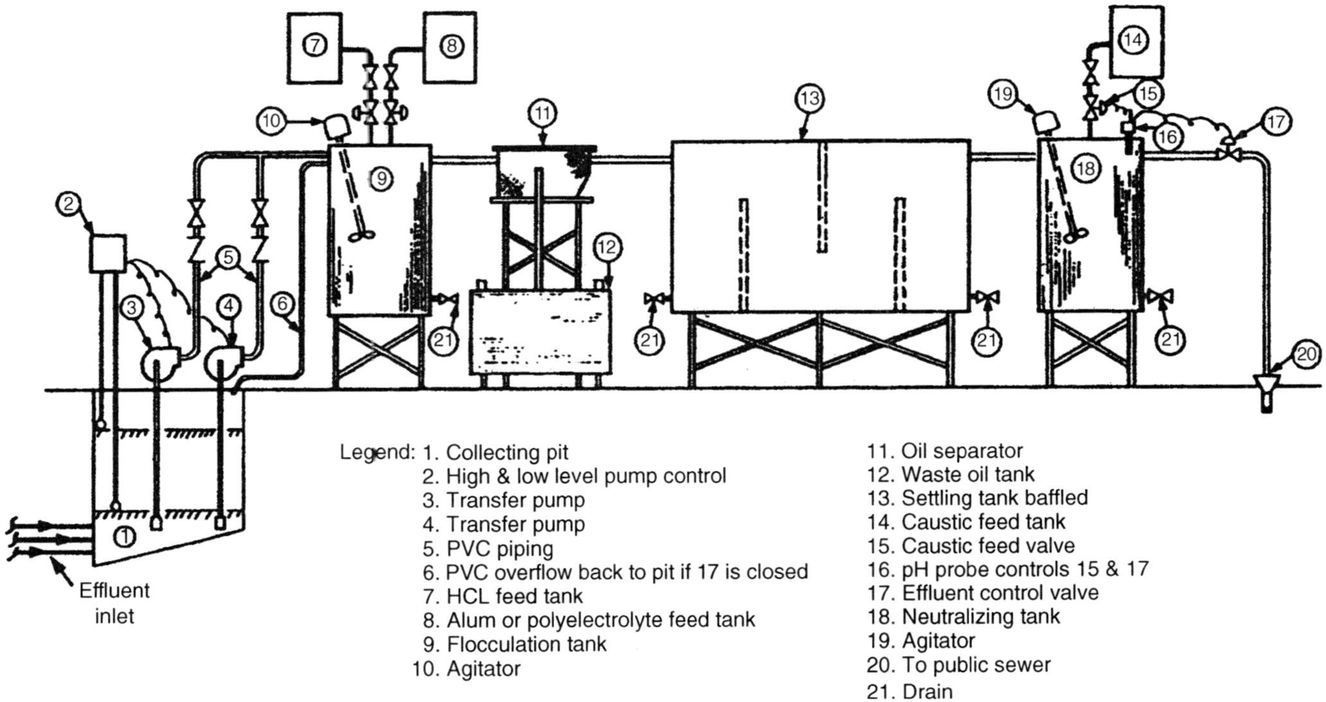

FIGURE 13.1 Typical large-scale waste neutralizer.

metal-bearing wastewaters. The cyanide is destroyed by oxidation with chlorine (or a sodium hypochlorite solution) at a pH of 9 to 11, and chromium VI is reduced to chromium III with sulfur dioxide or sodium bisulfate at a pH of 2. Various treatment methods are available.

For trace dissolved metals up to 1000 ppm, the use of ion-exchange vessels containing resin beads tailored to the application may be considered. Ion exchange is a reversible chemical reaction wherein an ion (an atom or a molecule that has lost or gained an electron and thus acquired an electric charge) from a wastewater solution is exchanged for a similarly charged ion attached to an immobile solid particle. These solid ion particles are either naturally occurring inorganic zeolites or synthetically produced organic resins. The synthetic organic resins are the predominant type used today because their characteristics can be tailored to specific applications. This process is similar to the process for purifying water for laboratory or process applications. The exact ion combinations are based upon the metals to be removed. The exchange vessels can be sized for the wastewater flow or multiple vessels manifolded together to handle higher wastewater flows or allow standby capacity. Automated controls and accessories can be furnished to backflush or regenerate them. The removed metals can then be reclaimed or safely disposed of. This technology should be evaluated against other technologies for overall operating cost based upon size and type of application.

Any suspended metals can be removed by filtration.

Oil–Water Separation

Oil–water separation may involve free-floating oils, mechanical emulsions or diversions, and chemical emulsions. All the oils are immiscible. The size of oil droplets in emulsions may range from below 20 microns (μ) to above 150 μ. In selecting separation equipment, the designer must consider the oil quantities, droplet size, presence of emulsifiers, temperature of the water and oil, difference in the specific gravity of the fluids, their viscosity, pH, and other wastewater constituents.

Separators are on the market to suit every situation, including simple settling tanks (e.g., American Petroleum Institute separators), tanks with coalescing media to promote the agglomeration of dispersed oils, and tanks with chemical pretreatment to promote the separation of chemically emulsified oils. Equipment options include sludge removal and automatic oil skimming. In special conditions, such as where oil is mixed with and adheres to suspended solids, dissolved-air flotation separators may be necessary. Figure 13.2 shows a unit designed to remove larger amounts of free and mechanically emulsified oil. For smaller amounts, refer to Chap. 9.

Biological Treatment

Wastewater intended to be discharged directly to surface or groundwater must be treated to remove organic compounds, which would deplete the dissolved oxygen in the receiving water. Typically, permits allow an average of 30 mg/L per liter of 5-day biological oxygen demand (BOD_5). Treatment is accomplished by processes that convert soluble organic compounds to biological cell mass, which can be separated from the effluent by gravity in a clarifier. Colloidal material and some nondegradable compounds are normally absorbed in the settled solids.

Biological wastewater treatment plants are expensive to build and operate, and specialized experience is necessary to produce a successful design. Consequently, an industry

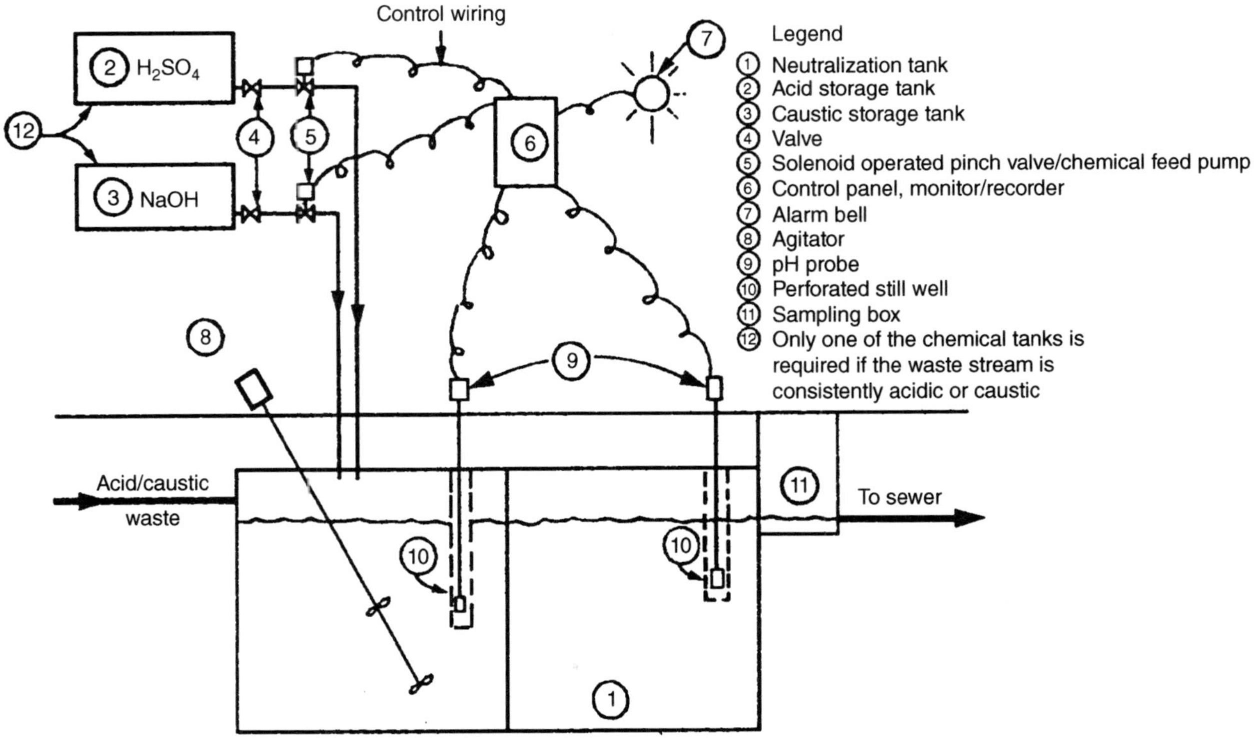

FIGURE 13.2 Large-scale acid caustic neutralization schematic.

faced with a biological treatment requirement should first compare the feasibility and economics of discharging to a municipal treatment system to building its own facility. If the build alternative is indicated, a choice must be made among the available application methods, including activated sludge, aerated lagoons, trickling filters and rotating filters, and the anaerobic process. Expert advice is strongly suggested.

Pipe Material and Joint Selection

Because of the vast diversity of manufacturing processes, it is impossible to make any general characterization of industrial wastewater. It is common to have various areas within a plant or industrial complex discharging different types of effluent with greatly varying characteristics.

A large quantity of effluent in an industrial facility originates from drains. Drains receive discharge from production equipment, floor washdown, process and production machines, and other equipment such as compressors and boilers. The floor drain and discharge pipe from the drain must be capable of resisting chemicals discharged from the production equipment. The selection of the most appropriate piping material can be accomplished only if the nature of the effluent, both present and future, is known and can be allowed for.

System Design Considerations

The design of the drainage system is dependent on the location, composition, and quantity of discharged effluent from all sources. The layout and engineering of a piping network require ingenuity and attention to detail.

The selection of the type and location of floor drains is a major aspect of drainage system design. The following are general guidelines for locating and selecting the drains:

1. Wet floors are to be avoided. Drains should be located next to equipment and be large enough to allow easy multiple discharges to spill over them without requiring a run of pipe over the floor or having to spill on the floor and run to the drain. If large flow rates are expected, select a large drain.
2. The use of long trench drains in areas where a number of pieces of equipment are placed will create an easy access to all the various drains from the equipment. This arrangement is usually less costly than multiple drains.
3. In many cases the discharge from equipment may be under pressure because of the head of water in the piece of equipment, such as a tank, being emptied. The drain should be large enough in physical size to accept the largest expected flow. The discharge pipe must be large enough to accept the maximum quantity flowing full by gravity without overflowing. An air gap shall be provided to prevent pressurizing the gravity drainage system.
4. To accept the largest number of multiple, small-size drainage lines from equipment, a funnel type of drain should be provided. The top of the funnel should be as close to the floor as reasonable in order for an air gap to be provided between the top of the floor drain and the end of the equipment drain. This air gap distance shall be twice the diameter of the drainage line. Provide adequate cleanouts in drain lines. In lines that are at the ceiling of high floors, extend the cleanouts to the floor above to avoid the need for maintenance personnel to climb ladders to clean stoppages.

5. The minimum-size drain line under the slab or underground should be 2 in (50 mm). To determine the size of drainage lines, refer to Table 19.2. Floor drains should be a minimum size of 4 in (100 mm).
6. Adequate venting of the drainage line must be provided to allow for smooth flow. These vents shall be connected to the top of the drain line in order to allow air at the top of the pipe either to be vented out (when there is a slug of liquid) or to admit air required by the flow of water or due to a partial vacuum created by the liquid flowing full. Vents shall be a minimum size of 2 in (DN 50). Larger vents shall be sized based on the local plumbing code.
7. Local regulations may require the use of double-contained piping to prevent potential leakage from discharging into the environment. A leak detection system for the double-contained piping should be provided that will allow leakage to be annunciated.

BIBLIOGRAPHY

Codes and Standards

ANSI/ASME B31.3, Process Piping. American National Standards Institute.

API Spec 12D, Specification for Field Welded Tanks for Storage of Production Liquids, American Petroleum Institute.

API Spec 12F, Specification for Shop Welded Tanks for Storage and Production Liquids, American Petroleum Institute.

API 650, Welded Steel Tanks for Oil Storage, American Petroleum Institute.

ASME Boiler and Pressure Vessel Code, American Society of Mechanical Engineers.

NFPA 30, Flammable and Combustible Liquids Code, National Fire Protection Association.

UL 142, Steel Aboveground Tanks for Flammable and Combustible Liquids, Underwriters Laboratories.

Government Publications

Activated Carbon Process for Treatment of Wastewater Containing Hexavalent Chromium, EPA-600/2-79-130, Environmental Protection Agency. Contains technical analysis and/or comparison of environmental problem solutions, equipment, and control strategies. It is available through the National Technical Information Service (NTIS) (see Resources below).

CFR Title 40, Protection of Environment U.S. Government Printing Office. The code is available from the U.S. Government Printing Office (see Resources below), local government bookstores, and most public libraries.

DeRenzo, D. J., *Corrosion Resistant Materials Handbook,* 4th ed., William Andrew Publishing/Noyes, Park Ridge, N.J., 1985.

Effluent Guidelines. Environmental Protection Agency.

Kirk-Othmer Encyclopedia of Chemical Technology, John Wiley & Sons, Inc., New York, 2004.

Perry's Chemical Engineers' Handbook, 8th ed. McGraw-Hill, New York, 2008.

Schweitzer, Philip A., *Handbook of Corrosion Resistant Piping,* 2d ed., Krieger Publishing.

Technology for the Storage of Hazardous Liquids: A State-of-the-Art Review. New York State Department of Environmental Conservation.

Treatment of Organic Chemical Manufacturing Wastewater for Reuse, EPA-600/2-79-184. Environmental Protection Agency. Available through the NTIS.

Wastech Controls & Engineering, *Ion Exchange for Heavy Metal Removal*, White Paper Resources.

Resources

American National Standards Institute, www.ansi.org
American Petroleum Institute, www.api.org
American Society of Mechanical Engineers, www.asme.org
National Fire Protection Association, www.nfpa.org
National Technical Information Service, www.ntis.gov
New York State Department of Environmental Conservation, www.dec.ny.gov
Underwriters Laboratories, www.ul.com
U.S. Government Printing Office, www.gpoaccess.gov

CHAPTER 14
FIRE SUPPRESSION WATER DRAINAGE

For smaller industrial and other similar facilities, the water from sprinkler and fire standpipe hose discharge used to suppress a fire could become contaminated with the stored products, products of combustion, and harmful raw materials it comes in contact with. Any water that has the possibility of being contaminated in this manner shall not be directly routed into the public sewer or allowed to be released into the environment unless the total discharge is extremely large. The smaller amounts of discharge should be routed to holding basins or tanks for analysis and removal, if in-house treatment is not possible or practical. The contaminated liquid shall be removed by approved competent contractors. If there is no material capable of causing contamination, no special consideration is necessary, except to protect other areas of the site and the facility from possible flooding. This may mean the water discharged shall be routed to a safe place outside and around the facility.

This chapter discusses the methods of disposal for contaminated sprinkler water discharge from smaller areas inside facilities. Major facilities that use fire hose or the public fire department for suppression, with their very large potential flow of water, are outside the scope of this handbook.

SYSTEM DESCRIPTION

The site-contaminated fire water collection system consists of three components: drains collecting contaminated water (most often sprinkler discharge from inside a fire area), a system of exterior main drain piping, and trenches that remove the sprinkler discharge from inside the facility and then route it to an exterior site collection basin.

Facility Construction

The room(s) in which the sprinkler water is intended to be collected must be provided with adequate methods of quickly removing sprinkler discharge. This is done by installing properly sized floor drains, trench drains, and/or scuppers. The floor shall be pitched to those locations. The scuppers are installed through the outside walls.

The doorways shall be protected with a threshold across the entrance. The threshold should be a minimum of 3 in (75 mm) above the high point of the floor. This depth will allow the discharged water to have enough momentary storage capacity to quickly flow completely down the drain. If there is a small area with the potential for high discharge rates, such as from a deluge system, the user shall be consulted.

For large buildings, such as a warehouse, manufacturing facility, or aircraft hangers, the floors should be pitched to the outside walls where drains or scuppers would be installed to quickly remove the water discharged by the sprinkler or hose system.

SYSTEM COMPONENTS

Drains

The drains could be conventional floor drains, trench drains, or scuppers. In general, conventional floor drains are not recommended because the limited flow into each drain would require a large number of them. The placement of trench drains shall be selected to intercept all the water discharged and sized to prevent it from damaging other parts of the facility, escaping from the property or into the soil.

Scupper drains located along the perimeter of the fire zone through the exterior walls are generally regarded as the best method to remove water from a room.

Fire Water Collection System

The fire water collection piping system (main drain) removes sprinkler discharge from interior drains placed in a fire area and routes this discharge into the exterior fire water collection system by gravity. An enclosed piping system or an open-trench system with a cover could be used to convey discharged water away from a fire area.

Each trench drain and scupper inside the building will have a separate branch outlet discharging directly into the main drain. If branches in the same fire area contribute flow into the main drain, it shall be increased in size as required by the entire calculated total flow at all points of connection.

The collection system shall be provided with a diverter system of an appropriate design with the outlet consisting of two possible routes. One is the normal setting to the storm water system for noncontaminated water, and the other is for contaminated water discharged to the fire water collection basin. This valve shall be activated by the fire alarm if a fire event occurs.

The outlet to the storm water system is used when the floor or equipment is washed, when the trench is cleaned, or when normal maintenance is performed within the fire area and there is no contamination.

Fire Snuffer Unit. The snuffer unit is required to extinguish a fire traveling on the surface of the collection trench along with the fire discharge. The fire marshal or other responsible official shall be consulted regarding the exact construction. One method consists of a baffle system which will skim off any floating material, preventing any flame from passing beyond the snuffer unit. A minimum of one snuffer unit shall be provided for each main building drain. Depending on the size of the building that the branch is serving, additional units may be required.

The snuffer unit shall be installed inside a vapor-sealed access manhole.

Fire Water Collection Basin. A holding basin or retention tank on site is necessary to contain all sprinkler discharge that may be contaminated by products of combustion and materials stored in the fire area. There shall be a provision for the testing of the discharged water to determine the level of contamination. If an in-house treatment system is provided, it shall be approved by the local AHJ in conformance with all regulations. If one is not provided, a contractor experienced in the removal of contaminated wastewater shall remove the water from the collection basin.

SYSTEM SIZING

General

The initial part of the design from an engineering standpoint is to first determine what areas within a facility would produce contaminated water as a result of products of combustion and a suppression system discharge. This would require consultation with the owner or operator to find out what materials will be stored in any of the areas within the facility and the potential contamination. It would then be necessary for the architect to arrange to have these areas located on the perimeter of the building.

The second item is to identify each independent fire zone and the required type of sprinkler coverage and density (e.g., ESFR, deluge, or standard sprinklers) in gallons per square foot (liters per square meter). This calculation would result in the expected flow into each individual drain and the total flow into the collection system.

Trench Drains

Trench drains will be selected with a configuration and length that will provide capacity for the required discharge. The total design discharge for the sprinkler system in each fire area shall be used to determine the configuration and number of drains required. The manufacturer shall be consulted regarding the design flow that each drain is capable of accepting. Traps shall be provided to minimize the possibility of spreading a fire from fire zone to fire zone.

Scuppers

They should be placed on the outside wall evenly along each perimeter of the fire zone in such a way that all water from each sprinkler zone shall be captured independent of other zones. Scuppers shall be located so as to prevent the spreading of a fire within the fire zone and shall be sized to discharge the calculated flow. A flap shall be installed over the rectangular slots which will prevent airflow through the slots into the room and the reverse flow of water from the outside. The floors shall be pitched to the scuppers, the degree of pitch determined after consultation with the general contractor. The bottom of each scupper shall be at the lowest level of the pitched floor. Each box shall be provided with a vent, and the vent shall have a flame arrester. The total quantity of drainage shall be the hydraulically calculated sprinkler flow only for the fire area under design. The flow of water through scuppers of various sizes is given in Table 14.1. The calculated discharge based on gallons per minute per foot squared (liters per minute per meter squared) shall be used to determine the size, number, and configuration of scupper drains required.

The discharge from the scupper drains on the ground floor shall be through a rectangular box on the outside wall which will receive the discharge exiting the building and feeding downward into the site collection system. The drainage from an upper floor shall be discharged vertically down to the lowest level through separate closed conduit into the site collection system located directly below.

TABLE 14.1 Discharge from Rectangular Scuppers, gpm

Head of water		Width of scupper, in							
mm	in	4	6	12	18	24	30	36	48
13	0.5	5	6	13	19	25	32	38	51
25	1	10	17	35	53	71	89	107	143
38	1.5	20	31	64	97	130	163	196	217
50	2	30	47	98	149	200	251	302	404
63	2.5	41	62	136	207	278	349	420	582
75	3	53	84	177	271	364	458	551	740
88	3.5	65	101	212	339	457	575	693	426
100	4	77	124	270	412	556	700	844	1135
150	6	124	211	476	471	1006	1270	1535	2067

Note: Head is measured 6'- 0" (2.9 m) back from opening.
1 in = 25 mm

Fire Water Collection System

The main drain piping shall be installed outside and parallel to the building wall (if reasonable) of the area being protected. A recommended minimum pitch of ¼ in/ft. (2.0 cm/m) shall be maintained in the entire collection system to prevent depositing solids on the bottom. A sufficient number of cleanouts in piping shall be installed for the main drain to be easily washed and cleaned. The trenches shall be covered with steel plates of adequate thickness and shall be capable of being removed for cleaning. The top of the trench shall be 1 in (25 mm) abovegrade to prevent the entrance of rainwater from a rainstorm. If other fire areas are protected in other parts of the facility, they shall be piped separately into the building drain. For each fire area, the main drain shall be sized for the total area contributing flow downstream. As each branch connects to the trench, the main pipe shall be increased in size as required.

Flow in open channels has been difficult to quantify. However, the calculation of gravity flow in piping has been well documented and been accepted by engineers. The only practical way, with an adequate safety factor, of avoiding difficulties is to design for a channel with partially full flow. The gravity flow calculations have resulted in formulas that are considered very accurate for piping. Many graphical and tabular solutions for flow have been found. One tabular method reference for regular pipe is Table 19.2, which establishes a pipe size based on pitch and volume for various depths of water. Experiments by engineers in the past, such as Colebrook and Moody, have led to considerable accuracy for trenches recognized by practicing engineers. These formulas would prove unwieldy to solve for low flow rates in the everyday calculations for 50 gpm (190 L/min) or less. To avoid such formulas, the following method will yield accurate results.

1. Calculate the flow from each fire area into each branch and the main drain.
2. Select a pitch. Use a minimum of ¼ in/ft (2.0 mm/m) pitch.
3. Refer to Table 19.2, three-fourths full column, and find the flow equal to or larger than that calculated.
4. Select the pipe size.
5. Refer to Table 14.2. Using the size pipe calculated, find the pipe area.
6. The area selected shall be used to configure the area of the trench.

TABLE 14.2 Characteristics of Pipe

Size		Inside		
in	DN	diameter, in	Radius, in	Area, in^2
6	150	6.52	3.25	33.17
8	200	8.64	4.32	59.00
10	250	10.66	5.33	39.00
12	300	12.32	6.16	119
14	350	14.79	7.40	172
15	375	15.36	8.18	210
16	400	16.85	8.43	223
18	450	18.92	9.46	281
20	500	20.98	10.50	333
24	600	25.07	12.53	493
30	750	31.15	16.08	812
36	900	37.36	18.18	843
42	1000	43.45	22.73	1620
48	1200	49.66	24.82	1963

7. Check Table 19.3 for the self-cleaning velocity to prevent buildup of deposits. The velocity shall be a minimum of 2 ft/s (0.6 m/s) to scour sand and grit from the trench or pipe bottom. The minimum size should be 6 in (DN 150).

8. To the trench area, adding a 10 percent safety factor is suggested.

9. For flow rates greater than 50 gpm (190 L/min), refer to Fig. 17.9, using the above conversion from area of piping to area of channels.

It is sound practice to avoid difficulties by designing a trench for a filling ratio of about 75 percent of the maximum design flow to make efficient use of the design capacity. Selecting a pipe size from the two-thirds column in Table 19.1 will satisfy this requirement.

The drainage collection system is sized on flow rate and pitch from one point in the facility to the detention basin. The flow rate of water required to be disposed of is determined by first calculating the actual sprinkler head discharge selected and the water density over the area used for hydraulic calculations. High velocity in the drainage pipe is not a major consideration. A shallow pitch will give a low velocity that may result in the deposit of some material that would have to be flushed out after the fire event. A high velocity will not affect the life of the collection system because of the relatively short amount of time the system is in operation. Each branch will be connected into a main drain.

Venting of the interior system is required to allow free flow of the effluent. Each branch where it connects to the main drain should have a loop vent of at least 2-in (50-mm) size. The vent could be connected to the sanitary vent system or carried through the roof independently. The pipe material selected shall be compatible with the potential chemicals it might carry. Because of these potential chemicals, double containment may be necessary. No venting is required in the channels.

Fire Water Collection Basin or Tank. The open-air collection basin may require a liner or other special protection to prevent the contaminated waste from infiltrating the surrounding soil. One possibility is that this liner should be sacrificial in nature. This will minimize the initial cost of the liner and, in addition, will allow the replacement cost to be considered part of the cleanup if a fire occurs. This will allow a mechanism by which a new liner could be easily installed if and when replacement is necessary. The reason for this philosophy is that it is expected that the liner will be removed and disposed of and a new liner installed. This procedure would eliminate the future treatment of stormwater passing through the basin after a fire event. The choice of the liner shall be determined by consultation with the AHJ and manufacturers. If a closed tank is preferred, it shall be constructed of material that is resistant to the chemicals expected to discharge into it.

The outlet system shall be routed primarily into the stormwater system, with a secondary route into the contaminated water collection basin if a fire event occurs. A valve-actuated alarm (or other device) connected to the fire alarm system shall be located on the main drain to the collection basin. A diverter valve shall be provided that will divert the entire flow automatically to the collection basin when the sprinkler or deluge system is activated. This diverter valve will remain closed to the stormwater discharge pipe until after the fire alarm system is reset and can only be manually reset after testing or treatment of the captured water. An overflow shall be provided that directs water onto the site in a safe area.

The basin might need a pump pad or depression to place a pump on (or into) if the disposal point is higher than the basin. A suitable disposal point design and location of access for a truck will be necessary.

The suggested basic criteria used for total containment sizing of the basin or tank will be to locate the tank invert below the incoming collection system invert and should be as follows:

1. Provide for the single largest creditable fire volume discharged from a contaminated source.
2. If the basin is exterior, the additional freeboard volume of water calculated from a 24-h, 100-y return period rainstorm shall be included. Refer to Table 17.12.
3. A safety factor of 30 percent shall be added to the factors above. Additional factors to be considered would be:
   a. A security fence
   b. The sideslope of the basin graded in such a manner that would permit mowing
   c. A clear area around the basin that will allow equipment to maneuver
   d. A subdrain provided to prevent groundwater from damaging the liner
   e. Provision of a convenient and accessible point for testing the liquid and integrity of the liner
   f. Provision of an emergency spillway to a protected area to guard against a catastrophic failure of the basin or surcharging by the site conveying system
   g. A monitor adjacent to the basin to control a fire if the liquid inside should reignite

Foam Discharge Disposal

In most areas of the United States there is no concern for the environmental impact of high-expansion foam. It is generally accepted that the foam should be allowed to dissipate. It is necessary to check with the local requirements during the design phase for foam solution containment and disposal. This lack of concern is not shared in some foreign locations.

If the foam is allowed to dissipate on its own, it will take several days. If there is no possibility of additional contamination, it could be done more quickly if there was a spray from a high-power hose stream or a lesser volume from a hose used by maintenance personnel. Since the foam is mostly made from a soap solution, the discharge, if contaminated, would be into the site collection system. If there is no contamination, it could be routed directly into the sewer. If contaminated, the residue shall be cleaned manually.

BIBLIOGRAPHY

Donnaway, S. S., "FPE Corner," *Plumbing Engineer.*

Standard design guides from various facilities.

Sturm, T. W., *Open Channel Hydraulics*, McGraw-Hill, New York, 2009.

CHAPTER 15
VOLATILE LIQUIDS: TREATMENT AND DISPOSAL

GENERAL

This chapter discusses methods that will protect the environment from an accidental spill or direct release of select volatile liquids. The treatment that eliminates or reduces the effluent to acceptable limits prior to either discharge or disposal is discussed. Some liquids are generated by industrial activity, and others exist in their natural state by the nature of their intended purpose. The liquids that are volatile and detrimental to the environment require the protection of double containment to both prevent and detect leakage into surrounding areas. Hydrocarbons, motor oil, and similar substances are discussed in Chap. 9.

This chapter will also discuss the potential leakage from underground liquid fuel piping and tanks into the soil. Because local and national codes require that the piping and tanks be double-contained, a spill is rare; but the design engineer must be aware of the leakage possibilities and methods of detection and disposal. The first section discusses gasoline service stations. The liquid stored is often referred to as *product*, which is the primary liquid that might leak. The remainder of this chapter considers the interception and disposal of other various volatile liquid waste products that a facility design engineer might encounter, such as metal cutting oil, motor lubrication oil, and waste oil storage systems.

The actual design of entire systems is outside the scope of this handbook, but can be found in books listed in the Bibliography. Items also not discussed are internal, non-product-containing vents (that are not subject to liquids) and all components concerned with potential discharges from any aboveground tank, piping, and product dispensers because leakage from aboveground piping and tanks is easily visible and contained. The detailed design of installed tanks and piping is discussed only as it pertains to a better understanding of how potential accidental spills or release of liquid fuel might occur or be prevented.

DOUBLE-CONTAINMENT PIPING

A double-containment piping system is an arrangement of liquid transport piping contained within a secondary housing. This provides a means of primary liquid product containment if the primary piping fails.

The primary piping shall be designed in such a manner that leakage will not occur in accordance with applicable codes for pressure, corrosion, and burial. The secondary containment shall be designed to meet or exceed the capabilities of the primary pipe, to contain a leak, and to withstand any possible effects of a process fluid release. Thermal expansion of the primary and secondary piping, if different, shall be taken into consideration. The two systems together are part of a complex, interacting assembly that is referred to as a *double-contained system.*

A method of automatic or manual leak detection or sensing of the interstitial space between the primary and secondary piping and tank systems is an important part of such a system.

Piping Materials

Piping aboveground from an aboveground storage tank (AST) is for the vent and product delivery. The most common piping material is A-53 steel with threaded joints and factory-applied corrosion protection. The pipe must be coated at the factory with an accepted and proven corrosion paint or coating. Note that Teflon tape should not be used on the pipe joints as it is incompatible with hydrocarbons. A common practice is to use a baked-on powder. Another material used where a high degree of corrosion protection and strength is required is stainless steel. Fiber-reinforced pipe (FRP) with ultraviolet protection added is another material often used. Galvanized steel pipe is not considered acceptable because of corrosion considerations. Adapters are used to connect aboveground steel pipe to double-contained underground plastic piping. Special connections are required.

For new and replacement underground storage tanks (USTs), interconnecting piping is almost exclusively plastic or FRP with the same material used for secondary containment. It is also common to use approved prefabricated rolls of flexible plastic piping. Requirements regarding approval of the specific piping material and connections selected and cathodic protection, if applicable, must be checked with the jurisdictional authority.

Plastic piping is commonly divided into two general types: flexible and rigid. Flexible pipe is generally manufactured from proprietary materials. If it is UL listed and/or FM Global approved, it is generally acceptable. In addition, the joints and connections should be selected to provide the greatest strength, ease of installation, and corrosion resistance. Flexible, plastic coaxial piping systems are a double-walled piping system installed as a single pipe.

Rigid FRP piping with an epoxy interior lining has been widely used and accepted. The primary pipe is assembled with socket-type fittings and epoxy cement. The outer (secondary containment) pipe is of the same material as the primary pipe and is manufactured in two half sections with a longitudinal flange. It is assembled after the primary pipe is tested, using cement placed on the adjacent flanges with nuts and bolts installed to hold the two half sections together until the cement dries.

Flexible pipe connectors are used to connect piping runs to sumps and manways to allow for settlement. In addition, because submersible pumps are often screwed into a tank connection, the product discharge will not always face the direction of the piping run to the dispenser when tightened. Flexible connectors are necessary inside manways to connect the submersible pump discharge to the dispenser supply piping.

For ASTs, because leakage is visible, double-contained piping is not required if the tank is located within a dike or inside a remote containment. If pipe is run underground or aboveground outside diked areas, it must be double-contained and provided with leak detection. If any liquid spills, it is allowed to evaporate into the air or removed manually.

Gasket materials must be either Buna-N (nitrile butadiene) or NBR (acrylonitrile butadiene rubber).

Design of Piping Components

For gravity discharge, the primary piping from most facilities would not be expected to have any pressure. If there is a pressure, the end user must be consulted. Plastic piping is most often used.

Thermal Expansion of Piping

When subject to a difference in temperature, piping materials will experience a change in dimension and volume. When the primary piping and secondary piping experience different magnitudes of expansion and contraction, it is called *differential expansion*. When they experience the same magnitude, it is called *nondifferential expansion*. For the piping that is expected to be used in the applications appearing in this book, the expansion problem is minimal. Most underground nonmetallic piping is subject to relatively low differences in temperature.

Nonpressure double-contained piping for underground installation is designed to drain chemical waste to a treatment facility by gravity. Both primary piping and secondary piping are designed to the same performance specifications. The most prevalent method of draining the annulus is to have the pipe enter a sump that will be large enough to contain any of the leakage, which would be emptied manually into a drain that discharges to a separator prior to discharge into a sewer. The recommended size of the containment pipe should be two sizes larger than the primary pipe.

Drainage Piping Installation

Underground piping should be installed in a trench, far enough below grade to prevent damage from surface earth loads. Initial backfill should be pea gravel, clean sand, or other acceptable material. A generally accepted minimum depth from grade surface to the top of the pipe where there is no slab is 18 in (45 cm), 12 in (30 cm) below asphalt slabs and 6 in (15 cm) under concrete slabs. It is also accepted practice, where electrical conduit is installed in the same trench as piping, to protect the electrical conduit with a concrete encasement and to provide 4-in (100-mm) separation.

Drainage Pipe Sizing

To size the drainage piping and trench, it is necessary to use Table 19.1. To accomplish this, first the viscosity of the oil or other liquids in Seconds Saybolt Universal (SSU) that are product must be determined. This is done by referring to a listing found in Table 15.1 that will give the viscosity of the liquid in question. If the liquid is not on the list and the only information available is the name, refer to Table 15.2 to find the specific gravity of the liquid. If the viscosity is found to be unfamiliar in units, refer to Table 15.3 to convert the listing to SSU viscosity.

The size of the drainage piping is found using Table 19.2. To use that table, the specific gravity of the liquid must be determined. If the specific gravity is different from that of water, a conversion must be made. This is calculated by dividing the specific gravity of the product found in Table 15.2 into 1 (the specific gravity of water). Multiply the product of this calculation by the gpm calculated for the system to find the actual gpm to be used in Table 19.2. Now it is possible to use Table 19.2, entering with the new calculated flow rate, using one-half full piping column and the proposed pitch.

TABLE 15.1 Viscosity of Various Hydrocarbons

	Viscosity, SSU						
	40°F	60°F	80°F	100°F	120°F	140°F	160°F
Liquid	4.4°C	15.6°C	26.7°C	37.8°C	40°C	60°C	71°C
Water	31.5	31.5	31.5	31.5	31.5	31.5	31.5
Gasoline	30	30	30	30	30	30	30
Jet fuel	35	35	35	35	35	35	35
Kerosene	42	38	34	33	31	30	30
Turpentine	34	33	32.8	32.6	32.4	32	32
Gasahol 15% gas	30	30	30	30	30	30	30
85% methanol	30	30	30	30	30	30	30
Fuel oil and diesel oil							
No. 1 fuel oil	40	38	35	33	31	30	30
No. 2 fuel oil	70	50	45	40	—	—	—
No. 3 fuel oil	90	68	53	45	40	—	—
No. 5A fuel oil	1,000	400	200	100	75	60	40
No. 5B fuel oil	1,300	600	490	400	330	290	240
No. 6 fuel oil	—	70,000	20,000	90,000	1,900	900	500
No. 2D diesel fuel oil	100	68	53	45	40	36	35
No. 3D diesel fuel oil	200	120	80	60	50	44	40
No. 4D diesel fuel oil	1,600	600	280	140	90	68	54
No. 5D diesel fuel oil	15,000	5,000	2,000	900	400	260	160

TABLE 15.2 Specific Gravities of Typical Oils and Other Liquids

Fuel oil	
#1 Fuel oil	0.79–0.85
#2 Fuel oil	0.81–0.92
#3 Fuel oil	0.82–0.95
#4 Fuel oil	0.88–0.97
#5A Fuel oil (Bunker A)	0.91–0.99
#5B Fuel oil (Bunker B)	0.91–0.99
#6 Fuel oil (Bunker C–low sulfur)	0.91–1.06
#6 Fuel oil (Bunker D–high sulfur)	0.93–1.07
#2 Diesel fuel oil	0.85
#2 Diesel fuel oil	0.82–0.95
#2 Diesel fuel oil	0.82–0.95
#2 Diesel fuel oil	0.82–0.95
Crankcase oils	
SAE 10	0.87
SAE 20	0.89
SAE 30	0.89
SAE 40	0.89
SAE 50	0.90
Machine tool and other industrial applications	
SAE 75	0.93
SAE 80	0.93
SAE 90	0.93
SAE 140	0.93
SAE 250	0.93

(Continued)

TABLE 15.2 Specific Gravities of Typical Oils and Other Liquids (*Continued*)

Aviation oils and fuels	
Jet fuels	0.74–0.85
Reciprocating engine fuels	<0.72
Reciprocating engine oils	0.88–0.89
Gear and transmission oils	
General purpose	0.88–0.92
Marine propulsion and stationary power turbines	
Light	0.87
Medium	0.87
Heavy	0.89
Miscellaneous oils	
Corn oil	0.93
Cottonseed oil	0.88–0.93
Turpentine	0.86–0.87
Linseed oil	0.93–0.94
Olive oil	0.91–0.92
Palm oil	0.92
Peanut oil	0.92
Rapeseed oil	0.92
Naphthalene	0.88–0.91
Crude oil	0.78–0.91

TABLE 15.3 Approximate Viscosity Conversions

Seconds Saybolt Universal (SSU)	Kinematic viscosity		Seconds Saybolt Furol (SSF)	Seconds Redwood 1 Standard	Seconds Redwood 2 Admiralty	Degrees Engler	Degrees Barbey
	cP	ft²/s					
31	1.0	0.00001076		29		1.00	6,200
31.5	1.13	0.00001216		29.4		1.01	5,486
32	1.81	0.00001948		29.8		1.08	3,425
32.6	2.00	0.00002153		30.2		1.10	3,100
33	2.11	0.00002271		30.6		1.11	2,938
34	2.40	0.00002583		31.3		1.14	2,583
35	2.71	0.00002917		32.1		1.17	2,287
36	3.00	0.00003229		32.9		1.20	2,066
38	3.64	0.00003918		33.7		1.26	1,703
39.2	4.00	0.00004306		35.5		1.30	1,550
40	4.25	0.00004575		36.2	5.10	1.32	1,459
42	4.88	0.00005253		38.2	5.25	1.36	1,270
42.4	5.00	0.00005382		38.6	5.28	1.37	1,240
44	5.50	0.00005920		40.6	5.39	1.40	1,127
45.6	6.00	0.00006458		41.8	5.51	1.43	1,033
46	6.13	0.00006598		42.3	5.54	1.44	1,011
46.8	7.00	0.00007535		43.1	5.60	1.48	885
50	7.36	0.00007922		44.3	5.83	1.58	842
52.1	8.00	0.00008611		46.0	6.03	1.64	775
55	8.88	0.00009558		48.3	6.30	1.73	698

(*Continued*)

TABLE 15.3 Approximate Viscosity Conversions (*Continued*)

Seconds Saybolt Universal (SSU)	Kinematic viscosity		Seconds Saybolt Furol (SSF)	Seconds Redwood 1 Standard	Seconds Redwood 2 Admiralty	Degrees Engler	Degrees Barbey
	cP	ft^2/s					
55.4	9.00	0.00009688		48.6	6.34	1.74	689
58.8	10.00	0.0001076		51.3	6.66	1.83	620
60	10.32	0.0001111		52.3	6.77	1.87	601
65	11.72	0.0001262		56.7	7.19	2.01	529
70	13.08	0.0001408		60.9	7.60	2.16	474
75	14.38	0.0001548		65.1	8.02	2.37	431
80	15.66	0.0001686		69.2	8.44	2.45	396
85	16.90	0.0001819		73.4	8.87	2.59	367
90	18.12	0.0001950		77.6	9.30	2.73	342
95	19.32	0.0002080		81.6	9.71	2.88	321
100	20.52	0.0002209		85.6	10.12	3.02	302
120	25.15	0.0002707		102	11.88	3.57	246
140	29.65	0.0003191		119	13.63	4.11	209
160	34.10	0.0003670		136	15.39	4.64	182
180	38.52	0.0004146		153	17.14	5.12	161
200	42.95	0.0004623		170	18.90	5.92	144
300	64.6	0.0006953	32.7	253	28.0	8.79	96
400	86.2	0.0009278	42.4	338	37.1	11.70	71.9
500	108.0	0.001163	52.3	423	46.2	14.60	57.4
600	129.4	0.001393	62.0	507	55.3	17.50	47.9
700	151.0	0.001625	72.0	592	64.6	20.44	41.0
800	172.6	0.001858	82.0	677	73.8	23.36	35.9
900	194.2	0.002099	92.1	762	83.0	26.28	31.9
1,000	215.8	0.002323	102.1	846	92.3	29.20	28.7
1,200	259.0	0.002786	122	1,016	111	35.1	23.9
1,400	302.3	0.003254	143	1,185	129	40.9	20.5
1,600	345.3	0.003717	163	1,354	148	46.7	18.0
1,800	388.5	0.004182	183	1,524	166	52.6	15.6
2,000	431.7	0.004647	204	1,693	185	58.4	14.4
2,500	539.4	0.005806	254	2,115	231	73.0	11.5
3,000	647.3	0.006967	305	2,538	277	87.6	9.6
3,500	755.2	0.008129	356	2,961	323	102	8.21
4,000	863.1	0.009290	408	3,385	369	117	7.18
4,500	970.9	0.01045	458	3,807	415	131	6.39
5,000	1,078.8	0.01161	509	4,230	461	146	5.75
6,000	1,294.6	0.01393	610	5,077	553	175	4.78
7,000	1,510.3	0.01626	712	5,922	646	204	4.11
8,000	1,726.1	0.01858	814	6,769	738	234	3.59
9,000	1,941.9	0.02092	916	7,615	830	263	3.19
10,000	2,157.6	0.02322	1,018	8,461	922	292	2.87
15,000	3,236.5	0.03483	1,526	12,692		438	1.92
20,000	4,315.3	0.04645	2,035	16,923		584	1.44

1 ft/s = 0.3 m/s

Source: Cameron hydraulic data.

LIQUID FUEL SUPPLY

Definitions and Liquid Fuel Classifications

This section is written for liquid fuel supply systems associated with automobile garages, emergency generators, and fire pumps. Liquid fuels are primarily gasoline, gasoline blends, or diesel fuel which are governed by the requirements of NFPA-30. In this standard, liquids are classified as either flammable or combustible based on their flash points. The *flash point* of a liquid is the temperature at which the liquid could give off vapor in sufficient concentration to form an ignitable mixture with air at or near the liquid's surface. In short, the flash point is the minimum temperature at which a fire or explosion could occur. Some frequently used definitions follow.

1. A *liquid fuel storage tank* is any stationary receptacle designed to contain an accumulation of regulated substances. Tanks can be constructed of materials such as steel, concrete, plastic, or various combinations that provide structural support.
2. An *underground storage tank* (UST) is required if the sum of the total tank volume of single or multiple tanks, including all the associated and interconnecting piping, is 10 percent or more below grade or covered with earth.
3. A *regulated substance* is any designated chemical that includes hydrocarbons derived from crude oil such as motor fuels, distillate fuel oils, residual fuels, lubricants, used oils, and petroleum solvents. Kerosene is also a regulated substance. Hydrocarbons are measured in parts per million (ppm) of total petroleum hydrocarbons (TPH). In future discussions, all these will be called *product*.
4. OSHA (29 CFR 1926) further defines storage tanks according to the following operating pressure ratings: atmospheric tanks, atmospheric pressure to 1.0 psig; low-pressure tanks, 1.0 to 15 psig; and pressure vessel, 15 psig or higher.
5. Liquids are classified by the NFPA as either flammable or combustible and are divided into classes I, II, and III. All *flammable liquids* under discussion are considered class I. They have a flash point below 100°F (37.8°C) and a vapor pressure no higher than 40 psia (2086 mmHg) at 100°F. Class I liquids are further subdivided into classes IA, IB, and IC. Class IA liquids have a flash point below 73°F (22.8°C) and a boiling point below 100°F (37.8°C); class IB includes gasoline and gasoline blends.
6. All *combustible liquids* are considered by the NFPA as class II and III liquids. They have a flash point at or above 100°F (37.8°C). Diesel fuel, light heating oil, and kerosene are class II flammable liquids with a flash point at or above 100°F (37.8°C) but below 140°F (60°C). Class III liquids are further subdivided into classes IIIA and IIIB. Class III liquids include motor lubrication and waste oil.
7. *Liquid petroleum* or *petroleum products* are defined by the NFPA as any hydrocarbon that is a liquid at atmospheric pressure and at temperatures between 20 and 120°F (–29 and 49°C), or discharged as a liquid at temperatures in excess of 120°F (49°C).
8. *Specific gravity* is the direct ratio of a liquid's weight to the weight of water at 62°F (37.8°C).
9. *Viscosity* is a measure of the internal friction between particles in a liquid that resists any force tending to produce flow. The higher the viscosity, the more slowly the liquid will flow under gravity conditions. Viscosity is obtained by measuring the time that a given quantity of liquid at a specified temperature takes to flow through an orifice. Viscosity is expressed in Seconds Saybolt Universal (SSU) for pump work and also kinematic viscosity centistokes (centipoise, cP), Seconds Saybolt Furol (SSF), and Seconds Redwood.

Conversion between these various units and others are given in Table 15.3. The specific gravity of various liquid fuels is given in Table 15.2.

CODES AND STANDARDS

The U.S. Environmental Protection Agency (EPA) has written basic, minimum regulations to protect the environment and people's health from the leakage of hydrocarbons and volatile organic compounds (VOCs) from USTs, ASTs, and associated piping. The basic purpose of these regulations is to ensure proper installation of the various system components, to prevent leaks or spills from occurring, and, if a leak or spill does occur, to ensure that the leak is quickly annunciated, found, and disposed of.

In almost all jurisdictions where these systems are installed, there exist specific requirements mandated by the local and state agencies. These requirements govern permits, registration, fees, and recordkeeping, as well as specific technical rules and regulations concerning system installation, materials, and leak detection. Specialists in this matter are highly recommended. Very often they may be more stringent than the federal EPA regulations cited here. A thorough code search will be necessary to ensure complete compliance with all applicable federal, state, and local regulations. The following is a list of commonly used codes, regulations, and guidelines. The guidance of specialists in this field are highly recommended. This list is not complete and must be verified in the locality where the project is constructed.

1. NFPA 30 and 30A
2. 40 CFR 112, Spill Containment Control and Countermeasures
3. Regional fire codes such as Building Officials Code Administrators (BOCA) International and Unified Facilities Criteria (UFC)
4. Underwriters Laboratories UL-142, Standard for Safety, and UL-2085, Fire Resistance
5. Resource Conservation and Recovery Act (RCRA), Subtitle 1 & C
6. Steel Tank Institute (STI)
7. Public Law (PL) 98 and 616
8. Clean Air Act Amendments (CAAA)
9. Superfund Amendments and Reauthorization Act (SARA), Title III, 1986

Liquid Fuel System Description

The liquid fuel storage and dispensing systems to be discussed are gasoline, gasoline blends and diesel oil. They are intended to store and dispense liquid fuel used to power internal combustion engines. Other required ancillary functions for installation include measurement of the level of liquid inside the storage tank, detection of any leakage from the storage tank and any leakage from the piping distribution system, and finally disposal of any leakage. A schematic detail of a typical underground tank and piping system using gasoline or diesel fuel for automobile facilities discussed in this chapter is shown schematically in Fig. 15.1.

System Components

Liquid fuel storage, whether in an aboveground storage tank or underground storage tank, requires many interrelated subsystems and components for proper operation and compliance

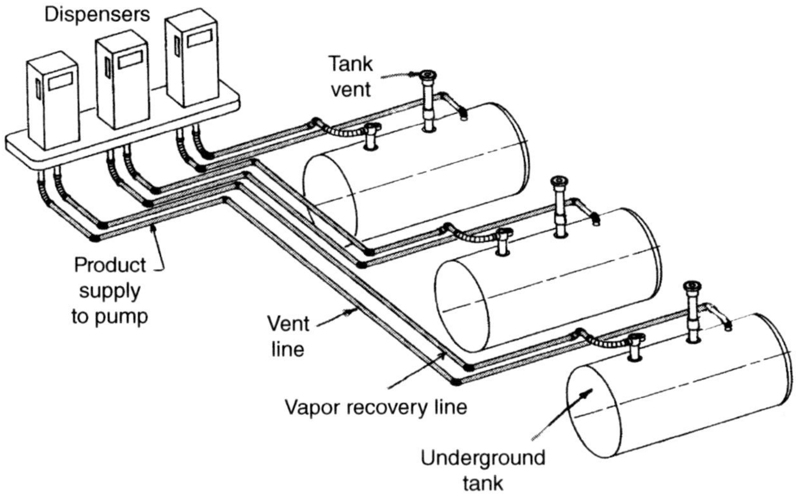

FIGURE 15.1 Typical underground tank and piping for automobile facilities.

with applicable codes and standards. The only systems discussed for drainage purposes are the following:

1. Storage tanks, including
   a. Tank filling and accidental spill containment
   b. Overfill protection
   c. Tank leakage
2. Product piping monitoring and leak detection

Distribution Systems

Storage Tanks

General. Storage tanks are designed and fabricated to prevent product releases due to structural failure and corrosion of the tank from the time of installation to the end of the expected useful life of the system. This requires that the tank manufacturer fabricate the tank in conformance with applicable codes and nationally recognized standards for structural strength and corrosion resistance. Since the tank must be installed in a manner that will prevent distortion and stress, the installation of the specific tank must be done by contractors trained and approved by the manufacturer. The tank foundation, bedding, and backfill shall be done only with material and methods approved by the manufacturer.

Spill Prevention. A tank is filled through a fill port on top of the tank. The purpose of the fill port is to have a safe filling method capable of catching spills from delivery hose disconnections. A typical 20-ft, 0-in length of 4-in delivery hose holds 15 gal. Product spilled must be prevented from entering the soil adjacent to the fill port. This is accomplished by providing safeguards that are code-mandated and recommended by good practice. A typical fill port is illustrated in Fig. 15.2.

Spill prevention is additionally accomplished by installing a below-grade containment basin with a capacity of 3.5 to 15 gal (13.5 to 57 L) to catch spillage of product from the

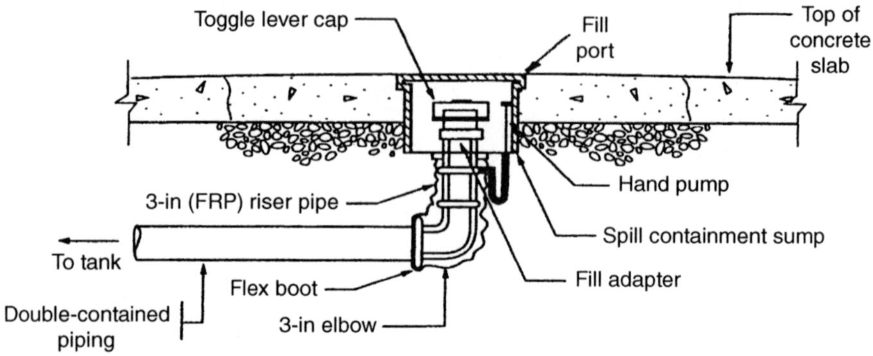

FIGURE 15.2 Typical fill port.

truck delivery hose. An optional device that could be included as part of the sump is a drain valve that, when opened, empties product in the sump, back into the fill line. Water in the sump is removed manually. A containment sump mounted on an underground tank, shown with included piping, is illustrated in Fig. 15.3.

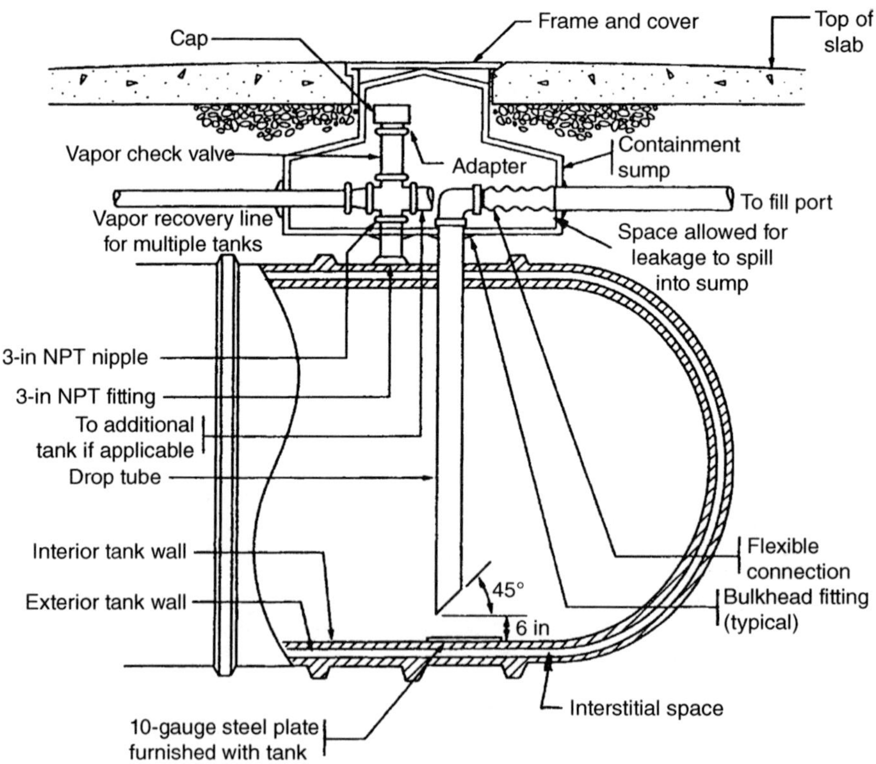

FIGURE 15.3 Typical containment sump, drop tube, and vapor recovery connection.

The fill port assembly is designed to accomplish the following:

1. Provide a watertight grade cover, allowing access to the fill hose connection.
2. Provide a fill hose connection for the tanker truck delivery hose. The truck hose has a standard end connection. This requires that an adapter be installed on the fill pipe leading to the tank. The gravity drop delivers about 200 gpm (750 L/min). The flow rate of a truck-mounted pump is generally 50 gpm (190 L/min).
3. Allow any fuel spillage from the fill hose to be contained and returned to the tank and any water to be removed. This is usually accomplished by installing a spill container around the fill port capable of containing 5 to 15 gal of product. A hand pump removes spilled fuel from the port bottom and returns the fuel into the fill line. Any remaining water is manually pumped out.
4. A dry disconnect coupling on the delivery truck hose could also be used to prevent spills.
5. Another, less commonly used fill method uses a coaxial hose to both fill the tank and recover vapor. This requires only one connection from the truck to the tank fill port and requires a different type of adapter than that used for a fill connection alone.

Overfill Prevention. All UST systems must be provided with overfill protection by installing one or more of the following devices or other methods accepted by the local authorities. The specific individual or combination of devices and methods accepted in any specific location shall be obtained from the local authorities.

1. A device that will alert the operator when the product level in the tank reaches 90 percent full by restricting the flow of product into the tank or alerting the operator by means of an audible alarm actuated by a high-level alarm probe
2. A device to shut down flow into the tank when the product level reaches 95 percent full
3. Equivalent devices or methods accepted and approved by the local implementing authority

The best method is the installation of an automatic device on the fill tube that reduces flow into the tank when it becomes 90 percent full and stops it entirely when the tank becomes about 95 percent full. There are a number of approved mechanical devices from different manufacturers that perform this function. Another method is to install an audible alarm actuated by the level gauge that automatically sounds when the liquid level in the tank reaches 90 percent full. Often a visible alarm is included. This alarm must be located in clear sight of the fill port.

In addition, tanks that do not require a vapor recovery system could have a floating-ball device installed on the atmospheric vent line that will close the vent when product reaches a predetermined point (usually at the 90 percent full level) such that additional filling may cause a spill. By closing, the air pressure will increase inside the tank and restrict the inflow, alerting the operator that the tank is approaching full. An extractor fitting is required that allows access into the line for removal of the float assembly.

LEAK DETECTION AND SYSTEM MONITORING

General

Leak detection for underground tank and piping is required by code. There are three basic requirements for leak detection:

1. Leakage must be capable of being detected from any portion of the tank or piping that routinely contains petroleum.

2. The leak detection equipment must meet performance requirements described in Federal Regulations.
3. Leak detection equipment must be installed, calibrated, operated, and maintained in accordance with the manufacturer's instructions.

The EPA has established various options, combinations of which must be used depending on project conditions, initial or long-term costs, and product. In addition, state and local requirements may differ in the number and application of these options. A schematic diagram indicating most required leak detection methods from tanks and piping is illustrated in Fig. 15.4. These options are:

1. Leakage from tanks
2. Manual tank gauging
3. Automatic tank gauging
4. Tank secondary containment interstitial monitoring
5. Vapor monitoring
6. Tank monitoring
   a. Liquid monitoring
   b. Hydrostatic monitoring
   c. Pressure monitoring

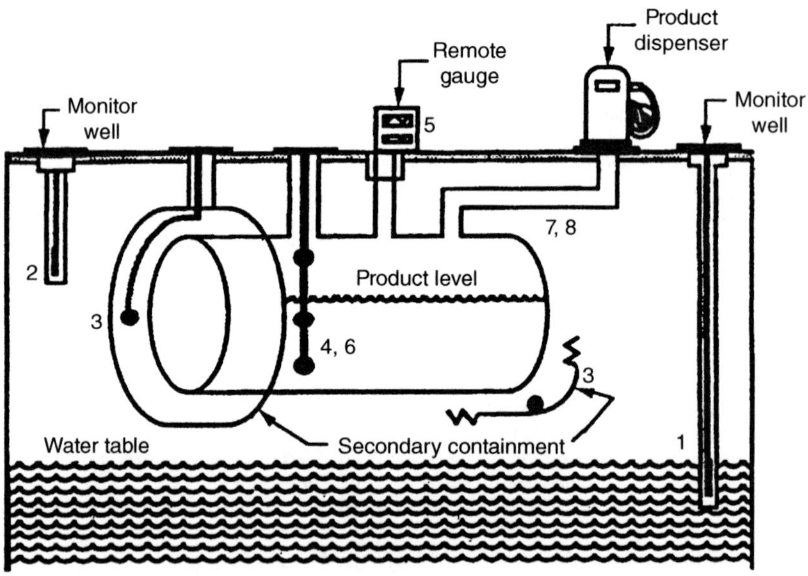

Leak detection methods for tanks and piping. (1) Groundwater monitoring. (2) Vapor monitoring. (3) Secondary containment with interstitial monitoring. (4) Automatic tank gauging. (5) Tank tightness testing with inventory control. (6) Manual tank gauging. (7) Leak detection for suction piping. (8) Leak detection for pressurized piping.

FIGURE 15.4 Leak detection for tanks and piping.

7. Leakage outside of tanks
   a. Groundwater monitoring
   b. Piping leak detection
   c. Tank tightness testing with inventory control

Leakage from Tanks

Leakage from tanks is measured by first gauging the level of product in a tank, then accounting for the amount of product dispensed, and again measuring the tank level. If the dispensed and remaining product figures agree with the amount of product delivered, there is no leakage.

Manual Tank Gauging. Referred to as *sticking*, manual gauging uses a long gauge stick calibrated to 1/8 in (3 mm) lowered directly into the tank until it rests on the bottom. This requires that a straight, direct access into the tank be provided. The wetted level leaves a mark on the stick that is read by the operator after removal from the tank. This method is generally limited to tanks of 2000 gal (7560 L) and smaller, and only in conjunction with tightness testing for tanks with a capacity larger than 550 gal (2080 L).

When this method is used, the tank being measured must be completely idle for at least 36 h. Two separate readings shall be taken at the beginning and end of that period and must be performed once per week. If the reading exceeds weekly or monthly standards, the tank is leaking.

It is common to provide manual tank gauging as a check against the mechanical or electronic methods and to allow measurement if electric power is out or there is a failure of the automatic devices.

Automatic Tank Gauging. Automatic tank gauging is accomplished by permanently installing a special probe, or monitor, into the storage tank. The probe could be mechanical, pneumatic, or electronic.

1. *Mechanical tank gauging*. The least costly tank gauging is a mechanical device, such as a float, that rides on the surface of the liquid. The movement of the float transmits the liquid level through a mechanical or other linkage to a remote indicator. General accuracy is about 2 percent.
2. *Pneumatic tank gauging*. This is a more accurate method that uses a bubbler pipe extending down into the liquid. A permanent source of compressed air or a hand-operated air pump forces air out of the bubbler pipe. The operating principle is to measure the pressure required to force air out of the bubbler pipe. The greater the pressure required to produce the bubbles, the greater the depth of the liquid.
3. *Electronic tank gauging*. The electronic tank gauging system consists of a probe mounted in a tank opening extending to the bottom of a tank and a remote panel that is microprocessor-controlled and could be programmable. The probe is capable of monitoring many parameters and extends from the primary tank bottom to a termination above the tank that both anchors the probe and acts as a junction box for the wiring. For USTs this point is below grade or the slab above the tank. Access to the box is through a small manhole in the slab. For ASTs, it terminates in a junction box on top of the tank.

   The advantage of the electronic gauging system is that it is capable of being programmed to automatically record many items. The probe will monitor such parameters as product and water level inside the tank and product temperature. In addition, probes extending from various locations can be electronically linked together to monitor vapor and liquid leakage from many sources, such as monitoring wells, piping, containment sumps, and tank interstitial spaces. Probes from multiple tanks can be linked to a single panel. Overfill and low-product levels can also be monitored.

The commonly available types of electronic probes are the magnetostrictive and ultrasonic. The magnetostrictive type uses a change in the magnetic field produced by movable product sensor floats, one for water and one for product. Each float has an integral magnet and is free to ride on the probe shaft as the product (and water) level changes. Signals are sent from a wire in the probe shaft and returned to a device on the top of the probe to gauge levels. The ultrasonic devices use ultrasound waves from the probe on the shaft to a receiver on the top of the probe to signal a change in product level. Capacitance probes are no longer recommended, having been replaced by less expensive and more accurate electronic probes.

Interstitial Monitoring. The interstitial space between the inner and outer tanks is used to contain and detect any leakage from the primary tank. Monitoring of the interstitial space is divided into two general categories: wet and dry. With dry methods, vapor is monitored for the presence of hydrocarbon vapors; with wet methods, liquid is monitored for the presence of liquid product. Wet methods are also used in pressurized systems using either air pressure or vacuum.

1. *Vapor monitoring*. Vapor monitoring is necessary to detect a leakage. This is achieved by placing a probe sensitive to product vapor into the dry interstitial space from a special tank opening. The probe is generally placed one-half the distance into the tank. If leakage occurs, then the vapor produced by the product is detected by the probe and a signal is given that leakage is present. This method has the advantage of not being affected by condensed water.
2. *Liquid monitoring*. Liquid monitoring is achieved by placing a probe at the bottom of the dry interstitial space that is capable of detecting liquid. If leakage occurs, it is detected by the probe and a signal is produced. A shortcoming of this method is that water condensation into the interstitial space is also detected as leakage.
3. *Hydrostatic monitoring system*. This is the most costly and most accurate method of interstitial monitoring. This is a wet system that uses a liquid, usually brine, installed at the factory that completely fills the interstitial space. The system is at atmospheric pressure with the liquid level carried above the tank into a reservoir that contains a level probe. Since normal product temperature differences will cause the liquid level to fluctuate, the level probe will annunciate only unacceptable changes.

   Water from a high water table leaking into the interstitial space from a hole in the secondary tank will cause the level to rise. Product leaking out of the primary tank into the interstitial space will also cause a rise in level. Conversely, a hole in the secondary tank will allow the brine to leak out of the system.

   To maintain atmospheric pressure, holes must be made in the standpipe cap. If there is a possibility of a high water table that will allow water to enter the holes, a separate vent line extended to a point safely above grade must be installed.
4. *Pressure monitoring*. Pressure monitoring of the interstitial is accomplished by injecting light air pressure (1.5 psig, 10 kPa) or creating a vacuum in the space. Any increase or reduction of pressure or vacuum indicates a leak that is sent to a remote panel for annunciation. Pressure systems are rarely used due to the increased maintenance required.

Leakage Remote from Tanks

Groundwater Monitoring. Groundwater monitoring senses the presence of product floating on the surface of groundwater. It can be used only where the products to be measured are immiscible. This method requires the installation of monitoring wells at strategic locations into the groundwater at several locations near the tank. The wells must be properly designed and sealed to eliminate contamination. Monitoring wells are discussed in Chap. 3.

They can be checked either manually or by automatic, electronic methods. Manual methods require the use of a "bailer" to collect liquid samples from inside the well and bring them to the surface. The samples must be sent for analysis once each month. Electronic methods use probes suspended in the well to continuously monitor for the presence of contamination in the groundwater. An alarm is annunciated at a remote panel. Monitor wells are limited to sites where the groundwater level is 20 ft (6.2 m) or less below the surface, with best results obtained from depths between 2 and 10 ft (0.60 and 3 m).

In-Ground Vapor Monitoring. In-ground vapor monitoring measures "fumes" in the soil around a storage tank to determine the presence of spilled product. This method requires the installation of monitoring wells similar to those used for groundwater monitoring. The wells can be checked by either manual or automatic, electronic methods. Manual methods require air sample gathering and submission to a laboratory for analysis once a month. Electronic equipment uses a probe suspended in the well and a remote, highly calibrated analyzer. Vapor monitoring should be considered only where backfill is sand, gravel, or other similar material through which vapor will readily move from the tank to the monitor. It is not recommended where a high groundwater level and excessive precipitation would interfere with the operation of this system for more than 30 days.

Leakage from Piping

Underground Pressurized Delivery Systems. Pressurized piping shall have one automatic leak detection method and one additional method, as follows:

1. Automatic leak detection consists of a permanently installed automatic flow restrictor or automatic flow shutoff device installed in the product discharge piping or a continuous alarm system. The continuous alarm could be a probe installed in the secondary pipe continuously monitoring liquid or vapor release of the secondary containment piping system. If the secondary containment terminates in a manway or containment sump, the lower end of the secondary containment pipe shall be open to the sump and allow any liquid to spill into the sump to be detected.
2. Additional methods include one of the following monthly requirements: groundwater monitoring, secondary pipe vapor monitoring, secondary pipe liquid monitoring, or annual pipe tightness testing. A continuous alarm to detect spillage as described above complies with this requirement as an additional method if installed in addition to an automatic detection method.

The following accuracy must be provided for automatic detection: a pressurized shutdown monitoring device that prevents flow of produce when a line tightness test indicates a minimum leakage of 0.10 gph (4.8 L/h). In addition, the device shall be capable of shutting down the system when a leak of 3 gpm (11.4 L/min) at 10 psi (68 kPa) is detected. One type of device commonly used is installed on the discharge of the submersible pump in the manway. It must act independently of any other pressurized shutdown monitoring device.

Underground Suction Delivery Systems. When the tank is below the dispenser, double-contained piping is not required for diesel fuel, but it is highly recommended. When the tank is higher than the dispenser, it is necessary to use double-contained piping. A sump shall be installed at the low point of the piping, and a probe is installed to detect leakage from the secondary pipe.

Tank Tightness with Inventory Control

This is a combination method using periodic tank tightness testing and a monthly inventory control. It is allowed only for the first 10 years of operation. Tank tightness requires that the tank be taken out of service and temporary equipment be installed. One method uses a volumetric test to exactly measure the change of level of product over a period of several hours. Another method uses ultrasound or tracer gas detection techniques to discover a leak in the tank wall. The monthly inventory requires that exact measurement be taken each month of the amount of product delivered compared to the amount stored in the tank and the amount dispensed. If the two figures balance, there is a leak.

Leakage into Secondary Containments

Any product leaking from the primary pipe will spill into the secondary containment pipe and flow by gravity to the low point. The piping will be pitched downward toward the tank, terminating in a *bulkhead fitting*. These fittings are used so that the secondary containment is open and free to have liquid spill into a sump to be detected. The product may be mixed with water and should not be returned to the tank but is removed manually into containers.

Leakage within containment sumps and manways is monitored by means of probes sensitive to liquid or vapor. They are suspended in or attached to the sides of the containment. The liquid probe level is adjusted to signal the presence of liquid. Probes are available to discriminate between water and petroleum products. They are connected by wires to a remote panel for annunciation. All leakage is disposed off manually.

Inductive sensors are also available that do not require penetration of the secondary containment pipe. This type of sensor is attached directly to the outside of the secondary pipe to detect leakage by interrupting the inductive path of the sensor.

Aboveground Storage and Distribution Systems

Storage Tanks

Tank Construction. ASTs are factory fabricated in both round and rectangular configurations. Primary tanks are often manufactured with compartments capable of storing different products. NFPA-30 limits motor fuel AST capacity to 6000 gal (22,680 L). All ASTs must be provided with some form of product leakage and overfill containment conforming to Spill Prevention Containment and Countermeasures (SPCC) regulations. This containment shall be either a dike or an impoundment capable of holding 110 percent of tank contents or with integral secondary containment of the primary tank. The leakage, in turn, is removed manually.

Most smaller tanks are provided with an integral secondary containment and an interstitial space. The width of the space varies among different manufacturers. There are several proprietary methods used to construct the secondary containment vault around the primary tank. An often used material is lined concrete, which provides a 2-h fire rating required by UL-2085 if sufficiently thick. Plastic-lined concrete with a wall thickness capable of providing a 2-h fire rating, as required by UL-2085, is generally used for an integral secondary containment. It is a code requirement that the outside of the concrete vault be protected against corrosion, weather, and sunlight. An external secondary containment vault of steel is also used. Insulation between the primary tank and secondary tank is provided by some manufacturers to protect the primary tank from temperature extremes. This may also be used to meet fire safety requirements for a 2-h rating.

Tank Connection and Access. Connections are located only on top of the tank, and they extend through the vault or secondary containment into the primary tank. There is usually no direct access for personnel entry into the primary tank, except for larger tanks. Standard connections include:

1. Tank vent
2. Emergency vent
3. Product dispenser outlet
4. Product fill (either coaxial or single as required)
5. Phase I vapor recovery (if necessary)
6. Tank gauging
7. Leak detection

Tank Filling and Spill Prevention. The tank is filled using a fill port assembly built into the tank encasement or containment at the top of the tank. The delivery truck must have a pump, and the operator is required to climb stairs to make the connection. The fill port cover could be locked if desired. To reach the top-mounted connection with the hose, stairs are provided as part of the installed tank. If a coaxial vapor recovery system is used, only one connection to the tank is required. If a two-point system is used, two connections are required. An optional, ground-level remote tank filling station is usually available as a separate piece of equipment to allow convenient filling without having to connect to the top of the tank. The remote fill station could have a self-contained pump or use the built-in fuel truck pump.

Spill Containment. For tanks with integral secondary containment, there is a containment sump surrounding the fill pipe. The size of the sump ranges from 5 to 15 gal (19 to 57 L). For tanks with external secondary containment, spills will enter the containment and be manually removed. The remote fill station could be provided with an integral spill containment sump to catch any hose spills. A small hand or electric pump could be provided to empty the containment sump back into the primary tank.

Overfill Prevention. Overfilling is prevented by automatic or manual means installed directly on the tank. Automatic overfill-prevention uses an overfill-preventing valve similar to that previously described in the UST subsection. Manual methods include a direct-reading level gauge installed in sight of the operator or an audible high-level alarm activated by a separate probe installed inside the tank. Alarms shall operate when the product level reaches 90 percent of capacity and are stopped when the level reaches 95 percent.

Leak Detection and System Monitoring

Leakage from Tanks. Aboveground tanks require a method of containing any possible product release and preventing contamination of the adjacent environment. Releases can result from small leakage from the tank or catastrophic failure of the tank. Required containment methods include providing a dike completely around the tank, remote secondary containment, or integral secondary containment. For ASTs without secondary containment, a dike shall be provided. Dikes are required to contain 110 percent of tank capacity and to be constructed of materials such as concrete, steel, or impermeable soil designed to resist the full head of water. They must be constructed in conformance with NFPA-30. Discharge from the dikes must have a separator along with the necessary control valves which may have to be self-actuating to conform with local codes. It is recommended that an additional impoundment basin be constructed at least 50 ft (16 m) from the AST and at a safe distance from other buildings or tanks. The purpose is to capture and isolate any flammable liquids released during a fire or tank failure and remove them to a safe distance away from the AST. For ASTs, dikes are seldom used because remote and integral secondary containment has a far looser initial cost.

Remote secondary enclosures are usually made of steel. They are of the totally enclosed type and are sealed in a manner so as to prevent the entrance of rainwater. They are required by code to provide 110 percent nominal capacity of the primary tank.

Tank integral secondary containment is achieved by enclosing the primary tank with an integral containment, usually steel or reinforced concrete. This type of tank has an interstitial space that is monitored for leakage in the same manner as the USTs discussed previously.

System Monitoring

System monitoring consists of product-level gauging and leakage annunciating. The AST systems can be monitored either manually or electronically.

Product-level gauging in the tank could be achieved by the use of a visual level gauge or an electrical gauge either mounted on or immediately adjacent to the tank or at a remote location. Level gauges similar to those installed in USTs can be used. Remotely mounted electronic gauges capable of recording and placing in memory many functions using probes similar to those installed in USTs are commonly used.

Leak detection for ASTs is much easier than for USTs because leakage from the tank can be easily observed manually. Automatic means of system monitoring include a stand-alone alarm panel or an alarm integral to an electronic panel used for product-level indication.

System Design Consideration

All piping and tanks installed underground require secondary containment and leak detection. Because of this, there is actually no drainage into the environment from piping associated with the fuel-dispensing system from a service station. If there is any spill from overfilling, it finds its way onto the ground, where it is quickly evaporated. The discharge piping from double-containment drains all leakage back into a manhole that is part of the storage tank, except for any water present. Any UST leakage is detected, and a signal is sent as an alert of a problem. If the leak is large enough to escape from the secondary containment, remote groundwater monitor wells are required that will detect any product or the fumes from product. No product is permitted to be spilled into the surrounding soil. The tank must be removed from service and raised aboveground to repair the leak.

DIESEL FUEL SYSTEMS

Fire Pump Fuel Storage

General. Diesel fuel is often used to supply fire pumps and emergency generator motors. This section discusses storage facilities for these purposes. The driver most often used to power fire pumps other than an electric motor is a diesel-fueled engine. This section discusses the storage of diesel fuel for that purpose.

Codes and Standards

The principal code for the fire pumps is NFPA-20. In future discussions, this will be referred to as *code*. Other codes may be local fire and building codes, as well as other local regulations governing the storage of diesel fuel.

General Design Considerations

The diesel fuel is stored in a tank whose bottom is elevated about 3 ft 0 in (1 m) above the floor. The fuel is fed by gravity to the diesel engine that has an integral fuel pump as part of the engine assembly.

The code recommends that diesel fuel be stored indoors adjacent to the fire pump if allowed by local regulations. The following discussions concern only an indoor tank. If the storage tank is a remote underground or aboveground tank, the requirements are the same as previously discussed for AST or UST installations. The aboveground tank shall be UL approved.

The minimum amount of actual storage mandated is 1 gal (3.8 L) per pump horsepower. Approximately 10 percent tank capacity should be added to the calculated storage to allow for freeboard.

The tank should be placed on legs to elevate it as high as reasonable to provide static height for gravity feed to the engine. A generally accepted distance is about 3 ft 0 in (1 m) above the floor. The entire tank shall be protected by a dike high enough to contain the entire contents with an additional allowance of several inches freeboard. It is recommended that a fire extinguisher be placed nearby for safety.

A fill line connection is factory installed at the top of the tank. Since the fill line shall extend outside the building to allow the pumper delivery truck to connect to it, the connection point will be placed lower than the top of the tank. Because of this, an antisiphon valve shall be placed on the high point of the pipe run to the fill line to prevent accidental emptying of the tank. A special "quick disconnect" connection shall be provided. If the fill line is in-ground, a check valve connection on the fill line shall be installed.

Other connections on top of the tank are for the vent and gauge. The vent, usually with 2-in size, must extend from the tank to a point 12 ft 0 in (3.6 m) above grade and in a safe location. A fuel-level gauge connection is provided that will permit a visual gauge to be mounted directly onto this connection. In this location the gauge must be read directly in front of the tank. If a more convenient gauge location is desired, a remote gauge can be placed on an adjacent wall near the fill line, if conditions allow. Interconnection tubing must be installed from the tank connection to the gauge. Although having a higher cost, the remote gauge, located at the fill connection if possible, will allow the delivery person to accurately observe the level of fuel in the tank to avoid overfilling.

Interconnecting fuel supply and return piping from the tank to the engine is generally made of copper tube.

Emergency Generator Fuel Storage

Emergency generators could have fuel stored in either AST or UST systems. When fuel is stored inside buildings, the capacity of the tank is limited to a maximum of 660 gal. This amount of fuel shall be stored in a double-contained tank either remotely or as an integral part of the base of the generator. It is common practice for integral tanks to have an interstitial space capable of containing 110 to 140 percent of the tank capacity. When a remote storage tank is used, it is a good practice to provide a day tank containing 2-h fuel supply immediately adjacent to the generator. If the storage containment is a remote underground or aboveground tank, the requirements are the same as previously discussed for AST or UST installations. The aboveground tank shall be UL approved. The installation must conform to all local rules and regulations previously discussed.

The fill port could be mounted at the generator or remotely. For a remote fill, the inlet should be in a locked box so that the public or a casual passerby could not have access to the inlet. On a private site, this will avoid tampering by employees. Another method is to place the fill port about 9 ft (3 m) abovegrade in an adjacent wall. Although this makes it difficult for delivery personnel to gain access, it also makes it much more difficult for the public to try to gain access to the inlet.

Spillage onto grade due to overfilling is prevented by installing visual and audible alarms adjacent to the fill port that signal at the 90 and 95 percent fill levels. The alarms are usually separate to have redundancy, but a single, dual alarm is available. Secondary protection is obtained by providing the nozzle on the delivery hose with an automatic shutoff similar to that on a gasoline-dispensing nozzle. Leakage into the interstitial space of integral storage tanks is usually signaled by a float arrangement or probes similar to the previously discussed alarms. The tank vent is routed away from the storage tank, 12 ft (4 m) abovegrade, either above the generator set or on an adjacent wall. If there is an overfill, the spill is removed manually. If desired, a drain could be placed adjacent to the fill connection and routed to an oil separator before discharge into a sanitary sewer.

The fill pipe and other fuel-containing interconnecting pipe should be double-contained to eliminate potential leaks. The line should be pitched to a low point. A tee at the low point, with a probe connected to an alarm, will detect any primary pipe leakage.

FLAMMABLE AND VOLATILE LIQUIDS

Federal, state, and local regulations have established standards for the discharge of volatile liquids, particularly oil, into stormwater and sanitary sewers. These standards vary among various locations, and the responsible enforcement and code authorities must be consulted to determine the level of treatment or separation required.

The most common flammable liquid is oil. The hazard created is either one of safety, since the vapors could create an explosive condition, a fire hazard where the oil that floats on water could be set on fire, or medical, where the breathing of the vapors is dangerous to health and toxic if ingested. The common characteristic of all volatile liquids discussed is that they are lighter than water and their removal is similar to that of oil. The viscosity of common volatile liquids is given in Table 15.1.

Oil in Water

Oil in water exists in several forms and is considered immiscible since it is in suspension and cannot be dissolved in water. These various forms are

1. Free oil.
2. Mechanically dispersed oil is fine droplets ranging in size from micrometers to fractions of a millimeter. They are stable due to electric charges and other forces but not due to the presence of surface-active agents.

3. Chemically stabilized emulsions are fine droplets that are stable due to surface-active agents.
4. Dispersed oil is suspended in such a small size (typically 5 μ or smaller) that ordinary filtration is not possible.
5. Oil-wet solids, which are particulates to which oil adheres.

Methods of Separation and Treatment

Larger Systems. Oil spills and leaks are best treated in their most concentrated state, which is at their source or as close as reasonable. The intention is to remove the oil to required levels prior to discharge. The primary methods used to separate and remove free oil and oil-wet solids are flotation, skimming, and separation. Secondary treatment, such as chemical treatment, coalescence, and filtration, is then used to break up oil–water emulsions and remove dispersed oil. Finally, tertiary treatment, such as ultrafiltration, biological treatment, and carbon adsorbtion will remove the final amount of oil present. The amount of treatment and filtration is governed by the local authorities as is the acceptable level of oil in wastewater. The AHJ must be consulted prior to selection of the method and criteria for equipment.

These are some of the areas or facilities where the use of a skimmer would be considered:

1. Where cutting oils or coolants would be present
2. Quench oil for heat treatment
3. Parts washing
4. Parking lots and garages
5. Wastewater sumps

The American Petroleum Institute (API) has established criteria for the large-scale removal of globules larger than 150 μ. In abbreviated form, they are as follows:

1. The horizontal velocity through the separator may be up to 15 times the rise velocity of the slowest-rising globule, up to a maximum of 3 ft (1 m).
2. The depth of flow in the separator shall be 3 to 8 in (75 to 200 mm).
3. The width of the separator shall be between 6 and 20 in (150 and 500 mm).
4. The depth-to-width ratio shall be between 0.3 and 0.5.
5. An oil retention baffle should be located no less than 12 in (300 mm) downstream from a skimming device.

Gravity Separators. Gravity separation is the primary and most often used separation method. It is based on the specific-gravity difference between immiscible oil globules and water. Since all these liquids are lighter than an equal volume of water, gravity separators operate on the principle of flotation. As the water and oil flow through the unit, the oil floats to the top and is trapped inside by a series of internal baffles. Since the oil remains liquid, it is easily drawn off.

Flotation Devices. For a larger-scale service, the flotation of oil and oil-wet solids to the top of the flotation chamber can be increased by the attachment of small bubbles of air to the surface of the slow-rising oil globules. This is done by adding compressed air to the bottom of the flotation chamber in a special manner that will create small bubbles that mix with, and attach themselves to, the oil globules.

Centrifugal Separators. For a larger-scale service, the centrifugal separator could be used. This device operates on the principle of inducing the combined oil–water mixture to flow around a circular separation chamber. The lighter oil globules will collect around the central vortex, which contains the oil removal mechanism, and the clear water will collect at the outer radial portion of the separation chamber. Methods have evolved that will produce effluent water with only 50 to 70 ppm of oil, and proprietary devices exist that will lower oil content to 10 ppm.

Filtration. Chemical methods used to break oil–water emulsions followed by depth-type filters to remove the destabilized mixture have proved effective in removing oil globules in a range of sizes between 1 and 50 μ. The velocity and the flow rate of the mixture must be carefully controlled to allow for optimum effectiveness of the system.

Skimmers. An oil skimmer is used when there is a significant amount of oil spilled in water and it floats to the top. This will provide basic information on how oil skimmers operate and the differences between them. Using an oil skimmer is often an inexpensive and effective way to remove free-floating oil. In many cases, an oil skimmer is all that is necessary to bring the normal discharge levels below the water/oil ratio to that which is necessary to release the drainage into a sewer.

Selection Considerations

1. A factor to be considered is the ambient atmosphere in which the skimmer will operate. Some hydrocarbons, such as kerosene and gasoline, give off vapors that may cause a hazardous atmosphere. In this situation, explosionproof components would probably be necessary.
2. The ambient temperature of the water is a factor. In areas where it is possible for either the water to freeze or the cold temperatures to cause the oil to congeal, a unit with a built-in heater will be required.
3. The manufacturers base their skimmer capacities on standard temperatures and viscosities. During the selection process, compare the ambient conditions and product viscosity.

Types of Skimmers

Belt Skimmers. Operation is based on a belt-shaped medium straight down through the surface of the product being skimmed via a motorized head pulley and is kept in a vertical position with a free-turning tail pulley that rides inside the belt. The oil is picked up by the belt and lifted up to the skimmer. It then passes through a set of wiper blades where it is wiped off. The waste oil then flows through a pair of troughs into an approved disposal container.

Tube Skimmers. The operation of a tube skimmer requires a small-diameter, special plastic tube running along the surface of where the product to be skimmed rests. The tube is rotated via a small drive unit, and the collected oil on the tube is scraped off by a set of ceramic wipers.

The advantage of the tube skimmer is that it can handle severe water-level fluctuations and that the unit can operate in shallow water until the bottom of the tube scrapes the bottom.

One disadvantage is that very little tube surface comes into contact with the oil, which requires a larger unit. Another is that the tubes are expensive to replace, and finally, a long tube is required to match the capacities of the belt skimmer.

Mop Skimmer. A mop skimmer uses oil-attracting fibers that are woven together into a rope. The rope is then guided across the surface of the water by the use of floating tail pulleys. The oil is removed by means of a set of motorized wringers.

Advantages include the ability for use in outdoor applications where no humanmade containment is present. It can be used in swamps, rivers, and other open water areas. They are capable of high pickups.

Disadvantages include high cost and greater operating space required. It loses effectiveness when used with high-viscosity oils. The skimming medium is costly to replace.

Smaller Systems

Oil separators for small flows usually take the form of a single unit consisting of a drain grating into which the effluent flows and inside which the oil remains to be drawn off manually. Another type of unit uses an overflow arrangement that sends the trapped oil to a remote oil storage tank.

Because there is the possibility that the vapor given off by the flammable liquid could also ignite, it is important to provide a separator vent that terminates in the open air at an approved location above the highest part of the structure. Some codes require that a flame arrestor be installed on the vent.

The most common material used for an oil interceptor is cast iron, although steel can be used for less severe service. Gratings must have the strength for the type of vehicle expected.

Refer to Fig. 15.5 for an illustration of a typical small oil interceptor. Figure 15.6 illustrates the installation of a typical oil interceptor with gravity oil draw-off for multiple floor drain inlets. Smaller amounts of spilled oil are discussed in Chap. 9.

Sand

Whenever a potential source of ordinary solids discharges into the drainage system, a sand interceptor, or trap, should be provided. A mechanical hazard could be created, since the sand could create a blockage in the piping system. A prime example of the use of a sand trap would be on the discharge of exterior footing drains.

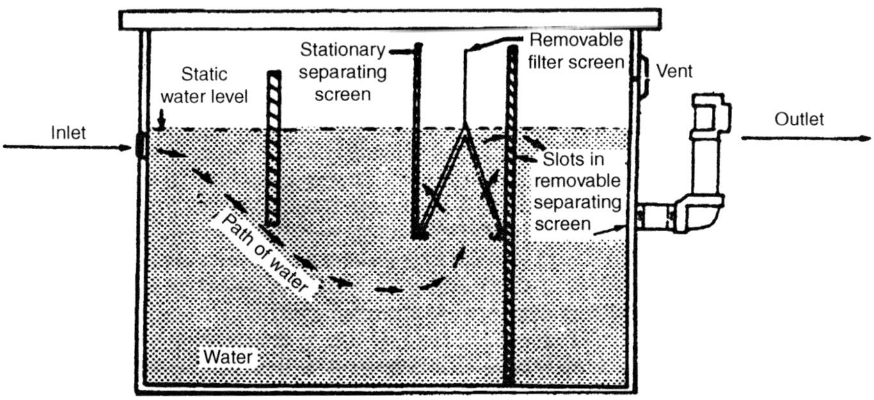

FIGURE 15.5 Typical oil interceptor. (*Courtesy of Rockford Co.*)

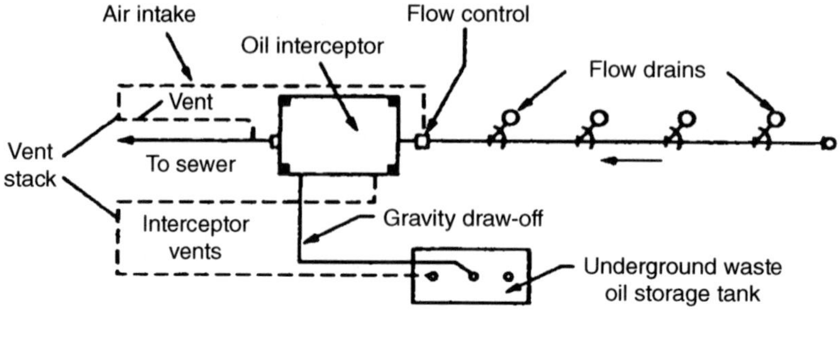

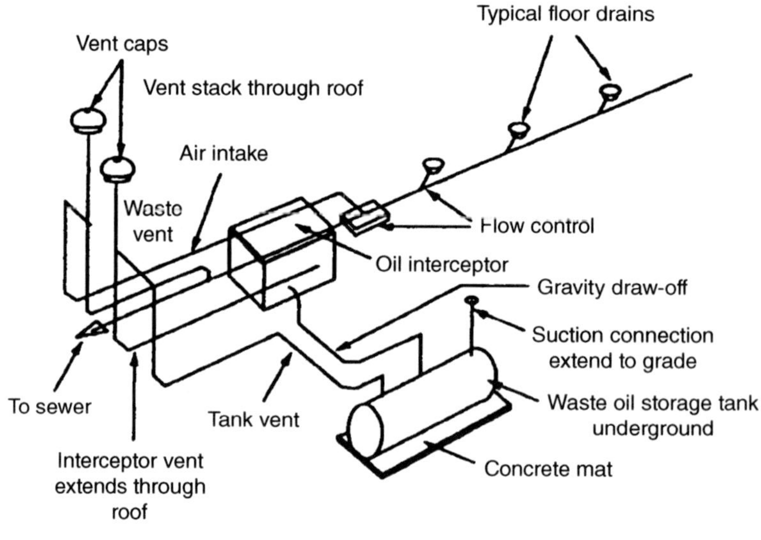

FIGURE 15.6 Typical oil gravity draw-off installation. (*a*) Plan, (*b*) isometric.

Since the solids are heavier than water, these traps operate on the principle of settlement, permitting the solids to accumulate at the bottom as the effluent flows through the device. The outlet from the sand trap should be located so that the accumulated material is prevented from being discharged. The solids must be removed from the trap by hand. Sufficient space must be available around the trap to make this easy to accomplish.

Sand traps are commonly constructed of masonry, but prefabricated units made of cast iron or steel are also used. Refer to Fig. 15.7 for an illustration of a typical sand interceptor.

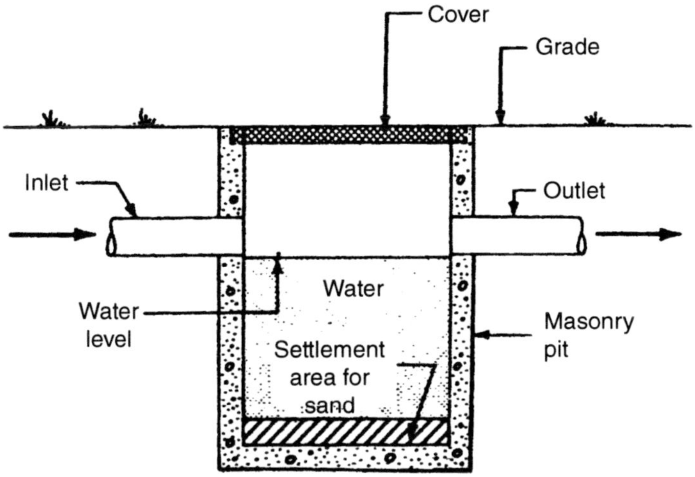

FIGURE 15.7 Sand interceptor.

BIBLIOGRAPHY

American Society of Plumbing Engineers, *Gasoline and Diesel Oil Systems*, Plumbing Engineers Design Handbook, vol. 3.

Floyd, C. M., "UST Leak Detection," *Environmental Protection.*

Gryer, W., "Bringing Storage Tank to the Surface," *Chemical Engineering.*

Leiter, J. L. (ed)., *Underground Storage Tank Guide.*

National Fire Protection Association, *Fire Protection Handbook*, NFPA, Quincy, Mass.

NFPA 30, *"Flammable and Combustible Liquids Code,"* NFPA, Quincy, Mass.

U.S. Environmental Protection Agency, *Musts for UST's.*

Ziu, C. G., *Handbook of Double Containment Piping Systems*, McGraw-Hill, New York.

CHAPTER 16
STORMWATER HARVESTING AND REUSE

This chapter discusses stormwater harvesting systems. It is intended only as a guide for various methods of collecting, storing, treating, and reusing precipitation (rainfall and snowmelt) that falls on the ground. It is not intended to be used for any specific design. The guidance of an experienced consultant is recommended.

Relatively clean surfaces such as the roofs of buildings, grounds, and surrounding areas of a facility are also included. The term *rainwater harvesting* is sometimes referred to as *catchment*, which means the collection, accumulation, recovery, treatment, and use of precipitation. Graywater is sometimes classified as light or dark, with dark water recovered from parking lots, etc. Both potable and nonpotable uses of harvested water are discussed.

It is well known that there is a shortage of potable water in many parts of the United States as well as a shortage of all types of water in many parts of the world. Because of this, finding a supplemental supply of water for such uses as drinking, agricultural, and industrial water is imperative. Rainwater is regarded as a valuable and available source of relatively clean nonpotable water. Of the total water used in United States, 85 percent does not require potable water. It is probable that rainwater harvesting systems will become much more popular and continue to grow. They will be used more as the cost premium comes down. Many owners will use this method because it is the right thing to do in spite of the fact that payback will not be short.

The codes throughout this country are taking an increasingly permissive view of using any alternative source of water and the reuse of rainwater as a safe and reliable water source. Rainwater harvesting is only one part of a larger story due to a broader group of systems that are becoming mainstream.

Other sources, such as graywater and blackwater, are much less clean and require a good deal more costly treatment to make them acceptable for human use or consumption. Graywater and blackwater are discussed in Chap. 2. A building with a rainwater harvesting system installed has the potential for earning LEED credits. For further discussion of LEED, refer to Chap. 23.

CODES AND STANDARDS

1. NSF International Protocol P151, Health Effects From Rainwater Catchment Systems
2. NSF/ANSI 6.1, Drinking Water System Health Effects
3. NSF—various standards for Drinking Water Treatment
4. National Weather Service, Hydro 35

5. NOAA, Technical Memorandums regarding rainfall
6. Applicable sections of model plumbing codes such as provisions of the UPC and IPC
7. Environmental Protection Agency, National Pollutant Discharge Elimination System
8. Rainwater Catchment Design and Installation Standards, ARSCA and ASPE
9. NSF Standard 53

Recent revisions of the model codes contain provisions establishing rainwater collected from roofs and other manmade areas that allow this water to be used for such nonpotable purposes as toilet flushing, trap priming, irrigation, water towers, and industrial processes if approved by the AHJ. The revisions are intentionally broad to include other uses such as car washes, automatic sprinklers, and other common uses. In addition, strict requirements are included that would permit the use of rainwater for potable use in residential applications.

BACKGROUND

Rainwater harvesting is an ancient practice. This has been used for collection and storage of freshwater for many centuries. The use of rainwater in the past includes potable, non-potable, fire protection, agricultural, and industrial purposes. This utilizes an otherwise wasted resource, reduces stormwater runoff and pollution, and helps reduce the demand on public water supplies.

The subject of rainwater harvesting was not mentioned in any major code until February 2010, when the International Association of Plumbing and Mechanical Officials (IAPMO) published the "Green Building Supplement." The water quality for nonpotable use is often not clear in the supplement. In many cases, the local authorities having jurisdiction are not familiar with rainwater harvesting and are undecided how to proceed as to what is acceptable or practical. The authority to allow the design and installation of a harvesting system depends on the willingness for acceptance of the local AHJ. There is still some opposition to the use of harvested rainwater. In Colorado, for instance, the capture of rainwater is not permitted for manufacturing projects because it will prevent the rain from reaching what would be its ultimate destination, such as waterways and groundwater replenishment. In a strange way, this might make sense. It is generally conceded that the biggest impediment to a national rainwater harvesting guideline is the western water rights.

Rainwater can be readily "harvested" primarily from rooftops or other elevated surfaces. Since rain has generally low concentrations of contaminants in the air, it requires less treatment (if at all) than other sources depending on its end use. However, it does pick up some contamination from bird droppings and other animals, as well as dust, dirt, leaves, organic droplets, and gases from the roofs, ground, and air. Rainwater frequency and quantities are difficult to predict. Another source of water is snowmelt from the area where large snowfall and melting systems are provided. Not only does precipitation represent a source of recoverable water, but also in the process of harvesting, it will reduce the frequency of sewage treatment plant overflows where systems are combined.

Why capture and reuse stormwater? There are many uses for harvested water. Nonpotable water uses are listed in Table 16.1. Potable water uses, where permitted, are listed in Table 16.2. The ultimate potential number of uses is limited only by the imagination of the designer.

TABLE 16.1 Nonpotable Water Uses

Landscape irrigation
Garden watering
Vehicle washing
Toilet and urinal flushing
Laundry washing
Building washing
Cooling tower makeup
Fire suppression
Household cleaning
Pond and pool filling
Boiler feedwater

Civil and plumbing engineers should work together during the design process to accomplish desired goals and to meet all the regulatory requirements. Water harvesting applications could use ponded and otherwise stored water directly for turf irrigation or for fire protection. Groundwater replenishment would also be a very important consideration. If the water were to be used for fixture flushing, a higher level of treatment might be necessary to remove turbidity and other odor-producing elements that the raw water contains. Potable water would have to be used as an auxiliary supply if there were insufficient precipitation. The clear separation and identification of potable from nonpotable water is mandatory.

TABLE 16.2 Major Potable Water Uses

Drinking water
Cooking culinary uses
Showering and bathing
Dishwashing

The basic components of stormwater harvesting would include a source of precipitation, a transport system to carry the precipitation to storage, usually a tank, a tank where the water can be collected and stored, a method of treating the water to whatever level is desired for its intended use, and the necessary piping and pumps to distribute the stored water to where it is needed, if it is not done by gravity.

Having decided to proceed, first we consider from where the rainwater supply could be collected for this project. It could be collected from any hard surface. Roofs and decks would be good examples as long as they are clear of any overhanging vegetation and airborne pollution sources. Since roofs are a primary collection point, a comparison of various roofing materials is given in Table 16.3.

TABLE 16.3 Comparison of Roofing Materials

Roofing material	Contaminants
Aluminum	Aluminum salts
Galvanized metal	Lead, cadmium, zinc salts
Sheet metal	Lead
Ceramic materials	Mold, algae, bacteria, moss
Composite materials	Mold, algae, bacteria, moss, dust, soot, petroleum compounds, gravel grit, copper, organics, fillers
Wood	Algae, bacteria, mold, moss, wood preservatives

Condensate and Cooling Tower Discharge

The use of clean condensate from HVAC air-cooling units could be considered in the total amount of water available. Its use and appropriate treatment would require the approval of the AHJ.

Rainwater Basics

There are three categories of harvesting.

1. Active systems utilize pumps, piping, and additional mechanical input.
2. Passive systems use gravity rather than pumps and may involve the use of topography, swales, ditches, and ponds which flow by gravity.
3. Combination systems may use both of the above.

Because regulations are still under development in the United States, most rainwater harvesting systems are used for nondrinking purposes, such as residential and commercial turf irrigation, crop watering, car washing and toilet flushing. Regulations in some local jurisdictions may require that some percentage of the rain obtained from rainwater harvested on site be used for landscape irrigation.

A typical rainwater harvesting system diverts, treats, and stores runoff from residential and commercial roofs and other areas. The most basic collection system for residential use is snow or rain falling on roofs being routed into barrels or small tanks. This has been increasing in use throughout the United States. Systems are made up of a collection surface (gutters, downspouts), first-flush devices, other filtration, a storage tank, sanitation if required, and a distribution system.

Simply put, the rain falls on roofs or other areas, is collected, and then is first conveyed through filters or the first diverting device into a storage tank. From the tank it is delivered to its final destination. Sometimes additional sanitation is required based on its end use. Although nonpotable water use requires much less treatment than that intended for potable use, even the most basic outdoor use of water requires some form of treatment. From a technical perspective, nonpotable water use would require much less rigorous treatment than potable water. Generally, nonpotable water would require only particulate removal for such applications as irrigation purposes, but when used for potable purposes, costly additional treatment to remove biological and organic impurities will be required.

Water Contaminants

It is possible to categorize water contaminant concentration in urban stormwater into plant water demand categories such as those given in Table 16.4. With the exception of water

TABLE 16.4 Typical Contaminant Levels and Classes in Urban Stormwater

Contaminant concentration in urban stormwater		
Contaminants found in runoff	Units	Average concentrations*
Total suspended solids	mg/L	80
Total phosphorus	mg/L	0.30
Total nitrogen	mg/L	2
Total organic carbon	mg/L	12.7
Fecal coliform bacteria	MPN/100 mL	3600
E. coli bacteria	MPN/100 mL	1450
Petroleum hydrocarbons	mg/L	3.5
Cadmium	µg/L	2
Copper	µg/L	10
Lead	µg/L	18
Zinc	µg/L	140
Chlorides (winter only)	mg/L	230
Insecticides	µg/L	0.1–2
Herbicides	µg/L	1–5

*Concentrations represent mean or median storm concentrations measured at typical sites and may be greater during individual storms. Also note that mean or median runoff concentrations from stormwater hotspots are 2 to 10 times higher than those shown here.

Units: mg/L = milligrams per liter
µg/L = micrograms per liter
MPN = most probable number

used for outdoor applications that cannot be inhaled by users, only suspended solids should be removed from the raw water. If the water is to be sprayed, some regulations require that a disinfectant residual be present. If the collected rainwater is to be used inside a building, much more stringent treatments are required even for nonpotable water use.

Biological contaminants are also present. The bacteria that cause disease along with their origin are listed in Table 16.5. The minimum permissible guidelines for bacteria in water for various uses are given in Table 16.6.

TABLE 16.5 Common Diseases and Their Origination

Disease	Bacteria name	Where it comes from	Method of water contamination
Listeriosis	*Listeria monocytogenes*	Fecal matter of infected animals, which can contaminate drinking water supplies and soil	Rodents leave droppings on roof or in catchment system
Leptospirosis	*Leptospira interrogans*	Urine of infected mammals	Rodents urinate on roof or catchment system
Campylobacteriosis	*Campylobacter jejuni*	Fecal matter of animals, domestic and feral; particularly carried by birds both domestic and wild	Birds or rats leave droppings on roof or in catchment system
Salmonellosis	*Salmonella gastroenteriti*	Fecal matter of animals, including birds and reptiles	Birds or geckos leave droppings on roof or in catchment system
Diarrhea	*Escherichia coli*	The fecal matter of all animals, but types pathogenic to humans are mostly found in human and cattle feces	Getting this disease via your water system is unlikely
Enteritis necroticans	*Clostridium perfingens*	Fecal matter of animals, domestic and feral	Animals leave dropping on roof or in catchment system

Preliminary Filtering

Precipitation is generally cleaner than surface water and groundwater, but it is far from being free from all contaminants. Before it contacts the catchment area, it picks up impurities such as gases, salts, and organic material from the air. Once water contacts a catchment surface, it picks up additional contaminants such as those from the surface material itself, leaves, sticks, insects, and wildlife droppings.

It is important to remember that the diversity and level of contaminants in collected rainwater are generally much less than those found in other natural sources. This is an incentive to consider properly treated rainfall as a resource to augment other sources used for residential, commercial, and industrial applications. In a well-designed rainwater harvesting system, emphasis should be placed on removal of contaminants from an initial rain. This will reduce the need for some additional treatment regardless of what the ultimate use of reclaimed water will be. Further treatment is dependent on the final use of the reclaimed water.

TABLE 16.6 Minimum Permissable Water Guidelines for Bacteria

Category of use	Rainwater quality for nonpotable indoor use	Rainwater quality for potable use
Single-family households	Total coliforms: <500 CFU/100 mL Fecal coliforms: <100 CFU/100 mL Water testing recommended annually	Total coliforms: 0 Fecal coliforms: 0 Protozoan cysts: 0 Viruses: 0 Turbidity: <1 NTU Water testing recommended every 3 months
Community or public water system	Total coliforms: <500 CFU/100 mL Fecal coliforms: <100 CFU/100 mL Water testing recommended annually	Total coliforms: 0 Fecal coliforms: 0 Protozoan cysts: 0 Viruses: 0 Turbidity: <0.3 NTU Water testing required monthly The water must meet all other applicable public water supply regulations and water testing requirements.

CFU: colony-forming units
NTU: nephelometric turbidity units
Source: Texas Commission on Environmental Quality.

SYSTEM COMPONENTS

General

All component materials used for the rainwater harvesting system shall be approved for the intended use by the local AHJ, or listed by ANSI/NSF as part of an accredited product certification program, if available.

The Catchment Surface

The roof of a building is the obvious first choice for catchment. Water quality from the material used for roofing is a function of the climatic conditions and the surrounding environment. Possible contaminants for roofing materials are given in Table 16.3. The materials used for potable and nonpotable water are different.

Surfaces for Potable Water

1. All components of the collection system shall be made of nontoxic materials approved for the intended use.
2. All surfaces shall be washed clean before they are placed into service.
3. Painted surfaces are acceptable only if the paint has been certified to ensure that the toxicity level of the paint is acceptable for drinking water. Lead-, chromium-, and zinc-based paints are not permitted. Also not permitted on the catchment surface are lead flashing and wood or cedar shake roofing.
4. Water collected from vehicular parking or similar surfaces is not permitted.

Surfaces for Nonpotable Water

1. Water collected from vehicular parking or similar surfaces is permitted.
2. The collection surface may be constructed of any suitable hard, impervious material.
3. Overhanging vegetation and proximity to airborne pollution sources are to be avoided.

Roof Surface Materials

1. *Metal.* This is a commonly used roofing material. The quality of rainwater collected is a function of the roof texture—the smoother, the better.
2. *Clay and concrete tiles.* These are both porous materials and are suitable for both potable and nonpotable use. They may contribute as much as a 10 percent loss of water due to texture, inefficient flow, and evaporation. To reduce this loss, tiles may be painted or coated with a special sealant or paint. This will also prevent bacterial growth.
3. *Composite or asphalt shingles.* Due to the leaching of toxins, these materials are not acceptable to collect for potable water use. They can be used for irrigation purposes. They may contribute as much as a 10 percent loss of water due to texture.
4. *Slate.* Its smoothness makes it ideal for a catchment surface. The initial high cost may preclude its use.
5. *Others.* Wood shingle, tar, and gravel used to harvest water make it usually suitable only for irrigation because of the leaching of toxins.

Gutters and Downspouts

Gutters and downspouts are installed to collect and route rainwater falling onto the roof to either a tank or a cistern close to the building or connect to piping, carrying it away from the building. The gutters are normally in the shape of half of a pipe that collects the runoff directly from the roof. The downspout conveys that water away from the building. The sizes of gutters and downspouts are given in the local Plumbing Code. Part of the installation is the outlet from the gutter and the fittings that allow the downspout to be fastened to the side of the building, including all the miscellaneous components used to allow the downspout to be installed close to the side of the building. The following precautions shall be observed.

1. For potable water, copper, zinc or wood shall not be used.
2. Sufficient pitch shall be maintained along the gutter to the downspout.

Roof Washers and Prefiltration

General. Contaminants swept along with the rain include dirt, leaves, other deposited foliage, pollen, insects, and wildlife droppings. If these items were to be simply deposited into a cistern or tank and left to sit, the general quality of the water inside would degrade. Those measures used to remove them include self-cleansing mesh screens and first-flush devices to reduce organics in the rainwater before the water enters a tank or cistern. One device is called a "roof washer," and the purpose of the roof washer is to eliminate accumulated debris discharged from the collection surface. This device does not waste any water except for the first 15 min of rainfall. Most of the debris carried off a roof during a storm (often sticks, branches, leaves, bird droppings, etc.) occurs during the early part of a rainfall event. The approximate amount of rainfall to be wasted shall be the first 15 min of rain. There are many variations of manufactured roof washers available; among them is a device called a *vortex filter*. These devices incorporate a fine filter in a housing that gives an unpressurized circular flow of water to clean it. It cleans the water of large material, and a finer filter removes smaller particulates. This allows debris to fall out of suspension and be discharged. One problem with the vortex washer is that 15 percent of the rainwater is continuously wasted. After the initial washing period, the clean water shall enter a tank.

Roof Washers. A roof washer eliminates larger debris from the rainfall. It is commonly called a *first-flush diverter*. Following is a list of considerations that should be used in selecting a roof washer.

1. The inlet to the roof washer shall be provided with a debris screen that protects the roof washer from the intrusion of waste and vermin. The debris screen shall be corrosion-resistant and shall have no openings larger than 0.5 in nominal. Prefilters that have a self-cleaning design are not required to have a debris screen.
2. Water containing debris drained from the first-flush diverter or prefilter shall be piped away from the storage tank and shall terminate in a location that will not cause damage to property or erosion.
3. If more than one storage unit is used, a screen, roof washer, or prefilter shall be provided for each separate cistern. Where cisterns are interconnected to supply water, a single prefilter can be used.
4. First-flush diverters and pre-filters shall be provided with an automatic means of self draining between rain events. A detail of a vortex roof washer and filter commercially available from one manufacturer is illustrated in Fig. 16.1. The vortex filter is a combination of a coarse screen and a fine filter that will remove a majority of particulates. The disadvantage of the vortex filter is that it continuously discharges 15 percent of all rainwater to drain.

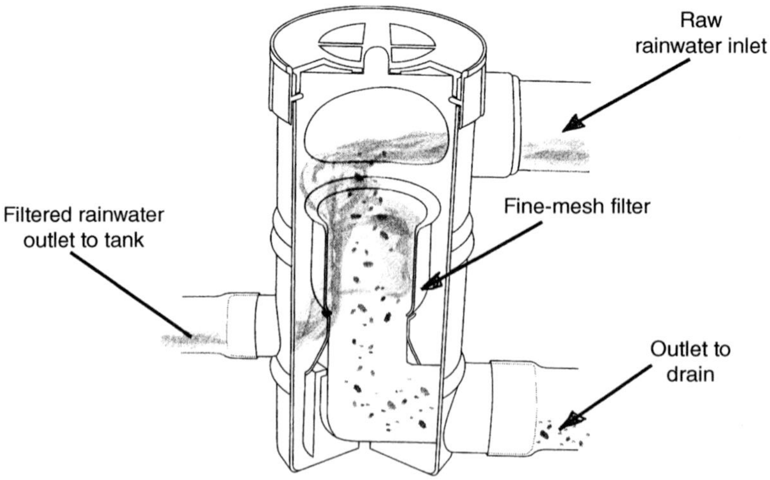

FIGURE 16.1 Vortex rainwater filter and roof washer.

Another type of fabricated first-flush diverter is available that consists of a tee pointing down in the supply line running to the cistern. As the tee is filled with water and debris from the first flush, a ball check valve on the tee prevents flush water from mixing with clean water. Clean water than flows to the cistern. The dead leg of the tee has small perforations at the base that will allow water to drain out during periods of no rain, but constant maintenance will be required to clean the diverter.

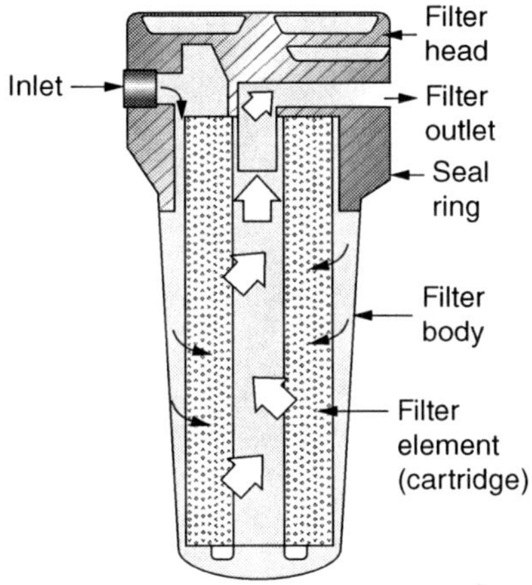

FIGURE 16.2 Cartridge filter.

Prefiltration Fine Filters. In addition to the roof washer which will capture and eliminate larger debris, a fine-mesh prefilter may be desired to eliminate smaller particles that pass through. If a fine filter is required, the use of a cartridge will be generally necessary. It consists of a granular material, such as sand, anthracite, or garnet, in a cartridge, which will retain the debris until it is taken off-line and backwashed. Such a cartridge filter is illustrated in Fig. 16.2.

Storage Tanks

General. The storage tank is usually the costliest part of the harvesting system aside from optional water treatment equipment. The size of the tank is dictated by variables such as the rainwater supply, intended usage, demand, projected length of dry spells, catchment area, personnel preference, and budget.

There are four basic methods needed for promotion of good, aerobic microbial processes:

1. The entering water needs to be oxygenated by encouraging mixing of the water with air.
2. The organic content of the entering water needs to be minimized.
3. A biofilm must be developed inside the tank.
4. The inlet and withdrawal of water from the tank shall avoid disturbing the sedimentation and biofilm.

A myriad of variations have been used in different eras and geographical areas. This discussion concentrates on the most commonly used today, easily installed, and readily available storage tank options.

Storage Tank Basics. In any rainwater harvesting application, it has been thought by many that the quality of water stored for reuse starts with the rainwater harvesting tank, or cistern.

The rainwater harvesting tank, or cistern, can be located either above or below grade. A modern cistern is defined as a small storage tank, usually the size of a wine barrel resting on the ground, although in the past, cisterns have been large storage areas belowground. The larger tank is the most costly component of the system. Aboveground tanks shall be opaque, shielded from direct sunlight, and shall be UV resistant. Below-grade tanks are typically manufactured from plastic or steel and shall be provided with a manhole. The tanks must have enough ballast to avoid floating away. An overflow outlet with a trap is required to carry floating debris out and to keep the tank healthy. A vent is also needed along with a screen to keep insects out.

Selecting the tank size is one of the keys for a successful, efficient, and cost-effective rainwater harvesting system. The expected demand should be based on the projected population and not on the maximum occupancy. If the maximum number were used, an oversize tank would result. As long as the rainwater harvesting system will be backed up by another source, such as a municipal system or a well, it should be sized on an average demand. In general, increasing the tank size leads to a greater ability to supply water because a larger tank can capture more water in a large event. However, going beyond a certain tank size can only lead to a marginal benefit. By comparing the advantages and disadvantages of a variety of tank sizes, the engineer can select a reasonable tank size.

The use of an additional treated water "day tank" in larger systems might be desirable if the whole system is to be utilized for more than one purpose. It would be used to store better-treated water than that used for a lesser purpose. The size would be governed by the expected peak demand rather than the average demand. Selecting a reasonable peak demand rather than a design peak load is important to keep the size of the tank small. While it may appear that it would be more expensive to have a system with two pumps, the multiple applications might be a more cost-effective approach because the pump's piping and treatment equipment for both systems are smaller than that for a single system.

Storage Tank Materials

1. Steel storage tanks are usually used only for the larger industrial and commercial projects where the primary purpose is to accumulate LEED points and to supply a larger demand. The steel tanks are usually large, quite strong, and intended for burial. Steel tanks are usually part of a complete recycling system consisting of treatment equipment, pumps, and a filtering system. They are custom-designed to provide nonpotable water for a variety of uses, all of which are intended to reduce dependence on potable water. A lining to the inside is often applied for corrosion protection. It is a costly installation but will have a return on investment that will prove attractive.
2. Wood tanks have been used for potable and nonpotable water for many years. It is a highly desirable choice when used in urban and suburban locations. These tanks are site-built by skilled technicians and provide a more pleasing appearance. Wood tank used for rainwater storage is considered more contaminated due to the bacteria present in rain rather than in the treated potable water. It is a costly installation.
3. Fiberglass tanks are available in vertical and horizontal configurations. Tanks for potable water use shall be made with a food-grade resin lining, and it should be opaque. Fittings for water connections shall be an integral part of the tank, eliminating the possibility of leaks.
4. Polypropylene tanks are commonly sold by suburban supply retailers and are suitable for smaller installations no larger than 10,000 gal (37,900 L). They are usually installed aboveground, but if they are to be used underground, reinforced tanks are recommended. The connections to piping are usually aftermarket modifications that might be subject to leakage.
5. Sheet metal tanks are galvanized and are available only in smaller sizes. They can be lined for potable water use, and if paint is used, it must be FDA and NSF approved. Most tanks are corrugated and galvanized for corrosion resistance.
6. Other materials, such as concrete and ferrocement, are available for smaller, residential applications.

The Presence of a Biofilm

There is a naturally occurring biofilm that forms in a tank after a period of time. This biofilm is recommended to be maintained and enhanced because it is beneficial for the tank and may remove additional bacteria and contaminants from the water. A small amount of bacteria and sediment will settle to the bottom of the tank and should be allowed to remain. This should be drained out periodically. It is similar to that used in water treatment, and it maintains water quality through the breakdown of organic matter and trapping of metals. If the tank is cleaned, care must be taken to avoid destroying the healthy biofilm.

Water Connections into a Storage Tank

How the water enters the storage tank is very important, and it should be made through a smoothing (stilling) inlet or above the high water level. The action of a smoothing inlet shall not allow water to disturb the bottom of the tank. It also oxygenates the water as well as exchanges the old water with new prefiltered water. It is directed upward to force the older water out the overflow, which should be the same size as the inlet. A detail of a typical stilling inlet is illustrated in Fig. 16.3.

The water discharged from the tank shall be done in a manner that will maintain the high quality of the water. It has been shown that water removed from the tank just below the surface (6 in, or 15 cm) is of the highest quality. This is accomplished by the use of a floating filter, which moves with the volume in the tank, keeping the outlet at the proper height. A floating filter is illustrated in Fig. 16.4.

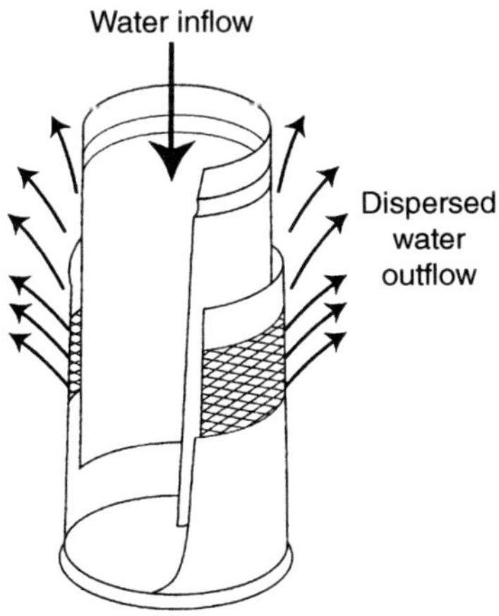

FIGURE 16.3 Typical stilling inlet.

FIGURE 16.4 Floating inlet filter.

To maintain a healthy water system, some overflow should occur. This helps by removing fine floating debris through a surface skimming action. This overflow is directed to a sanitary sewer or used as fertilizer for plants. If it is discharged to a sewer, a trap is required. A detail of a typical overflow with integral trap is illustrated in Fig. 16.5.

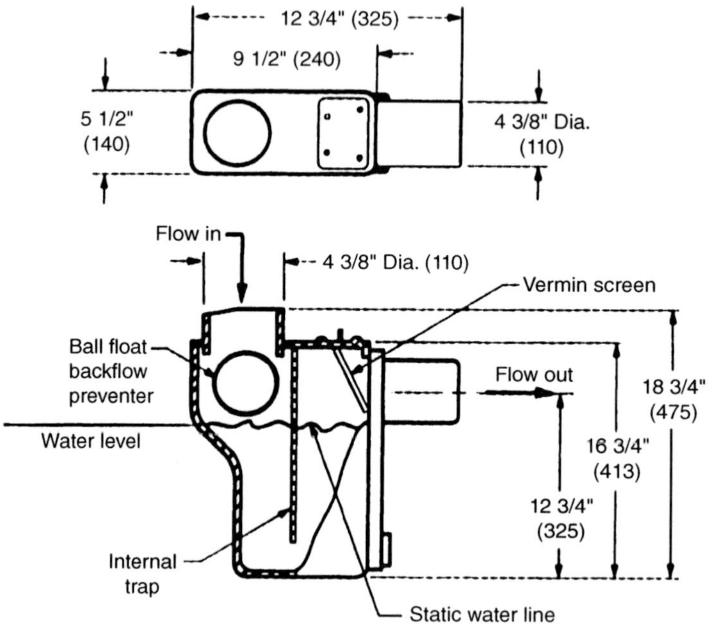

FIGURE 16.5 Tank over flow.

Tank and Water Conveying System Materials

The collection tank is the most expensive component and arguably the most critical to the successful operation of the system. The size of the tank is dictated by a number of factors. Tanks can be located either aboveground or belowground and could be made of a variety of materials. A comparison of possible tank locations is given in Table 16.7. The difference in characteristics of the different tank materials is given in Table 16.8.

TABLE 16.7 Water Tank Locations

Characteristic	Aboveground	Belowground
Installation cost	Low	High
UV-resistant material	Yes	No (exposed surfaces only)
Opaque Walls	Yes	No
Ease of inspection and maintenance	High	Low
Stored water temperature	Ambient	Cool

TABLE 16.8 Comparison of Tank Materials

	Tank materials	Strengths	Weaknesses	Relative cost*
Plastic	Fiberglass	Commercially available; minimal maintenance; lightweight	Must be sited on smooth, solid footing; UV degradable; relatively fragile	2
	Polyethylene and polypropylene	Commercially available; minimal maintenance	UV degradable; fittings may leak	3
Metal	Galvanized steel	Commercially available; lightweight	Possible corrosion (will not rust belowground)	1
Ceramic	Concrete	Durable, immovable; available poured in place of prefabricated; neutralizes pH	Potential to crack and leak	5
	Ferrocement	Durable; immovable; versatile; neutralizes pH	Potential to crack and leak; must be fabricated on side	4
	Concrete block and tile	Durable; immovable; keeps water cool in hot climates	Difficult to maintain; expensive to build; must be fabricated on site	6
Wood	Redwood, cedar, cypress, etc.	Aesthetically appealing	High maintenance; will rot belowground; must be fabricated on site	7

* Installed cost, with 1 being the lowest.

A conveying system comprises a flat area, usually a roof, drains, gutters, and piping that brings water to a harvesting tank. The approval of authorities shall be given if the water is to be used for potable purposes.

Piping

For potable water, the same piping used for any potable water system can be used. For nonpotable water, it is common to use PVC or CPVC piping of a color that will separate that pipe from any potable water. All piping systems used in the treatment and in the building supply system need to be colored blue or purple to identify them as a nonpotable source. It is not permitted by most jurisdictions to install fiberglass piping under the building or foundations. All protections for the possibility of having reverse flow and cross connections required by the AHJ shall be observed.

Labeling

1. At all rainwater supplied fixtures, a prominent label or sign indicating "NONPOTABLE—DO NOT DRINK" shall be affixed.
2. Nonpotable water piping shall be designated by solid color and colored bands as specified by the AHJ or national code. It shall be labeled "NONPOTABLE—RAINWATER."
3. Plumbing outlets and faucets served with harvested rainwater shall be easily recognized by symbol or color for nonpotable water.

Pumps

Water Pump. In practice, pumping water from a cistern or tank to the point of use is no different than that from a well or holding tank. The type and the size of the pump are determined by the peak demand and difference in elevations between the water storage low level and the height of the end use. Submerged pumps are often used and should have easily accomplished maintenance. Alarms shall be provided to shut down flow when the low level is exceeded.

Dye Metering Pump. Another pump is a dye metering pump that is used to inject color into water that is nonpotable, as required by local plumbing codes.

Backflow Prevention

It is very important that any use of rainwater that interconnects with the potable water supply inside the building have adequate protection from cross-flow by means of strict backflow prevention. Rainwater systems are not permitted to have any direct connection to potable water supply or any other alternative water source piping. Potable and other reclaimed water is permitted to be used as makeup water to the harvested system as long as the connection is provided with an air gap or reduced-pressure backflow preventer. This must be periodically tested to ensure that no cross-connection occurred.

SYSTEM SIZING AND COMPONENT DESIGN PARAMETERS

General

The following information for the following sizing and design examples is provided for illustrative purposes only and represents many methods and theories available from several sources. They are presented as methods that are believed to be credible and are not to be used without due consideration for any specific application. Comparison with other responsible information regarding applicability shall be thoroughly made before applying any formula or process for design purposes.

The following general principles shall be observed.

1. Separation shall be maintained between potable and nonpotable systems. Cross-connections are not permitted except with the proper protection as required by local code.
2. All materials used shall be approved and deemed suitable for the intended use.
3. Plastic pipe shall be protected from UV radiation by means of a factory-applied coating or manufacturing process.
4. Where rainwater harvesting and potable water piping are installed in the same trench, wall cavity, or other similar location, the potable piping shall be separated by a minimum distance of 12 in (300 mm) above or adjacent to the rainwater pipe.
5. Where potable water is used, all piping materials shall be nontoxic and approved by local authority.

Additional Water Treatment

When water leaves the tank, it may be treated again depending on its final intended use. The EPA as well as local authorities have established standards that require verification of

TABLE 16.9 Examples of Contaminants by Class

Class	Examples
Suspended solids	Dirt, clay, colloidal materials, silt, dust, insoluble metal oxides and hydroxides
Dissolved organics	Trihalomethanes, synthetic organic chemicals, humic acids, fulvic acids
Dissolved ionics (salts)	Heavy metals, silica, arsenic, nitrates, chlorides, carbonates
Microorganisms	Bacteria, viruses, protozoan, cysts, fungi, algae, molds, yeast cells
Gases	Hydrogen sulfide, methane, sulfur dioxide, carbon dioxide

suitability. Examples of various contaminants are shown in Table 16.9 and are important for appropriate treatment depending on the end use of the water. The Texas Commission on Environmental Safety has established minimum water treatment techniques, and their example could be used as a guideline for water quality. Refer to Table 16.10 for techniques of water treatment. If the water is to be used on the exterior of a facility, the

TABLE 16.10 Treatment Techniques

Treatment	Location	Result
Screening		
Leaf screens and strainers	Gutters and downspouts	Prevents leaves and other debris from entering tank
Settling		
Sedimentation	Within tank	Settles out particulate matter
Activated charcoal	Before tap	Removes chlorine*
Filtering		
Roof washer	Before tank	Eliminates suspended material
In-line/multicartridge	After pump	Sieves sediment
Activated charcoal	After sediment filter	Removes chlorine, improves taste
Slow sand	Separate tank	Traps particulate matter
Microbiological treatment and disinfection		
Boiling/distilling	Before use	Kills microorganisms
Chemical treatments (chlorine or iodine)	Within tank or at pump (liquid, tablet, or granular) Before activated charcoal filter	Kills microorganisms
Ultraviolet light	After activated charcoal filter, before tap	Kills microorganisms
Ozonation	After activated charcoal filter, before tap	Kills microorganisms
Nanofiltration	Before use; polymer membrane (pores 10^{-3} to 10^{-6} in)	Removes molecules
Reverse osmosis	Before use: polymer membrane (pores 10^{-9} in)	Removes ions (contaminants and microorganisms)

*Should be used if chlorine has been used as a disinfectant.
Source: Adapted from *Texas Guide to Rainwater Harvesting*, 2d ed., Texas Water Development Board.

purity requirements are much less than if the water was to be brought inside the building. Irrigation is the most probable outside use. Any use inside the building would require additional treatment. For outside use, the only treatment is a prefilter eliminating the suspended matter prior to the water entering the tank.

Although the harvested rainwater itself, after the first flush, is reasonably clean, it must be additionally treated when the water is brought for indoor use or is expected to come into any contact with people. With this additional treatment, the water will have additional purity to satisfy authorities, but it will still be classified as nonpotable.

1. For potable water final stage filtration
   - *a.* Carbon filtration could be provided for reduction of taste, organic matter, or odor.
   - *b.* All particulate filters shall be installed upstream of disinfection systems.
   - *c.* Disinfection and filtration shall be located as close to final use as possible.
   - *d.* All filters shall be adequately sized and be approved to be in conformance with NSF Standard 53.
   - *e.* Filtration shall be capable of producing water with 50 μm or less particulate size.
2. For nonpotable water filtration
   - *a.* Where water is used for exterior nonpotable use and noncritical installations such as irrigation, washdown, etc. A final-stage filter is not required.
   - *b.* Where water is used for interior nonpotable use such as for toilets and laundry, the water shall be filtered as a safeguard against sediments or discoloration that would interfere with valve operation. Some regulators require a disinfectant residual to be in any water that may come in contact with humans.
3. For potable water disinfection, regulators require a disinfectant residual to be in any water that may come in direct contact with humans.
   - *a.* Chlorination may be used with an automatic demand feed system, and if used, it shall ensure adequate contact time and residual according to local AHJ.
   - *b.* Ozone may be used with an approved zone system ensuring adequate time with the ozone. Provision must be made for off-gas ozone to be piped to a safe environment.
   - *c.* Ultraviolet disinfection may be used and shall be provided between filtration and the final point of use.

Following are details of a number of treatment technologies. A review appears in Table 16.10.

1. *Suspended solids removal.* This can be done with a cartridge-type filter that must be replaced when full. It is used when flow is 5 gpm (25 L/min) or less. Another type of filter is called a *media filter*, which contains sand, anthracite, or garnet and will remove particulates down to 10 to 20 μm with a flow between 5 and 20 gpm (25 and 90 L/min). Membrane technologies can also be used to remove particulates, but they are higher in initial cost. The waste solution is discharged into the sanitary sewer. Microfiltration is designed to remove particulate matter down to submicrometer range and capable of operating at 10 to 20 gpm (25 to 90 L/min). A cartridge filter is illustrated in Fig. 16.2.
2. *Dissolved organics removal.* This is done with specially prepared granular carbon media. It is also capable of removing unpleasant taste and odor. This medium is usually replaced twice a year.
3. *Dissolved salts removal.* The most common method for removal is reverse osmosis.
4. *Contamination.* Contamination from microorganisms represents the most troublesome of the contaminants. Disinfection is the process of eliminating or inactivating those microorganisms in the tank, if required by local authorities. Chemicals to reduce the microorganisms can be added to the water by pumping a solution or tablets of calcium

TABLE 16.11 Additional Treatment Options

Parking lots	• Higher UV treatment doses, combined with secondary chemical treatment • Requires oil and emulsion separation
Cooling tower blowdown	• Membrane filtration to remove TDS • Descaling agents
HVAC condensate	• Membrane filtration to remove TDS • Descaling agents
RO reject water	• Membrane filtration to remove TDS • Descaling agents
Graywater	• Higher UV treatment doses, combined with secondary chemical treatment • May require oil and emulsion separation • Slow sand filtration • Requires chemical removal (chemical destruct or oxidation) • Possible membrane treatment

hypochlorite and a variety of other disinfectants. The approval of the method used would be required from the AHJ.

5. *Salts removal.* The most common method for removal is reverse osmosis.

Table 16.11 has been prepared to address specific contamination that would result from water collected from nonstandard sources.

Pumps

Where a water pump is provided, it shall conform to the following:

1. If used for potable water, the pump and piping shall be listed with the appropriate authority.
2. The pump shall be capable of delivering the water with a minimum of 15 psig (100 kPa) residual pressure at the most hydraulically remote outlet. Maximum pressure shall not exceed 75 psig (500 kPa). Install a pressure-reducing valve if the pump is capable of producing higher than this pressure.

SYSTEM SIZING FOR TYPICAL PROJECTS

Simply put, the rainwater harvesting system has the following requirements:

1. The total demand for potable use only comes from the estimated plumbing use of water.
2. The total demand for industrial use only comes from the estimated use of industrial water.
3. The total demand for irrigation uses comes from the estimated water requirements and losses from the plants and soil.
4. The total demand for all uses comes from the estimated total amount of water used for domestic use, other commercial uses for facility purposes, and irrigation all added together.

5. Storage consists of consideration of the supply and demand for the total use over design time.

All of these subjects are discussed later in this chapter.

ESTIMATING WATER AVAILABLE FOR USE

Estimating Residential Water Demand

The judgment of the design engineer is necessary to determine the quantity of water needed for the intended use and the consequences of running out of water. It is necessary for the engineer to make the design calculations that reflect the actual plumbing fixtures, the number of occupants, and the specific usage patterns of the individual home. A list of typical residential potable water use is given in Table 16.12. The actual demand would be additive for the number of fixtures, the number of occupants, and the length of time the storage tank is sized for.

Estimating Industrial Water Demand

An estimate of potable water use for industrial plant toilet use is listed in Table 16.13. Any additional demand would be determined in consultation with the facility. The actual demand would be additive for the length of time the storage tank is sized for. Again, the storage tank size would be determined by using 1 month's water storage as long as the amount of rainfall is consistent. The 1-month amount of water stored is based on one-fourth of a consistent entire year's supply. If it is not consistent, there should be a connection to a public water supply to provide the difference based upon the actual supply.

TABLE 16.12 Residential Use of Potable Water

	Water consumption using water-conserving fixtures	Assumptions AWWA residential end-use study	Actual daily demand, gal
Toilets			
Ultra-low-flush	1.6 gal/flush	6 flushes/person/day	4.80
Dual flush	1 gal/flush (liquids) and 1.6 gal/flush (solid)	6 flushes/person/day	3.60
Baths and showers			
Showerhead	2.2 gpm	5 min/person/day	11
Bath	50 gal/bath	NA	50
Faucets			
Personal hygiene, cooking, and cleaning of surfaces	2.2 gal/faucet/min	5 min/person/day	11
Appliances or uses measured on a per-use, not per-person, basis			
Clothes washer, front loading, horizontal axis	18–25 gal/load	2.6 loads/week	63
Dishwasher	8 gal/cycle	0.7 cycle/day	5.60

*Appliances or use measured on a per-person basis

TABLE 16.13 Industrial Plumbing Use of Potable Water

	Gallons/use	Uses/person	Gallons/person/day
Toilet/urinal	1.6	4.0	6.4
Lavatory	1.0	4.0	4.0
Shower	25.0	0.4	10.0
	Total daily usage per person = 20.4 gal		

Estimating Irrigation Water Demand

Water for irrigation is considered for only estimated underground irrigation use. The reason for this is the additional treatment necessary if an aboveground sprinkler system will spray water that might be breathed in by people passing by.

Water used to irrigate landscaping often equals or exceeds the interior water use. Selecting native plants or plants that thrive in the similar climate where the facility is sited might be one answer necessary to reduce demand.

Because plant water needs vary greatly depending on soils, climate, plant size, etc., the engineer must choose one of several methods of estimating the water demand for irrigation. They all are based on a 1-month supply of water. The first method is calculated by the following formula.

$$G = R \times F \times A \times I \tag{16.1}$$

where G = gal/mo
R = rainfall, ft/mo (from Table 16.17)
F = 0.62 (in of rain/ft^2 converted to gallons)
A = area, ft^2
I = imperviousness of area (from Table 17.1)

The second method uses typical plant water demand based on area, and it is approximated in Table 16.14. The figure obtained would be multiplied by the amount of square feet in the area of planting.

TABLE 16.14 Total Plant Water Demand

Plant water demand category	Water use ft/ft^2
Very low water use	0.8
Low water use	1.7
Medium water use	2.9
High water use	4.2

The third method estimates the water for minimum water requirements over the square feet of irrigation area based on soil type where planting occurs. This is shown in Table 16.15.

Yet another method uses the *potential evapotransparation* (PET). This method estimates the amount of water evaporated from the soil in addition to the amount of water given off from plants, based on the humidity and temperature of the location. The PET is given in Table 16.16 for 1 day. The figure obtained would be multiplied by 30 days for 1 month.

TABLE 16.15 Soil Water Demand

Soil type	Minimum irrigation area per 100 gal of estimated graywater discharge per day, ft^2
Coarse sand or gravel	20
Fine sand	25
Sandy loam	40
Sandy clay	60
Clay with considerable sand or gravel	90
Clay with small amounts of sand or gravel	120

TABLE 16.16 Climate PET Table

Climate*	Inches daily	Gal/ft^2
Cool humid	0.10–0.15	0.062–0.09
Cool dry	0.15–0.20	0.09–0.125
Warm humid	0.15–0.20	0.09–0.125
Warm dry	0.20–0.25	0.125–0.156
Hot humid	0.20–0.30	0.125–0.187
Hot dry	0.30–0.45	0.187–0.218
	↑ Worst case	

*Cool is under 70°F as an average midsummer high. Warm is between 70 and 90°F as midsummer highs. Hot equals over 90°F. Humid is over 50% as average midsummer relative humidity (dry = under 50%).

Source: Rainbird, Inc.

Irrigation water requirements can be greatly reduced by the use of mulch, which is the choice of the owner to apply. The mulch retains water more efficiently than soil.

It is suggested that an irrigation consultant be contacted prior to the use of any method.

Estimating Rainfall Runoff

Estimating Rainfall, General. Two methods of determining system sizing use a monthly average rainfall (preferred) or the monthly median rainfall. Although the median rainfall is considered a more conservative method of determining storage, it will not be considered due to the lack of sufficient reliable information for the location. It might be available from local authorities.

Estimating Rainfall Runoff Using Rules of Thumb

1. English units. Rough rule of thumb for calculating rainfall runoff volume on a catchment surface: you can collect 600 gal (2300 L) of water per inch of rain falling on 1000 ft^2 of catchment surface.
2. You can collect 27,000 gal (102,600 L) of water per inch of rain falling on 1 acre of catchment surface.
3. You can collect 1000 L of water per 10 mm of rain falling on 100 m^2 of catchment surface.

4. You can collect 100,000 L of water per 10 mm of rain falling on 1 hectare of catchment surface.

Estimating Rainfall Volume Using Monthly Average Rainfall. The monthly average rainfall is obtained from the state or the closest city to the facility under design or from NOAA, using a report entitled "Climate Normals." A typical listing of monthly average rainfall is given in Table 16.17. The collection surface is often dictated by the architectural constraints, such as roof area. Equation (16.1) calculates the required facility catchment area based on the estimated rainfall, which shall require consultation with the architect if inadequate.

The amount of catchment area, based on the needed water demand per month, is found by the using the following formula:

$$A = \frac{D}{0.623} \times R \times E \tag{16.2}$$

where A = required surface area
D = demand per event, gal/mo, based on plants and planted area
R = rainfall for time period being considered, in/mo (obtained from Fig. 17.12)
E = efficiency (loss from leakage, evaporation, roof composition, etc., normally 0.85)

The number of gallons to be realized per rainfall event is described by the following formula:

$$G = R \times F \times E \tag{16.3}$$

where G = gallons (total of rainfall expected)
R = inches of rainfall (average)
F = factor (0.62 gal/in of rain per ft^2)
E = efficiency (loss from leakage, evaporation, roof composition, etc., normally 0.85)

Estimating Tank Capacity for a Project

The determination of the tank capacity is a very inexact calculation. The engineer's problem is to calculate an approximate water demand estimate and then estimate the project's available water. This would be coupled with the cost of the tank. Also to be considered is the length of time that rainwater is easily available, since it must store sufficient water to meet the engineer's requirements. It is not critical to collect water for tank replenishment if there is an alternative source of water, such as a well or municipal supply.

The information used to size the tank shall be used only as an initial guide. If the maximum volume is less than the annual water demand, than either the catchment area needs to be expanded or the demand shall be reduced.

The demand must take into account the many uses for the water discussed earlier. The capacity of the tank is based on several factors, including

1. Intended use of the water
2. Daily and instantaneous intended flow requirements
3. Rainwater frequency on a monthly basis
4. Rainwater volume (based on average conditions and daily rainfall)
5. Collection surface area

TABLE 16.17 Average Rainfall Data

	Years	Jan.	Feb.	Mar.	Apr.	May	Jun.	Jul.	Aug.	Sep.	Oct.	Nov.	Dec.	Annual
Birmingham AP, AL	30	5.45	4.21	6.10	4.67	4.83	3.78	5.09	3.48	4.05	3.23	4.63	4.47	53.99
Huntsville, AL	30	5.52	4.95	6.68	4.54	5.24	4.22	4.40	3.32	4.29	3.54	5.22	5.59	57.51
Mobile, AL	30	5.75	5.10	7.20	5.06	6.10	5.01	6.54	6.20	6.01	3.25	5.41	4.66	66.29
Montgomery, AL	30	5.04	5.45	6.39	4.38	4.14	4.13	5.31	3.63	4.22	2.58	4.53	4.97	54.77
Anchorage, AK	30	0.68	0.74	0.65	0.52	0.70	1.06	1.70	2.93	2.87	2.09	1.09	1.05	16.08
Annette, AK	30	9.67	8.05	7.96	7.37	5.73	4.72	4.26	6.12	9.49	13.86	12.21	11.39	100.83
Barrow, AK	30	0.12	0.12	0.09	0.12	0.12	0.32	0.87	1.04	0.69	0.39	0.16	0.12	4.16
Bethel, AK	30	0.62	0.51	0.67	0.65	0.85	1.60	2.03	3.02	2.31	1.43	1.37	1.12	16.18
Bettles, AK	30	0.84	0.61	0.55	0.38	0.85	1.43	2.10	2.54	1.82	1.08	0.90	0.87	13.97
Big Delta, AK	30	0.34	0.41	0.22	0.20	0.77	2.38	2.77	2.11	1.03	0.73	0.59	0.39	11.94
Cold Bay, AK	30	3.08	2.59	2.48	2.30	2.65	2.89	2.53	3.59	4.51	4.54	4.79	4.33	40.28
Fairbanks, AK	30	0.56	0.36	0.28	0.21	0.60	1.40	1.73	1.74	1.12	0.92	0.68	0.74	10.34
Gulkana, AK	30	0.45	0.52	0.36	0.22	0.59	1.54	1.82	1.80	1.44	1.02	0.67	0.97	11.40
Homer, AK	30	2.61	2.04	1.82	1.21	1.07	0.96	1.45	2.28	3.37	2.77	2.87	3.00	25.45
Juneau, AK	30	4.81	4.02	3.51	2.96	3.48	3.36	4.14	5.37	7.54	8.30	5.43	5.41	58.33
King Salmon, AK	30	1.03	0.72	0.79	0.94	1.35	1.70	2.15	2.89	2.81	2.10	1.54	1.39	19.41
Kodiak, AK	30	8.17	5.72	5.22	5.48	6.31	5.38	4.12	4.48	7.84	8.36	6.63	7.64	75.35
Kotzebue, AK	30	0.55	0.42	0.38	0.41	0.33	0.57	1.43	2.00	1.70	0.95	0.71	0.60	10.05
Mcgrath, AK	30	1.04	0.74	0.81	0.66	1.02	1.45	2.32	2.75	2.36	1.46	1.46	1.44	17.51
Nome, AK	30	0.92	0.75	0.60	0.65	0.74	1.14	2.15	3.23	2.51	1.58	1.28	1.01	16.56
St. Paul Island, AK	30	1.74	1.25	1.12	1.12	1.21	1.41	1.91	2.96	2.79	2.70	2.87	2.13	23.21
Talkeetna, AK	30	1.45	1.28	1.26	1.22	1.64	2.41	3.24	4.53	4.35	3.06	1.78	1.96	28.18
Unalakleet, AK	30	0.40	0.31	0.39	0.35	0.55	1.25	2.15	2.92	2.10	0.89	0.66	0.47	12.44
Valdez, AK	30	6.02	5.53	4.49	3.55	3.08	3.01	3.84	6.62	9.59	8.58	5.51	7.59	67.41
Yakutat, AK	30	18.18	10.99	11.41	10.80	9.78	7.17	7.88	13.27	20.88	24.00	15.17	15.85	160.38

Flagstaff, AZ	30	2.18	2.56	2.62	1.29	0.80	0.43	2.40	2.89	2.12	1.93	1.86	1.83	22.91
Phoenix, AZ	30	0.83	0.77	1.07	0.25	0.16	0.09	0.99	0.94	0.75	0.79	0.73	0.92	8.29
Tucson, AZ	30	0.99	0.88	0.81	0.28	0.24	0.24	2.07	2.30	1.45	1.21	0.67	1.03	12.17
Winslow, AZ	30	0.46	0.53	0.61	0.27	0.36	0.30	1.18	1.31	1.02	0.90	0.55	0.54	8.03
Yuma, AZ	30	0.38	0.28	0.27	0.09	0.05	0.02	0.23	0.61	0.26	0.26	0.14	0.42	3.01
Fort Smith, AR	30	2.37	2.59	3.94	3.91	5.29	4.28	3.19	2.56	3.61	3.94	4.80	3.39	43.87
Little Rock, AR	30	3.61	3.33	4.88	5.47	5.05	3.95	3.31	2.93	3.71	4.25	5.73	4.71	50.93
North Little Rock, AR	30	3.37	3.27	4.88	5.03	5.40	3.51	3.15	2.97	3.53	3.81	5.74	4.53	49.19
Bakersfield, CA	30	1.18	1.21	1.41	0.45	0.24	0.12	0.00	0.08	0.15	0.30	0.59	0.76	6.49
Bishop, CA	30	0.88	0.97	0.62	0.24	0.26	0.21	0.17	0.13	0.28	0.20	0.44	0.62	5.02
Eureka, CA	30	5.97	5.51	5.55	2.91	1.62	0.65	0.16	0.38	0.86	2.36	5.78	6.35	38.10
Fresno, CA	30	2.16	2.12	2.20	0.76	0.39	0.23	0.01	0.01	0.26	0.65	1.10	1.34	11.23
Long Beach, CA	30	2.95	3.01	2.43	0.60	0.23	0.08	0.02	0.10	0.24	0.40	1.12	1.76	12.94
Los Angeles AP, CA	30	2.98	3.11	2.40	0.63	0.24	0.08	0.03	0.14	0.26	0.36	1.13	1.79	13.15
Los Angeles C.O., CA	30	3.33	3.68	3.14	0.83	0.31	0.06	0.01	0.13	0.32	0.37	1.05	1.91	15.14
Mount Shasta, CA	30	7.06	6.45	5.81	2.65	1.87	0.99	0.39	0.43	0.87	2.21	5.08	5.35	39.16
Redding, CA	30	6.50	5.49	5.15	2.40	1.66	0.69	0.05	0.22	0.48	2.18	4.03	4.67	33.52
Sacramento, CA	30	3.84	3.54	2.80	1.02	0.53	0.20	0.05	0.06	0.36	0.89	2.19	2.45	17.93
San Diego, CA	30	2.28	2.04	2.26	0.75	0.20	0.09	0.03	0.09	0.21	0.44	1.07	1.31	10.77
San Francisco AP, CA	30	4.45	4.01	3.26	1.18	0.38	0.11	0.03	0.07	0.20	1.04	2.49	2.89	20.11
San Francisco C.O., CA	30	4.72	4.15	3.40	1.25	0.54	0.13	0.04	0.09	0.28	1.19	3.31	3.18	22.28
Santa Barbara, CA	30	3.57	4.28	3.51	0.63	0.23	0.05	0.03	0.11	0.42	0.52	1.32	2.26	16.93
Santa Maria, CA	30	2.64	3.23	2.94	0.91	0.32	0.05	0.03	0.05	0.31	0.45	1.24	1.84	14.01
Stockton, CA	30	2.71	2.46	2.28	0.96	0.50	0.09	0.05	0.05	0.33	0.82	1.77	1.82	13.84
Alamosa, CO	30	0.25	0.21	0.46	0.54	0.70	0.59	0.94	1.19	0.89	0.67	0.48	0.33	7.25

(Continued)

TABLE 16.17 Average Rainfall Data (*Continued*)

	Years	Jan.	Feb.	Mar.	Apr.	May	Jun.	Jul.	Aug.	Sep.	Oct.	Nov.	Dec.	Annual
Colorado Springs, CO	30	0.28	0.35	1.06	1.62	2.39	2.34	2.85	3.48	1.23	0.86	0.52	0.42	17.40
Denver, CO	30	0.51	0.49	1.28	1.93	2.32	1.56	2.16	1.82	1.14	0.99	0.98	0.63	15.81
Grand Junction, CO	30	0.60	0.50	1.00	0.86	0.98	0.41	0.66	0.84	0.91	1.00	0.71	0.52	8.99
Pueblo, CO	30	0.33	0.26	0.97	1.25	1.49	1.33	2.04	2.27	0.84	0.64	0.58	0.39	12.39
Bridgeport, CT	30	3.73	2.92	4.15	3.99	4.03	3.57	3.77	3.75	3.58	3.54	3.65	3.47	44.15
Hartford, CT	30	3.84	2.96	3.88	3.86	4.39	3.85	3.67	3.98	4.13	3.94	4.06	3.60	46.16
Wilmington, DE	30	3.43	2.81	3.97	3.39	4.15	3.59	4.28	3.51	4.01	3.08	3.19	3.40	42.81
Washington Dulles AP, DC	30	3.05	2.77	3.55	3.22	4.22	4.07	3.57	3.78	3.82	3.37	3.31	3.07	41.80
Washington Nat'l. AP, DC	30	3.21	2.63	3.60	2.77	3.82	3.13	3.66	3.44	3.79	3.22	3.03	3.05	39.35
Apalachicola, FL	30	4.87	3.76	4.95	3.00	2.62	4.30	7.31	7.29	7.10	4.18	3.62	3.51	56.51
Daytona Beach, FL	30	3.13	2.74	3.84	2.54	3.26	5.69	5.17	6.09	6.61	4.48	3.03	2.71	49.29
Fort Myers, FL	30	2.23	2.10	2.74	1.67	3.42	9.77	8.98	9.54	7.86	2.59	1.71	1.58	54.19
Gainesville, FL	30	3.51	3.39	4.26	2.86	3.23	6.78	6.10	6.63	4.37	2.50	2.17	2.56	48.36
Jacksonville, FL	30	3.69	3.15	3.93	3.14	3.48	5.37	5.97	6.87	7.90	3.86	2.34	2.64	52.34
Key West, FL	30	2.22	1.51	1.86	2.06	3.48	4.57	3.27	5.40	5.45	4.34	2.64	2.14	38.94
Miami, FL	30	1.88	2.07	2.56	3.36	5.52	8.54	5.79	8.63	8.38	6.19	3.43	2.18	58.53
Orlando, FL	30	2.43	2.35	3.54	2.42	3.74	7.35	7.15	6.25	5.76	2.73	2.32	2.31	48.35
Pensacola, FL	30	5.34	4.68	6.40	3.89	4.40	6.39	8.02	6.85	5.75	4.13	4.46	3.97	64.28
Tallahassee, FL	30	5.36	4.63	6.47	3.59	4.95	6.92	8.04	7.03	5.01	3.25	3.86	4.10	63.21
Tampa, FL	30	2.27	2.67	2.84	1.80	2.85	5.50	6.49	7.60	6.54	2.29	1.62	2.30	44.77
Vero Beach, FL	30	2.89	2.45	4.20	2.88	3.80	6.03	6.53	6.04	6.84	5.04	3.04	2.19	51.93
West Palm Beach, FL	30	3.75	2.55	3.68	3.57	5.39	7.58	5.97	6.65	8.10	5.46	5.55	3.14	61.39
Athens, GA	30	4.69	4.39	4.99	3.35	3.86	3.94	4.41	3.78	3.53	3.47	3.71	3.71	47.83
Atlanta, GA	30	5.03	4.68	5.38	3.62	3.95	3.63	5.12	3.67	4.09	3.11	4.10	3.82	50.20
Augusta, GA	30	4.50	4.11	4.61	2.94	3.07	4.19	4.07	4.48	3.59	3.20	2.68	3.14	44.58

Columbus, GA	30	4.78	4.48	5.75	3.84	3.62	3.51	5.04	3.78	3.07	2.33	3.97	4.40	48.57
Macon, GA	30	5.00	4.55	4.90	3.14	2.98	3.54	4.32	3.79	3.26	2.37	3.22	3.93	45.00
Savannah, GA	30	3.95	2.92	3.64	3.32	3.61	5.49	6.04	7.20	5.08	3.12	2.40	2.81	49.58
Hilo, HI	30	9.74	8.86	14.35	12.54	8.07	7.36	10.71	9.78	9.14	9.64	15.58	10.50	126.27
Honolulu, HI	30	2.73	2.35	1.89	1.11	0.78	0.43	0.50	0.46	0.74	2.18	2.27	2.85	18.29
Kahului, HI	30	3.74	2.36	2.35	1.75	0.66	0.23	0.49	0.53	0.39	1.05	2.17	3.08	18.80
Lihue, HI	30	4.59	3.26	3.58	3.00	2.87	1.82	2.12	1.91	2.69	4.25	4.70	4.78	39.57
Boise, ID	30	1.39	1.14	1.41	1.27	1.27	0.74	0.39	0.30	0.76	0.76	1.38	1.38	12.19
Lewiston, ID	30	1.14	0.95	1.12	1.31	1.56	1.16	0.72	0.75	0.81	0.96	1.21	1.05	12.74
Pocatello, ID	30	1.14	1.01	1.38	1.18	1.51	0.91	0.70	0.66	0.89	0.97	1.13	1.10	12.58
Chicago, IL	30	1.75	1.63	2.65	3.68	3.38	3.63	3.51	4.62	3.27	2.71	3.01	2.43	36.27
Moline, IL	30	1.58	1.51	2.92	3.82	4.25	4.63	4.03	4.41	3.16	2.80	2.73	2.20	38.04
Peoria, IL	30	1.50	1.67	2.83	3.56	4.17	3.84	4.02	3.16	3.12	2.77	2.99	2.40	36.03
Rockford, IL	30	1.41	1.34	2.39	3.62	4.03	4.80	4.10	4.21	3.47	2.57	2.63	2.06	36.63
Springfield, IL	30	1.62	1.80	3.15	3.36	4.06	3.77	3.53	3.41	2.83	2.62	2.87	2.54	35.56
Evansville, IN	30	2.91	3.10	4.29	4.48	5.01	4.10	3.75	3.14	2.99	2.78	4.18	3.54	44.27
Fort Wayne, IN	30	2.05	1.94	2.86	3.54	3.75	4.04	3.58	3.60	2.81	2.63	2.98	2.77	36.55
Indianapolis, IN	30	2.48	2.41	3.44	3.61	4.36	4.13	4.42	3.82	2.88	2.76	3.61	3.03	40.95
South Bend, IN	30	2.27	1.98	2.89	3.62	3.50	4.19	3.73	3.98	3.79	3.27	3.39	3.09	39.70
Des Moines, IA	30	1.03	1.19	2.21	3.58	4.25	4.57	4.18	4.51	3.15	2.62	2.10	1.33	34.72
Dubuque, IA	30	1.28	1.42	2.57	3.49	4.12	4.08	3.73	4.59	3.56	2.50	2.49	1.69	35.52
Sioux City, IA	30	0.59	0.62	2.00	2.75	3.75	3.61	3.30	2.90	2.42	1.99	1.40	0.66	25.99
Waterloo, IA	30	0.84	1.05	2.13	3.23	4.15	4.82	4.20	4.08	2.95	2.49	2.10	1.11	33.15
Concordia, KS	30	0.66	0.73	2.35	2.45	4.20	3.95	4.20	3.24	2.50	1.84	1.45	0.86	28.43
Dodge City, KS	30	0.62	0.66	1.84	2.25	3.00	3.15	3.17	2.73	1.70	1.45	1.01	0.77	22.35

(Continued)

TABLE 16.17 Average Rainfall Data (*Continued*)

	Years	Jan.	Feb.	Mar.	Apr.	May	Jun.	Jul.	Aug.	Sep.	Oct.	Nov.	Dec.	Annual
Goodland, KS	30	0.43	0.44	1.20	1.51	3.46	3.30	3.54	2.49	1.12	1.05	0.82	0.40	19.76
Topeka, KS	30	0.95	1.18	2.56	3.14	4.86	4.88	3.83	3.81	3.71	2.99	2.31	1.42	35.64
Wichita, KS	30	0.84	1.02	2.71	2.57	4.16	4.25	3.31	2.94	2.96	2.45	1.82	1.35	30.38
Greater Cincinnati AP	30	2.92	2.75	3.90	3.96	4.59	4.42	3.75	3.79	2.89	2.96	3.46	3.28	42.60
Jackson, KY	30	3.56	3.68	4.38	3.79	5.16	4.67	4.59	4.13	3.77	3.18	4.20	4.27	49.38
Lexington, KY	30	3.34	3.27	4.41	3.67	4.78	4.58	4.81	3.77	3.11	2.70	3.44	4.03	45.91
Louisville, KY	30	3.28	3.25	4.41	3.91	4.88	3.76	4.30	3.41	3.05	2.79	3.81	3.69	44.54
Paducah, KY	30	3.47	3.93	4.27	4.95	4.75	4.51	4.45	2.99	3.56	6.45	4.53	4.38	49.24
Baton Rouge, LA	30	6.19	5.10	5.07	5.56	5.34	5.33	5.96	5.86	4.84	3.81	4.76	5.26	63.08
Lake Charles, LA	30	5.52	3.28	3.54	3.64	6.06	6.07	5.13	4.85	5.95	3.94	4.61	4.60	57.19
New Orleans, LA	30	5.87	5.47	5.24	5.02	4.62	6.83	6.20	6.15	5.55	3.05	5.09	5.07	64.16
Shreveport, LA	30	4.60	4.21	4.18	4.42	5.25	5.05	3.99	2.71	3.21	4.45	4.68	4.55	51.30
Caribou, ME	30	2.97	2.06	2.57	2.64	3.28	3.31	3.89	4.15	3.27	2.99	3.12	3.19	37.44
Portland, ME	30	4.09	3.14	4.14	4.26	3.82	3.28	3.32	3.05	3.37	4.40	4.72	4.24	45.83
Baltimore, MD	30	3.47	3.02	3.93	3.00	3.89	3.43	3.85	3.74	3.98	3.16	3.12	3.35	41.94
Blue Hill, MA	30	4.78	4.06	4.79	4.32	3.79	3.93	3.74	4.06	4.13	4.42	4.64	4.56	51.22
Boston, MA	30	3.92	3.30	3.85	3.60	3.24	3.22	3.06	3.37	3.47	3.79	3.98	3.73	42.53
Worcester, MA	30	4.07	3.10	4.23	3.92	4.35	4.02	4.19	4.09	4.27	4.67	4.34	3.80	49.05
Alpena, MI	30	1.76	1.35	2.13	2.31	2.61	2.53	3.17	3.50	2.80	2.33	2.08	1.83	28.40
Detroit, MI	30	1.91	1.88	2.52	3.05	3.05	3.55	3.16	3.10	3.27	2.23	2.66	2.51	32.89
Flint, MI	30	1.57	1.35	2.22	3.13	2.74	3.07	3.17	3.43	3.76	2.34	2.65	2.18	31.61
Grand Rapids, MI	30	2.03	1.54	2.59	3.48	3.35	3.67	3.56	3.78	4.28	2.80	3.35	2.70	37.13
Houghton Lake, MI	30	1.61	1.25	2.05	2.29	2.57	2.93	2.75	3.72	3.11	2.26	2.14	1.75	28.43
Lansing, MI	30	1.61	1.45	2.33	3.09	2.71	3.60	2.68	3.46	3.48	2.29	2.66	2.17	31.53
Marquette, MI	30	2.60	1.85	3.13	2.79	3.07	3.21	3.01	3.55	3.74	3.66	3.27	2.43	36.31

Muskegon, MI	30	2.22	1.58	2.36	2.91	2.95	2.58	2.32	3.77	3.52	2.80	3.23	2.64	32.88
Sault Ste. Marie, MI	30	2.64	1.60	2.41	2.57	2.50	3.00	3.14	3.47	3.71	3.32	3.40	2.91	34.67
Duluth, MN	30	1.12	0.83	1.69	2.09	2.95	4.25	4.20	4.22	4.13	2.46	2.12	0.94	31.00
International Falls, MN	30	0.84	0.64	0.96	1.38	2.55	3.98	3.37	3.14	3.03	1.98	1.36	0.70	23.93
Minneapolis-St. Paul, MN	30	1.04	0.79	1.86	2.31	3.24	4.34	4.04	4.05	2.69	2.11	1.94	1.00	29.41
Rochester, MN	30	0.94	0.75	1.88	3.01	3.53	4.00	4.61	4.33	3.12	2.20	2.01	1.02	31.40
Saint Cloud, MN	30	0.76	0.59	1.50	2.13	2.97	4.51	3.34	3.93	2.93	2.24	1.54	0.69	27.13
Jackson, MS	30	5.67	4.50	5.74	5.98	4.86	3.82	4.69	3.66	3.23	3.42	5.04	5.34	55.95
Meridian, MS	30	5.92	5.35	6.93	5.62	4.87	3.99	5.45	3.34	3.64	3.28	4.95	5.31	58.65
Tupelo, MS	30	5.14	4.68	6.30	4.94	5.80	4.82	3.65	2.67	3.35	3.38	5.01	6.12	55.86
Columbia, MO	30	1.73	2.20	3.21	4.16	4.87	4.02	3.80	3.75	3.42	3.18	3.47	2.47	40.28
Kansas City, MO	30	1.15	1.31	2.44	3.38	5.39	4.44	4.42	3.54	4.64	3.33	2.30	1.64	37.98
St. Louis, MO	30	2.14	2.28	3.60	3.69	4.11	3.76	3.90	2.98	2.96	2.76	3.71	2.86	38.75
Springfield, MO	30	2.11	2.28	3.82	4.31	4.57	5.02	3.56	3.37	4.83	3.47	4.46	3.17	44.97
Billings, MT	30	0.81	0.58	1.12	1.74	2.48	1.89	1.28	0.85	1.34	1.26	0.75	0.67	14.77
Glasgow, MT	30	0.35	0.26	0.47	0.75	1.72	2.20	1.78	1.25	0.98	0.71	0.39	0.37	11.23
Great Falls, MT	30	0.68	0.51	1.01	1.40	2.53	2.24	1.45	1.65	1.23	0.93	0.59	0.67	14.89
Havre, MT	30	0.47	0.36	0.70	0.87	1.84	1.90	1.51	1.20	1.03	0.62	0.45	0.51	11.46
Helena, MT	30	0.52	0.38	0.63	0.91	1.78	1.82	1.34	1.29	1.05	0.66	0.48	0.46	11.32
Kalispell, MT	30	1.47	1.15	1.11	1.22	2.04	2.30	1.41	1.25	1.20	0.96	1.45	1.65	17.21
Missoula, MT	30	1.06	0.77	0.96	1.09	1.95	1.73	1.09	1.15	1.08	0.83	0.96	1.15	13.82
Grand Island, NE	30	0.54	0.68	2.04	2.61	4.07	3.72	3.14	3.08	2.43	1.51	1.41	0.66	25.89
Lincoln, NE	30	0.67	0.66	2.21	2.90	4.23	3.51	3.54	3.35	2.92	1.94	1.58	0.86	28.37
Norfolk, NE	30	0.57	0.76	1.97	2.59	3.92	4.25	3.74	2.80	2.25	1.72	1.44	0.65	26.66
North Platte, NE	30	0.39	0.51	1.24	1.97	3.34	3.17	3.17	2.15	1.32	1.24	0.76	0.40	19.66

(Continued)

TABLE 16.17 Average Rainfall Data (*Continued*)

	Years	Jan.	Feb.	Mar.	Apr.	May	Jun.	Jul.	Aug.	Sep.	Oct.	Nov.	Dec.	Annual
Omaha Eppley AP, NE	30	0.77	0.80	2.13	2.94	4.44	3.95	3.86	3.21	3.17	2.21	1.82	0.92	30.22
Omaha (North), NE	30	0.76	0.77	2.25	3.07	4.57	3.84	3.75	2.93	3.03	2.49	1.67	0.95	30.08
Scottsbluff, NE	30	0.54	0.58	1.16	1.79	2.70	2.65	2.13	1.19	1.22	1.01	0.80	0.56	16.33
Valentine, NE	30	0.30	0.48	1.11	1.97	3.20	3.01	3.37	2.20	1.61	1.22	0.72	0.33	19.52
Elko, NV	30	1.14	0.88	0.98	0.81	1.08	0.67	0.30	0.36	0.68	0.71	1.05	0.93	9.59
Ely, NV	30	0.74	0.75	1.05	0.90	1.29	0.66	0.60	0.91	0.94	1.00	0.63	0.50	9.97
Las Vegas, NV	30	0.59	0.69	0.59	0.15	0.24	0.08	0.44	0.45	0.31	0.24	0.31	0.40	4.49
Reno, NV	30	1.06	1.06	0.86	0.35	0.62	0.47	0.24	0.27	0.45	0.42	0.80	0.88	7.48
Winnemucca, NV	30	0.83	0.62	0.86	0.85	1.06	0.69	0.27	0.35	0.53	0.66	0.80	0.81	8.33
Concord, NH	30	2.97	2.36	3.04	3.07	3.33	3.10	3.37	3.21	3.16	3.46	3.57	2.96	37.60
Mt. Washington, NH	30	8.52	7.33	9.42	8.43	8.21	8.36	8.02	8.08	8.55	7.66	10.49	8.84	101.91
Atlantic City AP, NJ	30	3.60	2.85	4.06	3.45	3.38	2.66	3.86	4.32	3.14	2.86	3.26	3.15	40.59
Atlantic City C.O., NJ	30	3.44	2.88	3.79	3.25	3.16	2.46	3.36	4.16	3.02	2.71	2.96	3.18	38.37
Newark, NJ	30	3.98	2.96	4.21	3.92	4.46	3.40	4.68	4.02	4.01	3.16	3.88	3.57	46.25
Albuquerque, NM	30	0.49	0.44	0.61	0.50	0.60	0.65	1.27	1.73	1.07	1.00	0.62	0.49	9.47
Clayton, NM	30	0.30	0.27	0.62	0.99	2.08	2.21	2.81	2.69	1.56	0.74	0.54	0.32	15.13
Roswell, NM	30	0.39	0.41	0.35	0.58	1.30	1.62	1.99	2.31	1.98	1.29	0.53	0.59	13.34
Albany, NY	30	2.71	2.27	3.17	3.25	3.67	3.74	3.50	3.68	3.31	3.23	3.31	2.76	38.60
Binghamton, NY	30	2.58	2.46	2.97	3.49	3.55	3.80	3.49	3.35	3.59	3.02	3.32	3.03	38.65
Buffalo, NY	30	3.16	2.42	2.99	3.04	3.35	3.82	3.14	3.87	3.84	3.19	3.92	3.80	40.54
Isljp, NY	30	4.27	3.33	4.76	4.13	3.90	3.71	2.93	4.48	3.39	3.63	3.86	4.13	46.52
New York C.Park, NY	30	4.13	3.15	4.37	4.28	4.69	3.84	4.62	4.22	4.23	3.85	4.36	3.95	49.69
New York (JFK AP), NY	30	3.62	2.70	3.79	3.75	4.13	3.59	3.92	3.64	3.50	3.03	3.48	3.31	42.46
New York (Laguardia AP), NY	30	3.56	2.75	3.93	3.68	4.16	3.57	4.41	4.09	3.77	3.26	3.67	3.51	44.36
Rochester, NY	30	2.34	2.04	2.58	2.75	2.82	3.36	2.93	3.54	3.45	2.60	2.84	2.73	33.98

Syracuse, NY	30	2.60	2.12	3.02	3.39	3.39	3.71	4.02	3.56	4.15	3.20	3.77	3.12	40.05
Asheville, NC	30	4.06	3.83	4.59	3.50	4.42	4.38	3.87	4.30	3.72	3.18	3.82	3.40	47.07
Cape Hatteras, NC	30	5.84	3.94	4.95	3.29	3.92	3.82	4.95	6.56	5.68	5.31	4.93	4.56	57.75
Charlotte, NC	30	4.00	3.55	4.39	2.95	3.66	3.42	3.79	3.72	3.83	3.66	3.36	3.18	43.51
Greensboro-Winston-Salem, NC	30	3.54	3.10	3.85	3.43	3.95	3.53	4.44	3.71	4.30	3.27	2.96	3.06	43.14
Raleigh, NC	30	4.02	3.47	4.03	2.80	3.79	3.42	4.29	3.78	4.26	3.18	2.97	3.04	43.05
Wilmington, NC	30	4.52	3.66	4.22	2.94	4.40	5.36	7.62	7.31	6.79	3.21	3.26	3.78	57.07
Bismarck, ND	30	0.45	0.51	0.85	1.46	2.22	2.59	2.58	2.15	1.61	1.28	0.70	0.44	16.84
Fargo, ND	30	0.76	0.59	1.17	1.37	2.61	3.51	2.88	2.52	2.18	1.97	1.06	0.57	21.19
Grand Forks, ND	30	0.68	0.58	0.89	1.23	2.21	3.03	3.06	2.72	1.96	1.70	0.99	0.55	19.60
Williston, ND	30	0.54	0.39	0.74	1.05	1.88	2.36	2.28	1.48	1.35	0.87	0.65	0.57	14.16
Akron, OH	30	2.49	2.28	3.15	3.39	3.96	3.55	4.02	3.65	3.43	2.53	3.04	2.98	38.47
Cleveland, OH	30	2.48	2.29	2.94	3.37	3.50	3.89	3.52	3.69	3.77	2.74	3.38	3.14	38.71
Columbus, OH	30	2.53	2.20	2.89	3.25	3.88	4.08	4.62	3.72	2.92	2.31	3.19	2.93	38.52
Dayton, OH	30	2.60	2.29	3.29	4.03	4.17	4.21	3.75	3.49	2.65	2.72	3.30	3.08	39.58
Mansfield, OH	30	2.63	2.17	3.36	4.17	4.42	4.52	4.23	4.60	3.44	2.68	3.76	3.26	43.24
Toledo, OH	30	1.93	1.88	2.62	3.24	3.14	3.80	2.80	3.19	2.84	2.35	2.78	2.64	33.21
Youngstown, OH	30	2.34	2.03	3.05	3.33	3.45	3.91	4.10	3.43	3.89	2.46	3.07	2.96	38.02
Oklahoma City, OK	30	1.28	1.56	2.90	3.00	5.44	4.63	2.94	2.48	3.98	3.64	2.11	1.89	35.85
Tulsa, OK	30	1.60	1.95	3.57	3.95	6.11	4.72	2.96	2.85	4.76	4.05	3.47	2.43	42.42
Astoria, OR	30	9.62	7.87	7.37	4.93	3.28	2.57	1.16	1.21	2.61	5.61	10.50	10.40	67.13
Burns, OR	30	1.18	1.11	1.24	0.85	1.05	0.66	0.40	0.45	0.50	0.72	1.11	1.30	10.57
Eugene, OR	30	7.65	6.35	5.80	3.66	2.66	1.53	0.64	0.99	1.54	3.35	8.44	8.29	50.90
Medford, OR	30	2.47	2.10	1.85	1.31	1.21	0.68	0.31	0.52	0.78	1.31	2.93	2.90	18.37
Pendleton, OR	30	1.45	1.22	1.26	1.13	1.22	0.78	0.41	0.56	0.63	0.99	1.63	1.48	12.76

(Continued)

TABLE 16.17 Average Rainfall Data (*Continued*)

	Years	Jan.	Feb.	Mar.	Apr.	May	Jun.	Jul.	Aug.	Sep.	Oct.	Nov.	Dec.	Annual
Portland, OR	30	5.07	4.18	3.71	2.64	2.38	1.59	0.72	0.93	1.65	2.88	5.61	5.71	37.07
Salem, OR	30	5.84	5.09	4.17	2.76	2.13	1.45	0.57	0.68	1.43	3.03	6.39	6.46	40.00
Sexton Summit, OR	30	4.71	4.29	3.92	2.38	1.35	0.94	0.35	0.61	1.20	2.93	5.32	5.18	33.18
Allentown, PA	30	3.50	2.75	3.56	3.49	4.47	3.99	4.27	4.35	4.37	3.33	3.70	3.39	45.17
Erie, PA	30	2.53	2.28	3.13	3.38	3.34	4.28	3.28	4.21	4.73	3.92	3.96	3.73	42.77
Harrisburg, PA	30	3.18	2.88	3.58	3.31	4.60	3.99	3.21	3.24	3.65	3.06	3.53	3.22	41.45
Middletown/Harrisburg APT	30	3.18	2.88	3.58	3.31	4.60	3.99	3.21	3.24	3.65	3.06	3.53	3.22	41.45
Philadelphia, PA	30	3.52	2.74	3.81	3.49	3.89	3.29	4.39	3.82	3.88	2.75	3.16	3.31	42.05
Pittsburgh, PA	30	2.70	2.37	3.17	3.01	3.80	4.12	3.96	3.38	3.21	2.25	3.02	2.86	37.85
Avoca, PA	30	2.46	2.08	2.69	3.28	3.69	3.97	3.74	3.10	3.86	3.02	3.12	2.55	37.56
Williamsport, PA	30	2.85	2.61	3.21	3.49	3.79	4.45	4.08	3.38	3.98	3.19	3.62	2.94	41.59
Block Is., RI	30	3.68	3.04	3.99	3.72	3.40	2.77	2.62	3.00	3.19	3.04	3.77	3.57	39.79
Providence, RI	30	4.37	3.45	4.43	4.16	3.66	3.38	3.17	3.90	3.70	3.69	4.40	4.14	46.45
Interior PR	30	5.10	4.50	4.30	5.00	8.10	6.10	7.00	8.60	9.60	9.20	9.40	6.40	83.2
North Coastal PR	30	3.90	3.10	2.80	4.50	6.20	4.00	4.60	5.90	5.80	8.00	6.10	5.20	58.1
Charleston AP, SC	30	4.08	3.08	4.00	2.77	3.67	5.92	6.13	6.91	5.98	3.09	2.66	3.24	51.53
Charleston C.O., SC	30	3.62	2.62	3.83	2.44	2.77	4.96	5.50	6.54	6.13	3.02	2.18	2.78	46.39
Columbia, SC	30	4.66	3.84	4.59	2.98	3.17	4.99	5.54	5.41	3.94	2.89	2.88	3.38	48.27
Greenv'l-Spartanb'rg AP, SC	30	4.41	4.24	5.31	3.54	4.59	3.92	4.65	4.08	3.97	3.88	3.79	3.86	50.24
Aberdeen, SD	30	0.48	0.48	1.34	2.83	2.69	3.49	2.92	2.42	1.81	1.63	0.75	0.38	20.22
Huron, SD	30	0.49	0.57	1.67	2.29	3.00	3.28	2.86	2.07	1.80	1.59	0.89	0.39	20.90
Rapid City, SD	30	0.37	0.46	1.03	1.86	2.96	2.83	2.03	1.61	1.10	1.37	0.61	0.41	16.64
Sioux Falls, SD	30	0.51	0.51	1.81	2.65	3.39	3.49	2.93	3.01	2.58	1.93	1.36	0.52	24.69
Bristol-Johnson Cty TN	30	3.52	3.40	3.91	3.23	4.32	3.89	4.21	3.00	3.08	2.30	3.08	3.39	41.33
Chattanooga, TN	30	5.40	4.85	6.19	4.23	4.28	3.99	4.73	3.59	4.31	3.26	4.88	4.81	54.52

Knoxville, TN	30	4.57	4.01	5.17	3.99	4.68	4.04	4.71	2.89	3.04	2.65	3.98	4.49	48.22
Memphis, TN	30	4.24	4.31	5.58	5.79	5.15	4.30	4.22	3.00	3.31	3.31	5.76	5.68	54.65
Nashville, TN	30	3.97	3.69	4.87	3.93	5.07	4.08	3.77	3.28	3.59	2.87	4.45	4.54	48.11
Oak Ridge, TN	30	5.13	4.50	5.72	4.32	5.14	4.64	5.16	3.39	3.75	3.02	4.86	5.42	55.05
Abilene, TX	30	0.97	1.13	1.41	1.67	2.83	3.06	1.70	2.63	2.91	2.90	1.30	1.27	23.78
Amarillo, TX	30	0.63	0.55	1.13	1.33	2.50	3.28	2.68	2.94	1.88	1.50	0.68	0.61	19.71
Austin/City, TX	30	1.89	1.99	2.14	2.51	5.03	3.81	1.97	2.31	2.91	3.97	2.68	2.44	33.65
Austin/Bergstrom, TX	30	2.21	2.02	2.36	2.63	5.12	3.42	2.03	2.51	2.88	3.99	3.02	2.53	34.72
Brownsville, TX	30	1.36	1.18	0.93	1.96	2.48	2.93	1.77	2.99	5.31	3.78	1.75	1.11	27.55
Corpus Christi, TX	30	1.62	1.84	1.74	2.05	3.48	3.53	2.00	3.54	5.03	3.94	1.74	1.75	32.26
Dallas-Fort Worth, TX	30	1.90	2.37	3.06	3.20	5.15	3.23	2.12	2.03	2.42	4.11	2.57	2.57	34.73
Dallas-Love Field, TX	30	1.89	2.31	3.13	3.46	5.30	3.92	2.43	2.17	2.65	4.65	2.61	2.53	37.05
Del Rio, TX	30	0.57	0.96	0.96	1.71	2.31	2.34	2.02	2.16	2.06	2.00	0.96	0.75	18.80
El Paso, TX	30	0.45	0.39	0.26	0.23	0.38	0.87	1.49	1.75	1.61	0.81	0.42	0.77	9.43
Galveston, TX	30	4.08	2.61	2.76	2.56	3.70	4.04	3.45	4.22	5.76	3.49	3.64	3.53	43.84
Houston, TX	30	3.68	2.98	3.36	3.60	5.15	3.35	3.18	3.83	4.33	4.50	4.19	3.69	47.84
Lubbock, TX	30	0.50	0.71	0.76	1.29	2.31	2.98	2.13	2.36	2.57	1.70	0.71	0.67	18.69
Midland-Odessa, TX	30	0.53	0.58	0.42	0.73	1.79	1.71	1.89	1.77	2.31	1.77	0.65	0.65	14.80
Port Arthur, TX	30	5.69	3.35	3.75	3.84	5.83	6.58	5.23	4.85	6.10	4.67	4.75	5.25	59.89
San Angelo, TX	30	0.82	1.18	0.99	1.60	3.09	2.52	1.10	2.05	2.95	2.57	1.10	0.94	20.91
San Antonio, TX	30	1.66	1.75	1.89	2.60	4.72	4.30	2.03	2.57	3.00	3.86	2.58	1.96	32.92
Victoria, TX	30	2.44	2.04	2.25	2.97	5.12	4.96	2.90	3.05	5.00	4.26	2.64	2.47	40.10
Waco, TX	30	1.90	2.43	2.48	2.99	4.46	3.08	2.23	1.85	2.88	3.67	2.61	2.76	33.34
Wichita Falls, TX	30	1.12	1.58	2.27	2.62	3.92	3.69	1.58	2.39	3.19	3.11	1.68	1.68	28.83
Milford, UT	30	0.73	0.77	1.21	0.99	0.94	0.44	0.76	1.04	0.92	1.12	0.77	0.58	10.27

(Continued)

TABLE 16.17 Average Rainfall Data (*Continued*)

	Years	Jan.	Feb.	Mar.	Apr.	May	Jun.	Jul.	Aug.	Sep.	Oct.	Nov.	Dec.	Annual
Salt Lake City, UT	30	1.37	1.33	1.91	2.02	2.09	0.77	0.72	0.76	1.33	1.57	1.40	1.23	16.50
Burlington, VT	30	2.22	1.67	2.32	2.88	3.32	3.43	3.97	4.01	3.83	3.12	3.06	2.22	36.05
Lynchburg, VA	30	3.54	3.10	3.83	3.46	4.11	3.79	4.39	3.41	3.88	3.39	3.18	3.23	43.31
Norfolk, VA	30	3.93	3.34	4.08	3.38	3.74	3.77	5.17	4.79	4.06	3.47	2.98	3.03	45.74
Richmond, VA	30	3.55	2.98	4.09	3.18	3.96	3.54	4.67	4.18	3.98	3.60	3.06	3.12	43.91
Roanoke, VA	30	3.23	3.08	3.84	3.61	4.24	3.58	4.00	3.74	3.85	3.15	3.21	2.86	42.49
St Thomas VI	25	1.80	1.80	2.80	3.80	2.80	2.50	2.80	4.11	5.80	5.10	6.80	3.50	42.80
Olympia, WA	30	7.54	6.17	5.29	3.58	2.27	1.78	0.82	1.10	2.03	4.19	8.13	7.89	50.79
Quillayute, WA	30	13.65	12.35	10.98	7.44	5.51	3.50	2.34	2.67	4.15	9.81	14.82	14.50	101.72
Seattle C.O., WA	30	5.24	4.09	3.92	2.75	2.03	1.55	0.93	1.16	1.61	3.24	5.67	6.06	38.25
Seattle Sea-Tac AP, WA	30	5.13	4.18	3.75	2.59	1.78	1.49	0.79	1.02	1.63	3.19	5.90	5.62	37.07
Spokane, WA	30	1.82	1.51	1.53	1.28	1.60	1.18	0.76	0.68	0.76	1.06	2.24	2.25	16.67
Walla, WA	30	2.25	1.97	2.20	1.83	1.95	1.15	0.73	0.84	0.83	1.77	2.85	2.51	20.88
Yakima, WA	30	1.17	0.80	0.70	0.53	0.51	0.62	0.22	0.36	0.39	0.53	1.05	1.38	8.26
Beckley, WV	30	3.23	2.96	3.63	3.43	4.39	3.92	4.78	3.45	3.23	2.64	2.88	3.09	41.63
Charleston, WV	30	3.25	3.19	3.90	3.25	4.30	4.09	4.86	4.11	3.45	2.67	3.66	3.32	44.05
Elkins, WV	30	3.43	3.20	3.92	3.53	4.77	4.61	4.84	4.26	3.83	2.86	3.42	3.44	46.11
Huntington, WV	30	3.21	3.09	3.83	3.33	4.41	3.88	4.46	3.88	2.80	2.73	3.32	3.37	42.31
Green Bay, WI	30	1.21	1.01	2.06	2.56	2.75	3.43	3.44	3.77	3.11	2.17	2.27	1.41	29.19
La Crosse, WI	30	1.19	0.99	2.00	3.38	3.38	4.00	4.25	4.28	3.40	2.16	2.10	1.23	32.36
Madison, WI	30	1.25	1.28	2.28	3.35	3.25	4.05	3.93	4.33	3.08	2.18	2.31	1.66	32.95
Milwaukee, WI	30	1.85	1.65	2.59	3.78	3.06	3.56	3.58	4.03	3.30	2.49	2.70	2.22	34.81
Casper, WY	30	0.58	0.64	0.90	1.52	2.38	1.43	1.29	0.73	0.98	1.14	0.82	0.62	13.03
Cheyenne, WY	30	0.45	0.44	1.05	1.55	2.48	2.12	2.26	1.82	1.43	0.75	0.64	0.46	15.45
Lander, WY	30	0.52	0.54	1.24	2.07	2.38	1.15	0.84	0.57	1.14	1.37	0.99	0.61	13.42

Sheridan, WY	30	0.77	0.57	1.00	1.77	2.41	2.02	1.11	0.80	1.38	1.41	0.80	0.68	14.72
Guam, PC	30	5.58	5.11	4.24	4.16	6.39	6.28	11.66	16.17	13.69	11.88	9.34	6.11	100.61
Johnston Island, PC	30	1.64	1.29	2.01	1.86	1.14	0.87	1.40	2.07	2.46	2.78	4.78	2.70	25.00
Koror, PC	30	11.20	9.65	8.79	9.45	11.27	17.54	16.99	14.47	11.65	13.41	11.62	12.33	148.37
Kwajalein, Marshall Is., PC	30	5.12	3.73	3.82	7.63	8.62	8.86	10.24	10.42	11.82	11.46	10.74	7.94	100.40
Majuro, Marshall Is., PC	30	8.09	6.86	8.43	11.30	11.53	11.09	12.41	11.95	11.96	13.73	12.81	11.50	131.66
Pago, Amer Samoa, PC	30	14.02	12.14	11.15	11.16	10.43	5.94	5.76	6.43	7.36	10.03	11.16	13.38	118.96
Pohnpei, Caroline Is., PC	30	12.52	9.78	13.96	16.94	19.41	17.06	16.72	16.37	14.94	16.30	14.74	15.87	184.61
Chuuk, E. Caroline Is., PC	30	8.58	8.77	8.15	10.94	11.29	12.82	12.45	15.09	13.12	10.69	11.09	10.98	133.97
Wake Island, PC	30	1.40	1.89	2.38	2.11	1.70	1.95	3.44	5.62	4.82	4.27	2.78	1.87	34.23
Yap, W Caroline Is., PC	30	7.24	5.45	6.14	5.58	8.15	13.46	13.25	14.41	13.53	12.25	8.82	9.34	117.62
San Juan, PR	30	3.02	2.30	2.14	3.71	5.29	3.52	4.16	5.22	5.60	5.06	6.17	4.57	50.76

Abbreviation used for States.

AK	Alaska	LA	Louisiana	OK	Oklahoma
AL	Alabama	MA	Massachusetts	OR	Oregon
AP	Airport	MD	Maryland	PA	Pennsylvania
AR	Arkansas	ME	Maine	PR	Puerto Rico
AZ	Arizona	MI	Michigan	RI	Rhode Island
CA	California	MN	Minnesota	SC	South Carolina
CO	Colorado	MO	Missouri	SD	South Dakota
CT	Connecticut	MS	Mississippi	TN	Tennessee
DE	Delaware	MT	Montana	TX	Texas
FL	Florida	NC	North Carolina	UT	Utah
GA	Georgia	ND	North Dakota	VA	Virginia
HI	Hawaii	NE	Nebraska	VT	Vermont
IA	Iowa	NH	Near Hampshire	VI	Virgin Islands
ID	Idaho	NJ	New Jersey	WA	Washington
IL	Illinois	NM	New Mexico	WI	Wisconsin
IN	Indiana	NV	Nevada	WV	West Virginia
KS	Kansas	NY	New York	WY	Wyoming
KY	Kentucky	OH	Ohio		

TYPICAL TANK CALCULATIONS

Once demand is determined, the best way to size the storage tank can be calculated based on the input, on one hand, and the daily demand, on the other. For residential application, general assumptions for potential supply will be made to illustrate assumptions for the design. The figures are strictly made up to show an example of how a proposed design would proceed and should not be used for any actual facility. The following examples, applicable for both residential and commercial facilities, show different methods of approaching the system design.

The first calculation, Eq. (16.3), will be solved for a typical storage capacity and estimated tank size for 1 month. The assumptions made are arbitrary and intended to only show a method. The assumptions are as follows:

1. Demand = 3000 gal/mo

2. Imperviousness of collection surface = 85 percent

2. Rainfall rate = 0.62 gal/in-ft^2

4. Estimated preliminary tank capacity = 10,000 gal

5. Available existing storage = 1000 gal from previous month

6. Size of catchment area = 2500 ft^2

7. Rainfall = 1.97 in

$$G = R \times C \times F \times E \tag{16.4}$$

where G = gallons of storage estimated for 1 month
R = rainfall, in
C = catchment area, ft^2
F = factor giving gal/in-ft^2
E = efficiency of collection system

Solving for Eq. (16.3) gives

$$1.97 \times 2500 \times 0.62 \times 0.85 = 2595 \text{ gal available in any month}$$

To calculate the gallons in storage at the end of any particular month, add the volume of water in storage the prior month to the gallons collected in the present month in question and subtract the monthly demand. Therefore,

$$1000 + 2595 - 3000 = 595 \text{ gal available at end of 1 month}$$

A similar calculation should be continued for each additional month being considered. If the rain exceeds the storage capacity of the tank, the additional rain is lost to overflow and is not considered. If the collection surface is adequate to meet expected demand, there should be a positive storage remaining for the next month. If not, either the collection surface should be enlarged or the monthly demand should be reduced.

The second sample calculation requires information that will have to be researched and would be a more precise method of finding rainfall. A rule of thumb would be to use a 7-year monthly average at an area close to the project location. This information could be obtained from two excellent sources. One would be the National Oceanic and Atmospheric Administration National Climatic Data Center (www.nws.noaa.gov/oa/ncdc.html), and the other is the National Weather Service (www.nws.noaa.gov). Table 16.15 could be used for estimating the average rainfall in different parts of the United States if other information is not available. All the above values would be used.

Use the following formula to determine the required collection area.

$$S = 2 \times (G/R) \tag{16.5}$$

Use the following formula to determine the required volume of rainfall.

$$G = 2R \tag{16.6}$$

where S = surface collection area, ft²
G = amount of rainwater desired to be harvested, gal/mo
R = rainfall precipitation, in/h

This third example can be for a private residence and will use information that has been arbitrarily selected. An assumed average rainfall is 45 in/yr (1125 mm/yr), the roof area is 2500 ft^2 (2075 m^2), and a conservative figure (including service losses) is 0.50 for the volume of water collected per ft^2 (m^2) for each 1 in (25 mm) of rainfall of rain on 1000 ft^2 (836 m^2). Therefore, the number of gallons per/year available is

$$45 \times 2500 \times 0.50 = 56{,}250 \text{ gal/yr}$$

If the residence has four people, the average use of nonpotable water is assumed to be 50 gallons/day. The annual number of gallons required is

$$45 \times 50 \times 365 = 73{,}000 \text{ gal/yr required}$$

This underscores the need for additional water needs. The above figure would be divided by 12 if rainfall were about the same all year. If not, the number of months where there is rainfall would be used, and the remainder of the months require use of the alternative water supply.

If the only use is personal, 50 gal/person-day is considered a human consumption average. If used for irrigation, a pond should be considered using water from a single event for 1 day. Consult the local fire authorities for the minimum fire reserve. The amount of clothes washing and landscape irrigation will require a larger tank or storage area.

A fourth empirical formula has been developed using the daily rainfall data:

$$V = R \times A \times E \times F + D \tag{16.7}$$

where V = useful volume of tank for 1 month, gal
R = rainfall, in/day for an average 30 days
A = area of catchment, ft^2
E = efficiency of filter and runoff (estimated to be 0.30)
F = 0.62 factor (derived from converting inches per/square foot to gallons)
D = daily estimated demand × 30 days/mo

The storage tank volume would be controlled by the consequences of the tank running dry or the amount of time an empty tank would be acceptable, such as where there might be a continuous period without rain. Any seasonal rain pattern must be taken into consideration. Any available source of useful water is a factor. In some industrial applications, multiple storage tanks allow the use of different quality water and quantities for different purposes. Separate tanks would hold water for swimming pools, and other water would be used for turf irrigation.

Calculations for Irrigation Purposes Using Average Monthly Rainfall. To calculate the size of the tank, the tank has to be large enough to have the required volume of water stored for the average month. In addition, a greater storage volume of water should be considered to meet demand during the dryer months or through periods of little rainfall. These monthly rainfall figures are shown in Table 16.17.

Tank and Cistern Siting Considerations

General. The following general installation principles shall be observed.

1. Installation shall comply with the local AHJ.
2. Cisterns shall be installed on a flat and level foundation designed to support the weight.
3. Cistern inlets shall be installed to allow water to enter with a minimum of turbulence.
4. Overflows shall be provided with a means to prohibit the entrance of insects and vermin and be located in such a manner that will prevent erosion and are sized in conformance with gutter and downspout requirements.
5. Cisterns shall only connect with a public water supply through an approved backflow preventer or vacuum breaker installed.
6. Outlets shall be located at least 4 in (100 mm) from the bottom.
7. A schematic detail of a thought process for installation is illustrated in Fig. 16.6.

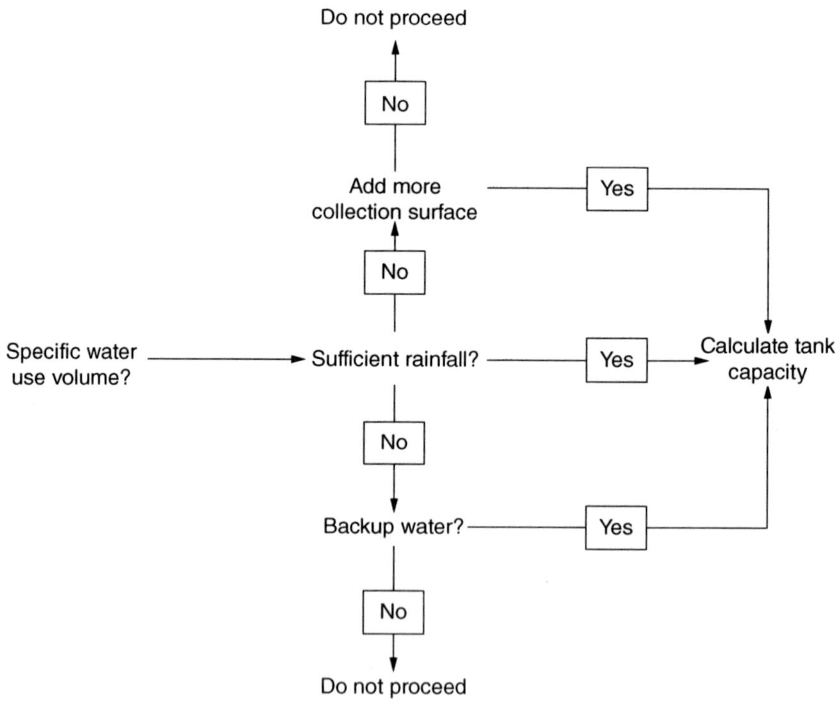

FIGURE 16.6 Thought process guiding tank capacity sizing.

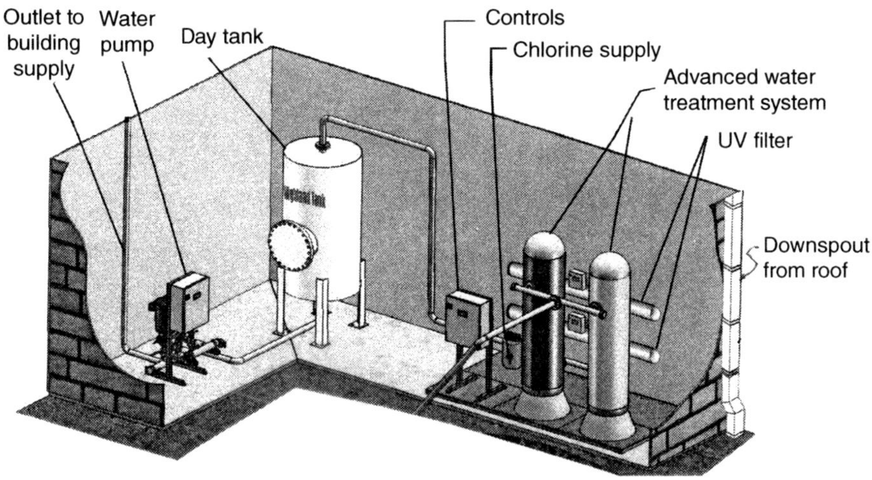

FIGURE 16.7 Typical treatment arrangement.

8. A typical detail of a large nonpotable water installation is illustrated in Fig. 16.7.
9. Storage tanks shall be protected from direct sunlight, if possible.
10. To ensure that water stored in underground tanks will be safe from contamination, tanks shall be located at least 50 ft (17 m) from a septic drain field.
11. Ample room and location shall be allowed for a water truck to supply the tank, if necessary.

Aboveground Installation

1. Above-grade installations shall be protected from direct sunlight by being opaque and UV resistant and shall utilize other means to prevent algae growth.
2. Where the tank is subject to freezing temperatures, it shall be protected from freezing.

Belowground Installation

1. Below-grade tanks shall have a manhole that extends a minimum of 4 in. (100 mm) abovegrade. The cistern shall comply with OSHA Part 1926, Subpart F.
2. If a situation exists where the soil surrounding an underground tank may become saturated, it shall be prevented from floating by the addition of ballast that exceeds the buoyant force which is equal to the weight of the tank filled with water plus a 10 percent safety factor.

A comparison of various materials used for tanks is shown in Table 16.11.

Schematic Diagrams of Storage Tank

When the design is started, a typical thought process might be as presented in Fig. 16.6. A schematic piping arrangement for a typical treatment layout is given in Fig. 16.7. A schematic tank detail is given in Fig. 16.8.

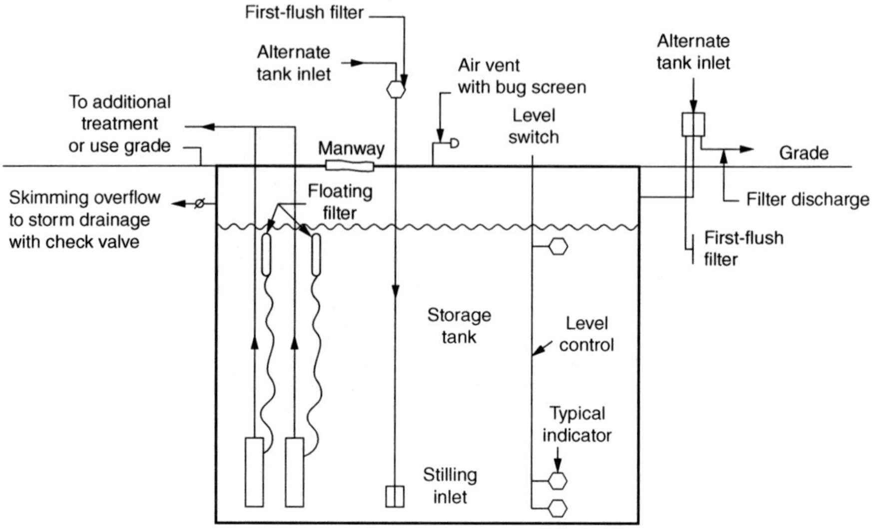

FIGURE 16.8 Typical storage tank schematic.

Inspections and Testing

1. Inspection of all elements shall be made before any part of the system is covered.
2. A flow test shall be conducted, first testing all elements. After this is completed, a hydrostatic pressure test shall be conducted and the system proved tight at a pressure of 50 psig (335 kPa). All parts of the system shall be proved watertight.
3. For potable water and after several rain event cycles, an initial sample of water shall be tested for compliance with local standards as listed in the "Standard Methods For Examination" of water and wastewater and objectionable bacteria. This shall be done once per year.
4. The system shall be tested annually.
5. If the system is to be abandoned, the rainwater piping shall be completely replaced with piping approved for potable water. Fixtures can then be reconnected.
6. Rainwater harvesting systems are required to be identified with an off-color piping (usually purple stripes or solid purple color).

BIBLIOGRAPHY

American Rainwater Catchment Systems Association, Basic Rainwater Catchment System.

Boulware, E. W., "Harvesting the Rain," *Plumbing Systems & Design*.

Boulware, E. W., Rainwater Catchment Design and Installation Standards, American Rainwater Catchment Systems Association and American Society of Plumbing Engineers.

Cartwright, P., "Rainwater Harvesting, Parts 1 & 2," *Plumbing Systems & Design*.

Crawfore, E., and S. Lawson, "Decentralized Rainwater Harvesting Systems," *PME Magazine*.

Darco Inc., *Rainwater Harvesting.*

Plumbing Engineering & Design Handbook, vol. 2, Chap. 6, "Storm Water Harvesting," American Society of Plumbing Engineers.

Texas Water Development Board, *Manual on Rainwater Harvesting*, 3d ed.

Water Control Inc., *Commercial Water Reclamation Systems.*

Waterfall, P., "Harvesting Rainwater for Landscape Use," University of Arizona.

CHAPTER 17
STORMWATER DRAINAGE AND DISPOSAL

This chapter is a guide for development of a stormwater drainage system for smaller, uncomplicated sites of less than 2 mi^2 (3.12 km^2). It is intended for use in designing the modest storm drainage systems associated with buildings and with industrial plants and their surrounding parking and lawn areas. Procedures follow established engineering principles for a conservative method of design and piping layout for small watersheds. Included in this chapter are general permitting requirements for runoff.

Stormwater runoff is defined as rain or snowmelt that flows over the ground, impervious surfaces, roofs, sidewalks, streets, and buildings from a point 5′-0″ (1.6 m) from the building wall and any surface on a facility site that prevents water from percolating into the ground. The selection and location of inlets, catch basins, manholes, and piping along with storm sewer design must be accomplished. The design of interior stormwater drainage is outside the scope of this book. The reuse of runoff is discussed in Chap. 16.

Pollutants conveyed in stormwater discharges from facilities (point source) involved with the manufacturing of chemical and allied products will require permits. It would be the responsibility of the design engineer (or a consultant) to research the necessity of having all the necessary stormwater permits required. Such permits are also discussed.

STORMWATER RUNOFF

General

The following investigations must be concluded before designing the proposed system:

1. The ultimate site stormwater discharge location or method of disposal, must be determined, such as public sewers, waterways, surface absorption, recharge basins, or drywells.
2. If surface waters are to be used as a final outfall, then local, state, or federal agencies must be contacted regarding permitted flow, outfall structures, and required permits. Such agencies may be the U.S. Army Corps of Engineers, EPA, National Pollutant Discharge Elimination System (NPDES) or the local sewer department. EPA permits are required if such a waterway is considered navigable. This is discussed later.
3. If a public sewer is to be the final outfall, public agencies must be contacted regarding permitted flow, connection details, plan submittal, and required approvals of AHJs.

4. If the stormwater is to be used for reuse on the site, refer to Chap. 16 for discussion.
5. If discharge is possible into waterways, high and low water levels of all streams and rivers should be known, including the history of floods.
6. If site stormwater will be discharged directly into smaller streams, it must be determined whether a possible flood condition downstream could be created.
7. Consideration should be given to any future site development.
8. The requirements for the Standard Urban Stormwater Mitigation Plan (SUSMP) must be checked for your project location.
9. Obstacles to be avoided, such as underground piping, trees, underground structures, and rock, should be located.

Layout of System

At this point, it is assumed a survey plan and a preliminary site plan with contours are available for a general layout of the system. Runouts from buildings should now be located, ultimate outfall of the storm sewer selected, and work started on a trial location of stormwater inlets. A trial layout of the site piping should be developed at this time and coordinated with water mains, electrical banks, sanitary sewers, gas mains, and other piping to ensure proper clearance. Allowance for proper and reasonable slope for the storm sewer lines should be made. Obstacles to be avoided, such as trees, underground structures, and rock, should be located. Recommendations for final grading could be made to the landscape architect to possibly reduce the cost of stormwater sewers by advantageous placement of inlets to minimize piping runs.

In general, the following should be considered when locating and designing inlets:

1. Where streets or roads intersect, inlets must be located upstream of traffic flow.
2. Where a series of inlets are located in a road with continuous slope, each intermediate inlet should be designed for approximately 75 to 90 percent of design flow, passing 10 to 25 percent on to the next downstream inlet. If the slope is shallow and flow is not great, 100 percent capacity should be considered. More than one inlet should be provided at the bottom of a slope to accept the additional flow.
3. Inlet capacity should be limited to approximately 5 cubic feet per second (cfs) (0.15 m^3/s) of water.
4. All site low points must be provided with an inlet if off-site gravity flow is not possible. Combination inlets are used for streets or roadways. Parking lot inlets should be flat grates only, even when located at a curb.
5. Distance between drainage inlets should be a maximum of about 300 ft (90 m) when other considerations have been considered.
6. In a roadway with a sag vertical curve, more than one inlet should be considered if a large quantity of runoff is anticipated. Additional inlets should be located on either side of the low point inlet to minimize flooding and sediment buildup resulting from large flows.
7. Manholes should be used at all changes of direction and slope of pipe and at multiple pipe connections. Drainage structures also may serve the same purpose. Distances between manholes should be 250 to 600 ft (75 to 190 m), depending on sewer size and site contours.
8. Steep slopes should have inlets located closer together than normally required to limit inflow.
9. A hooded catch outlet should be used as a final structure before connection to a combined sewer to prevent the passage of unwanted gases.

10. The efficiency of inlets in roadways depends on gutter flow, water velocity, roadway slope, and inlet depression, if any. In many cases, freedom from clogging or interference with traffic may take precedence over hydraulic considerations.

In general, the following should be considered:

1. The road crown should be as steep as practical.

2. Where traffic will not travel close to the curb and clogging is not a problem, use a depressed gutter inlet. Use a depressed curb or combination-type inlet where clogging may occur.

3. Where traffic moves close to the curb, use an undepressed inlet and gutter grate with longitudinal bars only.

Design Criteria

General. When a facility site requires storm drainage, such as commercial and industrial facilities and parking lots, the methods and tables found in plumbing codes for interior roofs are not applicable for site work. The solutions obtained using those methods for calculation would result in systems that are oversized for the flows involved, and would result in pipe sizes far too large to be economically feasible.

To correct this discrepancy, other methods have been developed. The two most common methods are the *unit hydrograph* for very large sites and the *rational method* for smaller sites of 2 mi^2 (4.68 km^2). The rational method, which is the only method to be discussed, uses a basic equation [see (Eqs. (17.1) and (17.2)] to calculate the flow quantity of surface runoff.

$$Q = AIR \qquad \text{IP units} \tag{17.1}$$

or

$$Q = \frac{AIR}{100} \qquad \text{SI units} \tag{17.2}$$

where Q = quantity of storm water runoff, cfs (1 cfs = 448 gpm, 1700 L/min)
A = area to be drained, acres, 43,560 ft^2 (4047 m^2)
I = imperviousness factor of surface comprising drainage area, dimensionless. If there are multiple types of surface areas, a weighted average should be used.
R = rate of rainfall, in/h (cm/h)

The use of the rational formula requires that several general concepts be understood:

1. The rainfall intensity is constant over a time period that equals the time of concentration of the area under design and is constant throughout the entire area.

2. The frequency distribution of the rainfall event and the peak runoff rate are each separately consistent for event-based calculations.

3. The time of concentration for an entire area is constant.

4. Discussions through the years have revealed that the percentage of rainfall converted to runoff can be reliably estimated.

5. The runoff coefficient estimate is accurate regardless of the season and the intensity of the rainfall.

To solve the rational formula, the determination of several interdependent criteria is necessary to calculate the peak flow:

1. The area contributing to the flow into one or many drainage inlets
2. The various different surfaces composing the contributing area
3. The rate of rainfall produced by the design storm

Because of the simplicity of the first two items to be inserted, they are discussed first. They are the area and imperviousness. The third criterion, R, is the maximum rate of rainfall, obtained by referring to the intensity–duration–frequency charts. This requires the determination of a design storm. First is a discussion of factors not affected by a storm.

Area. The first variable in the rational formula is A, which is the area, in acres, of the tributary area for any single drainage inlet or the entire contributing area to the design point. *Note* 1 acre = 43,560 ft^2 (4047 m^2).

Impervious Factor. The second variable is the *impervious factor I*, sometimes referred to as a *runoff coefficient*. This dimensionless coefficient is that portion of the total rainfall that is not lost to infiltration into the soil, ponding in surface depressions, water remaining on vegetation, or evaporation, compared to the total rainfall that falls on an area and flows over the ground to a drainage inlet. The coefficient depends on the slope of the land and the type of vegetation (if any) on it. The average figures for imperviousness of various surfaces are found in Table 17.1. For ease of calculations, use a factor of 1.0 for roofs, roads, etc.

Inspection, if possible, is the best way to determine the nature of an existing surface. In places where the exact nature of future development is uncertain, the figure selected should represent the least loss of precipitation that might be expected to the site, and should allow the largest quantity of stormwater to reach the drainage inlet. Where different types of surfaces are present on any particular area, a weighted coefficient should be used (see page 17.20).

Design Storm

The rational method reduces an inexact set of conditions to an exact formula. Such variables as rainfall rate, overland water flow, and the amount of stormwater that actually reaches the

TABLE 17.1 Imperviousness Factor for Surfaces

Surface	Flat slope, less than 2%	Average slope, 2.0 to 7.0%	Steep slope, 7.1% or more
Roofs and pavement (all types)*	0.95	0.95	0.95
Clay—sparse vegetation	0.40	0.55	0.70
Clay—lawn	0.15	0.20	0.30
Clay—dense vegetation	0.10	0.15	0.20
Clay—dense woods	0.07	0.12	0.17
Sand—sparse vegetation	0.20	0.30	0.40
Sand—lawn	0.07	0.12	0.17
Sand—dense vegetation	0.05	0.10	0.15
Sand—dense woods	0.03	0.08	0.13

*For ease of calculation, the factor for roofs and pavement could become 1.0 if excessive runoff will not result in a significant overdesign.

Source: Baltimore County Design Standards.

drainage inlet (DI) can never be exactly determined. Because of this unpredictability, it is important to realize that some judgment is required in applying the calculated and determined information in the design of a stormwater system.

The stormwater drainage system is generally designed to remove the maximum expected runoff as quickly as it falls, to avoid ponding or flooding. The ability to calculate the flow rate is complicated by not being able to accurately predict many of the factors affecting the actual amount of runoff resulting from any given storm. To calculate the estimated maximum runoff, an artificial "design storm" must be created. This storm will serve as a simulation capable of predicting runoff volume accurately enough to provide a basis for the design of the piping network.

The design storm is based on actual rainfall records, and it has been presented in a convenient form by the National Oceanic and Atmospheric Administration/National Weather Service (NOAA). Design storms are available as either intensity–duration–frequency curves or as charts and formulas appearing in several technical memoranda covering different areas of the United States. The intensity–duration–frequency charts are in a very convenient, easy-to-use form and are considered accurate for the watersheds that are the subject of this handbook. Charts for various cities throughout the United States are presented in NOAA publications. Typical intensity–duration–frequency charts are shown in Fig. 17.1

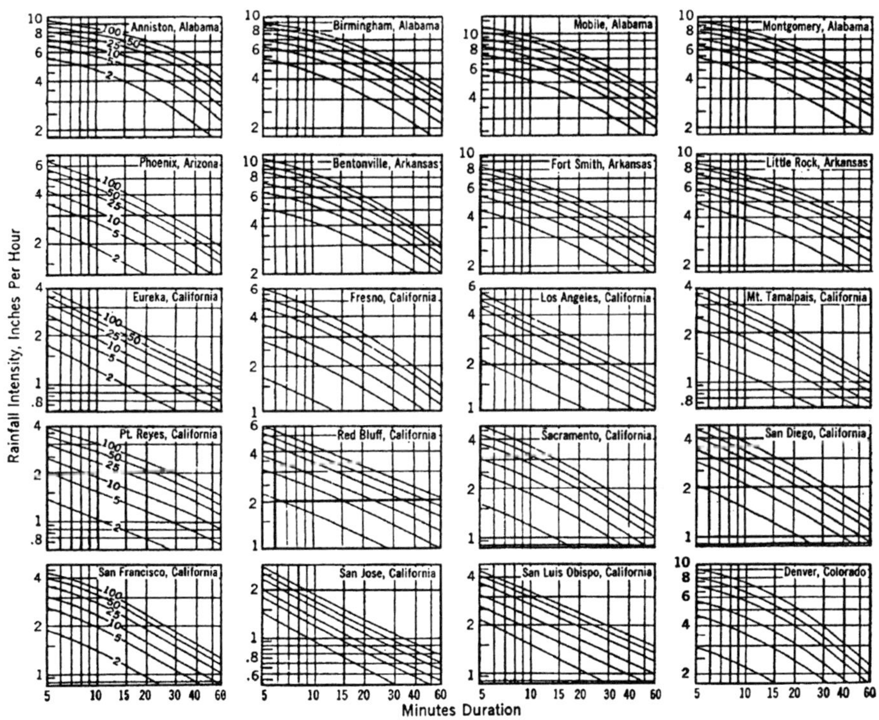

FIGURE 17.1 Rainfall intensity–duration–frequency charts.

The rainfall intensity–duration–frequency curves are abstracted from Technical Paper No. 25 of the Weather Bureau. Included are data for 200 selected stations in the United States and Puerto Rico. The data are substantially in the form found in the Technical Paper, except that the curves are cut off at the 60-min duration line, enabling data to be presented in six pages.

Rainfall data in this form are intended for use in designing the modest storm drainage system associated with buildings and with industrial plants and their surrounding parking and lawn areas.

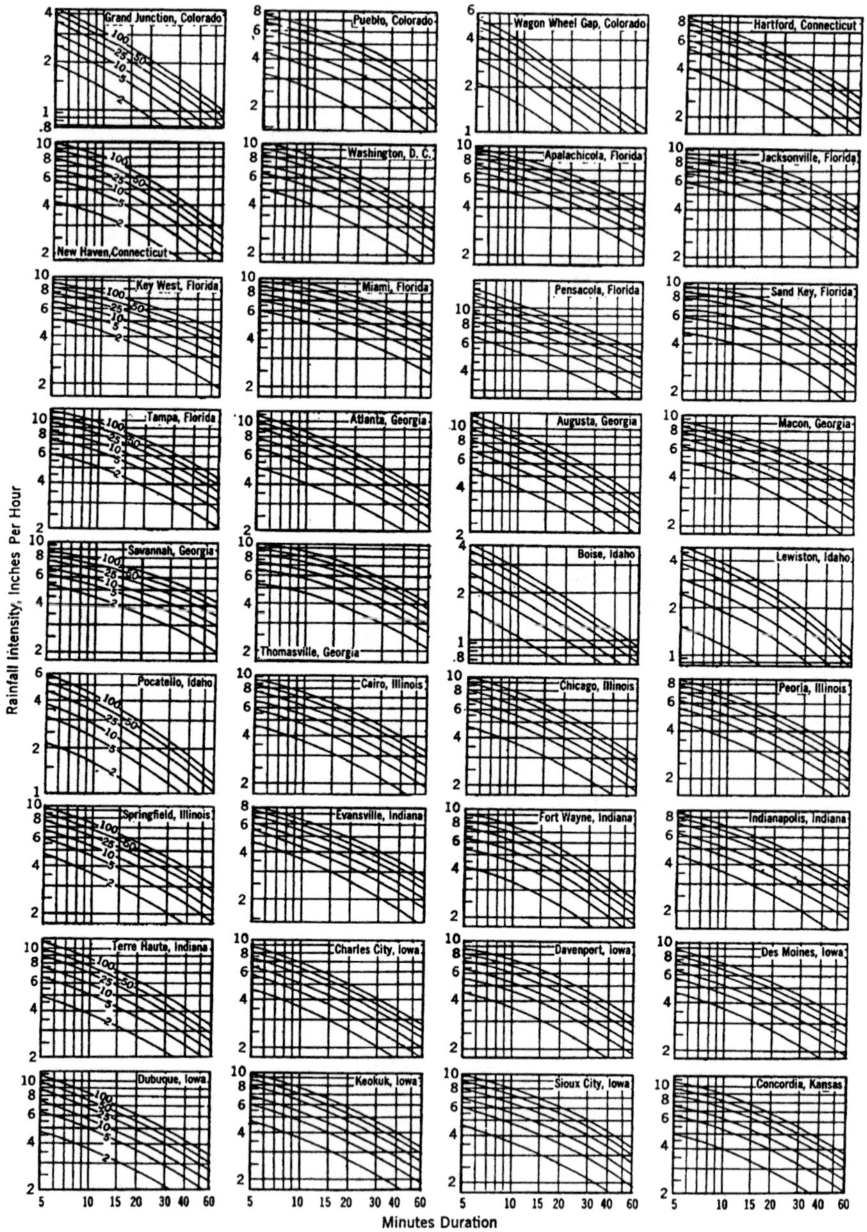

FIGURE 17.1 Rainfall intensity–duration–frequency charts. (*Continued*)

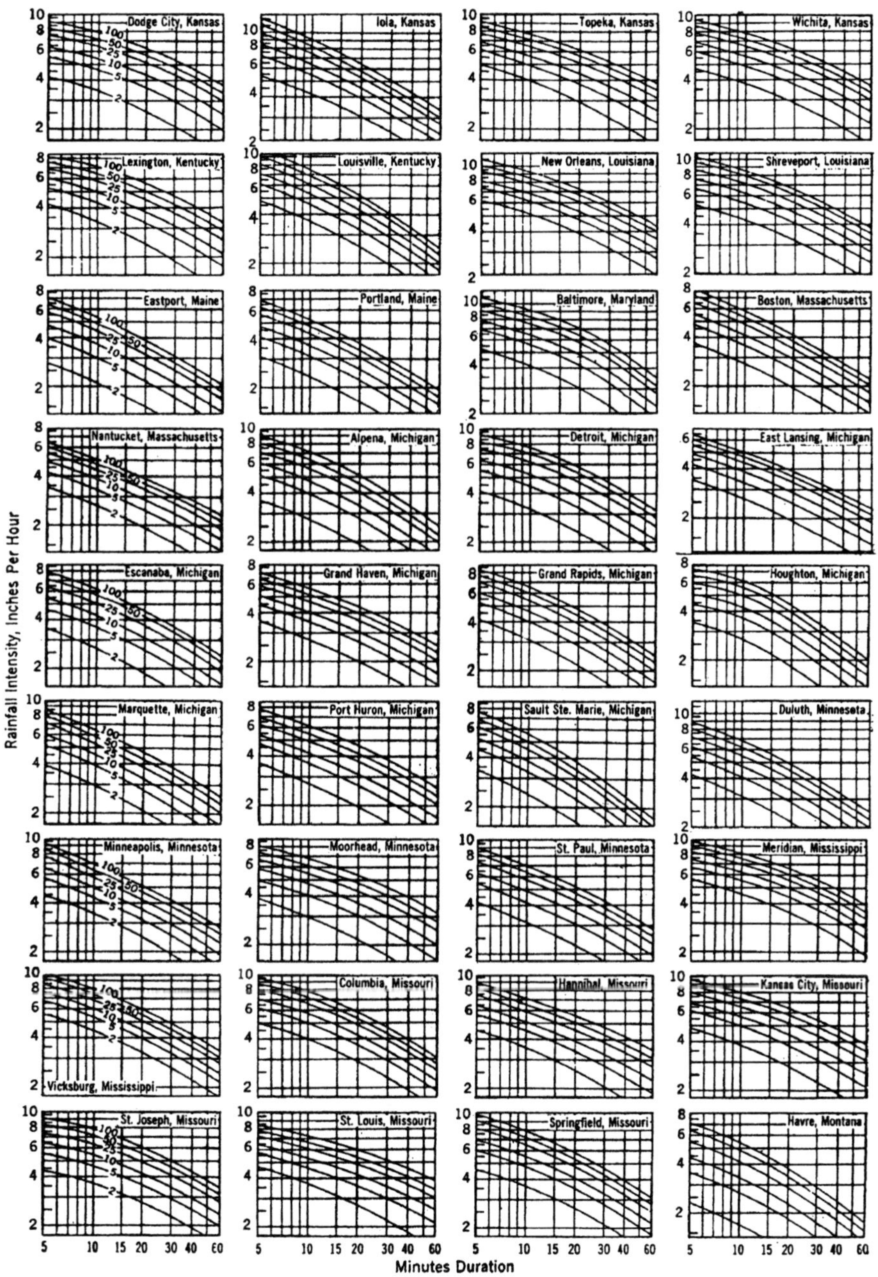

FIGURE 17.1 Rainfall intensity–duration–frequency charts. (*Continued*)

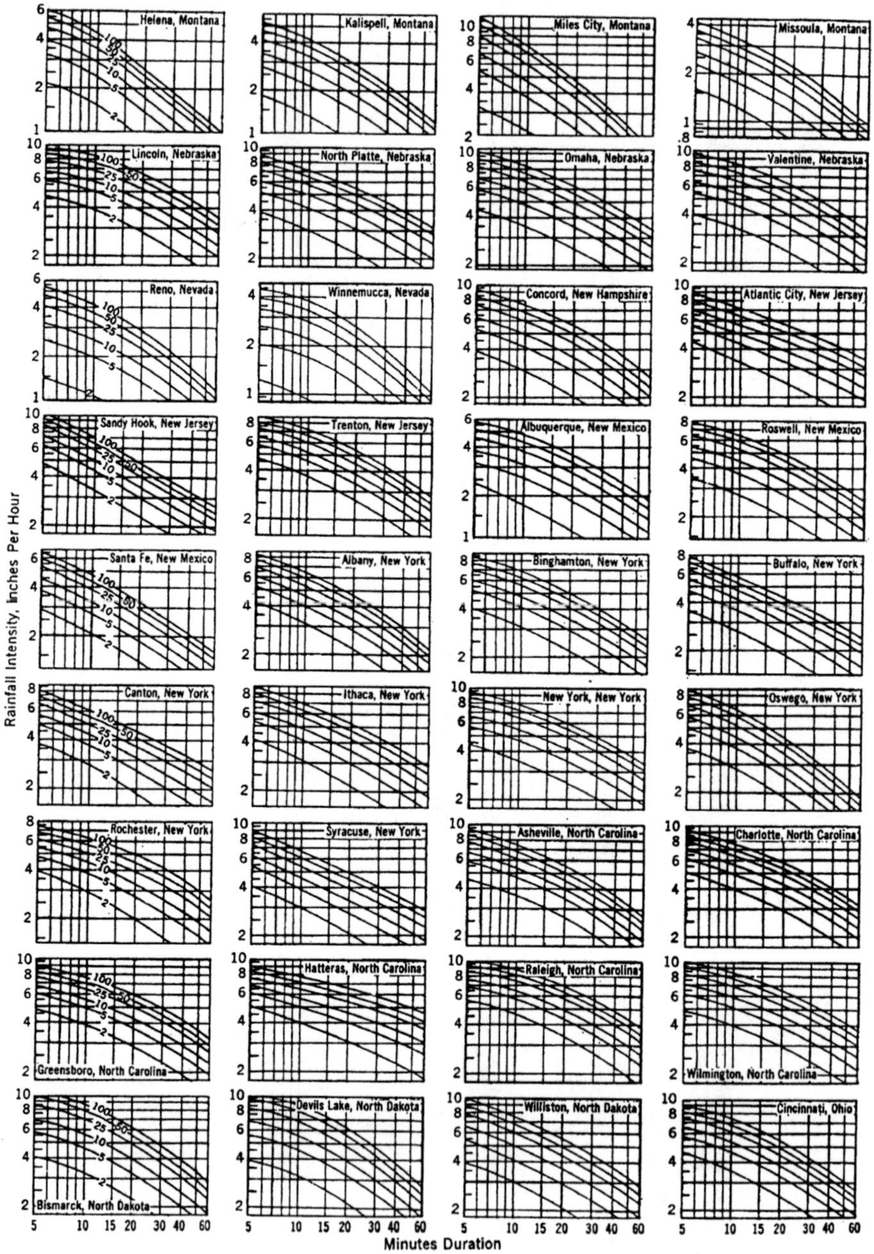

FIGURE 17.1 Rainfall intensity–duration–frequency charts. (*Continued*)

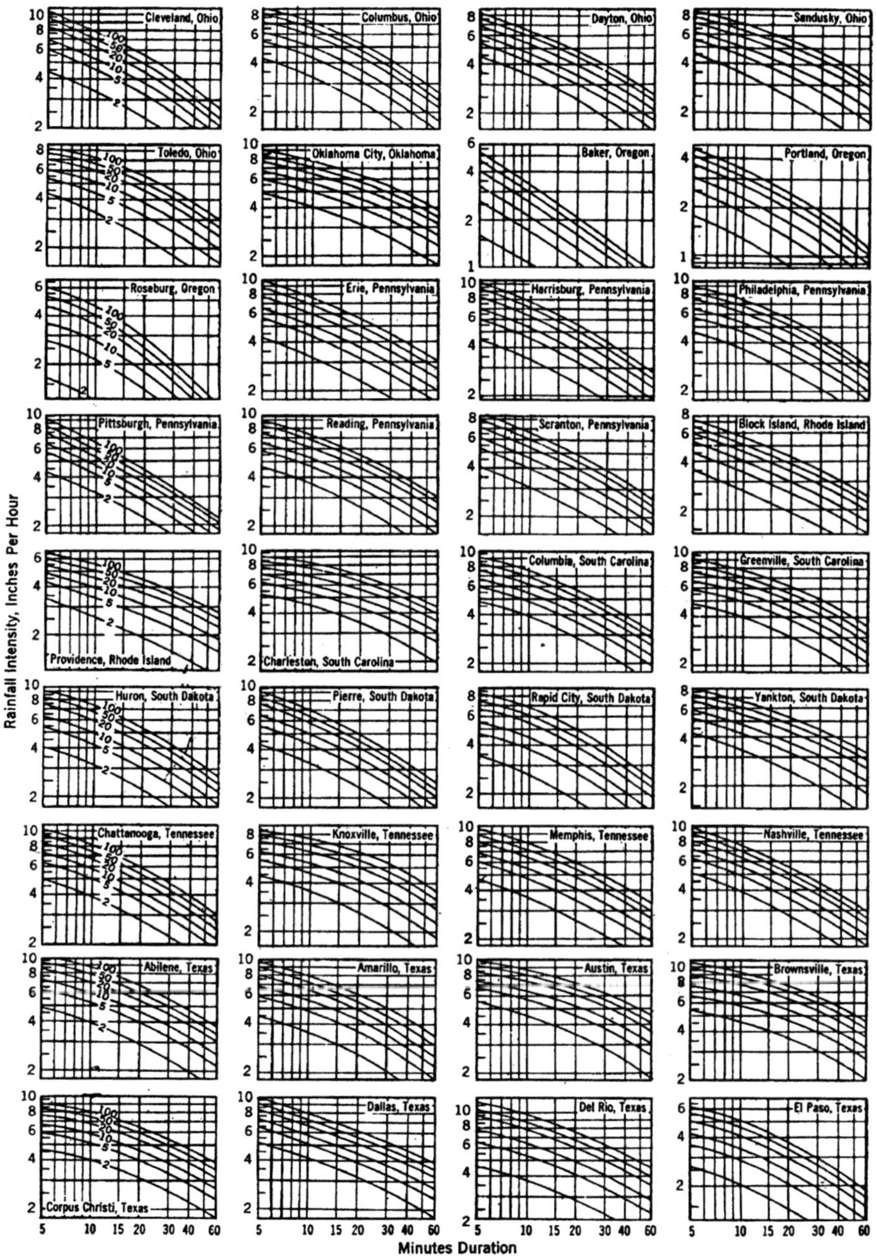

FIGURE 17.1 Rainfall intensity–duration–frequency charts. (*Continued*)

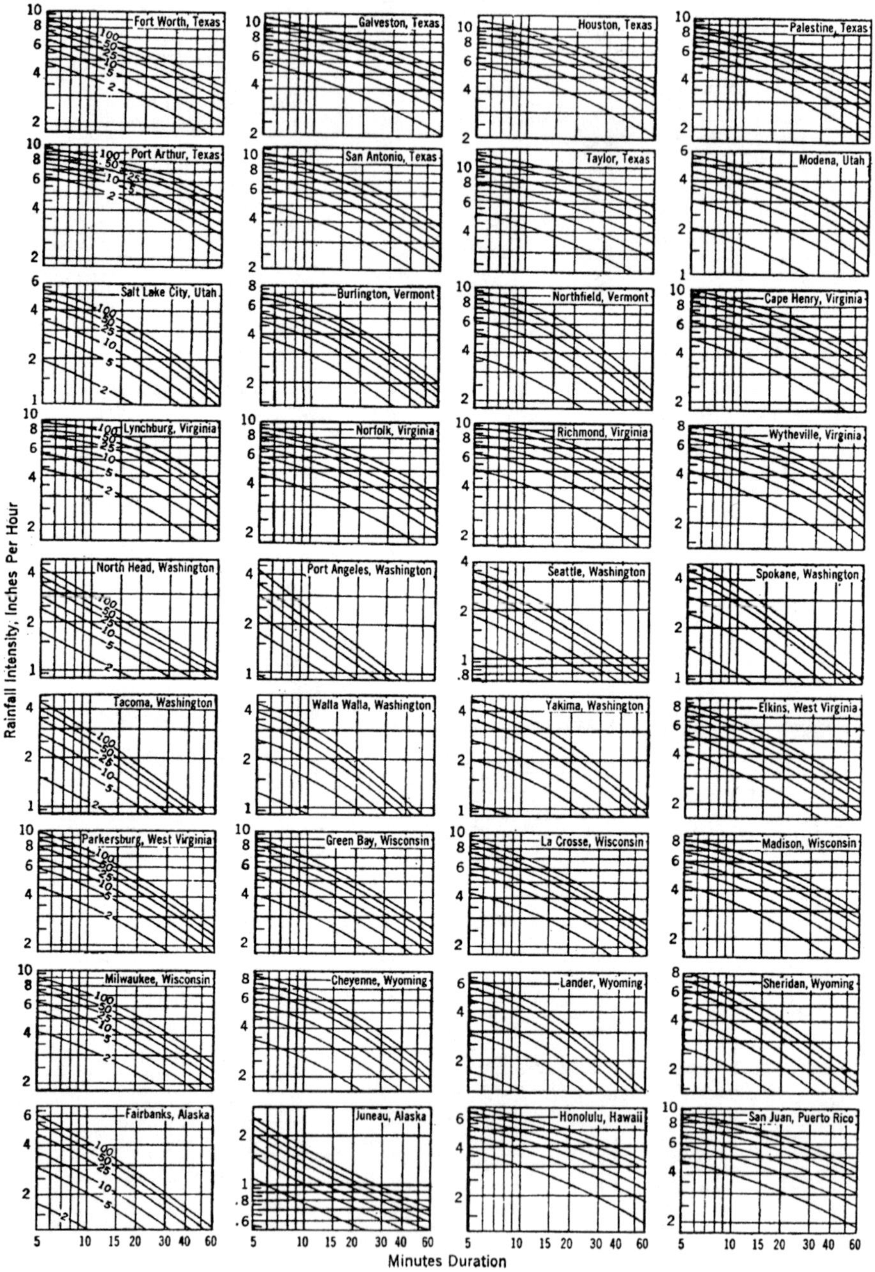

FIGURE 17.1 Rainfall intensity–duration–frequency charts. (*Continued*)

Determination of Design Storm

The design storm calculates the actual peak flow of rain that will be used for pipe sizing purposes along all points throughout the system. The underlying theory is that the maximum amount of rainfall will occur at the exact moment a drop of water from the farthest point of the tributary area is contributing to the flow in a particular section of the sewer. Therefore, calculations must establish the amount of time a drop of water takes to run from the farthest point of the site, into a DI, and then in the pipe to any design point of the sewer. This is necessary so that the shortest time of concentration is obtained which will result in using the largest hourly rainfall rate for design purposes. To use Fig. 17.1 and obtain the peak flow, the frequency (return period) with which a design rainstorm will recur must be selected.

Return Period (Frequency). The design storm return period, or frequency, is selected by the engineer. It is the statistical period of years that must elapse to produce the most severe design storm once in that period of time. The theory is that the worst storm that would ever be expected to statistically occur once in 100 years will be much more severe than the worst storm that would occur once in 2 years.

The curve for a given frequency is actually a plot of independent storms of varying lengths, each with a different duration, and not a single storm plotted against time. Therefore, a storm of 10-min duration is one that lasts for exactly 10 min and then stops. It is not an extension of any other plot. Any storm with a longer duration will have a lower instantaneous peak flow. Flooding would result if the design storm were exceeded; special consideration should be given to the degree of protection provided for the building and its contents by rapid removal of rainfall by the stormwater drainage system. This depends on the importance of the facility, how flooding may affect access, the importance of uninterrupted service, and the value of the equipment or material installed or stored. Since a severe thunderstorm or hurricane may produce rainfall rates greater than anticipated, the property value may require the selection of a frequency longer than the one originally chosen.

The return period selected could be based on client preference, code requirements, or the degree of safety desired by the engineer. Choice on the recurrence of the storm is based on economic considerations and risk tradeoff. The following criteria may he helpful as a guide. For average sites a design storm frequency of 10 years is generally used. Certain clients design their projects for 40 years of useful life, which may require a 50-year design storm if flooding might cause a problem. Many municipal authorities use a 2- or 5-year design storm for city sewers to keep the cost down. Some use a 10-year storm. The U.S. Army Corps of Engineers normally uses a 10-year storm, but for military installations a 25-year storm is used.

After the return period is selected, it is put aside for future use when the intensity is determined.

Rainstorm Duration (Time of Concentration). The final figure in the calculation of the rainfall rate is the determination of the length of time (duration) rain will fall during the design rainfall, measured in minutes. The reason for the importance of duration is that the peak flow of the rainstorm is assumed to occur the instant that a drop of water reaches the design point from the farthest part of the contributing area and then stops. The shorter the period of time; the more intense the rate of rain is assumed to be.

As an example, for New York City, the heaviest total amount of rain on record for a 5-min period is 0.75 in (17 mm). This figure multiplied out to determine the 1-h rate at the 5-min reading, gives an hourly figure of 9 in (225 mm) of rain. Yet the largest amount of rainfall ever recorded over a 1-h period is 3.11 in (85 mm). It can thus be seen that the rate of rainfall measured during a short period is much more intense than the rate measured over a longer time.

It is recommended that a 10-min duration be used as the minimum for typical design conditions. A very short duration will result in excessive runoff volume and excessive pipe sizes. Less than 10 min will not usually be justifiable even on a roof or parking lot.

We now calculate the amount of time it takes for a drop of water to reach any design point, commonly called the *time of concentration*. This is done by finding the overland flow time plus the time in pipe, if applicable.

Overland Flow Time (Inlet Time). The overland flow time, often called the *inlet time*, is the number of minutes it takes a drop of water to travel from the farthest point of the tributary area to a drainage inlet. For equal distances and slope, it takes longer for a drop of water to enter a DI from the farthest point of a lawn than from a paved parking lot. The slope of the land is also a factor, since for any given surface, the steeper the slope, the faster water will flow.

Figure 17.2 is a direct-reading chart, giving the overland flow time for a consistent grade surface when the distance from the most remote point to the DI, the average slope of grade, and type of surface are known. A special case is on pavement where the flow of water from

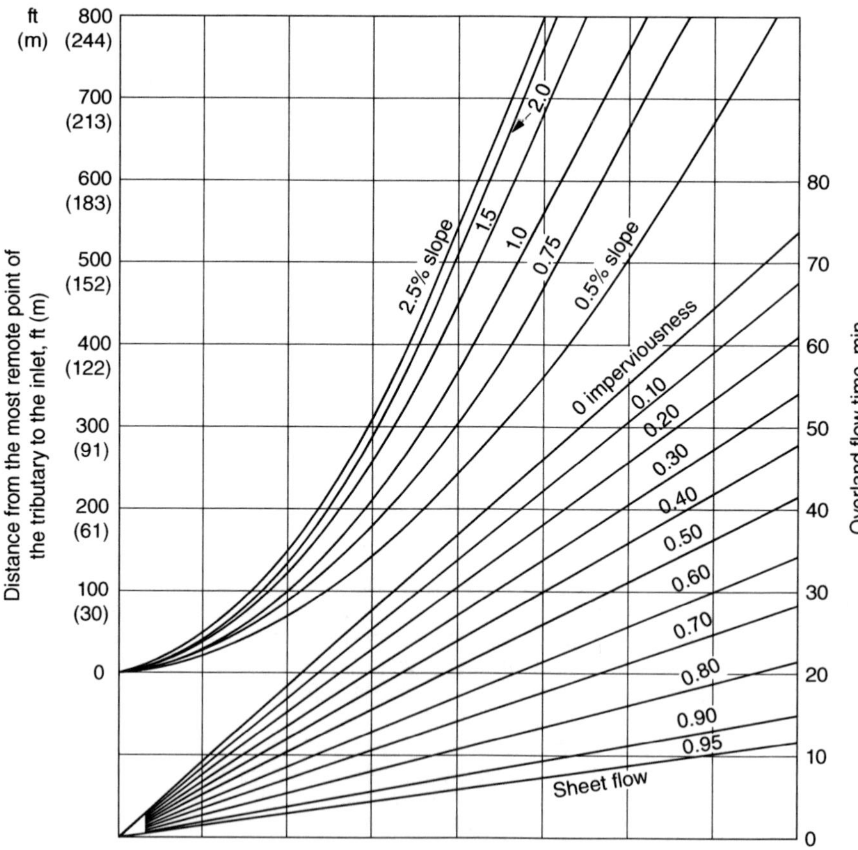

FIGURE 17.2 Overland flow time.

surfaces would be in the form of a very shallow sheet of water covering the surface of the ground. This phenomenon is known as *sheet flow* and has a high velocity. This is shown in the figure as sheet flow and has an imperviousness of 95. For mixed surfaces, see discussion under item 4, system design procedure.

Just as a point of interest, under normal conditions, it takes a rainfall of 2-min duration on turf before any excess rain would flow on the ground.

Time in Pipe. The *time in pipe* is the number of minutes the theoretical drop of water takes to flow from point *A*, (the outlet pipe of a DI or catch basin (CB) in question) to point *B*, the next design point, or from one design point to another. This requires that the pipe be sized and the velocity of the stormwater be known. Finding the velocity in feet per second and multiplying that figure by the actual pipe length, in feet, gives the time in pipe, in seconds. The velocity can be found in Table 19.1. Convert this to minutes. Note that a minimum velocity of 2 ft/s (0.6 m/s) shall be maintained in the stormwater system in order to maintain selfscouring properties.

Rate of Rainfall (Rainfall Intensity). The last variable for inclusion in the rational formula is *R*, the rate of intensity for rainfall. It is obtained from the rainfall intensity–duration–frequency curves, shown in Fig. 17.1, and is measured in inches per hour (millimeters per hour). The city closest to the local site should be selected for a preliminary figure, but consultation with the local officials will give the most accurate results.

Rainfall intensity is the amount of average precipitation, measured in inches per hour, in a given period of time. The determination of the intensity requires a combination of the duration of rainfall, in minutes, along with a set of curves for the frequency of occurrence to find the rainfall intensity. To enter Fig. 17.1, the return frequency of a design storm must have been selected and also the rainfall duration (which is the sum of the overland flow time and the time in pipe, if applicable) will have been calculated. To use the chart, first find the calculated duration at the bottom of the chart. Draw a vertical line up until it intersects the curve for the selected return frequency. From this intersection, draw a horizontal line to the left to read the rainfall rate. This is done for each individual point.

With the calculation of the rate of rainfall, it is now possible to substitute the appropriate figures into the rational formula necessary to calculate the flow and to size each design point on the site.

Storm Water Structure Selection

A drainage structure is an assembly outside of piping, built or installed belowground. Inlet structures are intended to collect surface stormwater and route the effluent into a piping system.

While the piping system is being designed, after the stormwater drainage inlet structures have been located, the type and size of inlets and grates must be selected. In most cases, the authority having jurisdiction has design standards available for such purposes. If no standards are available, the following can be used as a guide. Stormwater drainage inlet structures are classified according to their functions as follows:

Drainage Inlets. Drain inlets (DIs) are structures that admit stormwater into the piping system without restriction. Since all debris and sediment easily enter the piping system, they are generally located in areas that are reasonably free from sediment or debris such as paved, lightly vegetated, or unimproved areas. The bottom of a DI is leveled with the invert of the outlet pipe. A shallow DI is illustrated in Fig. 17.3, and a deep DI is illustrated in Fig. 17.4.

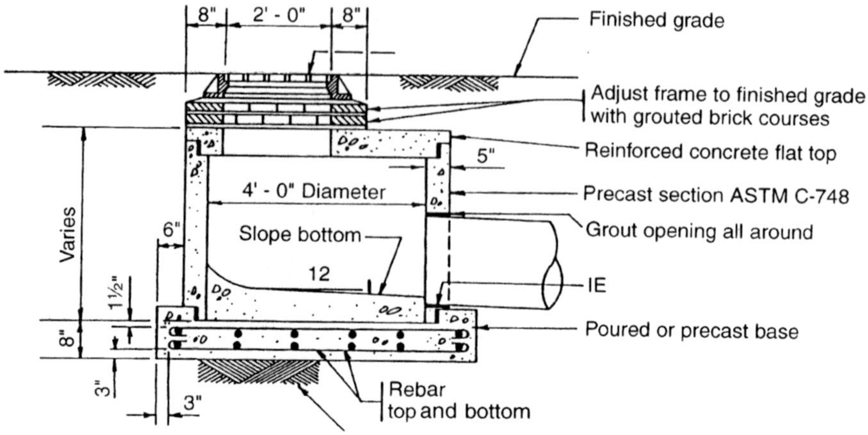

FIGURE 17.3 Typical shallow drainage inlet.

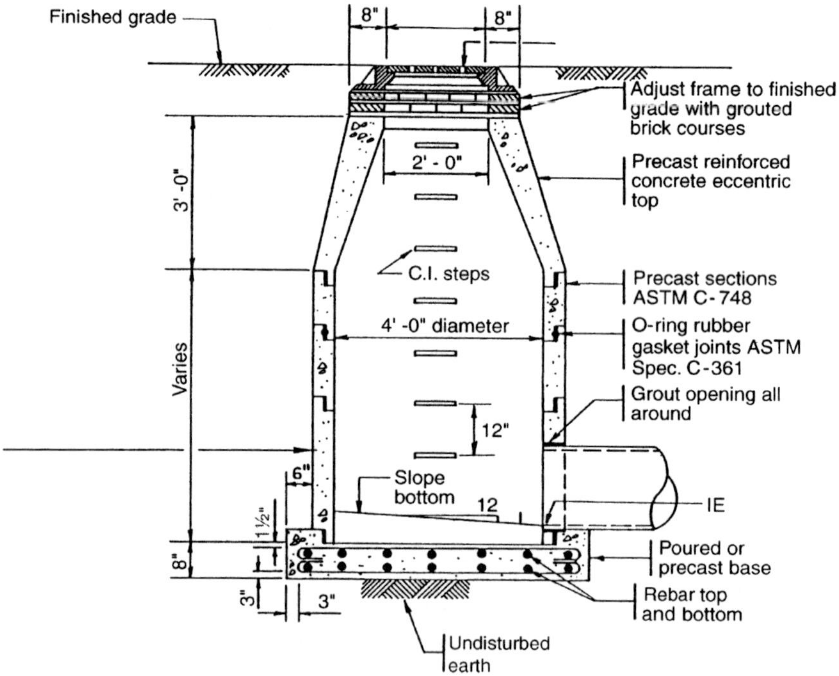

FIGURE 17.4 Typical deep drainage inlet.

Catch Basins. Catch basins (CBs) are similar to drainage inlets, except that there is a space below the inlet and outlet pipes for retention (catching) of debris or sediment. A CB is usually located in paved areas where debris can be easily washed in. Experience has indicated that inadequate maintenance negates their benefits, and CBs are not generally used unless good maintenance can be ensured. A typical CB is illustrated in Fig. 17.5.

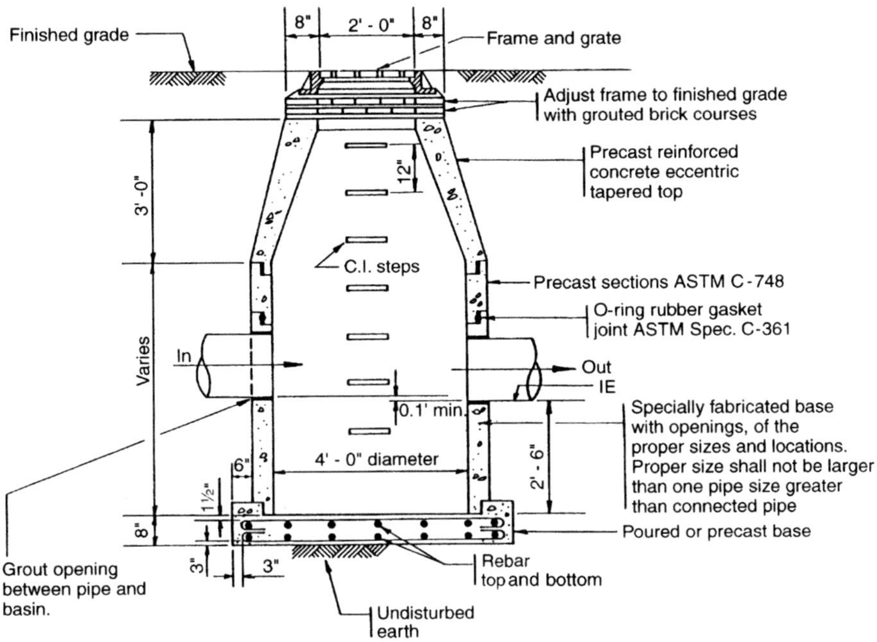

FIGURE 17.5 Typical catch basin.

Manholes. Manholes (MHs) are provided for ease of pipe connections and for cleaning purposes, but do not allow surface water to enter the cover. Drop manholes provide for pipe connections where a difference of more than 2 ft in elevation exists between inlet and outlet pipe. A precast MH is illustrated in Fig. 19.1, and a drop connection into a MH is illustrated in Fig. 19.3. Manhole covers and steps are illustrated in Figs. 19.4 and 19.5. Manholes have the same function for both the sanitary and stormwater systems.

Gratings. Gratings allow water to enter the drainage structure, and some gratings will prevent the entrance of debris into the piping system. Gratings are classified as plain, curb, or combination.

1. A plain grate (or gutter grate) is flat and allows water to enter from all unobstructed sides. Bars prevent the entrance of debris.
2. A curb grate has a single vertical opening in the side of a curb. Water enters only from the front. Debris can enter freely.
3. A combination grate is a plain grate and curb grate acting as a single unit with one placed in front of the other.
4. Multiple inlets are two or more closely placed inlets acting together.
5. A round cover cannot fall into a drainage structure.
6. Flat grates are assumed to be flush with grade and flooded.

Grates Installed in Flat Areas (1 Percent or Less). The following empirical formulas have been developed for the selection of gratings:

Plain grate

$$SF \times Q = CLH^{2/3} \tag{17.3}$$

where Q = quantity of water entering grate, cfs
C = constant of 3.0
L = total perimeter of grate, ft
H = depth of water, above level of road, ft
SF = safety factor

Discussion:

1. Usually H is 2 to 3 in, but never more than 6 in.
2. SF (the safety factor) is required to allow for blockage by debris. In paved areas, use 1.25. Other areas use 1.5 or 2.0, according to the conditions of possible debris accumulation. A safety factor of 2.0 is not unusual.
3. For the perimeter, where one side of a grate is against a curb, only three sides are used to compute the perimeter.
4. Figure 17.6 depicts a graphical solution of inlet capacity for a plain flooded grate, based on Eq. (17.3).

Use a manufacturer's catalog to select a standard size grate with the predetermined perimeter as a minimum starting point.

CURB GRATE. This type of inlet is rarely used alone unless conditions do not permit any other type and, in addition, favorable inlet depression and contour are possible. Figure 17.7 can be used to determine the flooded curb grating size.

COMBINATION GRATE. Experiments have shown that combination grates do not admit substantially more water than a flat grate; therefore, Eq. (17.3), for the plain grate, should be used as the primary method to determine the size. An additional 10 percent for paved areas and up to 75 percent for other areas should be deducted from the flow rate as safety factors plus an additional 25 to 75 percent as a debris factor. This type of grate is preferred where provision to collect all stormwater is required. Caution should be used in the application of the safety factor. The curb grate acts as a safety valve when a flat grate is clogged with debris.

Grates Installed in Sloped Areas (Greater Than 1%)

PLAIN GRATE. The size of a flat grate located in a sloping roadway can be determined from the trial length derived from the following formula:

$$L = \frac{V}{2}\sqrt{Y + D} \tag{17.4}$$

where L = minimum length of opening, ft
V = velocity of water in gutter, fps
Y = depth of water in curb, ft
D = thickness of grate, ft (from a manufacturer's catalog)

Discussion:

1. A minimum length of 3 ft (1 m) and a minimum width of 2 ft (0.6 m) are recommended.
2. Net opening of the grate should be 50 percent of width or greater.
3. The nomograph in Fig. 17.8 should be used to determine the water velocity in a gutter or channel.

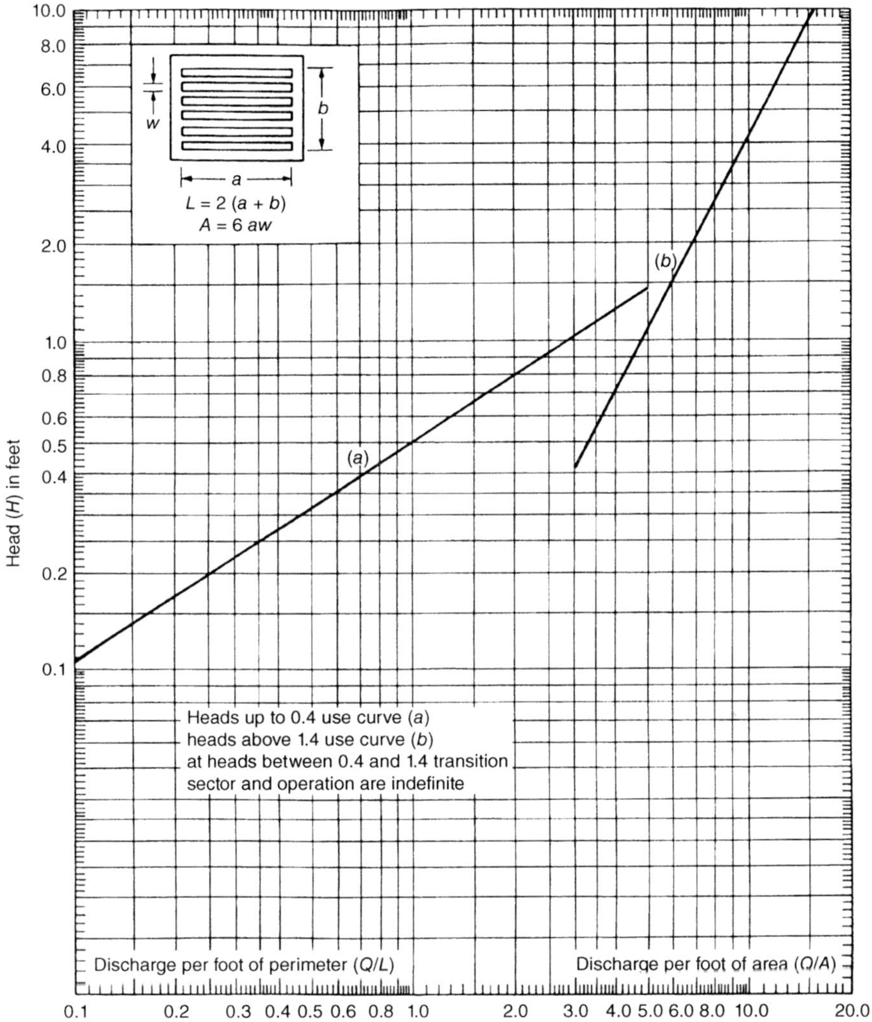

FIGURE 17.6 Graphical solution for flow into a plain flooded grate. (*Courtesy Bureau of Public Roads, Baltimore County.*)

CURB GRATE. This type of inlet is very inefficient, and a long length is usually required to collect the entire design flow. The capacity for complete interception is determined from the following formula:

$$Q = 0.7L\,(A + Y)^{1.5} \tag{17.5}$$

where Q = capacity of grate, cfs
L = length of clear opening, ft
A = depth of depression at inlet, ft (below roadway level)
Y = depth of water in gutter, ft (above roadway level)

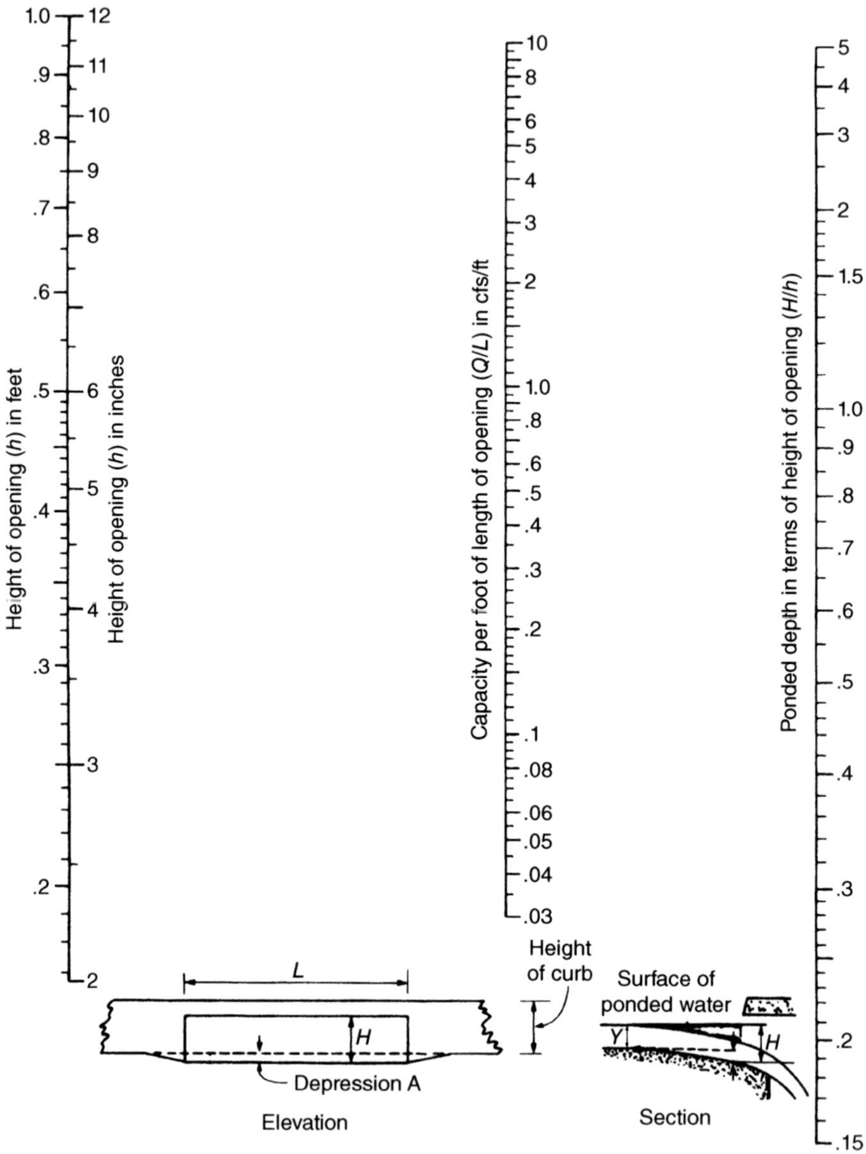

FIGURE 17.7 Graphical solution for flow into a curb grate.

COMBINATION GRATE. The capacity of combination inlets has been determined by a series of complex equations. Since actual experiments have proved that the capacity of combination inlets is not much greater than that of flat inlets, it is much easier to design the grate as if it were flat. Since some allowance can be made for the combination grate, the size of the flat grate portion can be reduced if the calculated size is larger than that of the standard size produced by a manufacturer.

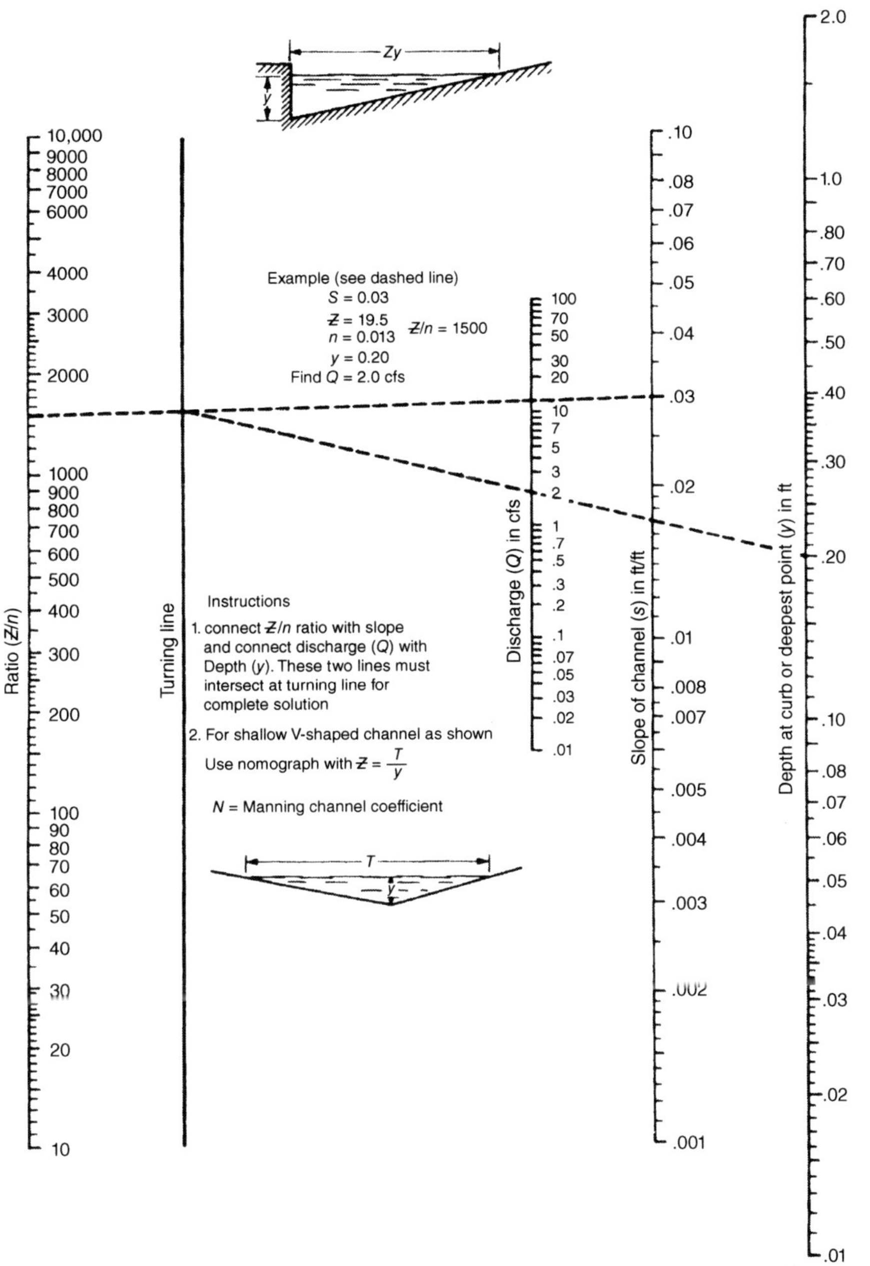

FIGURE 17.8 Water velocity in a gutter.

System Design Procedure

1. The drainage structures should be located and final coordination completed with the site planner. Select the type of drainage structures necessary, either DI or CB.
2. Select the return period for the design storm. If there is some concern of a recurrence in a short time, there is a new manner of determining the probability. The probability of any storm exceeding the rainfall rate to be used in the design of the stormwater system for any one year can be found by dividing the design time, in years, into 1.

   As an example, using a 25-year return period, what will be the probability P that this storm will be exceeded if a 1-year return period was selected? Solving for P gives

   $$P = \frac{1}{25} = 0.04$$

   In this example, the probability that a rainfall will occur in 1 year which exceeds a 25-year storm is 4 percent. Since the forces of nature are unpredictable, it is entirely possible for a specific rainfall rate to be exceeded at any time. With the selection of a higher average frequency of occurrence, such as 100 years, the probability of a similar storm occurring is considerably lower. Choices on the possible recurrence of the storm are based on economic considerations and risk tradeoff.
3. A site plan showing the location of the drainage structures and contour lines shall be used to determine the area draining into each DI and the distance from the farthest contributing point. The type of surface of the contributing area should be available or easily obtained. The contour lines will allow calculation of the slope of the land into each DI. A chart or tabulation in any convenient form should be prepared indicating the type of surface and the square feet (m^2) area of each type of surface draining into each DI.
4. Determine the imperviousness factor of the various surfaces in the contributing area of the site using Table 17.1, keeping in mind the average slope of the land and the type of surface or surface combinations. Use a weighted factor for areas having more than one surface type. The weighted factor is obtained by finding the area of each type of surface and multiplying the area by the imperviousness. This is done for each type of contributing area, adding them and dividing the sum by the whole area going into the DI. As an example, three separate tributary areas have been previously determined to be 0.5 acre of grass, 0.5 acre of pavement, and 0.2 acre of roof. This combination would be solved as follows:

   $$\frac{(0.5 \times 0.15) + (0.5 \times 0.90) + (0.20 \times 1.00)}{1.2} = 0.60 \text{ weighted factor}$$

5. Determination of the design storm will give the final criterion R necessary for the solution of the rational formula. To accomplish this, the duration of the design rainfall shall be calculated.
   - The first part is to find the overland flow time. This is done by using Fig. 17.2, which is a direct-reading chart that gives the overland flow time for various homogeneous surfaces and slopes.
   - Measure the distance from the farthest point of the tributary area into the DI.
   - Find or calculate the average slope of the ground from the farthest point of the tributary area.

- Determine the type or combination of surfaces. If the area consists of more than one surface, calculate the weighted coefficient.
- We now have the information necessary to use Fig. 17.1. Using the distance to the DI found above, find that distance on the left side of the diagram. Draw a line horizontally until it intersects the average slope found above. From there, draw a vertical line until it intersects the curve of the selected or calculated imperviousness coefficient. From this intersection, draw a line to the right to find the number of minutes that would be the overland flow time. The minimum overland flow time should be 10 min. Rarely will a lesser time be economically justifiable.

6. Utilizing the rainfall intensity–duration–frequency curves in Fig. 17.1 and utilizing the city closest to the project location, select the rainfall intensity using the return period selected and the time of concentration previously calculated. The time of concentration will be the overland flow time (found in Fig. 17.2) plus the time in pipe, if applicable (from the velocity found in Table 19.1). From data in the form just prepared and using the rational formula [Eq. (17.1)], calculate the total inflow to each individual DI. Place this information in the form for reference. The individual pipe from each DI only will be sized from this information using the quantity of rain along with the slope of the pipe.
7. With the total inflow in cubic feet per second to individual DIs now known, select the grate type and size. Consult the manufacturer's literature for the correct size. Add the grate type to the above form for reference. Adjust DI locations for flow requirement if necessary. Most manufacturers have developed inlet flowcharts for specific grates. Obtain from their catalog the size grate that can accept the calculated flow.
8. Lay out the piping system from the DI to the point of disposal. Locate MHs. Select the pipe material. The flow in stormwater sewers will have the same characteristics as the sanitary system. Refer to Chap. 18 for a complete discussion of sewers. To size the sewer with the flow calculated in cubic feet per second, use Fig. 17.9. To find the n factor (a coefficient for pipe roughness used in the Manning formula) for the pipe used, refer to Table 17.2.
9. Building roof drain runouts shall be connected to the stormwater drainage lines and their drainage area of the roof noted. Their size is governed by the local plumbing code. Once past 5′-0″ (1.6 m) from the building wall, the design of the sewer and size are selected by the engineer using the above procedures. Do not use a smaller size than that of the runout.
10. Select the critical inlet. The critical inlet is the one that produces the maximum combination of inlet time (overland flow time) plus flow time in sewer to the very first connection with any other branch. The inlet at the farthest end of the drainage system should be selected as the starting point. This point will be the longest in time, not necessarily in distance. This may require some iteration trial calculations of several DIs at various far ends of branches in order to determine which drainage inlet actually is critical. Distance will not be the only criterion. An area comprising asphalt which has a fast inlet time might be a much greater distance from an inlet than an area consisting of grass which would produce a longer inlet time. The slope and length of sewer pipe must also be considered.
11. With the selection of the critical inlet, the individual sewer pipe line can now be sized from the critical inlet up to the first point of intersection with any other contributing source of stormwater. The layout of the sewer system will establish the slope of the pipe, and the form will provide the flow rate of stormwater in cubic feet per second (cubic meters per second). The pipe material will have been selected, and the value of n for the pipe will be selected (Table 17.2). Entering Fig. 17.9 knowing the flow in cubic feet per second (cubic meters per second), draw a straight line from the flow rate to the slope of piping.

English/metric units

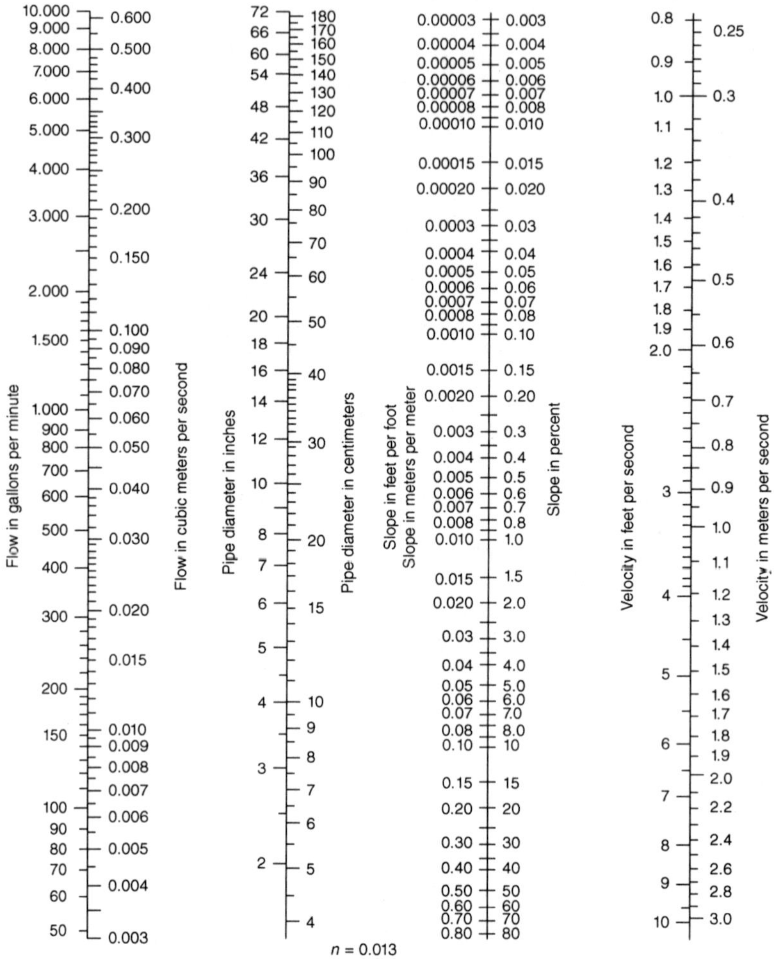

Value of n	0.008	0.010	0.011	0.012	0.013	0.015	0.019	0.021	0.024
Conversion factor for discharge and velocity	1.62	1.30	1.18	1.08	1.00	0.87	0.68	0.62	0.54

FIGURE 17.9 Pipe flowchart. Diagram for solution of gravity flow for Manning formula for circular pipes, ditches, and channels.

This will now establish the pipe size and water velocity. Adjust the line to intersect a standard size. If necessary, make the necessary adjustment as indicated by the conversion factor if a pipe material with different n from that of the chart. Multiply the desired figure by the conversion factor found at the bottom of Fig. 17.9 under the actual n. Generally accepted practice assigns an n value of 0.013 to most commonly used pipe materials. Another method of sizing the pipe would be by using Table 19.1 when a previously determined flow and slope are known. Good practice uses a one half-full pipe. If it is

TABLE 17.2 *n* Value of Pipe Used in the Manning Formula

Pipe material	Range of *n* values	Generally accepted value*
Asbestos-cement	0.011–0.015	0.013
Corrugated metal pipe	0.022–0.026	0.024
Cast iron	0.011–0.015	0.013
Concrete pipe	0.011–0.015	0.013
Ductile iron (cement-lined)	0.011–0.015	0.013
Plastic pipe, all kinds	0.010–0.015	0.011
Steel pipe	0.012–0.020	0.015
Vitrified clay	0.011–0.015	0.013

*Values will vary based on condition of pipe.

determined that the stormwater flow will be in ditches, refer to Chap. 18 for guidance in the design of such a system.

12. When the first junction with another contributing source has been reached, add the entire area of all contributing sources up to this first junction. Calculate the new time of concentration from the most remote inlet and the time in pipe to the design point of the storm sewer. Entering the intensity–duration–frequency chart to find the rainfall rate will now give all necessary information required for insertion into the rational formula [Eq. (17.1)] to find total (cubic meters per second). The resultant flow into the pipe along with the slope of the pipe is used to size that particular segment downstream from the design point. This is continued in succession to the end of the system, requiring a longer time of concentration each time a new design point is reached.

13. As progress is made along the route of the sewer, all branches to the main sewer line should be sized individually as their connection is reached. The branch is sized using the time to concentration required only for that branch. To size the main, add all the areas contributing to the flow at any particular junction and use the longer time of concentration determined from the initial overland flow time from the critical inlet added to the flow time of the water in the pipe.

System Design Considerations

1. The velocity in the piping system should not be less than 2 fps (0.6 m/s), to allow for the movement of sediment and maintain a self-cleaning velocity. Refer to Table 19.2 for the slope of the pipe that will give the desired velocity.
2. The maximum velocity should be limited to approximately 30 fps (9 m/s) to prevent erosion of the pipe interior. If large quantities of sand are present in the water, the scouring action will increase, and so the velocity should be reduced for this condition.
3. To limit velocity, the slope of the piping system should be decreased rather than use oversized pipe. There are other methods available also. If the slope of the ground is steep, a shallow slope along with a drop manhole should be used where the difference in elevation between inlet and outlet inverts is more than 2 ft (0.6 m).
4. A manhole should be located at every change in pipe size, slope, or direction. In many cases, a drainage structure can be used for this purpose if such placement is practical.

5. The minimum size of pipe for runouts from individual outlets should be 6 in (DN 150) because of the possibility of debris. For major branches, the size should be 12 in (DN 300) to reduce the possibility of stoppage by debris.
6. When a change in pipe size occurs, the pipe should be installed crown to crown to eliminate a surcharge of the upstream portion of the pipe.
7. Factors such as snowmelt may add additional, unexpected quantities of water. Snowmelt does not usually add a significant amount of inflow, but it may if the depth of snow is large. However, increasing one pipe size may lead to a problem by reducing velocity too much. In that case, it is advisable to increase the slope of the sewer line. As a general rule, one should be a little generous with pipe sizes. The slight increase in cost will be more than made up by the safety factor of additional capacity provided in the event of a larger than designed storm.
8. Exfiltration occurs when water leaks out of a pipe through bad joints or pipe cracks into the surrounding ground. When a pipe passes under a road or railroad, a joint should be chosen that will permit very little or no leakage, to avoid washing away the subgrade. Infiltration occurs through the same cracks and bad joints into an empty pipe. This is known as "dry flow."
9. Piping should be installed roughly parallel to the slope of the ground. Standard percentages of slopes, if possible, should be used to permit contractors to easily install the pipe.

STORMWATER PERMITTING

General

Stormwater collection systems contain many added substances in addition to stormwater. As stormwater flows over pavements, lawns, parking lots, and streets, it often collects considerable quantities of pollutants, such as oil, fertilizers, pesticides and metals. Active construction sites often discharge sediment into the storm drainage system. This section presents a simplified description of the permitting programs available to those involved in larger industrial construction activities. Nothing in this section should be construed as legal advice, which should be obtained only from people qualified to evaluate the specific project compliance and determine that there is a violation of any law or regulation. The information included is only to be used as a guideline, with specific criteria provided only by qualified authorities.

Stormwater permits are intended to achieve improvements in water quality by reducing or eliminating pollutant loading from some stormwater sources. The exact requirements for achieving this goal depend on the size of the project, type of permit, and the type of discharge, and the characteristics of water that receives the discharge should be handled by a qualified individual.

For a summary of typical industrial activity limitations for stormwater runoff, refer to Table 17.3. For a summary of pollutants that should be analyzed, refer to Table 17.4.

CODES AND STANDARDS

Historically, the National Stormwater Program is a federal government initiative that was developed by the Environmental Protection Agency (EPA). The most important development was the Clean Water Act (CWA). This, in turn, established the National Pollutant

TABLE 17.3 Industrial Activity for Stormwater Runoff

1. Facilities subject to stormwater effluent limitations guidelines, new source performance standards, or toxic pollutant effluent standards under 40 CFR, subchapter N.
2. Facilities classified as:

SIC 24	Lumber and wood products (except 2434)	SIC 32	Stone, clay, and glass products (except 323)
SIC 26	Paper and allied products (except 265 and 267)	SIC 33	Primary metal industries
SIC 28	Chemicals and allied products (except 283)	SIC 3441	Fabricated structural metal
SIC 29	Petroleum and coal products	SIC 373	Ship and boat building and repairing
SIC 311	Leather tanning and finishing		

3. Facilities classified as SIC 10 through 14 including active or inactive mining operations and oil and gas exploration, production, processing, or treatment operations, or transmission facilities that discharge stormwater contaminated by contact with or that have come into contact with, any overburden, raw material, intermediate products, finished products, by-products or waste products located on the site of such operations.

SIC 10	Metal mining	SIC 13	Oil and gas extraction
SIC 12	Coal mining	SIC 14	Nonmetallic minerals, except fuels

4. Hazardous waste treatment, storage, or disposal facilities, including those that are operating under interim status or a permit under subtitle C of RCRA.
5. Landfills, land application sites, and open dumps that receive or have received any industrial wastes, including those that are subject to regulation under subtitle D of RCRA.
6. Facilities involved in the recycling of materials, including metal scrapyards, battery reclaimers, salvage yards, and automobile junkyards, including but not limited to those classified as:

SIC 5015	Motor vehicle parts, used
SIC 5093	Scrap and waste materials

7. Steam electric power generating facilities, including coal-handling sites.
8. Those portions of transportation facilities that are involved in vehicle maintenance, equipment cleaning operations, or airport deicing operations, or which are otherwise identified as industrial activities.

SIC 40	Railroad transportation	SIC 44	Water transportation
SIC 41	Local and interurban passenger transit	SIC 45	Transportation by air
SIC 42	Trucking and warehousing (except 4221-25)	SIC 5171	Petroleum bulk stations and terminals
SIC 43	U.S. Postal Service		

9. Wastewater treatment works with a design flow of 1 Mgpd or more or required to have an approved pretreatment program under 40 CFR Part 403.
10. Construction activity including clearing, grading, and excavation activities except operations that result in the disturbance of less than 5 acres of total land area which are not part of a larger common plan of development or sale

TABLE 17.3 Industrial Activity for Stormwater Runoff (*Continued*)

11. Facilities where material handling equipment or activities, raw materials, intermediate products, final products, waste materials, by-products, or industrial machinery are exposed to stormwater.

SIC 20	Food and kindred products	SIC 31	Leather and leather products (except 311)
SIC 21	Tobacco products	SIC 323	Products of purchased glass
SIC 22	Textile mill products	SIC 34	Fabricated metal products (except 3441)
SIC 23	Apparel and other textile	SIC 35	Industrial machinery and equipment products
SIC 2434	Wood kitchen cabinets	SIC 36	Electronic and other electric equipment
SIC 25	Furniture and fixtures	SIC 37	Transportation equipment (except 373)
SIC 265	Paperboard containers	SIC 38	Instruments and related products and boxes
SIC 267	Misc. converted paper	SIC 39	Miscellaneous manufacturing industries products
SIC 27	Printing and publishing	SIC 4221	Farm product warehousing and storage
SIC 283	Drugs	SIC 4222	Refrigerated warehousing and storage
SIC 285	Paints and allied products	SIC 4225	General warehousing and storage
SIC 30	Rubber and misc. plastics products		

TABLE 17.4 Summary of Pollutants to Be Analyzed

	Industrial	
Pollutant	First-flush grab sample	Flow-weighted composite sample
Oil and grease	□	
pH	□	
Biological oxygen demand (BOD)	□	□
Chemical oxygen demand (COD)	□	□
Total suspended solids (TSS)	□	□
Total phosphorus	□	□
Nitrate and nitrite nitrogen	□	□
Total kjeldahl nitrogen	□	□
Any pollutant in the facility's effluent guideline	□	□
Any pollutant in the facility's NPDES permit	□	□
Any pollutant in EPA Form 2F Tables believed to be present	□	□

Discharge Elimination System (NPDES) which has focused primarily on the discharge of industrial process wastewater. Further legislation developed in the Water Quality Act established an approach for stormwater discharge regulation in the United States.

The EPA, under the Clean Water Act, has the primary responsibility for regulating and permitting stormwater discharge from any entity that could contribute a substantial amount of permits into the waters of the United States. The NPDES is the permitting mechanism used by many states. Still others use a third party for permitting. The state where the discharge is proposed shall have the ability to ensure that pollutant level discharge is reduced and controlled under all storm conditions. Where this is the case, all permit submissions must be made to the appropriate authority. In nonauthorized states, approval must be granted by the EPA.

Stormwater discharge permits are not required for construction sites under 5 acres (20,235 m^2). In addition, permits are not required for any facilities that have no material handling equipment or activities, final products, waste materials or by-products that are exposed to rainwater. These conditions account for a vast majority of facilities that a mechanical consulting design engineer will be responsible for. If there is a reasonable potential for items exposed to rainwater discharges associated with industrial activity, similar to leaks or spills discharging into stormwater sewers, a permit is required.

Guidelines for Industrial Discharge

For the design engineer involved with the design of an industrial facility, there are listings of facilities conforming to a specific "industrial activity" that requires permits for discharge either directly into U.S. waters or into separate municipal stormwater systems. There is a difference between industrial and municipal discharges. Also included in the list of industrial activities are construction sites and landfills associated with facilities.

A number of factors influence the extent to which industrial activities and significant materials can affect water quality and for which permits will be required. The exact factors can only be given by municipal or federal authorities.

1. Topography
2. Hydrogeology
3. Extent of impervious surfaces (e.g., concrete or asphalt)
4. Type of ground cover (e.g., vegetation, crushed stone, or dirt)
5. Outdoor activities (e.g., material storage, loading/unloading, vehicle maintenance)
6. Size of the operation
7. Type, duration, and intensity of precipitation events
8. Geographic location

The following are abbreviated definitions obtained from the EPA website to assist the reader in applying for a permit.

1. Point source discharges from other than facilities are confined and discrete conveyances such as pipes, ditches, channels, tunnels, and vessels from which pollutants will be discharged. They are required to be covered under NPDES.
2. Nonpoint sources are from streets, parking lots, and construction sites that are 1 acre or larger, that drain into surface water or groundwater.
3. The NPDES regulates stormwater from three sources: municipal storm sewer systems, construction sites, and industrial activities.

STORMWATER DISPOSAL

The traditional method of disposing of stormwater for small to medium-size projects designed by consulting engineers is to collect the surface runoff and route it into nearby municipal sewers. An urban area, complete with streets, parking areas, and roofs, is a prime example of this concept. When traditional methods are either not possible or not desired by the community, infiltration structures provide runoff volume control from structure roofs, parking buildings, and parking areas by diverting the runoff into an infiltration structure, thereby recharging the groundwater and, in many cases, preserving the base flow in nearby streams. They also eliminate the small amount of pollution that might flow into larger bodies of water.

Because of the range of criteria possible from the various authorities, the information discussed in this book is not intended to provide specific criteria, but rather to present guidelines and understanding that will allow the engineer to become familiar with general design criteria.

Infiltration Device Design Guidelines

The primary goal of small infiltration-limiting devices in stormwater management is to maintain the postdevelopment runoff characteristics at the source as nearly as possible compared to predevelopment characteristics. The multiple benefits make these devices suitable for meeting the intent and goals of maintaining the predevelopment runoff characteristics. These devices could be used either alone or in combination with one another.

Definitions

Stormwater infiltration devices to be discussed include these:

1. *Infiltration basin or pond* is an irregularly shaped, open excavated area used for stormwater collection, it is intended to allow rainwater to pond and infiltrate the soil below. It can be constructed by either excavation or embankment. It is designed to have a large surface area and a shallow depth. At the low point, an additional sand filter or dry well could be built to increase the infiltration capacity. An emergency outlet, such as an overflow, should be provided for exceptionally large storms. A typical infiltration basin is illustrated in Fig. 17.10.
2. *Infiltration trench* is an excavated trench for smaller paved areas with a width of about 3′-0″ to 10′-0″ (1 to 3 m) and dug to usually less than 10′-0″ (3 m) in depth. It is backfilled with stone aggregate.
3. *Dry wells.* A dry well is a hollow cylinder, either empty or filled with granular material, having porous sides that allow stormwater entering the dry well to disperse into the surrounding soil. Granular material should also be added around the outside of the dry well to allow easy dispersal. It is often constructed of precast concrete manhole segments stacked one on top of another downward from the surface. A typical dry well is illustrated in Fig. 17.11.

Infiltration Basins

An infiltration basin must have the capability to drain off and hold water between design storm events. The operational efficiency of the basin is dependent on the underlying soil conditions. Infiltration devices discussed here are not expected to contain large amounts, if any, of pollutants such as pesticides, manufacturing and industrial discharge, or combined

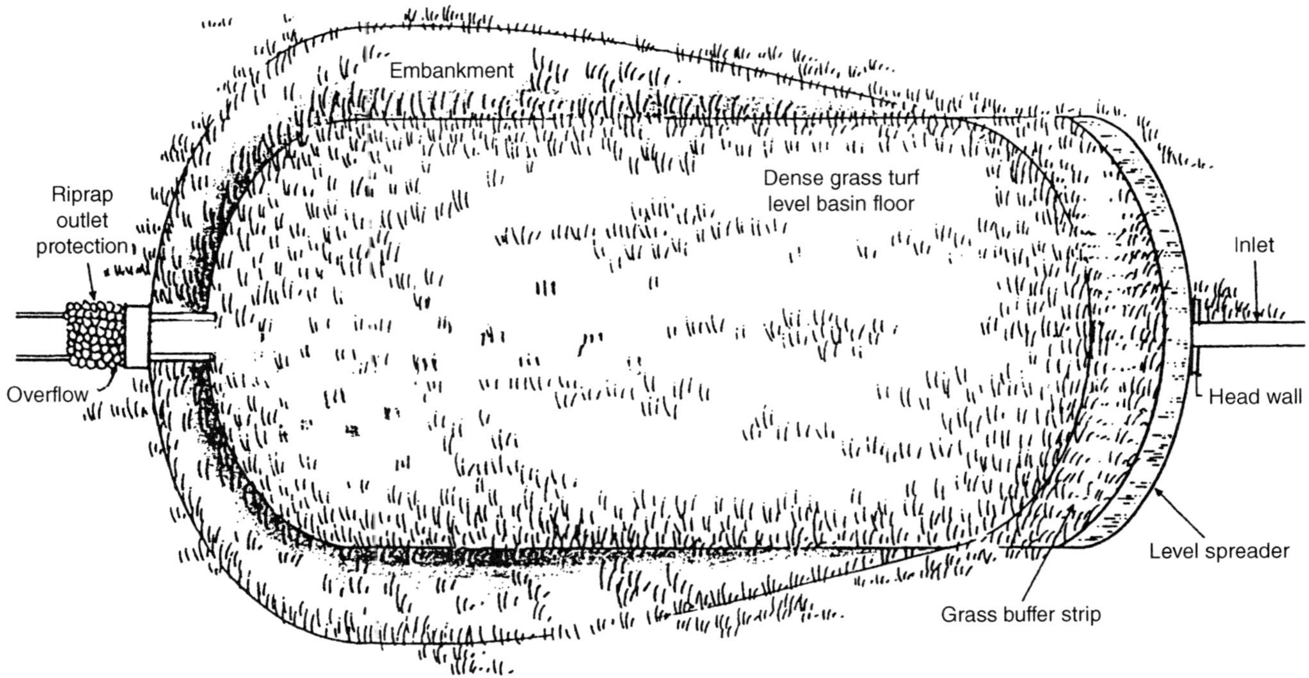

FIGURE 17.10 Schematic of a typical infiltration basin.

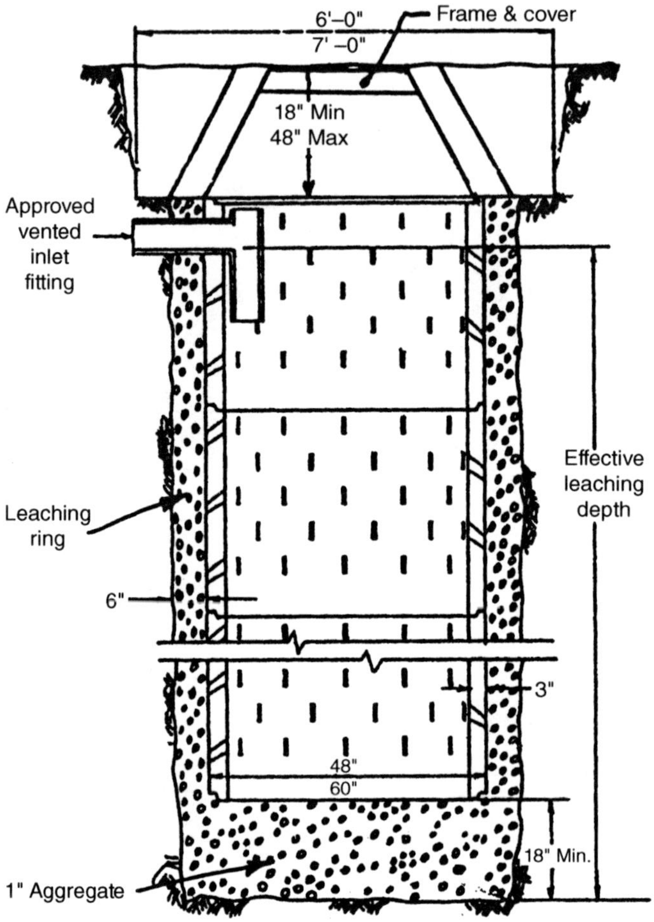

FIGURE 17.11 Detail of dry well.

sewer overflows. If these pollutants are expected, it would seriously defeat the possibility of successful installation.

Planning Guidelines

Building an infiltration basin is appropriate when groundwater recharge and pollution removal are a high priority for a small watershed. Since soil properties are critical factors in the design of basins, a preliminary screening of sites is necessary. Both soils with shallow groundwater or fractured bedrock and sandy soils with low absorption rates, areas of high loadings of pollutants, and areas where protection from possible pollutant contamination is undesirable are not suitable for basins. To limit seepage from the basin affecting groundwater quality, a minimum of 5-ft separation should be maintained to the top of the highest groundwater level found.

Infiltration basins are commonly used for drainage of areas from 5 to 10 acres, with surrounding land slopes to the basin less than 20 percent. The basin should be located greater than 50′-0″ (15 m) from slopes greater than 20 percent. Fill material and construction sites for pond location shall be avoided.

Design Guidelines

Site Evaluation. Soil borings should be conducted at the basin site for each 5000 ft^2 (4200 m^2) or where different soil groups are encountered. They must establish a minimum infiltration rate of at least 0.5 in/h (13 mm/h) and a minimum separation of 5-ft (1.5 m) to high groundwater or bedrock.

Since stormwater can carry pollutants, minimum setback distances from private water supply wells shall be at least 100′-0″ (30 m), and 1200′-0″ (360 m) from public walls. If the distance is closer than 1200′-0″ (360 m), a study of the groundwater flow in the area will be necessary. Basins must be located at least 10′-0″ (3 m) downslope and 100′-0″ (30 m) upslope from building foundations. Infiltration basins should not be located in a floodplain and shall conform to all regulations of the AHJ.

Treatment and Pretreatment of Runoff. Infiltration basins are not sediment control devices. Protective measures, should be installed such as a buffer strip of vegetation that will act as a pretreatment and collect any sediment that may be present. Pretreatment is a requirement for all basins that may receive stormwater that may contain particulate matter which might clog the infiltration basin. If necessary, a method of pretreatment, such as a grass filter to enable coarse sediment to be removed from the inflow and eventually cleaned up, should be installed at the inlet to the basin that would prevent premature clogging of the basin.

If excessive oil or sediment is expected, an oil interceptor or sediment trap shall be installed before water enters the basin. It is necessary to provide access to the pretreatment facilities because it will be necessary for these areas to be eventually cleaned and maintained to remove sediment buildup that occurs over the years.

It is expected that the stormwater runoff will contain pollutants that consist of small quantities of oil, dust, and biological sediments that would be a part of a normal commercial area.

Infiltration Time and Drain Time. To ensure the adequate treatment time, the total infiltration time in the basin should not be completed in less than 6 h or more than 72 h. This will provide the necessary time for adequate pollution treatment for groundwater protection. It is recommended that a minimum cation exchange capacity of 5 milliequivalents/100 g of dry soil be present to provide adequate treatment levels.

The actual design infiltration rate should be confirmed by field testing. The higher the infiltration rate, the longer the time that will elapse to when the basin will require possible maintenance.

Basin Shape

The basin shape can be any configuration that blends in with the surrounding site. All basins shall have a flat bottom. To enhance the infiltration, a large and shallow basin is preferred over a small and deep one. The basin floor shall be as flat as possible. Consider side slopes of 4 to 1 or flatter for ease of maintenance. A public right-of-way is necessary to grant access. The access routes should not be made on the basin floor. Fencing around the basin could serve as a safety feature, with the intent to deny public access to the basin.

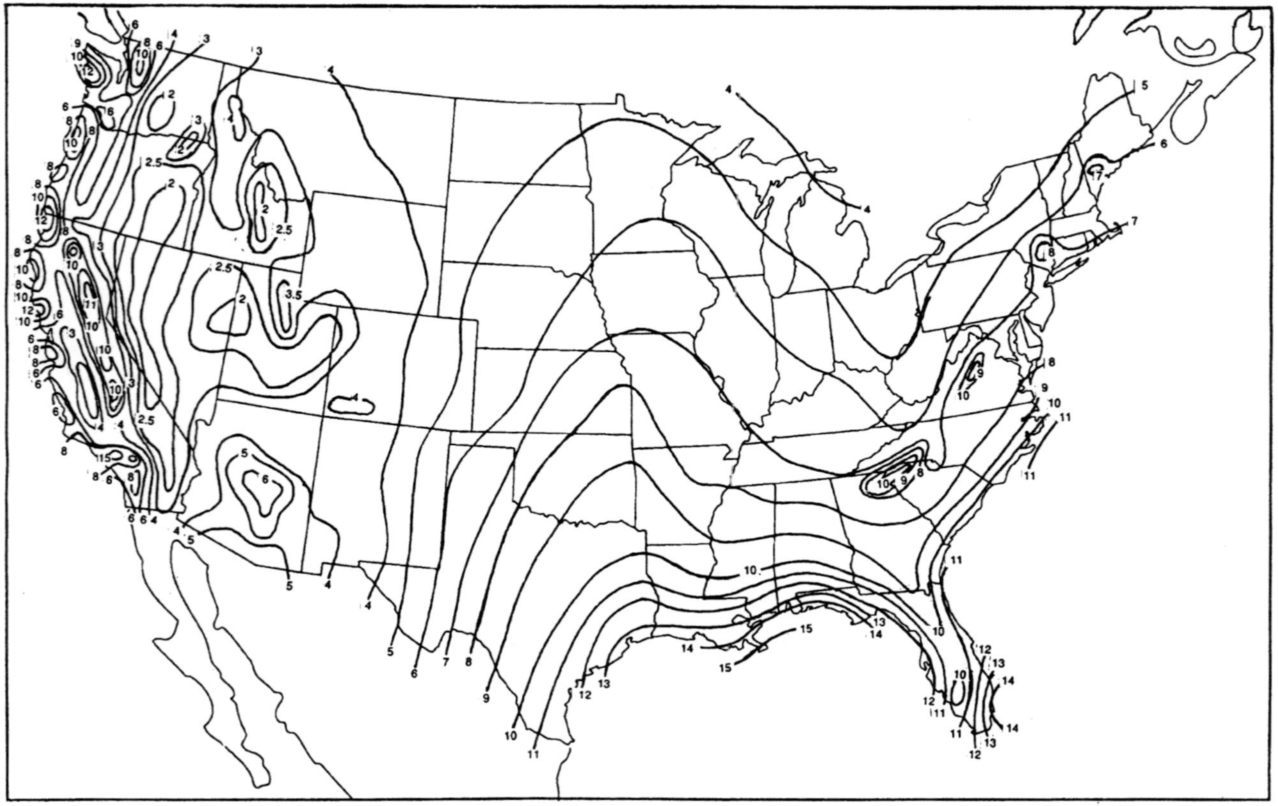

FIGURE 17.12 100-year, 24-h rainfall. (*Adapted from U.S. Weather Bureau map.*)

Vegetation. Plant a water-tolerant, fast-germinating, hardy grass on the bottom and side slopes. Generally, fertilizers should not be applied. A vegetative screen around the basin will restrict views from surrounding properties and enhance acceptance of the basin.

Basin Inlet. Erosion protection is required at the inlet to the basin. Riprap aprons, head walls, and a 20-ft (6-m) buffer strip of grass would help to dissipate excessive energy.

Winter Operation. When soil freezes, infiltration will stop. Therefore the basin could not be depended upon to treat rain or snowmelt during the winter.

Construction Guidelines. To avoid compaction and allow the basin to operate effectively, all heavy equipment shall be diverted away from the basin. Avoid bulldozers and front end loaders. Excavate during dry periods using only light earth moving equipment with over-size tires. After excavation, the bottom should be deeply tilled. Seed vegetation shortly after excavation with native fast-germinating grass.

Infiltration Basin Design. The size of the basin depends on the total volume of runoff from the tributary area and the infiltration rate of the soil. There is no national code that has been developed for the basin. The actual figure must be obtained from and approved by the local authorities and the owner. The design shall follow this progression.

1. Determine the extent of rainfall expected to fall on the project area. It is recommended that a 100-year, 24-h storm be selected so as not to overflow the basin. This is the largest creditable rain that would be expected. A shorter time may be selected with the approval of concerned parties. A map of the United States showing a 100-year storm lasting 24-h is illustrated in Fig. 17.12.
2. From the site examination, select the soil that will be on the bottom of the basin, and then determine the soil infiltration rate for that soil. Soil infiltration rates are given in Table 17.5. It is recommended that the engineer use one-half of the measured rate.

TABLE 17.5 Soil Infiltration Rates and Water Storage Capacities

Soil group	Infiltration rate, in/h	Water storage capacity Inch of water
Sand, open-structured	8.27	0.35
Loamy sand	2.41	0.31
Sandy loam	1.02	0.25
Loam	0.52	0.19
Silty loam	0.27	
Sandy clay loam	0.17	
Clay loam	0.09	
Silty clay loam	0.06	
Clay	0.02	

Source: Maryland Standards and Specification for Stormwater Management Infiltration Practices.

3. Select the length of time that the water in the basin will be given to fully infiltrate the soil. The length of time shall be from 6 to 72 h.
4. Determine the maximum basin depth for the design storm.
5. Calculate the design volume for the basin at maximum rainfall.

BIBLIOGRAPHY

"5 to 60 Minute Precipitation Frequency for Eastern and Central US," Technical Memorandum NWS Hydro 35, National Oceanic & Atmospheric Administration.

"5 to 60 Minute Precipitation Frequency for 11 Western States," NOAA Technical Memorandum, National Oceanic & Atmospheric Administration.

Davis, M. L., *Water and Wastewater Engineering*, McGraw-Hill, New York.

Fannon, W., "Storm Water Design," *Air Conditioning, Heating and Ventilating Magazine.*

Frankel, M., unpublished manuscript, 2009.

Mays, L. W., *Stormwater Collection Systems Design Handbook*, McGraw-Hill, New York.

"Rainfall Frequency Atlas of the United States," Technical Paper 40, U.S. Department of Commerce.

"Rainfall Intensity-Duration-Frequency Curves," Technical Paper 25, U.S. Department of Commerce.

CHAPTER 18
FLOW IN DITCHES AND OPEN CHANNELS

This chapter discusses the design requirements and flow measurement techniques for water in open channels and ditches. Open channel flow is the flow of liquids where there is only one liquid surface free of solid boundaries. A channel is a hard surface similar to pavement that is in contact with water. It is often referred to as a *conduit*. A ditch has a soft earthen configuration that contacts the water.

Ditch conditions are discussed because an engineer might be called upon to design a ditch containing stormwater flowing from a site to a point of disposal. Discussion is limited to only stormwater because sanitary drainage would not be permitted to flow in an open channel. Gravity, nonpressurized sanitary flow in pipes is discussed in Chap. 19.

The flow in open channels is classified into two types: steady and unsteady flow. Steady flow is a constant rate of discharge, and unsteady flow is characterized as variable rate of discharge over time. A flow is uniform if velocity and depth are mostly constant along the conduit. If velocity and depth change in the conduit, the flow is unsteady.

DETERMINING THE VALUE FOR n IN THE MANNING FORMULA

The value n is assigned for the roughness of the material that forms an open channel for calculation in the Manning formula. There has been a considerable amount of research done in the field of open-channel hydraulics. The determination of n has been established for a large variety of conditions, and the following represents a brief set of values for use in designing ditches to determine flow. Refer to Table 18.1. Additional information and values of n for a much more inclusive set of ditch and other conditions can be found in *Resistance in Fluids Flowing in Channels and Pipes*, authored by P. Ackers, London, England.

CALCULATING THE HYDRAULIC RADIUS OF AN OPEN CHANNEL

Often the hydraulic radius of the conduit must be found in order to calculate the flow rate. This is accomplished by first determining only the free surface area of the water flowing in the channel, in feet. Next, determine the total area of the conveyance channel itself (bottom and sides) that the flowing water is intended to flow in, in feet. To calculate the

TABLE 18.1 Values of *n* for Manning Formula in Open Channels

Surface	*n* Value		
	Min.	Avg.	Max.
A. Lined channels	0.011	0.012	0.014
1. Metal			
a. Smooth steel (unpainted)			
b. Corrugated	0.021	0.025	0.030
2. Wood			
a. Planed, untreated	0.010	0.012	0.014
3. Concrete			
a. Float, finish	0.013	0.015	0.016
b. Gunite, good section	0.016	0.019	0.023
c. Gunite, wavy section	0.018	0.022	0.025
4. Masonry			
a. Cemented	0.017	0.025	0.030
b. Dry rubble	0.023	0.032	0.035
5. Asphalt			
a. Smooth	0.013	0.013	
b. Rough	0.016	0.016	
B. Unlined channels			
1. Excavated earth, straight and uniform			
a. Clean, after weathering	0.018	0.022	0.025
b. With short grass, few weeds	0.022	0.027	0.033
c. Dense weeds, high as flow depth	0.050	0.080	0.120
d. Dense brush, high stage	0.080	0.100	0.140
2. Dredged earth			
a. No vegetation	0.025	0.028	0.033
b. Light brush on banks	0.035	0.050	0.060
3. Rock cuts			
a. Smooth and uniform	0.025	0.035	0.040
b. Jagged and irregular	0.035	0.040	0.050

Type of conduit	Wall and joint description	Manning's *n*
Concrete pipe	Good joints, smooth walls	0.011–0.013
	Good joints, rough walls	0.014–0.016
	Poor joints, rough walls	0.016–0.017
Concrete box	Good joints, smooth finished walls	0.012–0.015
	Poor joints, rough, unfinished walls	0.014–0.018
Corrugated metal pipes and boxes, annular corrugations	$2\frac{2}{3}$- by ½-in corrugations	0.027–0.022
	6- by 1-in corrugations	0.025–0.022
	5- by 1-in corrugations	0.026–0.025
	3- by 1-in corrugations	0.028–0.027
	6- by 2-in structural plate	0.035–0.033
	9- by 2½-in structural plate	0.037–0.033

TABLE 18.1 Values of n for Manning Formula in Open Channels (*Continued*)

Surface	n Value		
	Min.	Avg.	Max.
Corrugated metal pipes, helical corrugations, full circular flow	2 2/3- by ½-in corrugations, 24-in plate width		0.012–0.024
Spiral rib metal pipe	¾- by ¾-in recesses at 12-in spacing, good joints		0.012–0.013

Source: Data from Federal Highway Administration (2001).

hydraulic radius, divide the surface area of the water flowing in the channel by the length of contact surface wetted perimeter. The answer is the hydraulic radius in feet.

SUGGESTED DESIGN REQUIREMENTS FOR DITCH SYSTEMS

The following methods and dimensions are suggested if there is no code or requirements by facility standards or local authorities.

Cross Section

Recommended dimensions of ditches shall be trapezoidal in cross section with a minimum bottom width of 3 ft (0.67 m), although a smaller dimension will be acceptable for low flows. This is to avoid erosion of the ditch in the future. For ditch depths of 10 ft (3.3 m) or less, the dimensions will be as follows:

Asphalt-or concrete-lined	1.5 horizontal to 1 vertical
Grass-lined	3 horizontal to 1 vertical
Gravel-lined	2 horizontal to 1 vertical
Clay	2 horizontal to 1 vertical

For ditches over 10 ft (3.3 m) in depth, slope stability studies should be performed to determine the safe slope for the individual conditions. A maximum of 3 horizontal to 1 vertical shall be maintained. For all grass lined ditches, a maximum of 3 horizontal to 1 vertical shall be maintained to allow cutting of the grass. All gravel-lined ditches shall maintain a maximum of 2 horizontal to 1 vertical for stability. "V" ditches may be used for laterals not over 50 ft (17 m) in length.

The above dimensions shall be used in the absence of code or local authority requirements.

Velocity

Flow rate shall be determined from Fig. 17.9. To reduce erosion, the maximum velocity for ditches shall not exceed the figures shown in Table 18.2.

TABLE 18.2 Maximum Design Velocity of Ditches

Material	Maximum velocity	
	m/s	ft/s
Fine sand (noncolloidal)	0.16	2.5
Sandy loam (noncolloidal)	0.16	2.5
Silt loam (noncolloidal)	0.91	3.0
Ordinary firm loam	1.07	3.5
Fine gravel	1.52	5.0
Stiff clay (very colloidal)	1.52	5.0
Graded, loam to cobbles (noncolloidal)	1.52	5.0
Graded, silt to cobbles (noncolloidal)	1.68	5.5
Alluvial silts (noncolloidal)	1.07	3.5
Alluvial silts (colloidal)	1.52	5.0
Coarse gravel (noncolloidal)	1.83	6.0
Cobbles and shingles	1.68	5.5
Shales and hardpans	1.83	6.0

Design Depth and Slope

To reduce cost, ditches shall be designed to run full at the design discharge rate for ditches under 10 ft (3.3 m) with an additional allowable minimum freeboard of 6 in (150 mm). Preferred minimum freeboard for larger ditches is 12 in (300 mm). For ditches running parallel to public roads, the maximum design water surface shall not be higher than the base of the road. Low-capacity ditches shall have a minimum slope of 0.0002 ft/ft (0.01 m/m). Larger ditches will be hydraulically designed to set the ditch bottom slope.

Full Ditch Lining

Whether a ditch is to be lined depends on erosion characteristics and other factors. Ditches are normally unlined. A full pavement lining of concrete shall be used on ditches located in unstable soil. It shall also be used where the velocity of wastewater exceeds the velocities given above. Ditches used to convey contaminated fire water will be fully lined with concrete.

Partial Lining

Partial paved ditch sections are suggested for ditches under the following circumstances:

1. At the inlet and outlet of culverts
2. In a main unlined ditch where a culvert enters above the invert
3. At the confluence of main ditches
4. At the exit of the confluence

Culverts

A *culvert* is defined as any closed metallic structure with a clear opening of less than 20′-0″ (500 mm). For our purposes, it will be a smaller size used where very low, nuisance flows are routed under walkways etc. The recommended size of the culvert is to be a minimum 15-in (375-mm) diameter. Where the volume of flow would require pipes larger than 48 in, an improved open channel should be considered. The expertise involved in large culvert design involves consideration of hydraulics, imposed exterior loads, and local experience with erosion. This is considered far beyond the expertise of an average facility design engineer. The services of an experienced expert are strongly suggested. It is extremely unlikely that a large culvert would be used; therefore it will no longer be considered.

HEADWALLS AND WINGWALLS

The normal function of a headwall in low-to medium-flow conditions in a ditch is to control erosion and scour resulting from turbulence and higher velocities. If a headwall is not used, erosion protection such as riprap or sacked concrete with a cement/sand ratio of 9 : 1 is suggested around the outlet entrance. A typical headwall is illustrated in Fig 18.1.

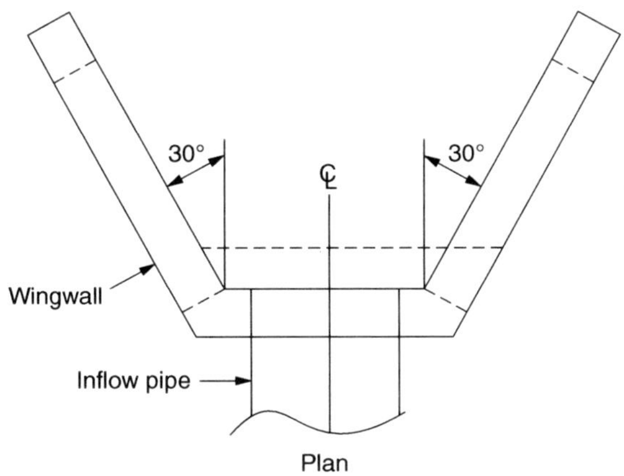

FIGURE 18.1 Typical headwall in ditch.

A wingwall is an extension of the discharge entrance and is suggested where the soil may be unstable. Headwall heights shall conform to structural requirements for strength. If a headwall is required for space, hydraulic, geometric or structural requirements, typical applications are as follows:

1. Straight headwalls are used for low approach velocity, light floating debris, broad or undefined approach channels, and where little change in alignment is present.
2. L headwalls are used where there are defined channels with low to moderate velocity where an abrupt change in alignment is required at the inlet.
3. Winged headwalls are used for channels with moderate velocity and medium floating trash. Wings are best set flush with the edges of the inlet aligned with the stream axis, and constructed with a flare angle of 18° to 45°.

Headwalls are usually constructed of plain or reinforced concrete and usually consist of either a straight headwall or a headwall with wingwalls as required by local conditions. Definite design criteria applicable to all conditions cannot be formulated, but the following considerations should ensure an efficient structure:

a. The primary problem is not one of hydraulics, but rather one of protecting the outfall against undermining bottom scour and perhaps degrading the downstream channel.

b. The structural integrity of the proposed structure shall consider soil, hydrostatic pressures, and frost protection for areas where frost penetration is possible.

c. The proposed structure shall be large enough to preclude the partial or complete stoppage of the drain by eating away of the adjacent soil.

d. The necessity for an apron (or stilling basin) is determined largely by the soil characteristics of the adjacent open channel and by the anticipated maximum turbulence and velocities at the drainage outlet. Most outlets operate under free-fall conditions, and the excess kinetic energy often must be dissipated to control damaging erosion.

e. Numerous endwall failures have resulted from an improper consideration of the relative elevation differences between the invert of the outlet pipe and the invert of the outlet channel. Select the apron elevation to ensure that sufficient depth of backwater occurs to prevent undesirable erosion.

f. A typical illustration of one type of headwall is shown in Fig. 18.1. If the flow is low, the wingwalls could be omitted. Local authorities must be consulted before selecting wingwalls if a headwall is necessary and the design.

Flume

A flume is a specially designed engineered open channel flow section that has restrictions in area that produce differences in velocity of the flowing fluid for the purpose of measuring flow. The most widely used flumes are the *Parchall* flume, illustrated in Fig. 18.2, and the *Palmer-Bowles* flume.

The flume measures water flow because the design alters the height of water in the converging section proportional to the flow through the flume. The increase in height is measured by instrument packages and converted to flow rates.

Flumes generally require relatively smooth flow of approximately 10 diameters upstream for proper operation. They should not be installed near a sharp change of slope or adjacent to obstructions in the flow path. They are selected where low head loss is important and where large amounts of suspended solids may be present.

Weirs

A *weir* is a dam or obstruction placed in the flow path of a partially filled ditch. The head of water is a function of flow velocity and, therefore, the flow rate through the weir. There is an opening that allows water to freely flow through it. A cross section through a typical weir is illustrated in Fig. 18.3. This opening has three shapes: rectangular, triangular, and Cipolletti. These shapes are illustrated in Fig. 18.4. The difference in geometry between the Cipolletti and the rectangular or triangular shape is that the Cipolletti weir has a 4-to-1 slope ratio of the sides. The advantage of the Cipolletti weir is that the discharge occurs as if there is no end contraction and no correction to the flow rate is necessary.

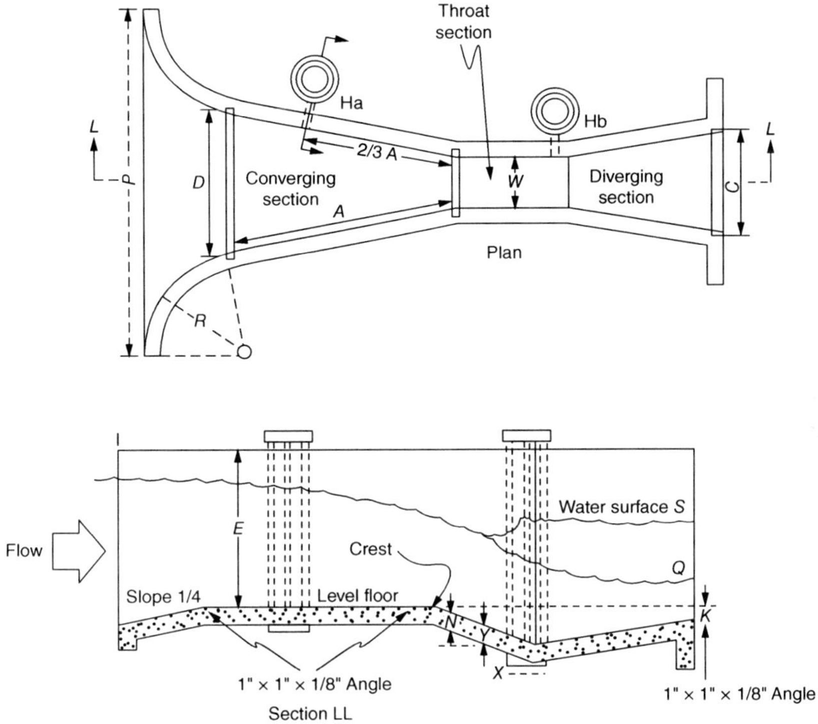

FIGURE 18.2 Plan and sectional views of a Parshall flume.

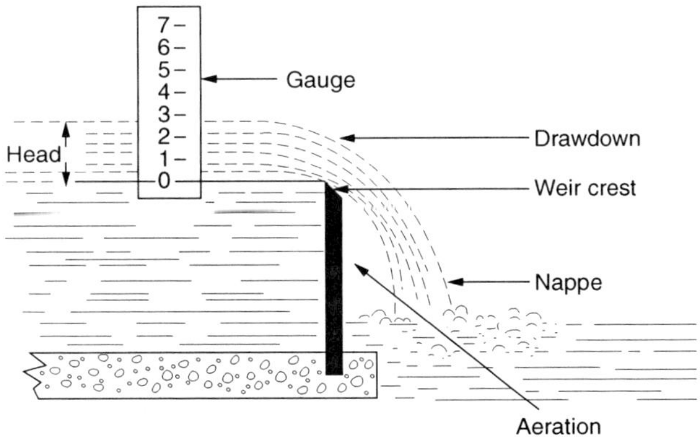

FIGURE 18.3 Cross section through a weir.

Discharge rates are found by obtaining the measurement of the vertical distance from the weir crest to the surface of the water. To ensure accuracy, the measurement must be taken some distance upstream of the weir. Generally accepted practice used for measuring the head of water places the staff gauge at distance of at least 4 times the head of water upstream of the weir crest. Debris and sediment should not be allowed to build up on the

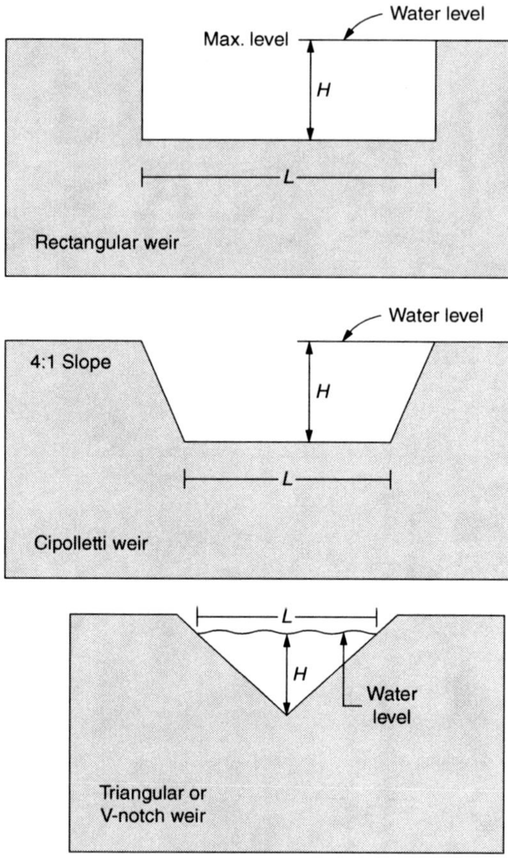

FIGURE 18.4 Elevations of sharp-crested weirs.

channel floor upstream of the plate. The basic formulas for flow through the various weirs are as follows:

$$\text{Rectangular or square weirs } Q = 3.33\ LH^{3/2} \tag{18.1}$$
$$\text{Cipolletti weirs } Q = 3.37\ LH^{3/2} \tag{18.2}$$
$$\text{Triangular weirs } (90°)\ Q = 2.44\ H^{3/2} \tag{18.3}$$
$$\text{Triangular weirs } (60°)\ Q = 1.41\ H^{3/2} \tag{18.4}$$

where Q = flow rate, cfs
L = for rectangular weir, ft, at base (for Cipolletti weir)
F = for Cipolletti weir, length at base, ft
F = for triangular weir, length at surface of flowing water above notch, ft
H = measured head, ft

Table 18.3 is a tabular solution for the discharge from a rectangular weir. Cipolletti weirs are calculated using Table 18.4 with a length equal to that at the bottom of the weir.

TABLE 18.3 Flow-through Rectangular Weirs, gpm

	Length *L* of weir, ft					Length *L* of weir, ft		
Head *H*, in	1	3	5	Additional gpm for each ft over 5 ft	Head *H*, in	3	5	Additional gpm for each ft over 5 ft
1	35.4	107.5	179.8	36.05	8	2,338	3,956	814
1¼	49.5	150.4	250.4	50.4	8¼	2,442	4,140	850
1½	64.9	197	329.5	66.2	8½	2,540	4,312	890
1¾	81	248	415	83.5	8¾	2,656	4,511	929
1	98.5	302	506	102	9	2,765	4,699	970
2¼	117	361	605	122	9¼	2,876	4,899	1,011
2½	136.2	422	706	143	9½	2,985	5,098	1,051
2¾	157	485	815	165	9¾	3,101	5,288	1,091
3	177.8	552	926	187	10	3,216	5,490	1,136
3¼	199.8	624	1,047	211	10½	3,480	5,940	1,230
3½	222	695	1,167	236	11	3,716	6,355	1,320
3¾	245	769	1,292	261	11½	3,960	6,780	1,420
4	269	846	1,424	288	12	4,185	7,165	1,495
4¼	293.6	925	1,559	316	12½	4,430	7,595	1,575
4½	318	1,006	1,696	345	13	4,660	8,010	1,660
4¾	344	1,091	1,835	374	13½	4,950	8,510	1,780
5	370	1,175	1,985	405	14	5,215	8,980	1,885
5¼	395.5	1,262	2,130	434	14½	5,475	9,440	1,985
5½	421.6	1,352	2,282	465	15	5,740	9,920	2,090
5¾	449	1,442	2,440	495	15½	6,015	10,400	2,165
6	476.5	1,535	2,600	528	16	6,290	10,900	2,300
6¼		1,632	2,760	560	16½	6,565	11,380	2,410
6½		1,742	2,920	596	17	6,925	11,970	2,520
6¾		1,826	3,094	630	17½	7,140	12,410	2,640
7		1,928	3,260	668	18	7,410	12,900	2,745
7¼		2,029	3,436	701.5	18½	7,695	13,410	2,855
7½		2,130	3,609	736	19	7,980	13,940	2,970
7¾		2,238	3,785	774	19½	8,280	14,460	3,090

Inches × 25 = mm
1 ft = 0.3 m
1 gpm = 3.8 L/min

TABLE 18.4 Flow-through Triangular Weirs

	Flow, gpm			Flow, gpm			Flow, gpm	
Head *H*, in inches	90° notch	60° notch	Head *H*, in	90° notch	60° notch	Head *H*, in	90° notch	60° notch
1	2.19	1.27	6¾	260	150	15	1912	1104
1¼	3.83	2.21	7	284	164	15½	2073	1197
1½	6.05	3.49	7¼	310	179	16	2246	1297
1¾	8.89	5.13	7½	338	195	16½	2426	1401
2	12.4	7.16	7¾	367	212	17	2614	1509
2¼	16.7	9.62	8	397	229	17½	2810	1623
2½	21.7	12.5	8¼	429	248	18	3016	1741
2¾	27.5	15.9	8½	462	267	18½	3229	1864
3	34.2	19.7	8¾	498	287	19	3452	1993
3¼	41.8	24.1	9	533	308	19½	3684	2127
3½	50.3	29.0	9¼	571	330	20	3924	2266
3¾	59.7	34.5	9½	610	352	20½	4174	2410
4	70.2	40.5	9¾	651	376	21	4433	2560
4¼	81.7	47.2	10	694	401	21½	4702	2715
4½	94.2	54.4	10½	784	452	22	4980	2875
4¾	108	62.3	11	880	508	23½	5268	3041
5	123	70.8	11½	984	568	23	4565	3213
5¼	139	80.0	12	1094	632	23½	5873	3391
5½	156	89.9	12½	1212	700	24	6190	3574
5¾	174	100	13	1337	772	24½	6518	3763
6	193	112	13½	1469	848	25	6855	3958
6¼	214	124	14	1609	929			
6½	236	136	14½	1756	1014			

Table 18.4 is the tabular solution for discharge from triangular weirs. Interpolation should be used to calculate discharges for intermediate dimensions.

SLUICE GATES

A *sluice gate* is a device used in controlling the flow in open channels, canals, or rivers. It consists of an opening installed in full obstruction in a channel that is controlled by an operable gate. The gate can be raised or lowered by means of a gear-operated drive that opens the gate from the bottom of the channel, allowing water to flow out.

Sluice Gate Discharge

To find the flow rate of water discharged through a sluice gate, the area of the opening necessary to discharge a given flow rate, or head of water above the discharge opening, refer to Fig. 18.5, entering the chart with any two values to find the third.

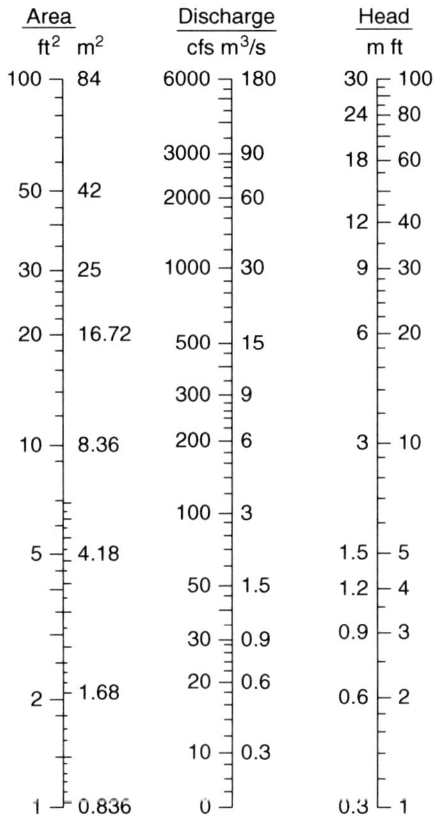

FIGURE 18.5 Sluice gate discharge.

Discharge of Water into Free Air

Use the illustration of free air sluice gates shown in Fig. 18.6 in conjunction with Fig. 18.5 to find the flow.

Submerged Discharge of Water

Use the illustration of submerged outlet, shown in Fig. 18.6, in conjunction with Fig. 18.5. The solution is similar to that of free air discharge except that the head of water used in the calculations is the difference of water levels as shown in the illustration.

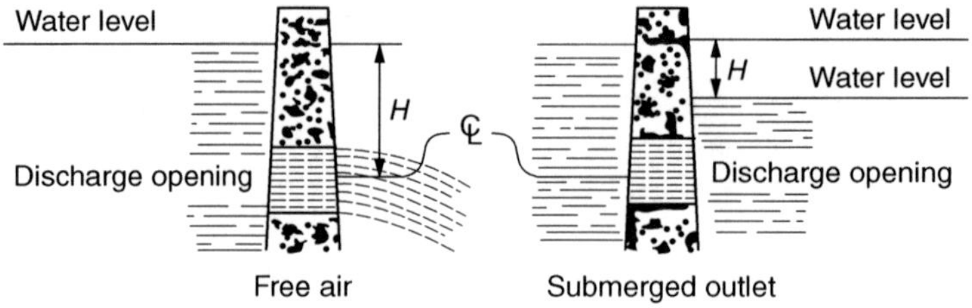

FIGURE 18.6 Free air sluice gates.

CALCULATING THE FLOW OF AN OPEN CHANNEL

A generally accepted method of finding the flow in ditches and open channels can be calculated using the Manning formula. The formula is simpler to solve and yields the same results as Kutter's formula. The Kutter formula would prove unwieldy to solve for engineers in everyday calculations. This Manning formula is used to find flow in pipes and is reproduced here for information only. The Manning formula is

$$Q = \frac{1.486}{n} \times R^{2/3} \times S^{1/2} \times A \tag{18.5}$$

where Q = capacity of ditch, cfs
A = cross-sectional area of flowing water, ft^2
R = hydraulic radius, ft (area of flow divided by wetted perimeter)
S = slope, ft/ft
n = Manning coefficient of channel roughness, dimensionless (see Table 18.1)

If it is necessary to find the loss of head from flowing water, refer to Fig. 18.7.

Finding the flow requires estimating the existing depth of water, the area of the ditch, the hydraulic radius, and the slope. The average velocity of the flow stream can be found by placing a floating object in the stream and observing the time it takes to move a given distance. Generally accepted practice requires that the surface velocity be reduced by 15 percent to give the average velocity.

A graphical solution of the Manning formula to find flow rates of 50 gpm (190 L/min) or greater in ditches is illustrated in Fig. 17.9, using the appropriate n factor found in Table 18.1 based on channel material and piping diameter. For flow rates less than 50 gpm (190 L/min), the following method will yield accurate results.

1. Calculate the anticipated maximum flow.
2. Select a pitch. Use a minimum of ¼ pitch (2.0 mm/m).
3. Refer to Table 19.1, three-quarters full column, and find the flow equal to or larger than that calculated.
4. Select the pipe size.
5. Refer to Table 14.2. Using the size pipe calculated, find the pipe area.
6. The area selected shall be used to configure the area of the ditch.

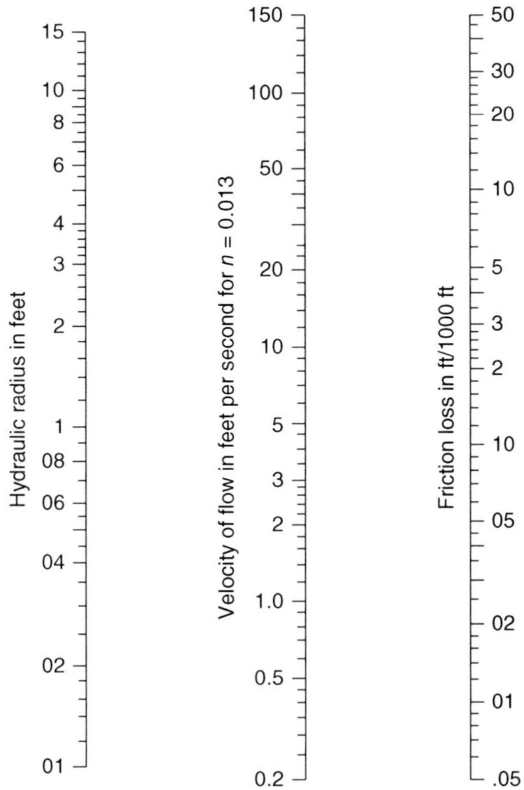

FIGURE 18.7 Chart to find head loss in Manning's formula.

7. Check Table 19.2 for the self-cleaning velocity to prevent buildup of deposits. The velocity shall be a minimum of 2 fps (0.6 m/s) to scour sand and grit from the trench or pipe bottom. The minimum size should be 6 in (DN 150).
8. To the ditch area, adding a 10 percent safety factor to the selected size is suggested.
9. Recommended maximum velocities for ditches will be found in Table 18.2.

BIBLIOGRAPHY

Blake, L. S., ed., *Civil Engineers Reference Book*, Butterworths, London.

Hicks, T. G., *Handbook of Civil Engineering Calculations*, McGraw-Hill New York.

Shun Dar Lin, *Water and Wastewater Calculations Manual*, McGraw-Hill, New York.

Simon, A. L., *Basic Hydraulics*, John Wiley & Sons, New York.

Sturm, T. W., *Open Channel Hydraulics*, 2d ed., McGraw-Hill, New York.

CHAPTER 19
SANITARY GRAVITY FLOW

GENERAL

This chapter discusses the sizing of gravity domestic sanitary and some nonchemical industrial discharge of facility wastewater from a point 5′-0″ beyond the building wall. Prior to this point it is considered part of the interior plumbing piping covered by the plumbing code and is outside the scope of this book. Site stormwater gravity flow, sizing, and disposal are discussed in Chap. 17. Sanitary and stormwater pumped discharge to the site is discussed in Chap. 20.

This chapter discusses gravity sanitary sewers and waste discharging from a point outside the building wall of facilities to disposal into a public sanitary sewer. Sanitary sewers from private residences are called *house sewers* and a *service lateral* for larger facilities.

If a public sewer is available, it is the least costly method of disposing sanitary waste. In urban areas, a sewer could be generally considered available if it is within 500 ft of the property line. This distance will vary based on local requirements. This may require that the house sewer travel a considerable distance, often in a public street. The alternative is a septic tank and system.

The plumbing code governs the size and installation of the house sewer, sometimes called the *building sewer* or *building lateral* inside a building. Another requirement is that if there is any chemical discharge into the public sewer, it must be treated so that the quality of the effluent is within guidelines established by the local authorities. These guidelines vary based on the municipal sewage treatment methods used. This requires that any properly treated laboratory or chemical waste exit separately from the building or facility site in order to permit pH or other measurements and grab samples to be taken. It could combine with the sanitary house sewer prior to discharge into the public sewer to dilute the discharge and to save installation costs. Waste that requires pH adjustments is discussed in Chap. 10.

If a public sewer is not available, the sanitary effluent must be treated by a septic tank or a private sewage treatment system to the extent required by the local authorities. The treated effluent will then be discharged to a sewer. The subject of septic tanks and sewage treatment is outside the scope of this book.

CODES AND STANDARDS

1. American Society of Sanitary Engineering (ASSE)
2. American Water Works Association (AWWA)

3. Cast Iron Soil Pipe Institute (CISPI)
4. NSF International (National Sanitation Foundation)
5. Plumbing and Drainage Institute (PDI)
6. American Society of Plumbing Engineers, (ASPE), *Plumbing Engineering Design Handbook*
7. Local plumbing code requirements

INFORMATION TO BE OBTAINED PRIOR TO DESIGN

The following information should be obtained prior to the design of any sewer:

1. Size, location, and depth of the existing public sewer. Is the public sewer a sanitary or combined sewer. Is the sewer considered overloaded. What is the pitch of the existing sewer.
2. The location and size of spurs and manholes in the public sewer.
3. Method of connecting to the public sewer.
   a. Do the local authorities require a manhole, or will a direct connection be acceptable. Will the local authority build the manhole.
   b. If a pipe is installed in a public street, are there special installation and other code requirements.
4. Topographical map along the proposed route of the new sewer.
5. Minimum depth of bury if mandated by the local authorities.
6. Obstacles to be avoided, such as underground piping, trees, underground structures, and rock, should be located.

GRAVITY SEWER COMPONENTS

Manholes

A manhole is an underground structure that provides access into the sewer for observation, maintenance, and minimization of hydraulic interference from connections and changes in direction. A precast manhole is commonly used due to cost, as illustrated in Fig. 19.1.

There are several different types of manholes, each used for a different purpose. A shallow manhole, illustrated in Fig. 19.2, is used when the depth from the invert of the sewer to grade is less then approximately 3′-0″ (1 m). A drop manhole, illustrated in Fig. 19.3, should be used when the invert of the incoming pipe is more than 2′-0″ (60 cm) higher than the discharge pipe. This is done to avoid having the inflow spill onto the channel of the manhole because it would quickly wear it away. Therefore, a drop pipe is provided to bring the effluent into the manhole to lessen the hydraulic impact. Two pipe entries into the manhole are provided in case the drop pipe has a stoppage. When the peak flow fills the pipe to more than three-fourths full, it is common practice to make the drop pipe one size larger than the incoming sewer line to minimize stoppages. Special-purpose manholes for pumped discharge are discussed in Chap. 20.

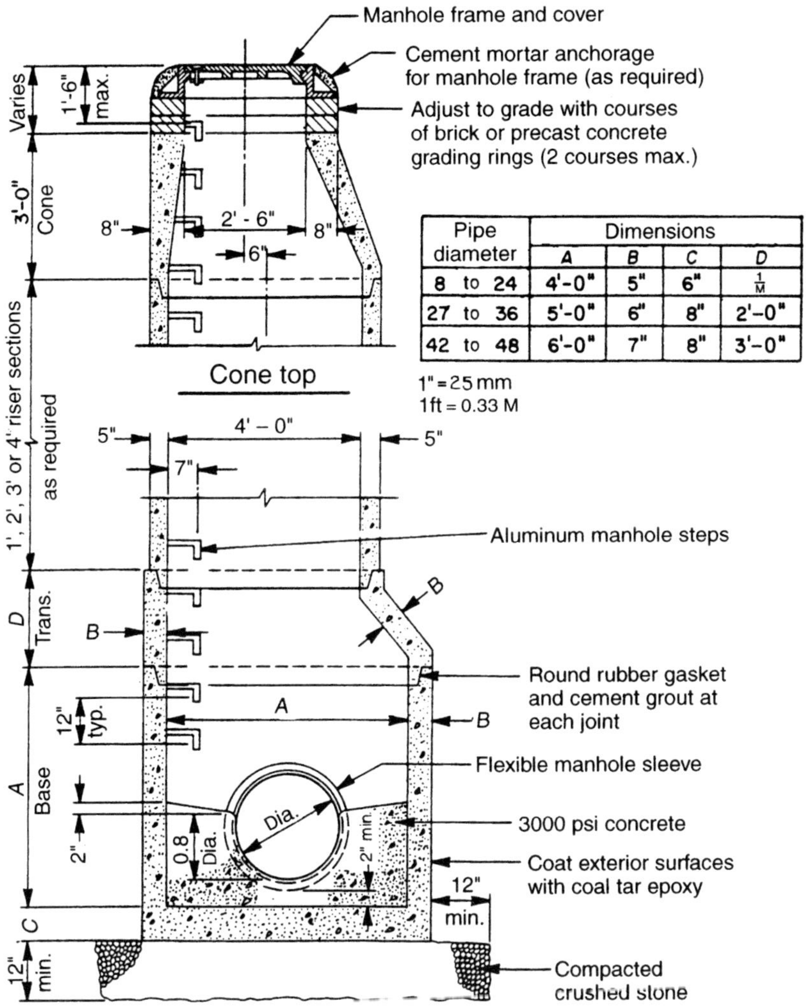

Pipe diameter	Dimensions			
	A	B	C	D
8 to 24	4'-0"	5"	6"	1/M
27 to 36	5'-0"	6"	8"	2'-0"
42 to 48	6'-0"	7"	8"	3'-0"

FIGURE 19.1 Typical precast manhole.

Manholes consist of the frame and cover, structure body, steps, sewage channel, and bottom slab.

Manhole Construction. Most manholes are circular, with the inside dimensions large enough for a worker to perform necessary functions without difficulty. The minimum recommended diameter of the major section is generally 3′-0″ (1 m), with 4′-0″ (1.25-m) inside diameter the most commonly used. It is common practice to use eccentric conical sections at the top to give a straight vertical side for the steps. If the cover is 2′-0″ (0.62 m), the clear inside dimension of the top section should be a minimum of 3′-0″ (1 m).

Manholes can be constructed of brick, precast concrete sections, poured (cast-in-place) concrete, and concrete block.

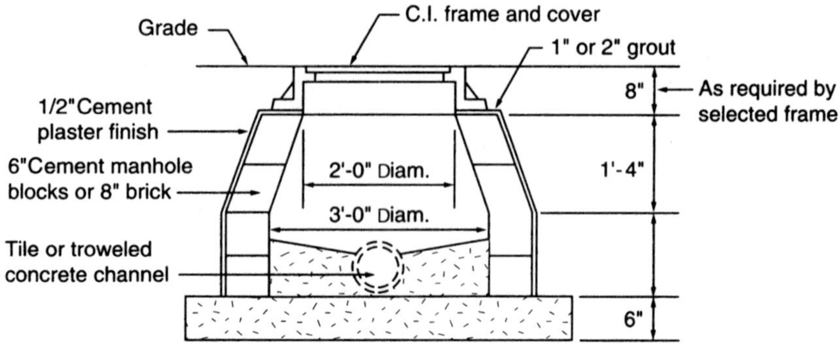

FIGURE 19.2 Detail of shallow manhole.

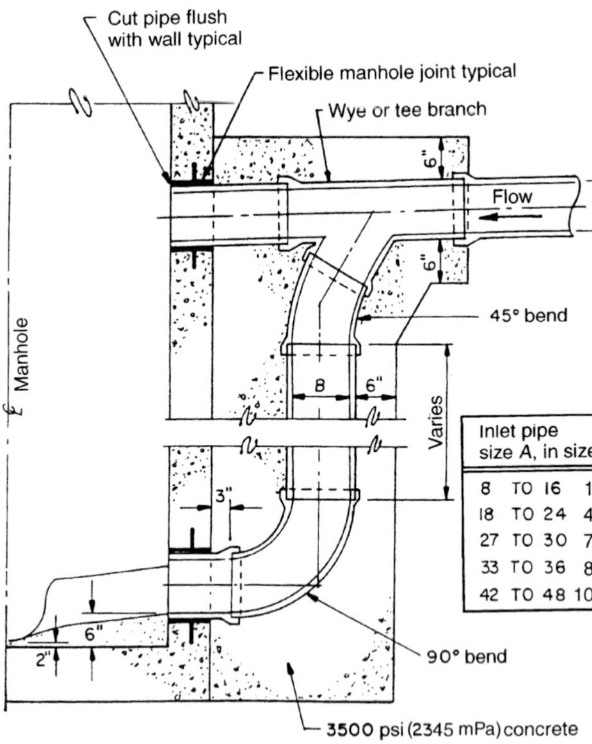

Inlet pipe size A, in size A DN		Drop connection size B, in DN	
8 TO 16	150 to 400	8	150
18 TO 24	450 to 600	12	300
27 TO 30	700 to 750	18	450
33 TO 36	800 to 900	24	600
42 TO 48	1050 to 1200	30	750

Note:
Drop pipe to be used in all cases where difference between inlet invert and lowest outlet invert is 2 ft (0.6 m) or greater.

FIGURE 19.3 Typical drop manhole connection.

The most common method of construction is with precast sections called *risers*. The typical precast manhole consists of a base, riser sections, and conical top section ready for installation. The sections fit together with tongue-and-groove joints with gaskets used in the joints to make it watertight. Install gaskets in accordance with manufacturer's recommendations regarding lubricants, cements, and other special installation requirements. These risers are available in a large variety of sizes and shapes.

Poured-in-place or precast reinforced concrete structures shall use concrete with a compressive strength of 4000 psi after 28 days in accordance with ASTM C-478. Install manhole steps into forms. Provide a concrete base under the structure. Rebars shall conform to ASTM A-615, grade 40 and wire fabric per ASTM A-185.

Brick and block manhole walls should be constructed with 8-in (20-mm) thickness up to 8′-0″ in depth and 12 in (30 mm) for greater depths. The outside of the manholes should be parged (coated) with at least one coat of cement mortar at least ½-in thick for protection against deterioration. Two coats are often used. For additional protection in wet soils, a coating of coal tar epoxy is often applied over the parging after it has been allowed to dry for 30 days.

Steps

Manhole steps are generally constructed of cast iron or aluminum. Common step diameters are ¾ in (19 mm) or 1 in (25 mm). Width is 16 in (40 cm) if it is desired to have two feet on one rung. The most common size rungs are 12 in (30 cm). Steps are spaced 12 in (30 cm) to 16 in (50 mm) apart, with 12 in the most common. Good practice is to have a distance of approximately 6½ in (16 mm) from the wall to the inside of the rung. Typical manhole steps are illustrated in Fig. 19.4.

Frame and Cover. Frames and covers are available in round and square configurations and in a wide variety of sizes. The advantage of a square cover is the additional space for entry into the manhole. The advantage of the round cover is that it will not fall into the structure. The frame and cover materials are normally cast iron, but ductile iron and aluminum are commonly available. They are also manufactured in light and heavy duty, based on the load that could be safely placed on it. A detail of a typical frame and cover is illustrated in Fig. 19.5.

It is common practice to install the top of the cover 1 in above the road surface or grade to keep infiltration of stormwater between the frame and cover to a minimum. If the distance from the frame to the top of the structure is greater than 3 in, use concrete rings or brick firmly bedded in mortar to raise the frame. Set casting to final grade elevation by

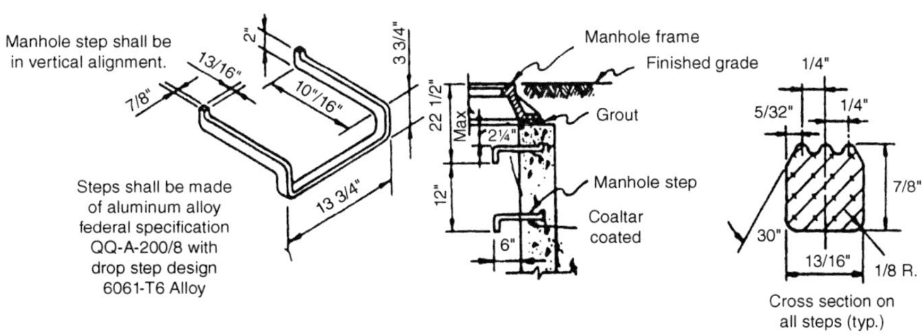

FIGURE 19.4 Typical detail of manhole step.

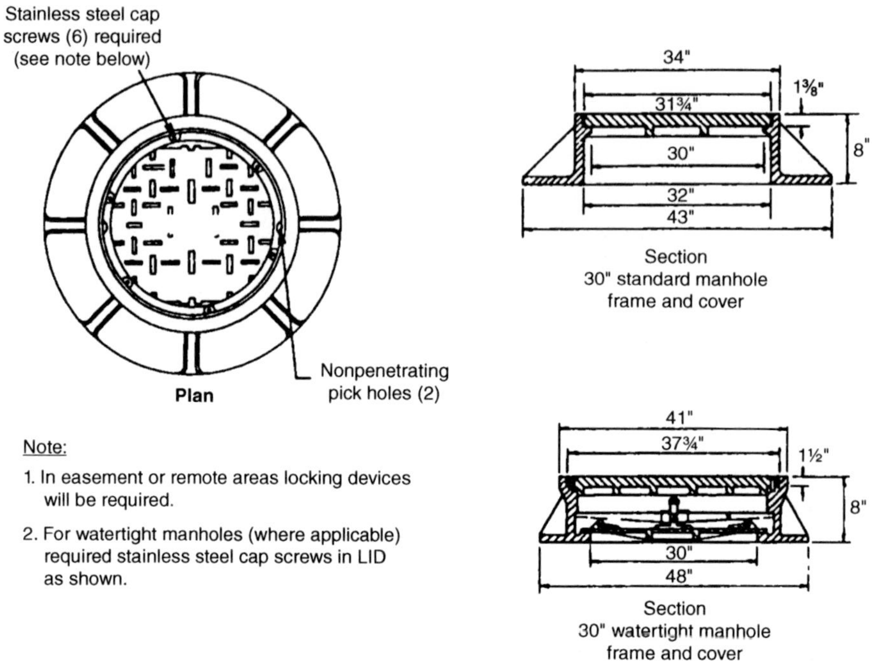

FIGURE 19.5 Standard or watertight manhole frame and cover.

means of mortar bed. Considerations in selecting frames and covers include a good fit to prevent rattling in traffic, reasonably tight closure, and resistance to unauthorized entry.

The size of the cover is determined by the intended use. For workers to enter, the minimum size is 2′-0″ (60 cm) round, which is considered small. The most commonly used size is 2′-6″ (75 cm).

It is common practice to have the words *Sanitary* or *Storm* cast into the cover as required for each different utility to easily identify the system.

Base and Channel. Structures are built on concrete bases or footings. For precast manholes, a recessed center is provided to receive the male tongue-and-groove joints of a riser section. The invert channel is formed by shaping poured concrete after the drainage lines have been installed. It also could be a cast iron pipe that has the top cut away after installation.

A channel is the lowest internal part of the manhole and provides a U-shaped open flow path for the liquid. It should be a smooth continuation of the pipe. Its height should be at least one-half of the pipe size for 12-in (30-cm) pipe and three-fourths of the height for 15-in (38-cm) and larger.

Spacing of Manholes. If manholes are installed in a public street, local authorities usually mandate the distance. There are usually no code requirements for the spacing of manholes on private property. When sewers are installed on private property, guidelines for placement and spacing of manholes for pipe sewers, including elliptical sewers, are as given in Table 19.1:

1. At all changes in direction
2. At changes of sewer alignment

TABLE 19.1 Manhole Spacing and Location of Pipe Sewers

A. Maximum Spacing of Manhole on Pipe Sewers

Pipe size, in (mm)	Recommended max. spacing, ft (m)	Absolute max. spacing, ft (m)
10 Dia. to 24 Dia. (254 to 610) Circular 14 H × 23 W (356 × 584) Horiz. elliptical pipe 23 H × 14 W (584 × 356) Vert. elliptical pipe	150(45.7)	200(61)
27 Dia. to 48 dia. (686 to 1219) circular 19 H × 30 W to 48 H × 76 W (483 × 762 to 1219 1930) horiz. elliptical pipe 30 H × 19 W to 49 H × 32 W (962 × 483 to 1245 813) vert. elliptical pipe	300(91.4)	400(121.9)
54 Dia. to 66 dia. (1372 to 1676) circular 53 H × 83 W to 68 H × 106 W (1346 × 2108 to 1727 × 2692) horiz. elliptical pipe 53 H × 34 W to 68 H × 43 W (1346 × 864 to 1727 × 1092) vert. elliptical pipe	400 (121.9)	500 (152.4)
72 Dia. (1829) and larger circular 72 H × 113 W (1829 × 2870) and larger horiz. elliptical pipe 76 H × 48 W (1930 × 1219) and larger vert. elliptical pipe	500 (152.4)	600 (182.8)

B. Manhole Location on Pipe Sewers

1. Change in grade or elevation
   (a) At all changes in grade or elevation for all sizes of sewers.
2. Change in alignment
   (a) At all changes in alignment for circular sewers up to and including 42-in (1067-mm) diameter.
   (b) At all changes in alignment for elliptical pipe sewers up to and including 43 in H × 68 in W (1092 mm × 1727 mm) and 42 in H × 27 in W (1067 mm × 686 mm) sizes.

3. Changes in the size of the sewer
4. Changes of grade
5. Intersections with other sewers
6. At the end of the sewer system

SANITARY SEWER DESIGN CONSIDERATIONS

Selecting the Pipe Material

The pipe material selected is based on many factors. Since no single material will meet all conditions, the material selected will be based on the most important characteristics. Some factors to consider are

1. Flow characteristics (friction coefficient)
2. Life expectancy
3. Resistance to scour

4. Chemical resistance to effluent and surrounding soil
5. Ease of handling and installation
6. Physical strength to resist soil loading
7. Type of joints and flexibility
8. Requirements of the local authorities
9. Cost

Vitrified Clay Pipe (VCP). This classic material has demonstrated its durability in use for the past 100 years. It has a high resistance to corrosion and abrasion but is very brittle. It is a gravity pipe, used in corrosive soil and for acid waste drainage and other corrosive waste. It is rarely installed nowadays.

Polyvinyl Chloride (PVC) Pipe. This pipe is made by the extrusion of PVC. It is available from 10 mm to 1.2 m and with a variety of connection ends. It is subject to attack by certain organic chemicals. It is the pipe of choice for gravity, low-pressure, and vacuum sewer systems. It is pressure-rated in its class at 23°C.

High-Density Polyethylene (HDPE). This pipe is available from 100 to 1200 mm. It is pressure-rated in its class at 23°C and transients not exceeding two times the nominal pressure class. The pipe is normally joined with thermal butt fusion, but it takes a higher skill lever to accomplish. Ductile iron fittings are commonly used. This pipe is corrosion-resistant, and no coating or lining is provided.

Ductile Iron Pipe (DIP). Its primary application is for force mains. Because wastewater is corrosive, it is common practice to line the pipe with cement and use an asphaltic outer coating. Manufacturers recommend that the pipe be encased in a loose-fitting, flexible polyethylene tube when installed in corrosive soils. It is available from 400 mm and smaller. It is also used for water distribution.

Reinforced Concrete Pipe (RCP). This pipe is available in diameters from 300 mm to 5.0 m and is made with integral bell and spigot ends. It is manufactured by techniques including centrifugal, vibration, packing, and tamping for consolidation of material when cast into forms. Adjustment of wall thickness, concrete strength, and reinforcement will produce a wide variety of strengths. This pipe is used for gravity systems usually in trunk lines and interceptor sewers.

Laterals and service connection runs for sanitary discharge are often extra-heavy cast iron with compression gasket joints. Another often used material is plastic, usually PVC with butt-fused joints.

Minimum and Maximum Effluent Velocity

The minimum velocity of the effluent should be sufficient to prevent depositing solids on the bottom of the pipe. This is called a *self-scouring velocity*. Tests have established this minimum velocity at 2 feet per second (fps) or 0.62 meter per second m/s when the pipe is flowing full. Accepted practice has a recommended minimum of 2.5 fps (0.77 m/s). Table 19.2 gives the slope necessary to maintain that minimum velocity. Since the pipes do not run full most of the time, it may be necessary to occasionally flush the sewer to remove any accumulated sediment.

In practice, sanitary sewers that have continuous high velocities and where grit is expected to be a problem should limit the highest velocity to 10 fps (3 m/s). For ordinary

TABLE 19.2 Gravity Drainage Pipe Sizing Tables
Discharge of circular sewer $n = 0.013$

Pipe size		Grade	½ Full			2/3 Full*			Full		
			Discharge		Velocity	Discharge		Velocity	Discharge		Velocity
in	in/ft	%	gpm	cfs	ft/s	gpm	cfs	ft/s	gpm	cfs	ft/s
2	1/8	1.0	7	0.02	1.3	10	0.02	1.5	13	0.03	1.3
	¼	2.1	9	0.02	1.8	14	0.03	2.0	18	0.04	1.8
	½	4.2	14	0.03	2.9	22	0.05	3.2	28	0.06	2.9
	1	8.3	20	0.05	4.0	32	0.07	4.5	40	0.09	4.0
3	1/8	1.0	18	0.04	1.7	29	0.06	1.9	36	0.08	1.7
	¼	2.1	26	0.06	2.4	41	0.09	2.7	51	0.11	2.4
	½	4.2	40	0.09	3.7	64	0.14	4.1	80	0.18	3.7
	1	8.3	57	0.13	5.3	90	0.20	5.9	114	0.25	5.3
4	1/8	1.0	39	0.09	2.0	61	0.13	2.2	77	0.17	2.0
	¼	2.1	55	0.13	2.8	87	0.20	3.1	110	0.25	2.8
	½	4.2	87	0.20	4.5	138	0.31	5.0	174	0.39	4.5
	1	8.3	123	0.28	6.3	194	0.44	7.1	245	0.55	6.3
6		0.5	79	0.18	1.8	124	0.28	2.0	157	0.35	1.8
	1/8	1.0	110	0.25	2.5	174	0.39	2.8	220	0.49	2.5
		1.5	135	0.30	3.1	213	0.47	3.5	269	0.60	3.1
	¼	2.0	157	0.35	3.6	248	0.55	4.0	314	0.70	3.6
		2.5	175	0.39	4.0	277	0.62	4.5	350	0.78	4.0
	3/8	3.0	193	0.43	4.4	305	0.68	4.9	386	0.86	4.4
		3.5	207	0.46	4.7	327	0.73	5.3	413	0.92	4.7
	½	4.0	225	0.50	5.0	355	0.79	5.6	449	1.00	5.0
	5/8	5.0	247	0.55	5.6	391	0.87	6.3	494	1.10	5.6
	¾	6.0	270	0.60	6.1	426	0.95	6.8	539	1.20	6.1
	7/8	7.0	292	0.65	6.6	461	1.03	7.4	583	1.30	6.6

(Continued)

TABLE 19.2 Gravity Drainage Pipe Sizing Tables (*Continued*)
Discharge of circular sewer n = 0.013

Pipe size		Grade	½ Full			2/3 Full*			Full		
			Discharge		Velocity	Discharge		Velocity	Discharge		Velocity
in	in/ft	%	gpm	cfs	ft/s	gpm	cfs	ft/s	gpm	cfs	ft/s
8		0.2	108	0.24	1.6	170	0.38	1.8	215	0.48	1.6
		0.4	153	0.34	2.0	241	0.54	2.2	305	0.68	2.0
		0.6	191	0.43	2.4	302	0.67	2.7	382	0.85	2.4
		0.8	236	0.53	2.9	372	0.83	3.2	471	1.05	2.9
	1/8	1.0	247	0.55	3.2	391	0.87	3.6	494	1.10	3.2
		1.5	303	0.68	3.8	479	1.07	4.3	606	1.35	3.8
	¼	2.0	348	0.78	4.5	550	1.22	5.0	696	1.55	4.5
		2.5	392	0.88	4.9	621	1.38	5.5	785	1.75	4.9
	3/8	3.0	427	0.95	5.4	674	1.50	6.0	853	1.90	5.4
		3.5	449	1.00	5.8	710	1.58	6.5	893	2.00	5.8
	½	4.0	494	1.10	6.2	780	1.74	6.9	987	2.20	6.2
		4.5	516	1.15	6.6	816	1.82	7.4	1,032	2.30	6.2
10		0.2	211	0.47	1.7	334	0.74	1.9	422	0.94	1.7
		0.4	303	0.68	2.4	479	1.1	2.7	606	1.35	2.4
		0.6	359	0.80	2.9	568	1.3	3.2	718	1.60	2.9
		0.8	438	0.98	3.5	692	1.5	3.9	875	1.95	3.5
	1/8	1.0	472	1.05	3.8	745	1.7	4.3	943	2.10	3.8
		1.5	561	1.25	4.5	887	2.0	5.0	1,122	2.50	4.5
	¼	2.0	651	1.45	5.3	1,029	2.3	5.9	1,302	2.90	5.3
		2.5	741	1.65	5.7	1,170	2.6	6.4	1,481	3.30	5.7
	3/8	3.0	808	1.80	6.4	1,277	2.8	7.2	1,616	3.60	6.4
		3.5	853	1.90	6.8	1,348	3.0	7.6	1,706	3.80	6.8
12		0.2	337	0.8	1.9	533	1.2	2.1	674	1.5	1.9
		0.4	472	1.1	2.7	745	1.7	3.0	943	2.1	2.7
		0.6	584	1.3	3.3	922	2.1	3.7	1,167	2.6	3.3
		0.8	718	1.6	4.1	1,135	2.5	4.6	1,436	3.2	4.1
	1/8	1.0	763	1.7	4.3	1,206	2.7	4.8	1,526	3.4	4.3
		1.2	831	1.8	4.7	1,313	2.9	5.3	1,661	3.7	4.7

		1.4	898	2.0	5.0	1,409	3.2	5.6	1,795	4.0	5.0
		1.6	965	2.2	5.3	1,525	3.4	5.9	1,930	4.3	5.3
		1.8	1,010	2.3	5.7	1,596	3.6	6.4	2,020	4.5	5.7
	¼	2.0	1,077	2.4	6.0	1702	3.8	6.7	2,154	4.8	6.0
		2.2	1,122	2.5	6.2	1,773	4.0	6.9	2,244	5.0	6.2
		2.4	1,167	2.6	6.6	1,844	4.1	7.4	2,334	5.2	6.6
14		0.1	382	0.8	1.6	603	1.3	1.7	763	1.7	1.6
		0.2	516	1.2	2.2	816	1.8	2.5	1,032	2.3	2.2
		0.3	651	1.5	2.7	1,029	2.3	3.0	1,302	2.9	2.7
		0.4	763	1.7	3.1	1,206	2.7	3.5	1,526	3.4	3.1
	1/16	0.5	853	1.9	3.5	1,348	3.0	3.9	1,706	3.8	3.5
		0.6	920	2.1	3.8	1,454	3.2	4.3	1,840	4.1	3.8
		0.7	1,010	2.3	4.1	1,596	3.6	4.6	2,020	4.5	4.1
		0.8	1,077	2.4	4.4	1,702	3.8	4.9	2,154	4.8	4.4
		0.9	1,145	2.6	4.6	1,816	4.0	5.2	2,299	5.1	4.6
	1/8	1.0	1,212	2.7	4.8	1,915	4.3	5.4	2,424	5.4	4.8
		1.1	1,257	2.8	5.1	1,986	4.4	5.7	2,513	5.6	5.1
		1.2	1,324	3.0	5.3	2,092	4.7	5.9	2,648	5.9	5.3
		1.3	1,369	3.1	5.5	2,163	4.8	6.2	2,738	6.1	5.5
		1.4	1,436	3.2	5.8	2,269	5.1	6.5	2,872	6.4	5.8
		1.5	1,481	3.3	5.9	2,340	5.2	6.6	2,962	6.6	5.9
		1.6	1,526	3.4	6.0	2,412	5.4	6.7	3,052	6.8	6.0
		1.7	1,571	3.5	6.3	2,483	5.5	7.1	3,142	7.0	6.3
15		0.1	439	1.0	1.6	694	1.6	1.8	878	2.0	1.6
		0.2	628	1.4	2.3	993	2.2	2.6	1,257	2.8	2.3
		0.3	763	1.7	2.7	1,206	2.7	3.0	1,526	3.4	2.7
		0.4	898	2.0	3.2	1,419	3.2	3.6	1,795	4.0	3.2
	1/16	0.5	1,010	2.3	3.6	1,596	3.6	4.0	2,020	4.5	3.6
		0.6	1,100	4.5	3.9	1,738	3.9	4.4	2,199	4.9	3.9
		0.7	1,190	2.7	4.3	1,880	4.2	4.8	2,379	5.3	4.3
		0.8	1,279	2.9	4.6	2,021	4.5	5.2	2,558	5.7	4.6
		0.9	1,347	3.0	4.8	2,128	4.7	5.4	2,693	6.0	4.8

(*Continued*)

TABLE 19.2 Gravity Drainage Pipe Sizing Tables (*Continued*)
Discharge of circular sewer $n = 0.013$

Pipe size		Grade	1/2 Full			2/3 Full*			Full		
			Discharge		Velocity	Discharge		Velocity	Discharge		Velocity
in	in/ft	%	gpm	cfs	ft/s	gpm	cfs	ft/s	gpm	cfs	ft/s
15	1/8	1.0	1,437	3.2	5.2	2,270	5.1	5.8	2,873	6.4	5.2
		1.1	1,481	3.3	5.3	2,340	5.2	5.9	2,962	6.6	5.3
		1.2	1,549	3.5	5.5	2,447	5.5	6.2	3,097	6.9	5.5
		1.3	1,616	3.6	5.8	2,553	5.7	6.5	3,231	7.2	5.8
		1.4	1,683	3.8	5.9	2,660	5.9	6.6	3,366	7.5	5.9
		1.5	1,751	3.9	6.2	2,766	6.2	6.9	3,501	7.8	6.2
		1.6	1,796	4.0	6.4	2,837	6.3	7.2	3,591	8.0	6.4
		1.7	1,863	4.2	6.6	2,943	6.6	7.4	3,725	8.3	6.6
16		0.1	539	1.2	1.7	851	1.9	1.9	1,077	2.4	1.7
		0.2	786	1.8	2.4	1,242	2.8	2.7	1,571	3.5	2.4
		0.3	967	2.2	2.9	1,528	3.4	3.2	1,933	4.3	2.9
		0.4	1,122	2.5	3.4	1,773	4.0	3.8	2,244	5.0	3.4
	1/16	0.5	1,257	2.8	3.8	1,986	4.4	4.3	2,513	5.6	3.8
		0.6	1,392	3.1	4.2	2,199	4.9	4.7	2,783	6.2	4.2
		0.7	1,504	3.4	4.5	2,376	5.3	5.0	3,007	6.7	4.5
		0.8	1,594	3.6	4.8	2,518	5.6	5.4	3,187	7.1	4.8
		0.9	1,683	3.8	5.1	2,660	5.9	5.7	3,366	7.5	5.1
	1/8	1.0	1,796	4.0	5.4	2,837	6.3	6.0	3,591	8.0	5.4
		1.1	1,885	4.2	5.6	2,980	6.6	6.3	3,770	8.4	5.6
		1.2	1,975	4.4	5.8	3,121	7.0	6.5	3,950	8.8	5.8
		1.3	2,020	4.5	6.1	3,192	7.1	6.8	4,040	9.0	6.1
		1.4	2,110	4.7	6.3	3,333	7.4	7.1	4,219	9.4	6.3
		1.5	2,177	4.9	6.6	3,440	7.7	7.4	4,354	9.7	6.6

18		0.1	719	1.6	1.8	1,136	2.5	2.0	1,437	3.2	1.8
		0.2	1,010	2.3	2.6	1,596	3.6	2.9	2,020	4.5	2.6
		0.3	1,257	2.8	3.3	1,986	4.4	3.7	2,514	5.6	3.3
		0.4	1,459	3.3	3.7	2,306	5.1	4.1	2,918	6.5	3.7
	1/16	0.5	1,616	3.6	4.1	2,554	5.7	4.6	3,232	7.2	4.1
		0.6	1,796	4.0	4.5	2,837	6.3	5.0	3,591	8.0	4.5
		0.7	1,953	4.4	4.8	3,085	6.9	5.4	3,905	8.7	4.8
		0.8	2,043	4.6	5.2	3,228	7.2	5.8	4,085	9.1	5.2
		0.9	2,155	4.8	5.5	3,405	7.6	6.2	4,309	9.6	5.5
	1/8	1.0	2,244	5.0	5.7	3,546	7.9	6.4	4,488	10.0	5.7
		1.1	2,357	5.3	6.2	3,724	8.3	6.9	4,713	10.5	6.2
		1.2	2,469	5.5	6.8	3,901	8.7	7.6	4,937	11.0	6.8
		1.3	2,581	5.8	7.3	4,078	9.1	8.2	5,162	11.5	7.3
		1.4	2,693	6.0	7.8	4,255	9.5	8.7	5,386	12.0	7.8
20		0.1	965	2.2	2.0	1,525	3.4	2.2	1,930	4.3	2.0
		0.2	1,392	3.1	2.8	2,199	4.9	3.1	2,783	6.2	2.8
		0.3	1,706	3.8	3.5	2,695	6.0	3.9	3,411	7.6	3.5
		0.4	1,975	4.4	4.0	3,121	7.0	4.5	3,950	8.8	4.0
	1/16	0.5	2,200	4.9	4.5	3,476	7.7	5.0	4,400	9.8	4.5
		0.6	2,401	5.4	4.8	3,794	8.5	5.4	4,802	10.7	4.8
		0.7	2,648	5.9	5.3	4,184	9.3	5.9	5,296	11.8	5.3
		0.8	2,805	6.3	5.6	4,432	9.9	6.3	5,610	12.5	5.6
		0.9	2,918	6.5	6.0	4,610	10.3	6.7	5,835	13.0	6.0
	1/8	1.0	3,150	7.0	6.3	4,977	11.1	7.1	6,300	14.0	6.3
		1.1	3,254	7.3	6.5	5,142	11.5	7.3	6,508	14.5	6.5
		1.2	3,366	7.5	6.8	5,319	11.9	7.6	6,732	15.0	6.8

(Continued)

TABLE 19.2 Gravity Drainage Pipe Sizing Tables (*Continued*)
Discharge of circular sewer n = 0.013

Pipe size		Grade	1/2 Full			2/3 Full*			Full		
			Discharge		Velocity	Discharge		Velocity	Discharge		Velocity
in	in/ft	%	gpm	cfs	ft/s	gpm	cfs	ft/s	gpm	cfs	ft/s
21		0.1	1,100	2.5	2.1	1,738	3.9	2.4	2,200	4.9	2.1
		0.2	1,571	3.5	2.9	2,483	5.5	3.2	3,142	7.0	2.9
		0.3	1,930	4.3	3.6	3,050	6.8	4.0	3,860	8.6	3.6
		0.4	2,200	4.9	4.1	3,476	7.7	4.6	4,400	9.8	4.1
	1/16	0.5	2,491	5.6	4.6	3,936	8.8	5.2	4,982	11.1	4.6
		0.6	2,693	6.0	5.0	4,257	9.5	5.6	5,386	12.0	5.0
		0.7	2,918	6.5	5.5	4,610	10.3	6.2	5,835	13.0	5.5
		0.8	3,150	7.0	5.8	4,977	11.1	6.5	6,300	14.0	5.8
		0.9	3,299	7.4	6.2	5,212	11.6	6.9	6,597	14.7	6.2
	1/8	1.0	3,479	7.8	6.5	5,497	12.2	7.3	6,957	15.5	6.5
		1.1	3,658	8.2	6.7	5,780	12.9	7.5	7,316	16.3	6.7
24		0.05	1,122	2.5	1.6	1,773	4.0	1.8	2,244	5.0	1.6
		0.1	1,616	3.6	2.3	2,554	5.7	2.6	3,232	7.2	2.3
		0.2	2,244	5.0	3.2	3,546	7.9	3.6	4,488	10.0	3.2
		0.3	2,805	6.3	3.9	4,432	9.9	4.4	5,610	12.5	3.9
		0.4	3,187	7.1	4.5	5,035	11.2	5.0	6,373	14.2	4.5
	1/16	0.5	3,591	8.0	5.1	5,673	12.6	5.7	7,181	16.0	5.1
		0.6	3,927	8.8	5.5	6,205	13.8	6.2	7,854	17.5	5.5
		0.7	4,264	9.5	6.0	6,738	15.0	6.7	8,528	19.0	6.0
		0.8	4,488	10.0	6.4	7,092	15.8	7.2	8,976	20.0	6.4
27		0.05	1,482	3.3	1.7	2,341	5.2	1.9	2,963	6.6	1.7
		0.1	2,110	4.7	2.5	3,333	7.4	2.8	4,219	9.4	2.5
		0.2	3,130	6.8	3.5	4,787	10.7	3.9	6,059	13.5	3.5
		0.3	3,703	8.3	4.3	5,851	13.0	4.8	7,406	16.5	4.3
		0.4	4,309	9.6	4.8	6,808	15.2	5.4	8,617	19.2	4.8
	1/16	0.5	4,713	10.5	5.5	7,446	16.6	6.2	9,425	21.0	5.5
		0.6	5,162	11.5	6.0	8,156	18.2	6.7	10,323	23.0	6.0
		0.7	5,610	12.5	6.5	8,864	19.8	7.3	11,220	25.0	6.5

30		0.05	2,020	4.5	1.9	3,192	7.1	2.1	4,040	9.0	1.9
		0.1	2,805	6.3	2.7	4,432	9.9	3.0	5,610	12.5	2.7
		0.2	4,040	9.0	3.8	6,383	14.2	4.3	8,079	18.0	3.8
		0.3	4,938	11.0	4.6	7,802	17.4	5.2	9,876	22.0	4.6
		0.4	5,610	12.5	5.3	8,864	19.8	5.9	11,220	25.0	5.3
	1/16	0.5	6,508	14.5	6.0	10,282	22.9	6.7	13,016	29.0	6.0
		0.6	7,181	16.0	6.5	11,346	25.3	7.3	14,362	32.0	6.5
33		0.05	2,581	5.8	2.0	4,078	9.1	2.2	5,162	11.5	2.0
		0.1	3,703	8.3	2.8	5,851	13.0	3.1	7,406	16.5	2.8
		0.2	5,162	11.5	4.0	8,156	18.2	4.5	10,323	23.0	4.0
		0.3	6,512	14.5	4.9	10,289	22.9	5.5	13,023	29.0	4.9
		0.4	7,406	16.5	5.6	11,701	26.1	6.3	14,811	33.0	5.6
	1/16	0.5	8,303	18.5	6.4	13,119	29.2	7.2	16,606	37.0	6.4
		0.6	9,201	20.5	7.0	14,537	32.4	7.8	18,401	41.0	7.0
36		0.05	3,254	7.3	2.1	5,142	11.5	2.4	6,508	14.5	2.1
		0.1	4,713	10.5	3.0	7,446	16.6	3.4	9,425	21.0	3.0
		0.2	6,733	15.0	4.3	10,638	23.7	4.8	13,465	30.0	4.3
		0.3	8,303	18.5	5.2	13,119	29.2	5.8	16,606	37.0	5.2
		0.4	9,425	21.0	6.0	14,892	33.2	6.7	18,850	42.0	6.0
	1/16	0.5	10,772	24.0	6.8	17,019	37.9	7.6	21,543	48.0	6.8

*The depth of flow is equal to two-thirds the pipe diameter.
Source: After Chezy.

sewers without grit and only occasional high periods of flowing full, a maximum velocity of 20 fps (7 m/s) would be considered acceptable.

For large facilities, if the house sewer is sized for the peak flow rate, the velocity during the peak flow rate will be adequate to flush the pipe clean of any deposits left during periods of lesser flow, such as during the night.

Cleanouts shall be provided to allow rodding of a house sewer. Such a cleanout is illustrated in Fig. 19.6.

A complete discussion of Manning's formula along with a broad list of values for roughness required for the sizing of gravity piping using this formula, is presented in Chap. 17.

Connections to the Public Sewer

Connecting the private sewer to the public sewer is governed by the local sewer authority and is usually done through a manhole. The local authority often has regulations for sewers that are run in public streets.

Most public sewers in a public street have smaller built-in connections, called spurs, installed during their construction. The location, size, and invert of spurs are often available on utility survey maps. If the size of the lateral is small enough to use a spur, it should be used.

Local authorities usually require approval of the connection to public sewers. If the spurs are large enough, connections are preferred to spurs. If the public sewer is very large compared to the house sewer, a special method of connection is required to pierce the main sewer and install the new pipe. One such suggested method is illustrated in Fig. 19.7.

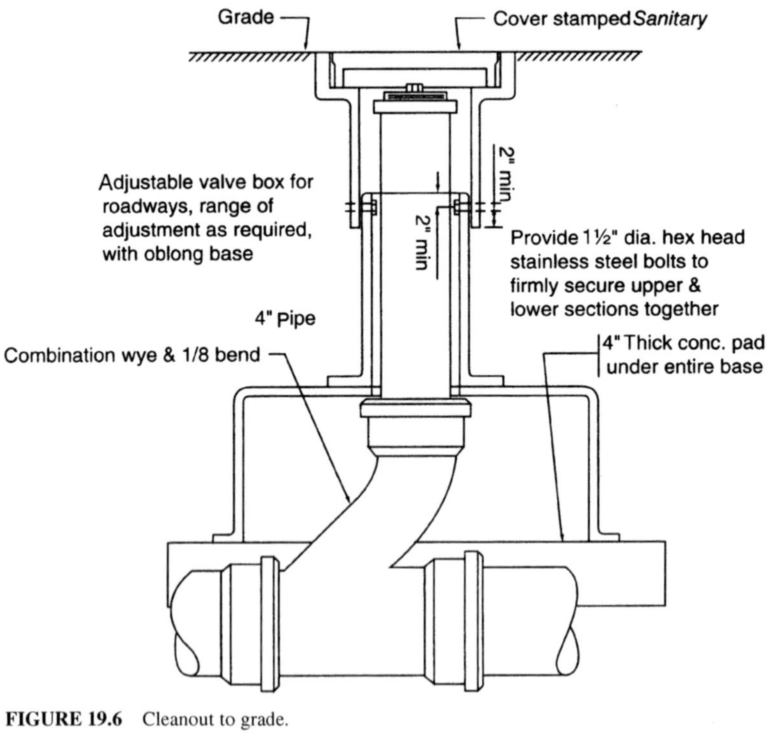

FIGURE 19.6 Cleanout to grade.

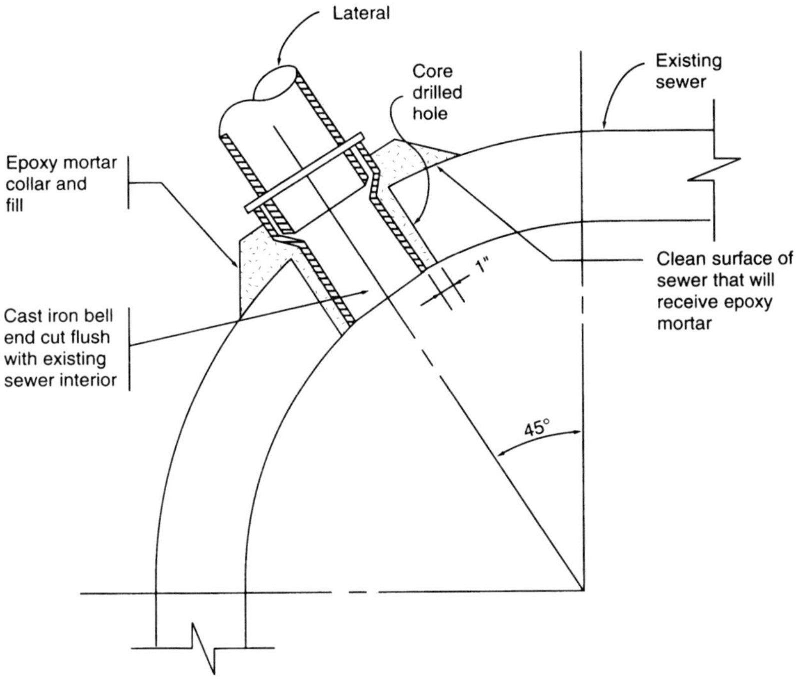

FIGURE 19.7 Connection to large sewer.

Depth of Bury

There are no hard-and-fast rules used to determine the depth to which sanitary sewers should be buried. When sewers are run when designing for a specific facility, that facility may have standards that must be followed. Often, the local authorities have established this depth. When no guidance exists, a reasonable starting point would be 3′-0″ of cover, with a depth of 2′-0″ in areas where the pipe will not be disturbed. Consideration should be given to the depth necessary to resist possible pipe breakage by traffic or other vehicles passing over the pipe if it is installed in such an area. When crossing other utilities, the sewer should be placed below them whenever possible. Since it is good practice to allow building laterals to have a minimum of ¼ in pitch when connecting to a sewer, the sewer should be deep enough to accommodate this pitch. A complete discussion of depth of bury is given in Chap. 20.

Depth of Flow

Sewers serving facilities should be designed to carry the peak design flow rate, including the allowance for the future, flowing about one-half to two-thirds full. This practice allows a safety factor for peak flow rates higher than originally designed for. Another recommendation for this practice is that the airspace above the flowing effluent allow for ventilation of the piping network.

Separation and Clearances from Drainage Piping

Clearance is the distance between the sewer pipe and any other pipe and is measured from exterior to exterior. Clearances are usually mandated by either code requirements or local authorities. There are different dimensions required both horizontally and vertically.

The most stringent requirements are between potable water and the sewer. Generally accepted practice is to have a minimum 10′-0″ (3-m) separation between water and sewer lines horizontally, with the sewer line below the water line. If that distance cannot be maintained, the sewer line should be considerably lower than the water main, with an often recommended minimum distance of 4′-0″ to 6′-0″ (1.25 to 1.85 m) below the water line. When crossing, the sewer should be a minimum of 1′-0″ (0.30 m) below the water main; in addition, the sanitary pipe shall be encased in concrete for a distance of 1′-0″ (0.30 m) past the water main.

For other utilities, 6-in (150-mm) clearance is adequate, and encasement is not necessary.

SIZING THE SEWER

Peak Flow Rate Determination

Four factors must be considered when calculating the peak flow:

1. Peak sanitary discharge and flow from fixtures
2. Peak drainage and estimated leakage flow from all process, utility, and manufacturing sources; also air-conditioning condensate, blowdown, and other similar sources
3. Allowance for future
4. Infiltration for long runs of piping

Peak Sanitary Discharge

Peak sanitary discharge is calculated using fixture units in accordance with the local plumbing code. This figure is available from the facility interior plumbing calculations. A method of converting fixture units into gpm (L/min) is presented in Fig. 19.8. Private sewers serving multiple buildings are also sized in the same manner until the fixture unit count becomes too large. After the code fixture unit value is exceeded, the drainage flow rate figure is calculated by using the facility water flow rate demand.

Other Peak Discharge

The process, utility, and manufacturing discharge can only be obtained from the user or by a study of the entire facility and a determination of the total gpm expected to be discharged from all sources. It is doubtful that all discharges will occur at the same time; therefore some diversity should be used to reduce the peak flow. If there is any doubt, use the highest figure. The owner is to be consulted regarding discharge. Peak discharge includes boiler blowdown, cooling tower discharge, and other HVAC use.

To estimate a preliminary size before the plan for the entire job is started, determine the total gallons per day (liters per day) for the facility. Divide that figure by the number of hours the facility is in operation to give an average hourly flow, in gallons. Double the hourly flow rate and divide by 60 to find the average peak flow rate in gallons per minute

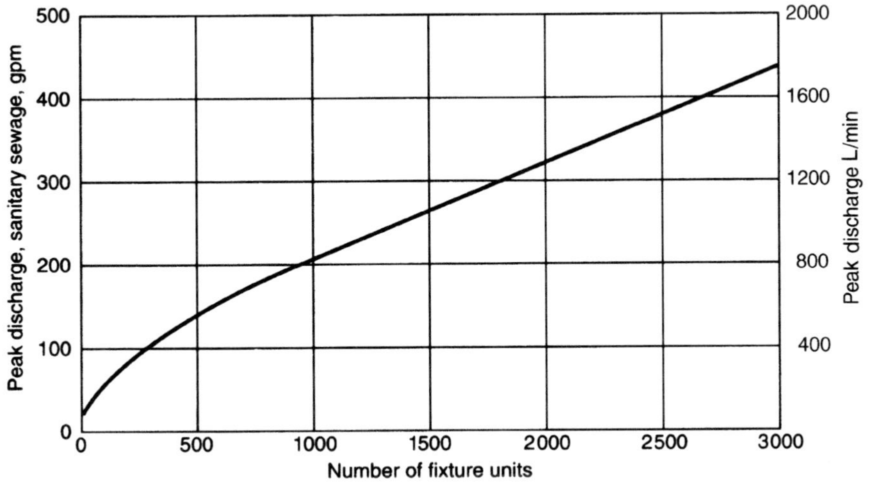

FIGURE 19.8 Fixture unit Conversion.

(liters per minute). This is only an average figure. In the determination of the pipe size, allow an additional 10 percent safety factor for the peak flow rate.

Slope of the Piping

The pipe slope is generally determined by both the topography where the sewer is being installed and the elevation of the outfall where the sewer line will be discharged. Often, it is a combination of both. Keeping a fairly uniform depth of bury will establish the general slope, with adjustment made to keep the slope uniform in different sections of the run. If the grade is flat, establish the slope based on the optimum velocity desired, if possible. If the outfall elevation is the controlling factor, there is little choice in the slope. If it is found that an adequate slope to provide the necessary velocity cannot be maintained, it will be necessary to use a lift station and pump the effluent. The slope of the line shall be steep enough to provide a minimum scouring velocity of 2 fps (0.6 m/s). Table 19.2 gives the minimum slope to maintain such a velocity.

Allowance for the Future

Allowance for the future should be calculated and added to the peak discharge.

Infiltration

Infiltration is the amount of groundwater or stormwater entering the sewer piping network from faulty joints in underground piping, leakage around manholes and manhole covers, and cracked pipe. Another major contributor to infiltration is from poorly installed laterals. Infiltration should be considered only for a long run from the building to the sewer or in a large network serving multiple facilities on a single site. Existing sewers are frequently found to be quite leaky.

TABLE 19.3 Slopes of Cast Iron Soil Pipe Sanitary Sewers Required to Obtain Self-Cleansing Velocities of 2.0 and 2.5 fps (0.6 and 0.75 m/s) (Based on Manning's Formula with $n = 0.012$)

Pipe size		Velocity		¼ Full		½ Full		¾ Full		Full	
in	DN	fps	m/s	Slope, ft/ft	Flow, gpm	Slope, ft/ft	Flow, gpm	Slope, ft/ft	Flow, gpm	Slope, ft/ft	Flow, gpm
2.0	50	2.0	0.6	0.0313	4.67	0.0186	9.34	0.0148	14.09	0.0186	18.76
		2.5	0.75	0.0489	5.84	0.0291	11.67	0.0231	17.62	0.0291	23.45
3.0	80	2.0	0.6	0.0178	10.11	0.0107	21.46	0.0085	32.23	0.0107	42.91
		2.5	0.75	0.0278	13.47	0.0167	26.82	0.0133	40.29	0.0167	53.64
4.0	100	2.0	0.6	0.0122	19.03	0.0073	38.06	0.0058	57.01	0.0073	76.04
		2.5	0.75	0.0191	23.79	0.0114	47.58	0.0091	71.26	0.0114	95.05
5.0	125	2.0	0.6	0.0090	29.89	0.0054	59.79	0.0043	89.59	0.0054	119.49
		2.5	0.75	0.0141	37.37	0.0085	74.74	0.0067	111.99	0.0085	149.36
6.0	150	2.0	0.6	0.0071	43.18	0.0042	86.36	0.0034	129.54	0.0042	172.72
		2.5	0.75	0.0111	53.98	0.0066	107.95	0.0053	161.93	0.0066	215.90
8.0	200	2.0	0.6	0.0048	77.20	0.0029	154.32	0.0023	231.52	0.0029	308.64
		2.5	0.75	0.0075	96.50	0.0045	192.90	0.0036	289.40	0.0045	385.79
10.0	250	2.0	0.6	0.0036	120.92	0.0021	241.85	0.0017	362.77	0.0021	483.69
		2.5	0.75	0.0056	151.15	0.0033	302.31	0.0026	453.66	0.0033	604.61
12.0	300	2.0	0.6	0.0028	174.52	0.0017	349.03	0.0013	523.55	0.0017	678.07
		2.5	0.75	0.0044	218.15	0.0026	436.29	0.0021	654.44	0.0026	872.58
15.0	375	2.0	0.6	0.0021	275.42	0.0012	550.84	0.0010	826.26	0.0012	1101.68
		2.5	0.75	0.0032	344.28	0.0019	688.55	0.0015	1032.83	0.0019	1377.10

The allowance for infiltration is based on the sewer system when it is reaching the end of its useful life, not when it is new. The types of joints and the pipe material used have an effect of the figure to be selected. Common figures used for private sewer systems are 500 gallons per day per mile (gal/day/mi) for smaller sewers and 1000 gal/day/mi for larger sewer lines. The figure selected for infiltration does not relate to the infiltration allowance used as a test for sewer acceptance.

For a single facility with a short run to a public sewer, no allowance should be made. For longer lengths of run shorter than 1500 ft (450 m), some consideration should be made depending on the local authorities. This will require a determination of the stability of ground conditions and the pipe jointing methods selected. If the run is more than about 1500 ft (450 m), allowance for infiltration should be made.

Sewers are sized using Kutter's or Manning's formula. Both will yield the same results, although of the two, Manning's formula is much less cumbersome to use. The easiest method, rather than to use the actual equation, is to find readily available prepared charts. Refer to Table 19.2 which is a direct reading of the size based on gpm flow rate, pitch, and depth of flow. Refer to Table 19.3 to ensure that the slope and velocity are sufficient for self-scouring purposes.

Determining the Size of the Sewer

To calculate the sewer size:

1. Select the pipe material and the n value to be used for sizing.
2. Determine the slope of the sewer to its final outfall.
3. Calculate the peak flow that is expected for the sewer.

With the above information selected, use Table 19.2 with the flow and the slope to find the appropriate pipe size. To allow a safety factor, use two-thirds full discharge, and select a figure that is equal to or greater than the expected flow or velocity to determine the size. The velocity shall be selected that will give a figure of at least 2 fps (0.6 m/s) to maintain a scouring velocity.

Table 19.2 is based on an n value of 0.013, which is used for cast iron pipe. If the n value is selected for a different pipe material, use the conversion factor found at the bottom of Fig. 17.9 to adjust the flow and/or velocity.

BIBLIOGRAPHY

Corbitt, R. A., ed., *Standard Handbook of Environmental Engineering*, McGraw-Hill, New York.

Design and Construction of Sanitary and Storm Sewers, American Society of Civil Engineers.

Plumbing Engineering Design Handbook, American Society of Plumbing Engineers.

Sewer Manual for Corrugated Steel Pipe, National Corrugated Steel Pipe Association.

CHAPTER 20
PUMP DISCHARGE SYSTEMS

This chapter discusses methods used to pump sewage and clear water discharged from industrial and commercial facilities installed on the site. When it is due to topography or elevation, if the ultimate point intended for disposal of sanitary or clear stormwater effluent from a facility or site is higher than the initial discharge point, pumping up to that ultimate point of disposal is required. When an interior sump or ejector pump is located outside a building on a site, it is commonly called a *lift station* or *pumping station*. There are also many occasions where these pumps are located inside a building for discharging to a higher site sewer. They should still be referred to as lift stations, and the discharge pipe to the disposal point is called a *force main*. Obstacles to be avoided, such as underground piping, trees, underground structures, and rock, should be located.

There are several possible classifications of sewage pump station design: wet well/dry well, submersible, suction lift, and screw pump. Only submersible pump stations are discussed here because they are by far the most common.

CODES AND STANDARDS

There are no requirements in specific model and plumbing codes that apply to sewage pumping installations outside the building. The specific appropriate code for any pumped discharge of stormwater for the local area where the facility is located must be followed. Refer to Chap. 17 for further discussion of applicable stormwater codes. If the discharge is routed outside the building into public streets, special codes for the pipe in public streets will be required from the local authorities.

SYSTEM COMPONENTS

General

There are three major components of the pump system: the pump, the basin, and the discharge piping routed to the public sewer or point of disposal, along with ancillary components such as valves, controls, and alarms. The pump discharge line is often referred to as a force main.

Pumps

There are two classifications of pumps used: sewage ejectors and sump pumps. The difference in terminology between a sewage ejector and a sump is sometimes blurred. A sewage ejector is used for the removal of sanitary sewage with large solids, and a sump pump is for the discharge of clear water with few solids. The impeller of the ejector pump is designed to pass solids and transport sanitary waste with larger solids suspended in the effluent. The ejector designation is reserved for sanitary pumps. Pumps designed to transport large quantities of nonsanitary water with larger suspended solids are sometimes referred to as *trash pumps.*

Basins

Basins receiving sanitary effluent shall be provided with a gasketed cover that has an atmospheric vent. The location shall be such that an odor emanating from the basin would be easily dissipated before causing any discomfort to the occupants. The cover shall be easy to remove so that the pumps can be raised to the surface for servicing. Where submersible pumps are selected, guide rails are attached to the inside of the basin for this purpose. A hose bibb or lawn hydrant should be provided adjacent to the basin to wash off the pump.

Discharge Piping

A force main is a pipeline designed to carry wastewater under pressure from its source to the point of disposal. Sewage and waste effluent has the same hydraulic characteristics as water; therefore, the pump discharge piping should be sized using the same criteria as for potable water. The force main includes valves, cleanouts, air and vacuum release valves, and thrust restraints. The stopping of a pump will sometimes lead to water hammer. If this is a concern, a slow-closing check valve should be considered. The pump discharge piping shall have a control valve and a check valve on the outlet. The check valve will prevent the effluent in the discharge piping from draining back into the basin.

The piping material should be selected primarily based on the soil, effluent corrosion resistance, and pressure rating. The piping suitable for force pumps would be similar to that suitable for water distribution. Other considerations are the strength of pipe material on the site and what materials are allowed by the local authorities. The most common pipe material in use is ductile iron pipe, covered in Chap. 19. When there is a choice among several different materials, the total cost of installation will be the deciding factor.

COMPONENT SELECTION

General

The design of the complete pumping station as a whole is an iterative one, where selection of each component is somewhat dependent on other information regarding size and capacity.

Pump Type Selection

There are three pump systems generally used, and all could be considered for installation either inside the building or outside on the site. These systems are categorized as follows:

1. Submersible pumps in a receiver (see Fig. 20.1)
2. Wet-pit receiver with vertical lift pumps (see Fig. 20.2)
3. Wet-pit receiver with cantilever type pumps (see Fig. 20.3)

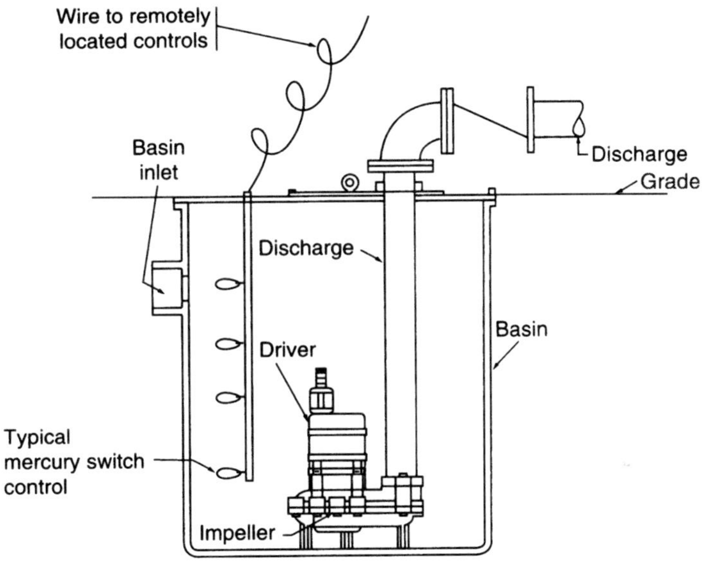

FIGURE 20.1 Submersible pump system.

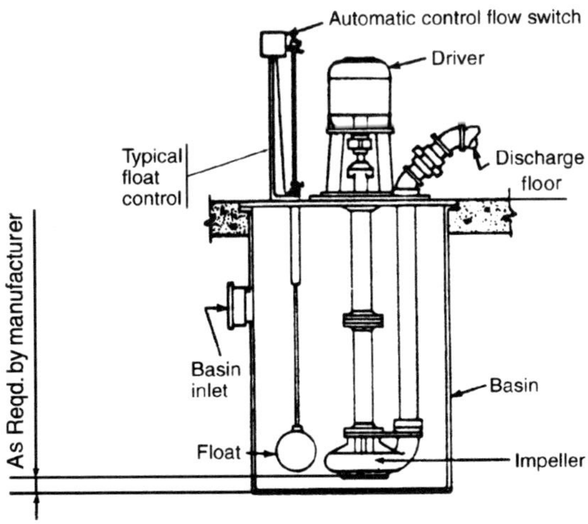

FIGURE 20.2 Vertical lift submerged pump system.

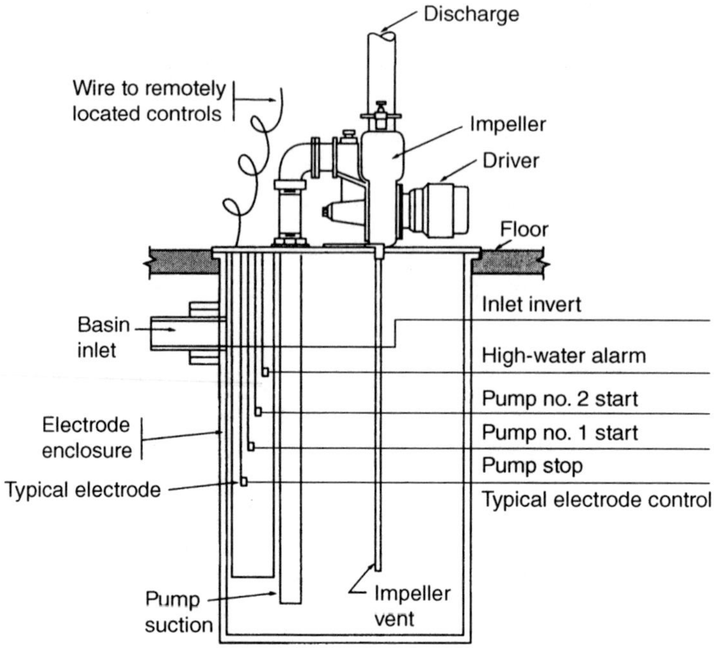

FIGURE 20.3 Cantilever-type system.

Each system has specific advantages and disadvantages that must be evaluated based on the following major considerations:

1. Range of head and capacity for motor and impeller combinations
2. Initial cost
3. Floor space requirements
4. Materials available for pump and bearing construction
5. Headroom required to remove pump and impeller
6. Ability to pump abrasive and corrosive effluent

Following is a discussion of the individual differences between system types.

Submersible Type. The submersible pump system utilizes a centrifugal pump that has a close-coupled motor and impeller combined in a single pump housing that is totally submerged in the effluent within the receiver. The pump can be directly joined to the discharge pipe or, by using a proprietary coupling, connected to it by gravity, thus permitting the entire assembly to be brought up to floor level for servicing. The discharge piping could be installed underground in a pit.

The operating cycle is controlled by level-sensing devices immersed in the effluent, starting and stopping the pumps when the liquid reaches the set point for the various operations.

The advantages to using the submersible pump are

1. Wide hydraulic range of pumps
2. Low initial cost
3. Minimum floor space requirements
4. Very deep receivers possible
5. Little maintenance required

Disadvantages are

1. Few materials available for pump construction
2. Limited acceptability for pumping hot liquids

Consult with manufacturers for specifics.

Vertical Lift Type. The vertical lift system utilizes a vertical shaft centrifugal pump installed above the basin floor and a separate driver mounted on the basin cover. The pump is immersed in the effluent and supported from under the cover by a special column that encloses the connecting shaft. The motor is independently supported on top of the cover directly over the pump. The discharge piping could be installed underground in a pit.

The operating cycle is controlled by level-sensing devices immersed in the effluent, starting and stopping the pumps when the liquid reaches the set point for the various operations. The advantages of using the vertical shaft pump, wet pit system are

1. Widest selection of materials for pump
2. Wide hydraulic range of pump
3. Wide selection of bearing materials and shaft lubrication methods
4. Underground discharge possible
5. Driver aboveground
6. Wide choice of impeller types
7. Available as complete assembly of receiver and cover with premounted equipment

Disadvantages are

1. Headroom greater than pit depth required to remove pump assembly for maintenance
2. Least efficient of all system types

This system is the most widely used of all the types.

Cantilever Type. The cantilever, or wet-pit, system utilizes a horizontal centrifugal pump and close-coupled driver mounted on the receiver cover. A suction pipe is extended (or cantilevered) from the pump down into the receiver, so that the effluent can be lifted up to the level of the impeller intake.

The pump selected most often is the submersible type, because of its low initial cost, wide range of capacities, and tolerance of many starts. Another advantage is that a smaller basin can be used because additional height below grade is not required to house the motor of a conventional vertical, submerged ejector pump. Pump casings, impellers, and other components are available in a wide variety of materials to resist chemical corrosion, abrasion, and suspended solid size.

The operating cycle is controlled by level-sensing devices immersed in the effluent, starting and stopping the pumps when the liquid reaches the set point for the various operations. The advantages of using the cantilever pump, wet-pit system are

1. No bearings suspended in the effluent
2. Handles highly abrasive and corrosive liquids and slurries
3. Pump and driver easily maintained

The disadvantages are

1. High initial cost
2. Limited length of suction pipe
3. Possibility of losing prime

Pump Control

Control of the operating cycle used to start and stop the pumps is based on the level of water in the basin. Alarms of the following types are also included in the control panel.

1. *Mercury switch.* This is a sealed, buoyant device containing mercury and integral contacts, attached at one end only and suspended at various levels inside the basin. As the effluent level rises and falls, so does the orientation of this device. The mercury then makes or breaks contact. Experience dictates that no less than 6 in (150 mm) between levels should be used.
2. *Float.* This type of device uses a rod with an attached float on the surface of the effluent level, with a set point on the rod operating an exposed switch on the basin cover.
3. *Pneumatic.* This alarm is an inflated device suspended below the level of the effluent. The difference in pressure of the effluent on the inflated device is used to make or break remotely located contacts.
4. *Electrode type.* This alarm uses separate electrodes suspended in the effluent. As the level of effluent rises and falls, different sets of electrodes are immersed and send a signal to remotely located controls.
5. It is good practice to have these on a separate circuit. In addition, another alarm that is highly recommended is a *pump failure to start alarm light* for duplex installations. In many cases, if the first pump does not function, the second pump will automatically start and satisfy system requirements. Operating personnel will not know that the first pump has broken until the second pump breaks down and the entire pumping system does not function.
6. The relative positions of controls and alarms applicable to all pumps are illustrated in Fig. 20.3.

Basin Material Selection

Basins, often referred to as *receivers, sumps*, or *wet wells*, are typically manufactured from cast iron, fiberglass, poured-in-place concrete, and steel. Cost will also be a factor if the basin is large.

Force Main Piping Material Selection

The discharge pipe, commonly called a force main, is sized to convey the effluent economically from the pump to its ultimate point of discharge. The pipe material selected is based on the flow rate, friction loss of the fluid in the pipe, chemical resistance to the effluent and soil, and fluid velocity. A minimum size of 4 in (100 mm) is highly recommended to allow solids to pass easily through the pipe with little chance of producing stoppages.

The selection of pipe material depends on the type of effluent expected and the strength of the pipe for underground burial, based on its installed condition. Commonly used materials include carbon steel, ductile iron, PVC, CPVC, FRP, and PE. For sanitary effluent, pressure-rated PVC or PE is often selected. Soil contamination may not permit plastic to be used. When the force main is to be routed in public streets, the local authorities may have pipe material and specific installation requirements. Where strength of pipe material is a factor (e.g., burial close to the surface of a public road), ductile iron should be considered. It is common practice to coat the exterior of the pipe or protect it with a film of PE in corrosive or contaminated soils. Several methods of installation are described in ANSI/AWWA A-21.3.

COMPONENT SIZING

General—Pump Sizing

In the general conditions for which pumps are sized, one pump (primary) of a multiple-pump system is running, which is the normal condition. The second, arrangement has all the pumps running simultaneously (secondary condition) at the same time if the level of water in the basin rises too high. This is considered an emergency condition.

A sanitary pump is sized on the discharge of the sanitary piping from the building. The means used to convert the drainage fixture units to gpm is given in Fig. 20.4. Add 10 percent extra to that figure for unexpected large flows. This figure will not supersede the actual calculation of that made for the interior plumbing system, but rather would be used for preliminary purposes.

Required Pump Capacity

There are several considerations when selecting the pump capacity. The first is based on the sanitary effluent discharge from the facility. For a facility sewage pump, it is accepted practice to have a pump capacity equal to the highest instantaneous flow rate expected from that facility. The reasoning is that a single pump must be capable of discharging all possible effluent from all sources. It is also critical that there be at least a duplex set of pumps, each with the same capacity, to ensure that the facility will be kept in operation if one pump is out of service. If there is the possibility of a wide range of instantaneous flow rates, then a three-pump system, with each pump sized at 75 percent of maximum flow, should be considered. One pump would be on standby. For conversion of sanitary fixture units into gpm, refer to Fig. 20.4.

The capacity of the pump could also be based on a maximum desired number of starts per hour. Recommended practice is to have an average of 6 or 7 starts per hour up to a maximum of 12 starts per hour. Another consideration is to allow a minimum running time of 1 min.

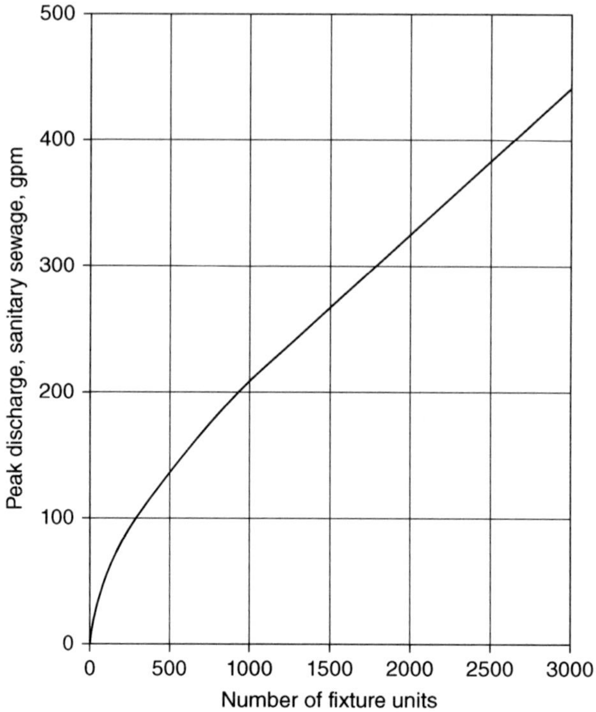

FIGURE 20.4 Fixture unit conversion to gpm.

For a pump that includes stormwater, it must be capable of discharging the heaviest rainfall. Refer to Chap. 17 for the selection of the design rainfall. The pump should be duplex, with one pump operating and one pump on standby.

Within these parameters, and with consultation of the pump manufacturer, the pump capacity can be selected.

Required Pump Head

The required system head is calculated by first determining the static height (in feet) from the bottom of the basin to a point 1 ft above the point of discharge. Next is added the total pressure loss for the equivalent length of run from friction of the water (in feet of head) of the liquid flowing through the pipe based on the flow of two pumps.

Pressure Loss of System

The pressure loss shall be found by using the maximum flow with the maximum number of pumps running. Use standard friction loss charts for water piping, based on the pipe material selected, flow rate, and its size. This figure will allow the calculation of the actual pressure loss in the run of pipe. The friction loss is obtained from standard engineering texts, using water as the liquid for sanitary effluent, such as *Cameron Hydraulic Data* or the ASPE *Book of Tables*. For more viscous liquids, appropriate charts or a conversion factor shall be used.

Velocity

The velocity in the pipe should be within a range between a minimum of 2 fps (0.6 m/s) and the generally accepted maximum of about 10 fps (3 m/s). The low flow velocity is based on the flow rate when one pump is running, and the highest velocity occurs when all pumps are running together. These figures can be obtained from the above references. Since one rarely needs the flow rate for all pumps running, a slightly higher velocity should be considered acceptable if the possibility of the development of excessive water hammer does not arise. One recommendation is to use a slow-closing check valve to limit the generation of excessive pressures every time the pumps stop.

Basin Capacity

Since the exact inflow into the basin in any time period is uncertain, a minimum running time shall be selected based on the number of gallons stored; and if additional flow into the basin occurs, the running time of the pump is extended, which is an added benefit.

The storage capacity of the basin is based on a single primary pump discharge flow rate and the desired number of starts per hour. Recommended practice is to have an average of 6 or 7 starts per hour up to a maximum of 12 starts per hour, and a minimum running time of 1 min. Within these parameters, and with consultation of the pump manufacturer, the storage capacity can be selected.

Large basin capacities require that the design of the basin include principles important to proper pump functioning. These principles are as follows:

1. The flow of effluent from the basin inlet should be directed to the pump inlet in a manner that will reduce swirl and hydraulic loss.
2. Water depth at low level shall be the minimum established by the manufacturer to avoid surface vortices. The flow shall be stopped to prevent a pump from running dry when water reaches this level.
3. Excessive turbulence should be avoided. A small amount of turbulence shall be limited to that required to prevent stagnation of stored water.
4. To prevent accumulation of sedimentation and solids, the floor of the basin should be sloped to a small lower sump in a corner depression that will allow a temporary pump to completely empty the basin. A 6-in (150-mm) space below the pump inlet is recommended to allow space (mud bottom) for sediment to accumulate on the basin floor.
5. Where the inflow into the basin is at a high elevation, the falling water entrains air as it spills onto the surface of water in the sump. This entrained air should be allowed to rise to the surface before reaching the pump. This is accomplished by providing a baffle wall in the sump and/or a dissipation arrangement at the inflow. If the basin is large, there should be sufficient distance from inlet to the basin to the point of inlet to the pumps that will permit air to dissipate.

The standard basin sizes are designed to accommodate the number and size of the pumps selected; the inlet and discharge pipes that are preinstalled in the basin; and the depth necessary to contain the desired quantity of liquid. The complete basin is installed as a single, integrated unit. The manufacturer should be consulted regarding the size of basins in relation to the pumps selected.

TABLE 20.1 Capacity of Basins per Foot of Depth

Circular				Square			
Diameter		Capacity		Side		Capacity	
ft	m	gal	L	ft	m	gal	L
2	0.6	23.50	90	2	0.6	30.00	114
2½	0.75	36.72	140	2½	0.75	45.00	171
3	0.9	52.88	200	3	0.9	67.50	257
3½	1.0	79.91	275	3½	1.0	90.00	342
4	1.2	94.00	357	4	1.2	120.00	456
4½	1.35	110.32	420	4½	1.35	149.60	570
5	1.5	146.89	560	5	1.5	187.00	710
6	1.8	158.64	600	6	1.8	270.00	1026
7	2.1	170.00	646	7	2.1	365.50	1400
8	2.4	181.00	680				
9	2.7	193.00	733				
10	3.3	204.00	775				

The actual size may be limited by space conditions where the basin is located. Additional depth may be used to obtain the desired volume if the length or width cannot be adjusted. To size the basin, refer to Table 20.1 for capacities in gallons (liters) of various basin sizes.

For a method of determining the basin depth, use the following approximate dimensions as a guide, assuming duplex pumps and starting from the invert of the inlet pipe. The actual dimensions for a specific facility will vary depending on installation requirements.

1. From the invert of the inlet pipe, allow 6 in (150 mm) to the high-water alarm.
2. From the high-water alarm, allow 6 in (150 mm) to pump no. 2 start.
3. From pump no. 2 start, allow a dimension to pump no. 1 start as determined by the pump capacity to accommodate the volume of the basin.
4. Below pump no.1 start, allow 6 in (150 mm) to pump stop. The dimension also depends on the pump capacity.
5. Allow 6 in (150 mm) from pump stop to the top of the basin inlet.
6. Allow 1 ft (0.33 m) from the inlet invert of the basin to the basin bottom. This dimension shall be confirmed with manufacturers.

Pipe Sizing

The flow rate and friction loss used to size the force main are based on the single condition of having the maximum number of pumps running. This maximum flow rate will result in greater friction loss. The force main pipe size is based on considerations of flow rate, velocity, and friction loss.

After the pump capacity is chosen, the only criterion that can be adjusted is the pipe friction loss. A balance must be made between a large pipe size with low friction loss and a small pipe size with large friction loss. A small pipe size may also require a high velocity, which should be avoided. With different pipe sizes, at some point the friction loss will change the horsepower of the pump. The optimum economical point is reached when a larger pipe size will no longer reduce the horsepower of the pump(s). It is also possible that for a long pipe run, a smaller pipe size will justify a larger pump size. Standard engineering references, such as *Cameron Hydraulic Data,* the ASPE *Design Handbook,* and the ASPE *Book of Tables,* shall be used to determine friction loss depending on the pipe size and flow rate.

SYSTEM DESIGN CONSIDERATIONS

Lift Station Location

If outdoors, the basin should not be located at a point on the site that is subject to flooding unless it is the only possible location. Borings of the area should be made to determine if the area has high groundwater or other underground obstacles that would interfere with the installation of the basin. The basin shall be installed with its top several inches above grade so that stormwater will not be directed toward the basin. It also must be located on stable soil. Soil-bearing loads are given in Table 20.2.

Layout of Piping at Pumps

Although not shown on details, the discharge piping should have a control valve and check valve installed as close as practical to the pump in a pit or exposed. A slow-acting check valve shall be used where there is a danger of water hammer due to the shutdown of the pumps. If the installation is outdoors, these valves shall be installed in a pit, allowing access for maintenance.

Flotation

Any structure extending below the groundwater table, or extending into a soil that may become saturated, must be checked to ensure that the buoyant force will not cause the basin to float. To determine the buoyant force, calculate the upward and downward forces. If the sum of the upward forces exceeds the downward force, the structure will float.

The downward forces are the sum of the equipment and basin weight, the weight of the concrete base slab under the basin, if present, and the dry weight of the soil on top of the basin, if any. The dry weight is calculated by determining the wet weight minus the weight of the water. For typical soils, refer to Table 21.1 to obtain the weight per cubic foot (cubic meter), and subtract the weight of a cubic foot (or meter) of water, which is 62.4 lb/ft^3 (922 kg/m^3). The effective weight of concrete is 87 lb/ft^3 (1392 kg/m^3) in water, or 150 lb/ft^3 (2400 kg/m^3) of concrete minus 63 lb/ft^3 (922 kg/m^3) for water. For displaced saturated

TABLE 20.2 Soil-Bearing Loads

Soil	Bearing load lb/ft^2	Bearing load kg/m^3
Muck	0	0
Soft clay	1,000	16,000
Silt	1,500	24,000
Sandy silt	3,000	48,000
Sand	4,000	64,000
Sandy clay	6,000	96,000
Hard clay	9,000	144,000

soil, the figure used is 50 lb/ft^3 (800 kg/m^3), with the assumed weight of soil of 100 lb/ft^3 (1600 kg/m^3).

The buoyant force is the volume of displaced weight of groundwater measured to the highest level possible or the weight of the displaced saturated soil to that of the empty basin. The buoyancy of an empty basin is given in Table 20.3.

The above calculations do not consider additional restraining forces such as skin friction between the basin and the soil and the shear between overburden and adjoining soil, which may have to be considered for large basins. Common practice regards these figures as optional safety factors for smaller basins, which adds a slight safety factor to the calculations.

If it is determined that the basin will float, the most common method to offset floating is to use a concrete slab under the basin to provide the necessary added weight. The basin must be anchored to the slab. It is common practice to use a 25 percent safety factor when calculating the slab weight.

TABLE 20.3 Approximate Buoyant Force of Empty Tanks

Tank capacity, gal	Tank diameter	Tank length, ft	Tank weight lb	Wall thickness	Net upward force (net buoyancy) totally submerged in water, lb*
560	4′	6′	480	12 gauge	4,190
560	4′	6′	828	7 gauge	3,842
1,000	5′4″	6′	875	10 gauge	7,475
1,000	4′	11′	968	10 gauge	7,382
1,000	5′4″	6′	1,180	7 gauge	7,459
1,500	5′4″	9′	1,575	7 gauge	11,200
2,000	5′4″	12′	1,980	7 gauge	14,720
2,500	5′4″	15′	2,375	7 gauge	18,479
3,000	5′4″	18′	2,765	7 gauge	22,235
3,000	6′	14′	2,565	7 gauge	22,150
4,000	5′4″	24′	3,560	7 gauge	29,790
4,000	6′	19′	3,310	7 gauge	30,040
4,000	7′	14′	3,150	7 gauge	30,200
5,000	6′	24′	5,400	¼ in	36,300
5,000	8′	13′6″	4,775	¼ in	36,925
6,000	6′	29′	6,380	¼ in	43,720
6,000	8′	16′	5,370	¼ in	44,630
7,500	8′	20′	6,500	¼ in	56,238
8,000	8′	21′6″	7,000	¼ in	59,750
8,000	10′6″	12′5″	6,470	¼ in	60,980
10,000	8′	27′	8,360	¼ in	74,356
10,000	10′6″	15′6″	7,600	¼ in	75,019
12,000	10′6″	18′6″	8,400	¼ in	91,440
15,000	10′6″	23′3″	12,600	5/16 in	112,833
20,000	10′6″	13′	15,900	5/16 in	151,345
25,000	12′	30′	20,875	3/8 in	190,509
30,000	10′6″	46′3″	27,000	3/8 in	222,518
30,000	12′	36′	24,294	3/8 in	229,329

*lb × 2.2 = kg
1 ft = 3.3 m

Pipe Restraints

Mechanical joints and push-on joints seal when the elastomeric gasket becomes compressed in the space between the spigot end of the pipe and the bell. These joints are leaktight when installed on a straight line. When a fitting introduces a change of direction, the flowing water exerts a pressure on that fitting. Pressures are listed in Table 20.4. Locations where thrust forces commonly occur are:

Bends	Tees	Wyes	Dead ends
Reducers	Valves	Offsets	Hydrants

Some joints are not capable of resisting this pressure; therefore a suitable means must be provided to restrain the joint from coming apart. The most common methods are integral joint restraint, clamps and tie rods, and thrust blocks.

Plastic piping with butt-fused joints has the same resistance as the piping and does not require additional restraint. Split couplings with grooved pipe are also capable of resisting this pressure without restraint, if couplings designed for this purpose are used. This jointing method is generally recommended for new underground pipe installation.

Clamps and Tie Rods. Resistance is provided by clamps attached to the pipe at each side of the joint and rods connecting the clamps across each joint. Standard charts are available giving the required restraint for different size pipes, depth of bury, number of joints to be restrained, and different fittings.

Integral Joint Restraint. This type of joint uses integrally cast glands for mechanical joint pipe and internally locked, grooved, and keyed joints for push-on joints.

TABLE 20.4 Pressure Exerted at Joints by Flowing Water. *Total pounds resultant thrust at fittings at 100 psi water pressure*

Nom. Pipe						
dia., in	DN	Dead end	90° bend	45° bend	22½°	11½° bend
4	100	1,810	2,559	1,385	706	355
6	150	3,739	5,288	2,862	1,459	733
8	200	6,433	9,097	4,923	2,510	1,261
10	250	9,677	13,685	19,353	3,776	1,897
12	300	13,685	19,353	10,474	5,340	2,683
14	350	18,385	26,001	14,072	7,174	3,604
16	400	23,779	33,628	18,199	9,278	4,661
18	450	29,865	42,235	22,858	11,653	5,855
20	500	36,644	51,822	28,046	14,298	7,183
24	600	52,279	73,934	40,013	20,398	10,249
30	750	80,425	113,738	61,554	31,380	15,766
36	900	115,209	162,931	88,177	44,952	22,585
42	1,050	155,528	219,950	119,036	60,684	30,489
48	1,200	202,683	286,637	155,127	79,083	39,733
54	1,350	256,072	362,140	195,989	99,914	50,199

Note: To determine thrust at pressures other than 100 psi, multiply the thrust obtained in the table by the ratio of the pressure to 100. For example, the thrust on a 12-in pipe with a 90° bend at 125 psi is 19,353 × 125/100 = 24,191 lb.
1 lb = 0.45 kPa

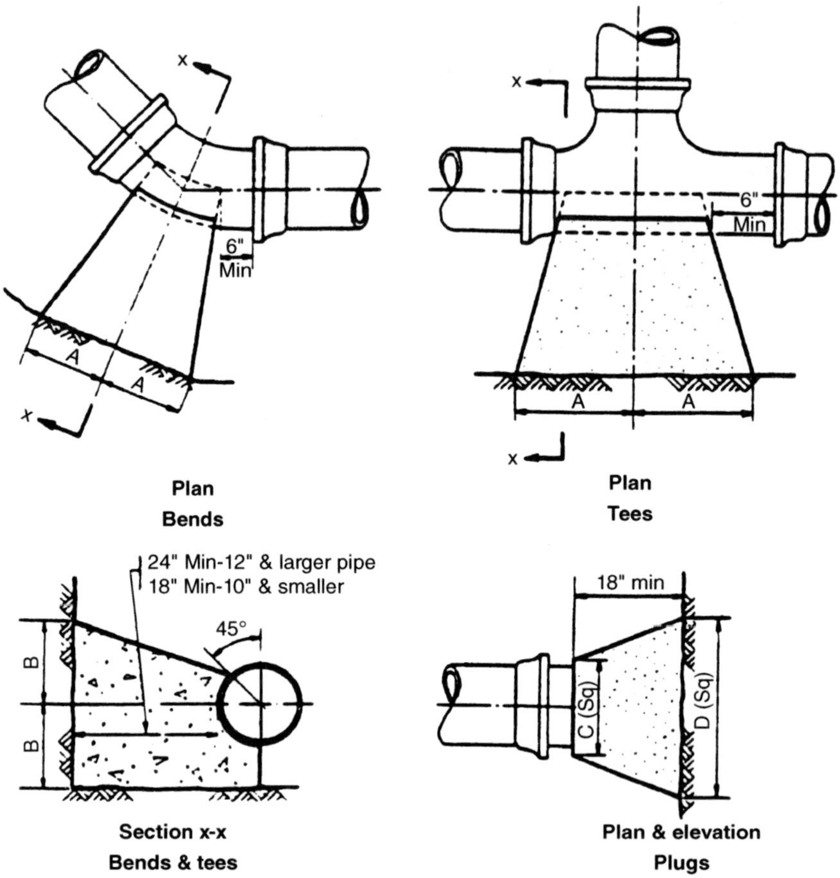

Type	Size	1/4 Bends		1/8 Bends		1/16 Bends		Tees		Plugs	
		A	B	A	B	A	B	A	B	C	D
Type I 4000 PSF soil	6"	8"	10"	6"	8"	3"	8"	8"	8"	10"	15"
	8"	12"	12"	8"	10"	5"	9"	9"	12"	12"	20"
	10"	16"	14"	10"	12"	6"	10"	11"	14"	14"	25"
	12"	19"	16"	12"	14"	8"	11"	14"	16"	16"	30"
	14"	23"	18"	14"	16"	10"	12"	16"	18"	18"	34"
	16"	26"	20"	16"	18"	11"	13"	18"	20"	20"	38"
Type II 2000 PSF soil	6"	16"	10"	9"	10"	6"	8"	10"	12"	10"	21"
	8"	22"	13"	12"	13"	8"	10"	13"	16"	12"	29"
	10"	26"	17"	14"	17"	10"	13"	16"	20"	14"	36"
	12"	29"	21"	16"	21"	11"	16"	18"	24"	16"	41"
	14"	35"	24"	19"	24"	12"	20"	22"	27"	18"	48"
	16"	38"	27"	21"	27"	12"	24"	24"	30"	20"	54"

Note: Based on 100 psi static pressure plus AWWA water hammer.
All bearing surface to be carried to undisturbed ground.

FIGURE 20.5 Typical thrust block dimensions.

Thrust Blocks. Thrust blocks consist of cast-in-place concrete blocks, the size and weight of which depend on the pipe size, water pressure (including surge loads), and safe soil-bearing load. Recommended locations for thrust blocks are illustrated in Fig. 20.5. These blocks must rest on undisturbed soil in order to resist the force exerted. Average safe soil-bearing loads are given in Table 20.2. These loads are general and must be verified for specific projects. The pressure exerted on joints is given in Table 20.4.

To determine the size of a thrust block, the thrust developed at the joint must be divided by the soil resistance. This will give the required area to be provided by the thrust block. Typical dimensions of thrust blocks for different fittings are given in Fig. 20.5.

In general, thrust blocks are not recommended for the following reasons:

1. The blocks are comparatively large.
2. Often, undisturbed soil is not present.
3. The cost is high.
4. Space for the blocks may not be available due to the proximity of adjacent site utilities.
5. The installation details for various fittings are shown in Fig. 20.6.
6. Thrust blocks are the least recommended of joint restraints.

Ancillary Equipment Considerations

Vacuum and Pressure Venting. It is good practice to provide an air pressure vent at the high point of the piping run. If the high point of the piping system occurs prior to the point

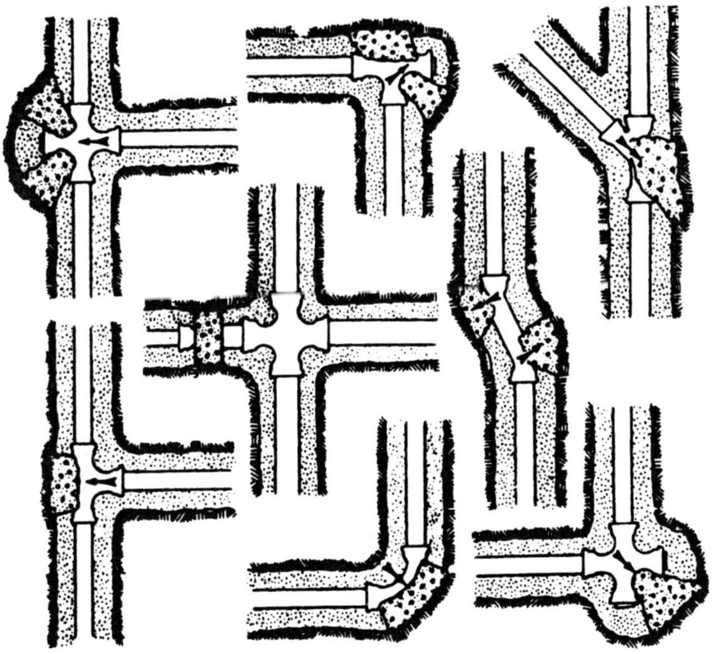

FIGURE 20.6 Locations for thrust blocks.

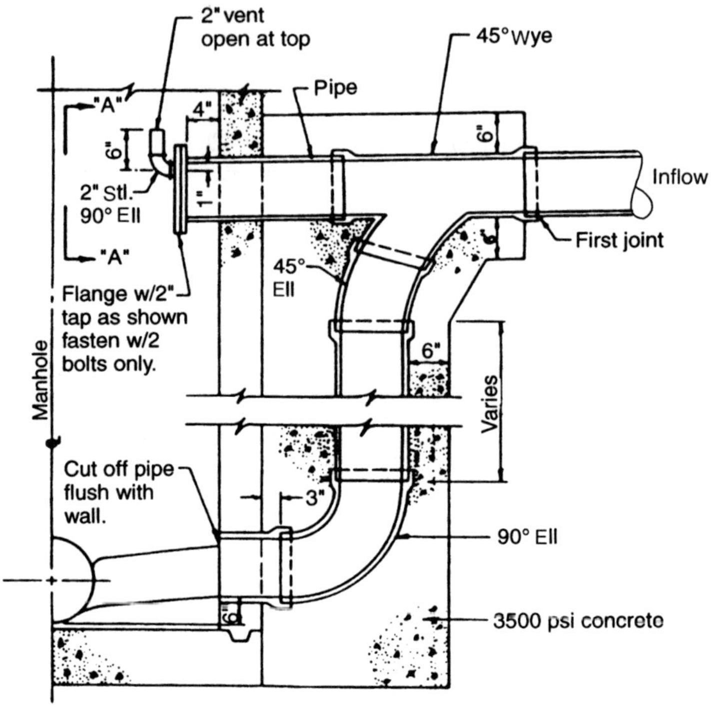

FIGURE 20.7 Force main termination.

of discharge, it is desirable to place a combination pressure/vacuum device in the pipe to allow accumulated air at the high point to be eliminated. Since the remainder of the run is downhill, a vacuum may be produced by the flowing liquid that will have to be broken to allow free flow of the liquid. Since there is a probability that air relief devices will discharge some water, the device is usually put in a large valve box with a gravel bottom to adsorb the liquid. Some authorities require that air release devices be installed in manholes. Ease of maintenance is the most important consideration. A detail of a typical device in a manhole is shown in Fig. 6.12.

Termination of Force Main. For the termination of a force main, a provision shall be made to eliminate any air entrapped in the effluent. A detail of such an air elimination method at the final discharge into a public sewer is illustrated in Fig. 20.7.

BIBLIOGRAPHY

American Society of Plumbing Engineers, *Book of Tables*.

Blake, L. S., ed., *Civil Engineers Handbook*, Butterworths, Boston.

Mackenzie, L. D., *Water and Wastewater Engineering*, McGraw-Hill, New York.

Mays, L. W., *Water Distribution Handbook*, McGraw-Hill, New York.

Westway, C. R., and A. W. Loomis, *Cameron Hydraulic Data*. Ingersol Rand.

CHAPTER 21

UNDERGROUND PIPING DESIGN

This chapter describes the criteria and methods used for the installation of underground piping in a site, starting 5′-0″ (1.5 m) from the outside wall of the facility. The loads placed on the pipe, the method of calculating these loads, and determination of the required supporting strength of the pipe are discussed. Underground double-wall pipe is covered in Chap. 23.

GENERAL

All buried pipe is subject to stresses imposed upon it by the nature of burial. Such factors as the pipe itself, bedding methods, type of backfill, material, type and shape of trench, loading from live loads, and storage of material over the trench will each contribute to the total load transmitted to the pipe and the ability of the pipe to support such loads. The pipe selected, combined with the method of installation, must have the strength to withstand the entire load placed upon it without crushing, cracking, or deforming. Obstacles to be avoided, such as underground piping, trees, and underground structures, should be located.

CODES AND STANDARDS

1. ASTM and ANSI standards for pipe and installation
2. AWWA standards for pipe, installation, and disinfection
3. NFPA-20 standard for installation of private water mains

PIPE AND INSTALLATION CLASSIFICATIONS

Pipe Installation

Underground pipe installation is divided into three general classifications:

1. *Pipe in a tunnel.* This is a classification for pipe forced through the earth by a ram or other means that leaves the surrounding earth undisturbed. This type of installation is outside the scope of this handbook.

2. *Pipe in an embankment condition.* This is a classification for pipe installed on the surface of the ground (positive projection) or in a trench (negative projection) with a layer of earth placed above the original ground surface.
3. *Pipe in a trench or ditch.* This is a classification for pipe installed in an excavation below grade, with the trench backfilled approximately to the original ground line or higher.

Pipe Classifications

Piping is divided into two classifications: rigid and flexible. Rigid pipe material is concrete, composite, metal, or clay. Rigid pipe fails when it breaks under a three-edge-bearing test. Concrete pipe fails when the three-edge-bearing test produces a crack 0.01 in (0.25 mm) wide and 2 in (50 mm) long. Flexible pipe is a plastic, composite, or corrugated steel. Flexible piping is considered to fail when loads produce a vertical deflection of 5 percent or more of pipe diameter.

Definitions

The earth load is the weight of all earth backfill over the pipe.

The superimposed load is produced by either a moving object (live load) passing over the pipe or a uniformly distributed (dead) load placed at ground level over the pipe in addition to the earth load.

A uniformly distributed load is an additional load, such as fill or lumber, stored at grade on top of the buried pipe.

A live load consists of the actual weight of any vehicle passing over the pipe.

The impact load places additional stress on the pipe due to the fact that a load is moving.

An incomplete projection occurs when a horizontal plane above a pipe installed in either a ditch or an embankment condition causes the soil to settle equally. This is the normal condition.

The transition width is the widest dimension that needs to be considered, beyond which there is no longer any additional load added to the buried pipe when computing the earth load.

The maximum trench width at the top of the pipe is the only dimension that needs to be considered in calculating the earth load.

Impact factor is a dynamic load caused by a moving object.

The surface load factor is the weight transmitted to the pipe from a truck on a road passing over the pipe.

Bedding is the contact between the pipe and earth.

Rigid pipe is any pipe that has been tested for strength in the laboratory by means of a three-edge-bearing test, such as metal or clay.

Flexible pipe is any pipe where internal and external forces have to be evaluated, such as plastic or corrugated steel pipe.

Radial deflection is a change in pipe diameter caused by an exterior load on a portion of the pipe.

Longitudinal deformation is a change along the length of a pipe that causes the pipe to bend. Often it is referred to as *sagging*.

The load factor for a bedding condition is used to determine the actual supporting strength of a pipe.

The Safety factor is an additional load to allow for the possibility of all extreme conditions occurring simultaneously.

LOADS ACTING ON A PIPE

General

The total load on the underground pipe under varying conditions compared to the allowable pipe supporting load that the pipe is designed to accept under those same varying conditions will determine the suitability of that particular pipe for installation. All the following discussion will determine that suitability.

Earth Loads

General. The earth load is the vertical force that earth transmits to a buried pipe. The pipe is subject to a very complex relationship among many other factors. The type of fill and the depth and width of a trench must all be considered. Original research at Iowa State College under the direction of Prof. Anson Marston produced a formula for the determination of earth loads. This, and additional research, resulted in Eqs. (21.1) and (21.2).

Earth Load Calculation. The earth load is obtained by selecting the lesser result obtained from the calculation of either Eq. (21.1) or (21.2) Actual tests have proved that the loads for a positive projection condition should be used when the result is the lesser of the two equations, even if the pipe is placed in a trench. The equations for earth load are

$$L_e = C_p\,WD \qquad \text{positive projection condition} \tag{21.1}$$

$$L_e = C_d\,WB \qquad \text{ditch condition} \tag{21.2}$$

where L_e = earth load on the pipe, lb/ft (kg/m) of pipe length
C_d = coefficient for load calculation of ditch condition
C_p = coefficient for load calculation of positive projection condition
W = weight of backfill, lb/ft^3 (kg/m^3)
B = width of trench above pipe, ft (m); see Fig. 21.1 for various conditions
D = outside diameter of pipe, ft (m)

Discussion:

1. To determine the coefficient C_d or C_p, first refer to Fig. 21.1 to obtain the appropriate letter for the specific ditch condition in use. The letters are

H = height of backfill over top of pipe, ft
D = outside diameter of pipe, ft
B = width of trench at top of pipe, ft

Then refer to Fig. 21.2 or 21.3 to obtain the coefficient for substitution in Eq. (21. 1) or (21.2).

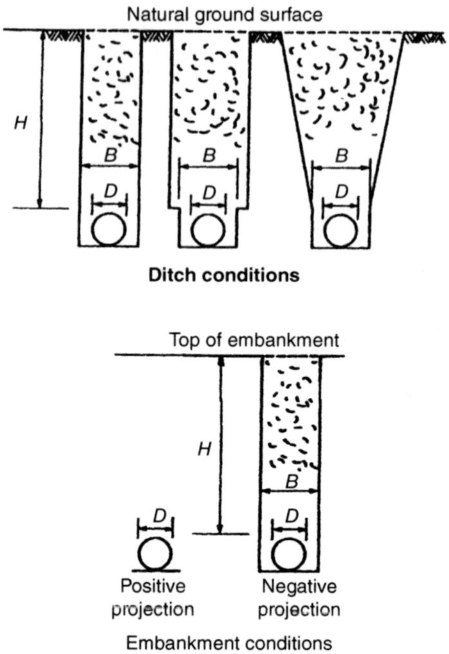

FIGURE 21.1 Installation conditions.

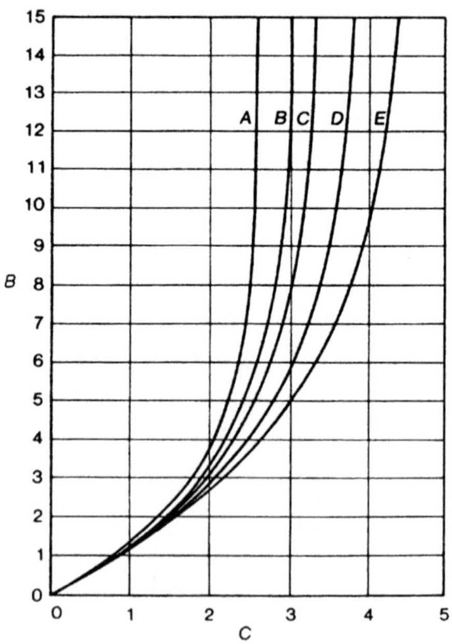

FIGURE 21.2 Ditch condition coefficient load calculation. (*Courtesy: ANSI 21.1.*)

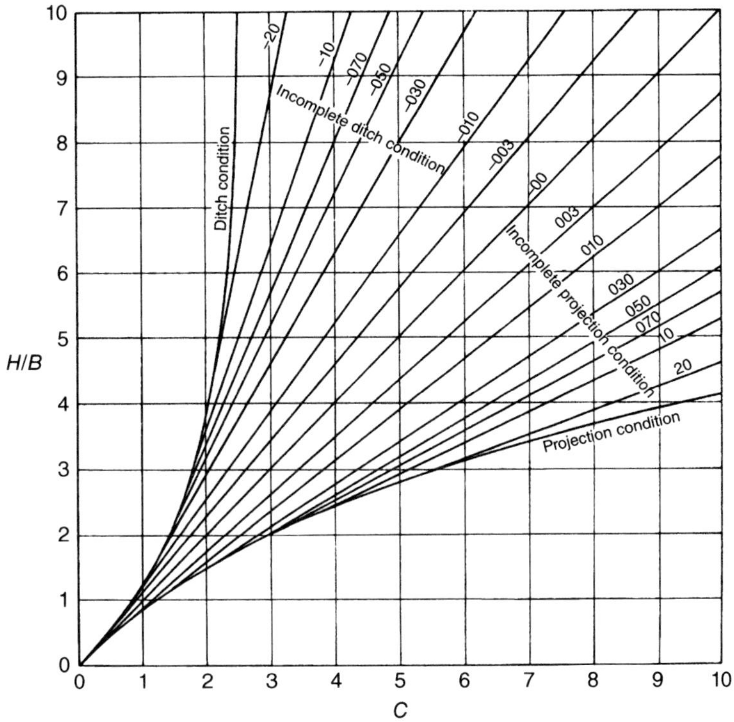

FIGURE 21.3 Negative and positive projection coefficient for load calculation. (*Courtesy ANSI 21.1.*)

2. To use Fig. 21.2, the letters represent the following curves:

   A = sand and sandy loam
   B = sand and gravel
   C = saturated topsoil
   D = average clay
   E = saturated clay

3. To use Fig. 21.3, first calculate H over B. With the result, draw a horizontal line to the intersection with the 0.70 line in the incomplete projection condition; then read down to find C for Eq. (21.1) or (21.2).
4. To determine the weight of soil W, refer to Table 21.1.
5. Trench width has an important influence on the soil load transmitted to the pipe. Any trench must provide sufficient working space for the installation with tolerance allowed for accepted construction practices and field conditions. Experiments have proved that only the trench width at the top of the pipe needs to be considered in calculating the earth load. If the trench is widened above the top of the pipe, it does not contribute any additional load to the buried piping. If sheeting, is used, the width of the trench is

TABLE 21.1 Weight of Soil

Type of soil	Weight kg/m³	lb/ft³	Correction factor	Average moisture content, %
Sand and sandy loam	1600	100	0.83	15
Sand and gravel mix	1760	110	0.90	20
Saturated topsoil	1840	115	0.95	25
Average clay	1920	120	1.00	30
Wet clay, peat	2080	130	1.10	100

computed to the inside of the sheeting if, it is to be removed. It is recommended that the sheeting remain, if possible.

6. Solving Eq. (21.1) or (21.2) (depending on conditions) will give the earth load.

At any given depth, for a particular pipe size and type of soil, a trench width is reached beyond which there is no longer any additional load added to the buried pipe. This is called the *transition width,* and it is the widest dimension that needs to be considered to compute the earth load.. Table 21.2 gives the transition width for various combinations of pipe size and depth of burial. It is common practice to allow 1 ft (0.3 m) on either side of a pipe in a trench for working space.

The most often used pipe burial is in a ditch, as shown in Fig. 21.1. For this condition, Fig. 21.4 provides direct determination of earth load without consideration of any equation. If soil weight is other than 120 lb/ft^3 (1920 kg/m^3), proportionally decrease or increase the resulting weight according to the correction factor in Table 21.1.

Superimposed Loads

Live Loads. Whenever pipe is buried in the ground, there is always the possibility of heavy equipment or trucks riding over the buried pipe during or after construction. When pipe is placed under a road or railroad, this is a normal occurrence. This load must be added to the earth load to determine the total force exerted on the buried pipe. The general contractor should be consulted to determine if piping will be buried to its full depth during construction or if additional cover will be added later.

A live load consists of two separate types of loads: the actual weight of any vehicle passing over the pipe and the impact load that places additional stress on the pipe because the load is moving.

Live Load from Trucks. The type of road surface over the pipe may have a reducing effect on the amount of live load reaching the pipe. A flexible surface such as light-duty asphalt will have little or no reducing effect. A concrete road or heavy-duty asphalt will greatly reduce the load intensity. Railroad tracks are constructed over a standard bed consisting of rock ballast and ties, which gives a uniform distribution of load to the piping.

For calculating the live load transmitted by a truck to a pipe under a road, the following formulas are used.

Unpaved or light-duty pavement:

$$L_t = CRPF \tag{21.3}$$

TABLE 21.2 Transition Width for Trenches

Cover over top of pipe*		Pipe size, DN												
		250	300	350	375	400	400	500	525	600	675	750	825	900
		Pipe size, in												
ft	m	10	12	14	15	16	18	20	21	24	27	30	33	36
2'-0"	50	1'-9"	1'-11"	2'-1"	2'-2"	2'-3"	2'-5"	2'-8"	2'-9"	3'-1"	3'-5"	3'-7"	3'-10"	4'-3"
2'-6"	70	1'-10"	2'-0"	2'-3"	2'-4"	2'-5"	2'-8"	2'-10"	2'-11"	3'-3"	3'-6"	3'-10"	4'-1"	4'-4"
3'-0"	75	1'-11"	2'-3"	2'-5"	2'-6"	2'-7"	2'-10"	3'-0"	3'-2"	3'-5"	3'-8"	3'-11"	4'-3"	4'-7"
4'-0"	100	2'-0"	2'-5"	2'-8"	2'-10"	2'-11"	3'-1"	3'-4"	3'-6"	3'-9"	4'-1"	4'-5"	4'-6"	4'-11"
5'-0"	125	2'-2"	2'-7"	2'-11"	3'-1"	3'-2"	3'-5"	3'-8"	3'-10"	4'-1"	4'-6"	4'-8"	5'-1"	5'-3"
6'-0"	150	2'-5"	2'-9"	3'-1"	3'-2"	3'-4"	3'-8"	3'-11"	4'-1"	4'-5"	4'-8"	5'-0"	5'-5"	5'-8"
7'-0"	175	2'-5"	2'-10"	3'-2"	3'-4"	3'-6"	3'-10"	4'-1"	4'-4"	4'-8"	5'-1"	5'-5"	5'-8"	6'-0"
8'-0"	200	2'-6"	2'-11"	3'-3"	3'-6"	3'-7"	4'-0"	4'-3"	4'-6"	4'-10"	5'-3"	5'-7"	6'-1"	6'-2"
9'-0"	225	2'-8"	3'-0"	3'-5"	3'-7"	3'-9"	4'-1"	4'-5"	4'-7"	5'-0"	5'-6"	5'-10"	6'-3"	6'-8"
10'-0"	250	2'-9"	3'-1"	3'-6"	3'-8"	3'-11"	4'-3"	4'-6"	4'-9"	5'-2"	5'-7"	6'-1"	6'-5"	6'-10"
12'-0"	300	3'-0"	3'-4"	3'-7"	3'-10"	4'-0"	4'-5"	4'-10"	5'-0"	5'-5"	6'-0"	6'-5"	6'-10"	7'-4"
14'-0"	350	3'-1"	3'-5"	3'-11"	4'-0"	4'-2"	4'-7"	5'-1"	5'-3"	5'-8"	6'-3"	6'-9"	7'-2"	7'-8"
16'-0"	400	3'-2"	3'-8"	4'-0"	4'-3"	4'-5"	4'-9"	5'-2"	5'-5"	5'-11"	6'-6"	7'-0"	7'-5"	7'-11"
18'-0"	450	3'-3"	3'-10"	4'-3"	4'-5"	4'-7"	5'-0"	5'-4"	5'-6"	6'-1"	6'-8"	7'-3"	7'-9"	8'-2"
20'-0"	500	3'-5"	4'-1"	4'-4"	4'-6"	4'-9"	5'-2"	5'-7"	5'-8"	6'-2"	6'-10"	7'-5"	7'-11"	8'-6"

*1 ft = 0.3 m.

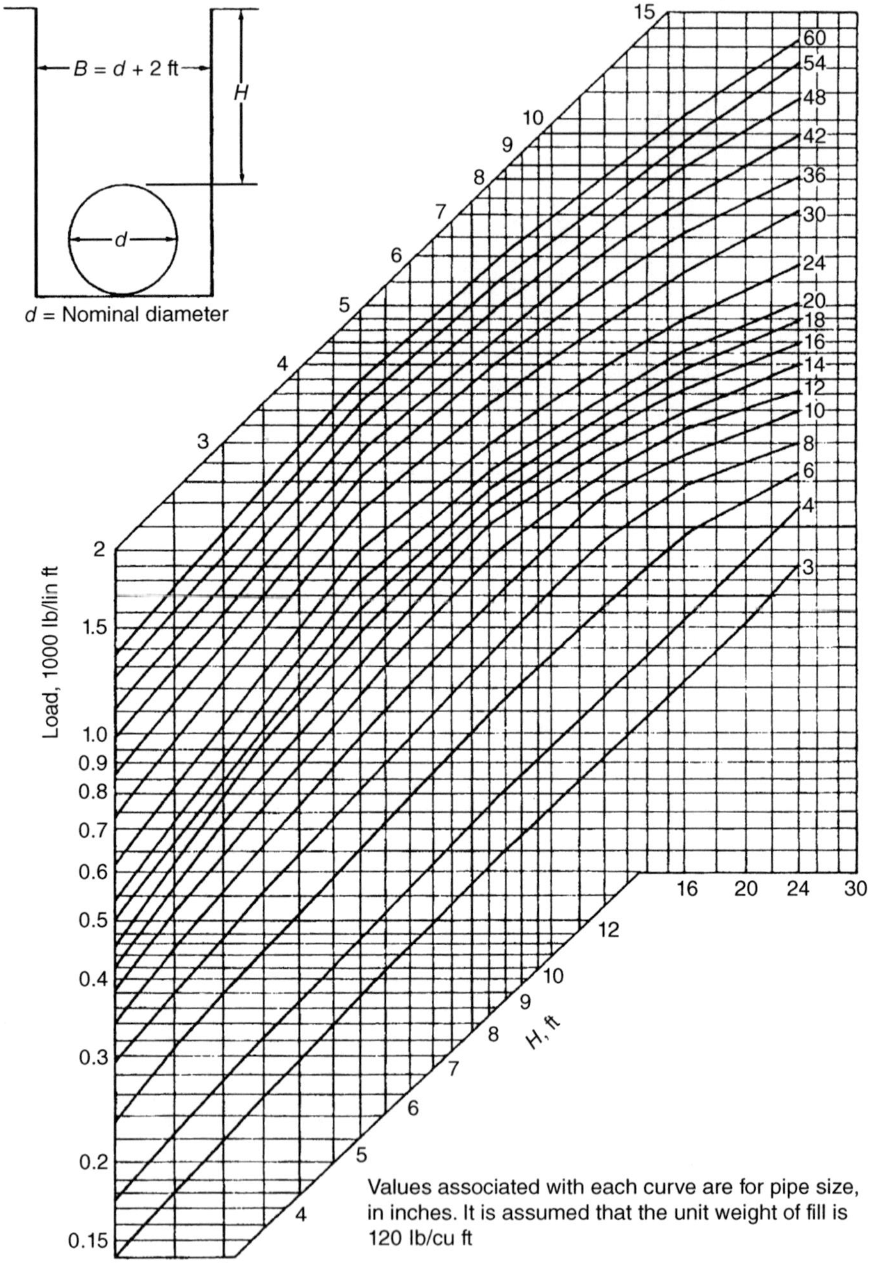

FIGURE 21.4 Direct-reading chart to determine earth load on pipe in trench. (*Courtesy: ANSI 21.1.*)

Rigid or heavy-duty pavement:

$$L_t = CBPF \tag{21.4}$$

where L_t = truck superload, lb/ft of pipe
C = surface load factor
R = reduction factor
P = wheel load
F = impact factor
B = outside diameter of pipe, ft

The surface load factor is the weight transmitted to the pipe from a truck on a road passing over the pipe. Tables have been prepared for both one truck and two trucks passing over the pipe simultaneously. The proper chart should be consulted based on road construction and the number of trucks. Always assume the worst case. Refer to Table 21.3 (one truck on flexible pavement), Table 21.4 (two trucks on flexible pavement), or Table 21.5 (one or two trucks on rigid pavement). The factors shown in the tables shall be multiplied by the wheel load to give the load transmitted to the pipe.

The reduction factor allows for the fact that adjacent portions of pipe that are not directly under the load assist in carrying some portion of the live load. Refer to Table 21.6.

Wheel load is the actual weight of the truck and its cargo on a wheel passing over the pipe. Refer to Table 21.7 to find the wheel load. The H-20 loading is most commonly used.

The impact factor is a dynamic load caused by a moving object. Refer to Table 21.8 to determine the factor based on depth of burial.

Table 21.9 gives the outside diameter of pipe, based on pipe size and material.

Live Load from Trains

For calculating the live load transmitted to a circular pipe under a railroad, refer to Table 21.10. Recommendations of the American Railway Engineering Association are used. This figure should be added to the weight of the track structure. The following criteria are used:

1. Copper E72 railroad design load is assumed.
2. The weight of track structure and ballast is assumed to be 200 lb/ft.
3. An impact factor of 1.4 is used, decreasing to 1.0 with 10 ft of cover or more. The impact factor has been included in the calculations used to prepare Table 21.10.

Loads from Other Equipment. For calculating the live load transmitted to a pipe from any other heavy equipment (crane, bulldozer, or others), the total weight of the vehicle must be found. The weight on each of its tracks or wheels is calculated by dividing the number of wheels or tracks into the vehicle weight. With the track or wheel weight and the pipe depth of burial known, refer to Table 21.11 for the percentage of load transmitted to the pipe. The track or wheel weight is then multiplied by the percentage of load transmitted to calculate the actual load on the pipe. Regardless of how slowly the vehicle is capable of moving, an impact factor must be used. The calculated actual load on the pipe is then multiplied by the impact factor obtained from Table 21.8. The result of this calculation is the total actual live load in pounds per foot on the buried pipe.

TABLE 21.3 Surface Load Factor—One Truck on Flexible Pavement*

Pipe size		Depth of cover, ft†												
DN	in	2	2½	3	3½	4	5	6	8	10	12	16	20	24
80	3	0.028	0.020	0.014	0.011	0.009	0.006	0.004	0.002	0.0015	0.001	0.0006	0.0004	0.0002
100	4	0.034	0.024	0.017	0.013	0.011	0.007	0.005	0.003	0.002	0.0015	0.0008	0.0005	0.0003
150	6	0.048	0.034	0.025	0.020	0.015	0.010	0.007	0.004	0.003	0.002	0.001	0.0007	0.0004
200	8	0.062	0.044	0.033	0.026	0.020	0.013	0.009	0.006	0.0035	0.0025	0.0013	0.0008	0.0005
250	10	0.074	0.054	0.040	0.031	0.025	0.016	0.012	0.007	0.004	0.003	0.0016	0.001	0.0006
300	12	0.087	0.063	0.048	0.036	0.030	0.019	0.014	0.008	0.005	0.0035	0.002	0.0012	0.0007
350	14	0.099	0.072	0.055	0.042	0.034	0.022	0.016	0.010	0.006	0.004	0.0025	0.0015	0.0008
400	16	0.110	0.082	0.061	0.047	0.038	0.025	0.018	0.011	0.007	0.005	0.003	0.0017	0.001
450	18	0.122	0.090	0.068	0.052	0.042	0.028	0.020	0.012	0.008	0.0055	0.0035	0.002	0.0012
500	20	0.132	0.098	0.075	0.058	0.046	0.031	0.022	0.013	0.009	0.006	0.004	0.0025	0.0015
600	24	0.150	0.113	0.087	0.068	0.054	0.037	0.026	0.015	0.010	0.007	0.0045	0.003	0.0017
750	30	0.171	0.132	0.102	0.081	0.065	0.045	0.031	0.019	0.012	0.009	0.005	0.0035	0.002
900	36	0.188	0.148	0.117	0.093	0.076	0.052	0.037	0.022	0.015	0.010	0.006	0.004	0.0025

*These factors are for a single concentrated wheel load centered over an effective pipe length of 3 ft.
†1 ft = 0.3 m.
Source: ANSI 21.1.

TABLE 21.4 Surface Load Factor—Two Trucks on Flexible Pavement*

Pipe size		Depth of cover, ft†												
DN	in	2	2½	3	3½	4	5	6	8	10	12	16	20	24
80	3	0.019	0.012	0.008	0.006	0.005	0.004	0.0035	0.003	0.0025	0.002	0.001	0.007	0.0005
100	4	0.032	0.022	0.016	0.012	0.009	0.006	0.005	0.004	0.0035	0.003	0.002	0.0013	0,0010
150	6	0.058	0.042	0.032	0.024	0.020	0.014	0.010	0.007	0.006	0.005	0.003	0.0019	0.0015
200	8	0.076	0.058	0.044	0.036	0.030	0.022	0.017	0.011	0.009	0.007	0.004	0.0027	0.0021
250	10	0.092	0.072	0.056	0.046	0.039	0.028	0.021	0.014	0.011	0.008	0.005	0.0033	0.0026
300	12	0.108	0.086	0.070	0.056	0.047	0.034	0.027	0.018	0.012	0.009	0.006	0.0039	0.0030
350	14	0.122	0.098	0.078	0.065	0.055	0.040	0.031	0.020	0.014	0.010	0.007	0.0046	0.0036
400	16	0.136	0.110	0.090	0.074	0.062	0.046	0.036	0.024	0.016	0.012	0.009	0.0060	0.0046
450	18	0.149	0.122	0.101	0.084	0.070	0.052	0.041	0.027	0.019	0.014	0.010	0.0066	0.0051
500	20	0.162	0.136	0.115	0.096	0.080	0.060	0.048	0.032	0.022	0.016	0.012	0.0079	0.0061
600	24	0.185	0.152	0.126	0.106	0.091	0.067	0.053	0.036	0.026	0.119	0.013	0.0086	0.0067
750	30	0.212	0.176	0.146	0.124	0.107	0.080	0.064	0.044	0.032	0.024	0.016	0.0106	0.0081
900	36	0.235	0.202	0.169	0.146	0.127	0.095	0.075	0.052	0.038	0.028	0.019	0.0125	0.0097

*These factors are for two trucks with 6-ft rear wheel spacing passing with inside rear wheels 3 ft apart. Effective pipe length is 3 ft coinciding with the distance between the adjacent inside wheels.

†1 ft = 0.3 m.

TABLE 21.5 Surface Load Factor—One or Two Trucks on Rigid Pavement*

Depth of cover, ft‡	Pavement thickness, in†							
	One truck				Two passing trucks			
	4	6	8	10	4	6	8	10
2	0.0244	0.0149	0.0101	0.0076	0.0410	0.0263	0.0186	0.0142
2½	0.0213	0.0139	0.0097	0.0072	0.0364	0.0246	0.0177	0.0136
3	0.0186	0.0126	0.0090	0.0070	0.0333	0.0228	0.1067	0.0129
3½	0.0164	0.0114	0.0085	0.0066	0.0290	0.0206	0.0156	0.0122
4	0.0144	0.0102	0.0079	0.0061	0.0262	0.0187	0.0146	0.0117
5	0.0114	0.0084	0.0066	0.0054	0.0210	0.0156	0.0123	0.0102
6	0.0093	0.0071	0.0057	0.0047	0.0170	0.0133	0.0107	0.0088
8	0.0065	0.0052	0.0043	0.0036	0.0114	0.0097	0.0081	0.0069
10	0.0046	0.0039	0.0033	0.0029	0.0080	0.0070	0.0062	0.0055
12	0.0034	0.0030	0.0026	0.0023	0.0059	0.0054	0.0049	0.0045
16	0.0022	0.0019	0.0017	0.0016	0.0034	0.0032	0.0030	0.0028
20	0.0013	0.0011	0.0010	0.0009	0.0024	0.0023	0.0022	0.0021
24	0.0008	0.0007	0.0006	0.0005	0.0015	0.0014	0.0013	0.0012

*These factors were computed by the methods explained in "Vertical Pressure on Culverts under Wheel Loads on Concrete Pavement Slabs," Portland Cement Assn., Chicago, IL, in Bulletin ST65.
†1 in = 25 mm.
‡1 ft = 0.3 m.
Source: ANSI 21.1.

TABLE 21.6 Reduction Factor

Pipe size, in	Depth of cover, ft*			
	2½–3½	4–7	8–10	>10
3–12	1.00	1.00	1.00	1.00
14	0.92	1.00	1.00	1.00
16	0.88	0.95	1.00	1.00
18	0.85	0.90	1.00	1.00
20	0.83	0.90	0.95	1.00
24–30	0.81	0.85	0.95	1.00
36–60	0.80	0.85	0.90	1.00

*1 ft = 0.3 m.
Source: ANSI 21.1.

TABLE 21.7 Truck Wheel Load

AASHTO truck	Gross weight, tons	Wheel load P, lb*
H-10	10	8,000
H-15	15	12,000
H-20	20	16,000

*1 lb = 2.2 kg.

TABLE 21.8 Truck Impact Factor

Depth of cover*	Impact factor F
Up to 1 ft 0 in	1.30
1 ft 1 in–2 ft 0 in	1 20
2 ft 0 in–2 ft 11 in	1.10
3 ft 0 in or more	1.00

*1 ft = 0.3 m.

Uniformly Distributed Load

When an additional load, such as fill or lumber, is stored at grade on top of the buried pipe, this is called a *uniformly distributed load.* This load is added to the earth load over the pipe. The formula for calculating the uniform load is

$$L_s = CBW \tag{21.5}$$

where L_s = superimposed load on pipe, lb/ft of pipe
C = coefficient, dimensionless
B = width of trench at top of pipe, ft
W = weight of superimposed load, lb/ft^2

Discussion:

1. The additional load L_s is limited in dimension when the length of pipe is used.
2. To find C, refer to Table 21.12 to determine the coefficient for vertical load on pipe installed in trench.

Table 21.12 is entered with the following values.

D is either length or width of material at grade, ft.

L is either length or width of material at grade, ft.

H is the height of backfill over top of pipe, ft, prior to placement of new load.

Bedding

General. Bedding is the contact between pipe and earth and is used when pipe is installed in a ditch or embankment condition, prior to the major backfilling operation. The type of bedding used has an important influence on the total load any pipe can support.

Bedding Methods

Rigid Pipe. The four most popular bedding methods for rigid piping are shown in Fig. 21.5. The load factors are indicated in each detail. The least desirable is class D, which is not recommended. The bell holes are dug out prior to placement of pipe except for class D.

The most common burial method is class B. Figure 21.6 indicates a detail enlarging on this method showing a more accurate detail of burial.

Flexible Pipe. The recommended installation method for flexible pipe in a trench is shown in Fig. 21.7.

TABLE 21.9 Pipe Outside Diameter

Nominal size		Inside	Nominal	Outside diameter, ft						
DN	in	area, ft^2	ID ft	Cast iron	Ductile iron	Plain concrete	Reinforced concrete	Clay	Corrugated steel, 1/8 in	Composite
250	10	0.545	0.84	0.89	0.93	1.04	—	1.04	0.92	0.94
300	12	0.785	1.00	1.06	1.10	1.29	1.33	1.25	1.08	1.13
350	14	1.18	1.18		1.28		—	—		1.33
375	15	1.227	1.25	1.32		1.56	1.62	1.54	1.33	1.44
400	16	1.53	1.33	—	1.46	—	—	—	—	1.52
450	18	1.767	1.50	—	1.62	1.87	1.92	1.86	1.58	1.68
500	20	2.26	1.70	—	1.82	—	—	—	—	1.87
525	21	2.405	1.75	—		2.20	2.20	2.14	1.83	1.98
600	24	3.142	2.00	—	2.12	2.62	2.62	2.39	2.08	2.27
675	27	3.976	2.25	—		2.92	2.92	2.77	2.33	2.55
750	30	4.909	2.50	—	2.66	3.20	3.20	3.08	2.58	2.84
825	33	5.940	2.75	—		3.50	3.50	3.31	2.83	3.14
900	36	7.069	3.00	—	3.19	3.79	3.79	3.66	3.08	3.39

1 in = 25 mm.

TABLE 21.10 Railroad Loads on Circular Pipe, lb/lin ft*

Pipe size		Height of fill H above top of pipe																
		0.3	0.6	0.9	1.2	1.5	1.8	2.1	2.4	2.7	3.3	3.6	4.2	4.8	5.4	6.6	8.2	9.0
in	DN	1	2	3	4	5	6	7	8	9	10	12	14	16	18	20	25	30
12	300	3,270	3,059	2,750	2,431	2,109	1,808	1,550	1,330	1,138	982	788	646	528	442	380	259	185
15	375	3,986	3,729	3,352	2,964	2,571	2,204	1,890	1,622	1,388	1,198	960	788	644	540	463	315	226
18	400	4,702	4,400	3,955	3,497	3,033	2,599	2,229	1,913	1,637	1,413	1,133	930	759	636	546	372	266
21	525	5,416	5,067	4,555	4,027	3,493	2,904	2,568	2,204	1886	1,627	1,305	1,071	874	733	629	428	307
24	600	6,132	5,738	5,158	4,560	3,955	3,390	2,908	2,495	2,135	1,842	1,478	1,212	990	830	712	485	348
27	650	6,849	6,408	5,760	5,093	4,417	3,786	3,247	2,786	2,384	2,058	1,650	1,354	1,106	927	796	542	388
30	750	7,562	7,075	6,360	5,623	4,877	4,180	3,586	3,077	2,633	2,272	1,822	1,495	1,221	1,024	879	598	428
33	825	8,279	7,746	6,963	6,156	5,339	4,576	3,925	3,368	2,882	2,487	1,995	1,637	1,336	1,120	962	655	469
36	900	8,995	8,416	7,565	6,689	5,801	4,972	4,265	3,660	3,132	2,702	2,167	1,778	1,452	1,217	1,045	711	510
42	950	10,420	9,754	8,768	7,752	6,724	5,763	4,943	4,422	3,630	3,132	2,512	2,061	1,683	1,411	1,211	824	591

*Cooper E72 design loading consisting of four 72.000-lb axles spaced 5 ft c/c. Locomotive load assumed uniformly distributed over an area 8 × 20 ft. Weight of track structure assumed to be 200 lb/lin ft. Impact included. Height of fill measured from top of pipe to bottom of ties. Interpolate for intermediate pipe sizes and/or fill heights.

1 lb/ft = 0.45 kg/m.

Source: *Concrete Pipe Design Manual.*

TABLE 21.11 Percentage of Wheel Loads Transmitted to Pipes*

Tabulated figures show percentage of wheel load applied to one linear foot of pipe.

Depth of backfill over top of pipe		Pipe size, in													
		6	8	10	12	15	18	21	24	27	30	33	36	39	42
		Pipe size, DN													
ft	m	150	200	250	300	375	400	525	600	650	750	825	900	975	1050
1	0.3	12.8	15.0	17.3	20.0	22.6	24.8	26.4	27.2	28.0	28.6	29.0	29.4	29.8	29.9
2	0.6	5.7	7.0	8.3	9.6	11.5	13.2	15.0	15.6	16.8	17.8	18.7	19.5	20.0	20.5
3	0.9	2.9	3.6	4.3	5.2	6.4	7.5	8.6	9.3	10.2	11.2	11.8	12.5	12.9	13.5
4	1.2	1.7	2.1	2.5	3.1	3.9	4.6	5.3	5.8	6.5	7.2	7.9	8.5	8.8	9.2
5	1.5	1.2	1.4	1.7	2.1	2.6	3.1	3.6	3.9	4.4	4.9	5.3	5.8	6.1	6.4
6	1.8	0.8	1.0	1.1	1.4	1.8	2.1	2.5	2.8	3.1	3.5	3.8	4.2	4.3	4.4
7	2.1	0.5	0.7	0.8	1.0	1.3	1.6	1.9	2.1	2.3	2.6	2.9	3.2	3.3	3.5
8	2.4	0.4	0.5	0.6	0.8	1.0	1.2	1.4	1.6	1.8	2.0	2.2	2.3	2.5	2.6

*These figures make no allowance for impact.
1ft = 0.3 m.
Source: *Clay Pipe Engineering Manual.*

TABLE 21.12 Vertical Load Coefficients

	L/2H													
D/H	0.1	0.2	0.3	0.4	0.5	0.6	0.7	0.8	0.9	1.0	1.2	1.5	2.0	5.0
0.1	0.019	0.037	0.053	0.067	0.079	0.089	0.097	0.103	0.108	0.112	0.117	0.121	0.124	0.128
0.2	0.037	0.072	0.103	0.131	0.155	0.174	0.189	0.202	0.211	0.219	0.229	0.238	0.244	0.248
0.3	0.053	0.103	0.149	0.190	0.224	0.252	0.274	0.292	0.306	0.318	0.333	0.345	0.355	0.360
0.4	0.067	0.131	0.190	0.241	0.284	0.320	0.349	0.373	0.391	0.405	0.425	0.440	0.454	0.460
0.5	0.079	0.155	0.224	0.284	0.336	0.379	0.414	0.441	0.463	0.481	0.505	0.525	0.540	0.548
0.6	0.089	0.174	0.252	0.320	0.379	0.428	0.467	0.499	0.524	0.544	0.572	0.596	0.613	0.624
0.7	0.097	0.189	0.274	0.349	0.414	0.467	0.511	0.546	0.584	0.597	0.628	0.650	0.674	0.688
0.8	0.103	0.202	0.292	0.373	0.441	0.499	0.546	0.584	0.615	0.639	0.674	0.703	0.725	0.740
0.9	0.108	0.211	0.306	0.391	0.463	0.524	0.574	0.615	0.647	0.673	0.711	0.742	0.766	0.784
1.0	0.112	0.219	0.318	0.405	0.481	0.544	0.597	0.639	0.673	0.701	0.740	0.774	0.800	0.816
1.2	0.117	0.229	0.333	0.425	0.505	0.572	0.628	0.674	0.711	0.740	0.783	0.820	0.849	0.868
1.5	0.121	0.238	0.345	0.440	0.525	0.596	0.650	0.703	0.742	0.774	0.820	0.861	0.894	0.916
2.0	0.124	0.244	0.355	0.454	0.540	0.613	0.674	0.725	0.766	0.800	0.849	0.894	0.930	0.956

Source: *Clay Pipe Engineering Manual.*

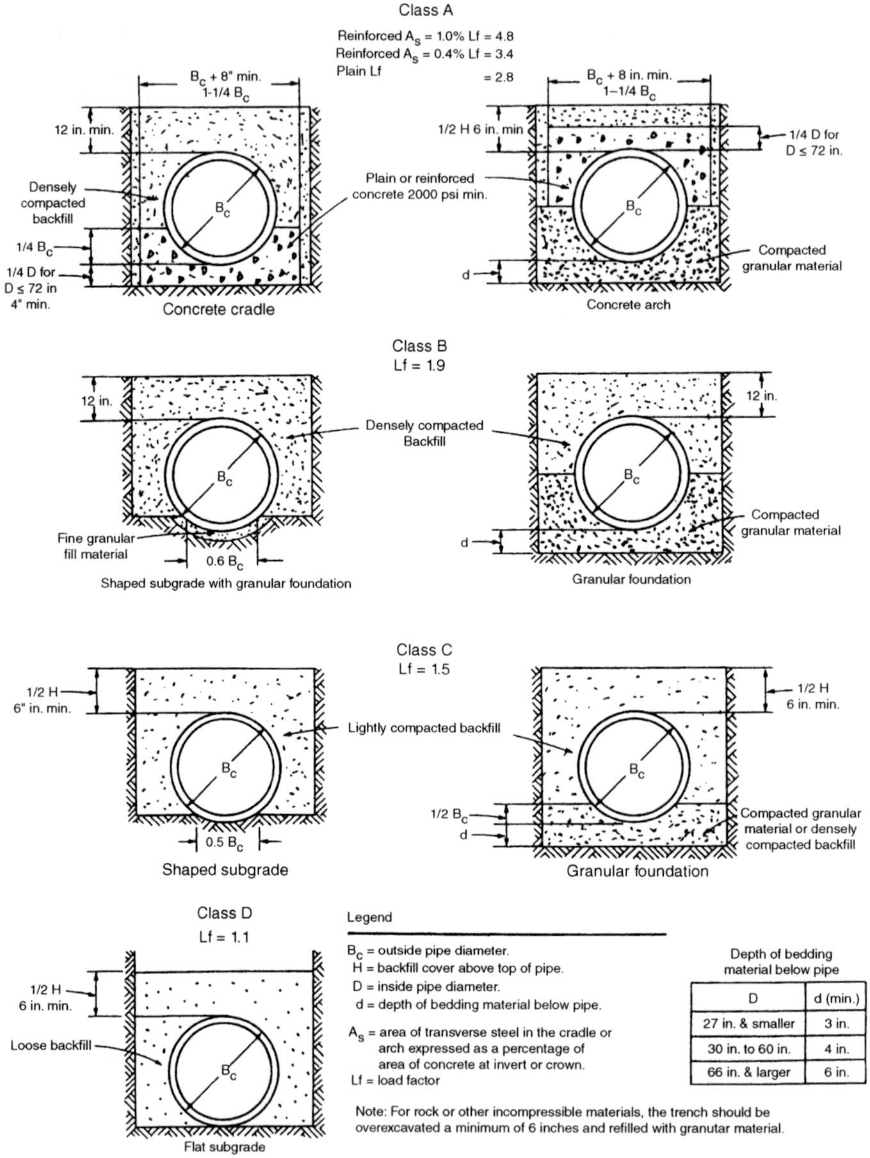

D	d (min.)
27 in. & smaller	3 in.
30 in. to 60 in.	4 in.
66 in. & larger	6 in.

FIGURE 21.5 Rigid pipe bedding methods. (*Courtesy: ASPE Data Book.*) *Note:* 1 in = 25 mm.

Pipe Bedding Material Classification

ASTM D 2321 presents a method of classifying soils and aggregates that are used as foundation, bedding, and backfill around pipes. Refer to Table 21.13 for the classification of embedment and backfill material. Table 21.14 gives the recommendations for use of various soil classifications.

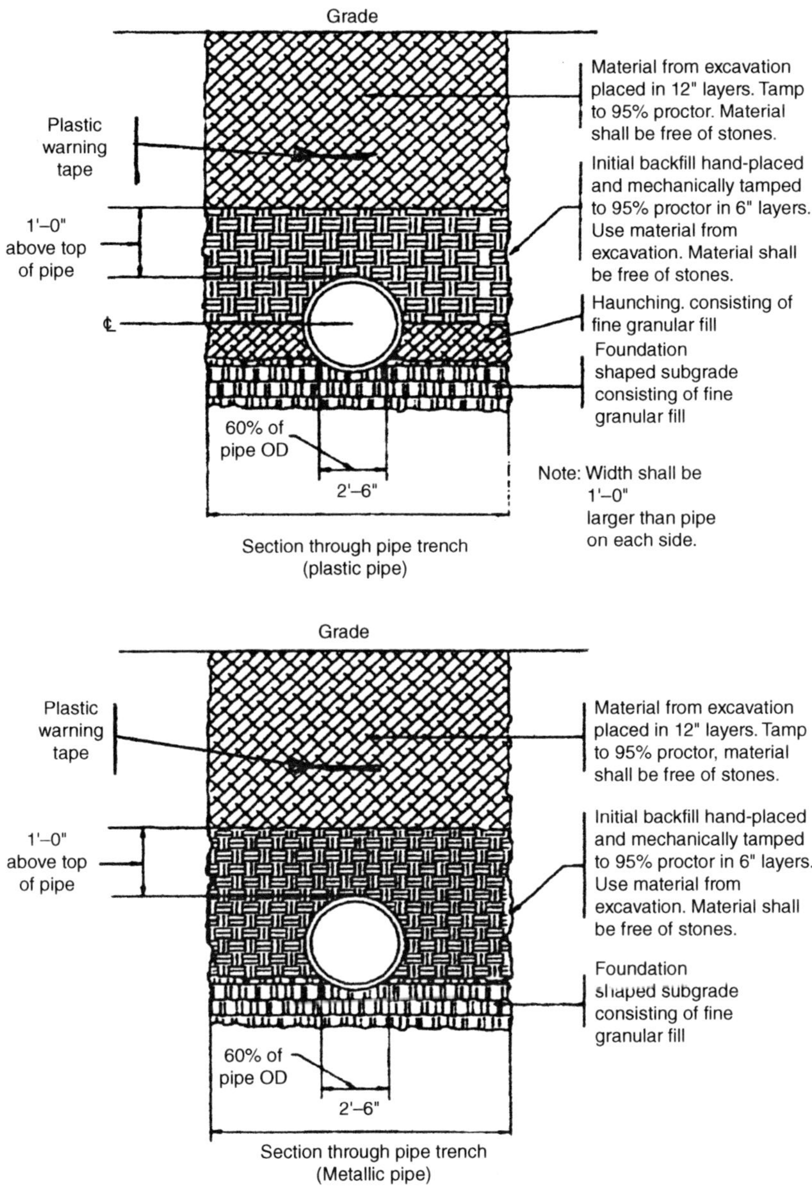

FIGURE 21.6 Complete installation of rigid pipe in trench.

Rigid Pipe Load Factor. The load factor for a bedding condition is used to determine the actual supporting strength of a pipe. As will be seen, the load factor increases the total load a pipe can support. The laboratory-calculated pipe strength multiplied by a load factor will give the *field supporting strength* of a pipe. These load factors have been determined experimentally for the various bedding methods and are indicated in Fig. 21.5

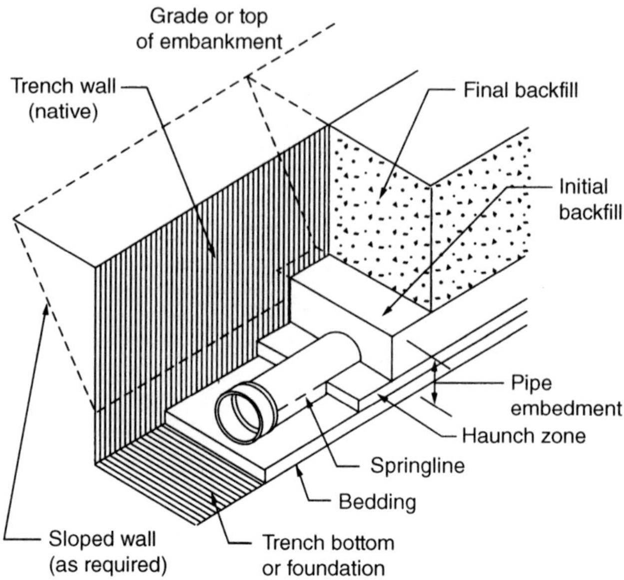

FIGURE 21.7 Flexible pipe bedding.

in each bedding diagram. However, they do not contain a safety factor. The result of the calculations will determine whether the field supporting strength of the pipe is great enough to resist the imposed load.

Flexible Pipe Load Factors

Pipe Deformation (Deflection)

1. *Radial deflection.* Radial deflection is a change in pipe diameter caused by an exterior load on a portion of the pipe. This will occur in buried pipe from a load on the top portion of the pipe if the support of the surrounding soil is not enough to resist the force.
2. *Longitudinal deformation.* Often referred to as *sagging*, longitudinal deformation is a change along the length of a pipe that causes the pipe to bend.
3. *Deformation at structures.* Not limited to flexible pipe, this type of deformation is caused by uneven settlement at wall penetrations and manholes.

The structural behavior of flexible pipe is not the same as that for rigid pipe. Flexible pipe does not have the structural strength and stiffness to resist traffic and backfill loads without support from surrounding embedment. Being flexible, the pipe deforms (or deflects) under load. This generally takes the form of radial elongation. This elongation structurally stresses the surrounding embedment material. The resistance to this stress is required for long-term support for the pipe. To achieve long-term resistance, proper placement of acceptable material around the pipe is required.

Strength of Materials

Rigid Pipe. All rigid pipe has been tested for strength in the laboratory by means of a three-edge-bearing test. To simplify selection of some piping, a method was developed to

TABLE 21.13 Classification of Soils for Embedment and Backfill

Class	Type	Soil group symbol D2487	Description	Percentage passing sieve sizes			Atterberg limits		Coefficients	
				1½ in (40 mm)	No. 4 (4.75 mm)	No. 200 (0.075 mm)	LL	PI	Uniformity C_u	Curvature C_u
IA	Manufactured aggregates: open-graded, clean	None	Angular, crushed stone or rock, crushed gravel, broken coral, crushed slag, cinders or shells; large void content, contain little or no fines.	100	< 10	< 5	Nonplastic			
IB	Manufactured, processed aggregates; dense-graded, clean	None	Angular, crushed stone (or other Class 1A materials) and stone/sand mixtures with gradations selected to minimize migration of adjacent soils; contain little or no fines (see X1.8).	100	< 50	< 5	Nonplastic			
II	Coarse-grained soils, clean	GW	Well-graded gravels and gravel-sand mixtures; little or no fines.	100	< 50 of "coarse fraction"	< 5	Nonplastic		> 4	1 to 3
		GP	Poorly-graded gravels and gravel-sand mixtures; little or no fines.		> 50 of "coarse fraction"				< 4	< 1 or > 3
		SW	Well-graded sands and gravelly sands; little or no fines.						> 6	1 to 3
		SP	Poorly graded sands and gravelly sands; little or no fines.						< 6	< 1 or > 3
	Coarse-grained soils, borderline clean to w/fines	e.g., GW-GC, SP-SM	Sands and gravels which are borderline between clean and with fines.	100	Varies	5 to 12	Nonplastic		Same as for GW, GP, SW, and SP	

(Continued)

TABLE 21.13 Classification of Soils for Embedment and Backfill (*Continued*)

Class	Type	Soil group symbol D2487	Description	Percentage passing sieve sizes			Atterberg limits		Coefficients	
				1½ in (40 mm)	No. 4 (4.75 mm)	No. 200 (0.075)	LL	PI	Uniformity C_u	Curvature C_c
III	Coarse-grained soils with fines	GM	Silty gravels, gravel-sand-silt mixtures.	100	< 50 of "coarse fraction"	12 to 50		< 4 or < A line		
		GC	Clayey gravels, gravel-sand-clay mixtures.					< 7 and > A line		
		SM	Silty sands, sand-silt mixtures.		> 50 of "coarse fraction"			> 4 or < A line		
		SC	Clayey sands, sand-clay mixtures.					> 7 and > A line		
IVA*	Fine-grained soils (inorganic)	ML	Inorganic silts and very fine sands, rock flour, silty or clayey fine sands, silts with slight plasticity.	100	100	> 50	< 50	< 4 or < A line		
		CL	Inorganic clays of low to medium plasticity, gravely clays, sandy clays, silty clays, lean clays.					> 7 and > A line		
IVB	Fine-grained soils (inorganic)	MH	Inorganic silts, micaceous or diatomaceous fine sandy or silty soils, elastic silts.	100	100	> 50	> 50	< A line		
		CH	Inorganic clays of high plasticity, fat clays.					> A line		
V	Organic soils	OL	Organic silts and organic silty clays of low plasticity.	100	100	> 50	< 50	< 4 or < A line		
		OH	Organic clays of medium to high plasticity, organic silts.				> 50	< A line		
	Highly organic	PT	Peat and other high organic soils.							

*Includes Test Method D 2487 borderline classifications and dual symbols depending on plasticity index and liquid limits.
Note: "Coarse fraction" as used in this table is defined as material retained on a No. 200 sieve.
Source: Copyright ASTM. Reprinted with permission.

TABLE 21.14 Recommendations for Installation and Use of Soils and Aggregates for Foundation, Backfill, and Embedment

Use	Soil class (see Table 21.13)*				
	Class IA	Class IB	Class II	Class III	Class IV-A
General recommendations and restrictions	Do not use where conditions may cause migration of fines from adjacent soil and loss of pipe support. Suitable for use as a drainage blanket and underdrain in rock cuts where adjacent material is suitably graded.	Process materials as required to obtain gradation which will minimize migration of adjacent materials. Suitable for use as drainage blanket and underdrain.	Where hydraulic gradient exists check gradation to minimize migration. "Clean" groups suitable for use as drainage blanket and underdrain.	Do not use where water conditions in trench may cause instability.	Obtain geotechnical evaluation of proposed material. May not be suitable under high earth fills, surface-applied wheel loads, and under heavy vibratory compactors and tampers. Do not use where water conditions in trench may cause instability.
Foundation	Suitable as foundation and for replacing over-excavated and unstable trench bottom as restricted above. Install and compact in 6-in maximum layers.	Suitable as foundation and for replacing over-excavated and unstable trench bottom. Install and compact in 6-in maximum layers.	Suitable as a foundation and for replacing overexcavated and unstable trench bottom as restricted above. Install and compact in 6-in maximum layers.	Suitable as foundation and for replacing overexcavated trench bottom as restricted above. Do not use in thicknesses greater than 12 in total. Install and compact in 6-in maximum layers.	Suitable only in undisturbed condition and where trench is dry. Remove all loose material and provide firm, uniform trench bottom before bedding is placed.
Bedding	Suitable as restricted above. Install in 6-in maximum layers. Level final grade by hand. Minimum depth 4 in (6 in for rock cuts).	Install and compact in 6-in maximum layers. Level final grade by hand. Minimum depth 4 in (6 in for rock cuts).	Suitable as restricted above. Install and compact in 6-in maximum layers. Level final grade by hand. Minimum depth 4 in (6 in for rock cuts).	Suitable only in dry trench conditions. Install and compact in 6-in maximum layers. Level final grade by hand. Minimum depth 4 in (6 in for rock cuts).	Suitable only in dry trench conditions and when optimum placement and compaction control is maintained. Install and compact in 6-in maximum layers. Level final grade by hand. Minimum depth 4 in (6 in for rock cuts).

(Continued)

TABLE 21.14 Recommendations for Installation and Use of Soils and Aggregates for Foundation, Backfill, and Embedment (*Continued*)

	Soil class (see Table 21.13)*				
Use	Class IA	Class IB	Class II	Class III	Class IV-A
Haunching	Suitable as restricted above. Install in 6-in maximum layers. Work in around pipe by hand to provide uniform support.	Install and compact in 6-in maximum layers. Work in around pipe by hand to provide uniform support.	Suitable as restricted above. Install and compact in 6-in maximum layers. Work in around pipe by hand to provide uniform support.	Suitable as restricted above. Install and compact in 6-in maximum layers. Work in around pipe by hand to provide uniform support.	Suitable only in dry trench conditions and when optimum placement and compaction control is maintained. Install and compact in 6-in maximum layers. Work in around pipe by hand to provide uniform support.
Initial backfill	Suitable as restricted above. Install to a minimum of 6 in above pipe crown.	Install and compact to a minimum of 6 in above pipe crown.	Suitable as restricted above. Install and compact to a minimum of 6 in above pipe crown.	Suitable as restricted above. Install and compact to a minimum of 6 in above pipe crown.	Suitable as restricted above. Install and compact to a minimum of 6 in above pipe crown.
Embedment compaction†	Place and work by hand to ensure all excavated voids and haunch areas are filled. For high densities use vibratory compactors.	Minimum density 85% Std. Proctor.‡ Use hand tampers or vibratory compactors.	Minimum density 85% Std. Proctor.‡ Use hand tampers or vibratory compactors.	Minimum density 90% Std. Proctor.‡ Use hand tampers or vibratory compactors. Maintain moisture content near optimum to minimize compactive effort.	Minimum density 95% Std. Proctor.‡ Use hand tampers or impact tampers. Maintain moisture content near optimum to minimize compactive effort.
Final backfill	Compact as required by the engineer.	Compact as required by the engineer.	Compact as required by the engineer.	Compact as required by the engineer.	Suitable as restricted above. Compact as required by the engineer.

*Class IV-B (MH-CH) and Class V (OL, OH, PT) materials are unsuitable as embedment. They may be used as final backfill as permitted by the engineer.

†When using mechanical compactors, avoid contact with pipe. When compacting over pipe crown, maintain a minimum of 6-in cover when using small mechanical compactors. When using larger compactors, maintain minimum clearances as required by the engineer.

‡The minimum densities given in the table are intended as the compaction requirements for obtaining satisfactory embedment stiffness in most installation conditions.

eliminate any difference due to pipe size. Therefore, a D load has been developed by dividing the actual laboratory strength by the pipe size, in feet. In cases where different sizes of pipe have different strengths, this is not possible. Table 21.15 gives the laboratory design loads for various pipe material.

Flexible Pipe. Flexible pipe is defined as any pipe where internal and external forces have to be evaluated, such as plastic or corrugated steel pipe. The failure of this type of piping is considered to occur when the total calculated load on the pipe produces a radial deflection of 5 percent or more of the pipe diameter. Refer to Table 21.16 for the maximum allowable deflection of various size pipes.

Total Load on Buried Pipe

To calculate the actual load used to select a rigid pipe, the following formula should be used. The result of this calculation is the actual number that will be compared to the actual load for the pipe (Table 21.15).
For pipe using the D load:

$$L_d \times F = \frac{Lt}{D} \times \text{SF} \tag{21.6}$$

For pipe using the ordinary load:

$$L_a \times F = L_t \times \text{SF} \tag{21.7}$$

where L_d = maximum allowable D load on pipe, lb/ft (from Table 21.15)
L_a = maximum allowable load on all other rigid pipe, lb/ft (from Table 21.15)
L_t = total calculated combined load (earth, superimposed, railroad, if any), lb/ft
D = outside diameter of pipe, ft (from Table 21.9)
F = load factor (from Fig. 21.5)
SF = safety factor

Discussion:

1. Here L_a or L_d is the allowable load for all types of rigid pipe.
2. Here L_t is the actual total load on pipe after determining the earth load, superimposed load, and/or railroad load.
3. Here F is the load factor determined from the pipe bedding method (Fig. 21.5).
4. SF is the safety factor. A practice normally followed is to allow for the possibility of all extreme conditions occurring simultaneously, thereby producing maximum loading on the buried pipe. Therefore, a safety factor is added to the calculated loads on the buried pipe. This factor ranges between 1.25 and 1.50. How closely and competently the excavation and backfill specification is both written and supervised will determine the number selected. When using reinforced concrete pipe, with the 0.01 crack as the failure criterion, no safety factor is required. This is the only exception. An SF of 1.5 is to be used unless close supervision of the job will be maintained and a very descriptive and tight specification is written. If such is the case, a 1.25 SF can be used. In addition, a 1.25 SF should be used for cast iron pipe in all conditions.
5. The allowable load (or field supporting strength) must be greater than the actual load so that the pipe will not fail under design conditions.

TABLE 21.15 Pipe Design Loads, lb

Pipe size								D loads		
								Reinforced concrete pipe		
DN	in	Cast iron	Clay pipe, regular strength	Clay pipe, extra strength	Concrete pipe, standard strength	Concrete pipe, extra strength	Composite pipe, strength class	Strength class	0.01 Crack	Ultimate
250	10	2265	1600	2400	1400	2000	1500	I	800	1200
300	12	2231	1800	2600	1500	2250	2400	II	1000	1500
375	15	2302	2000	2900	1750	2750	3300	III	1350	2000
450	18		2200	3300	2000	3300	4000	IV	2500	3000
525	21		2400	3850	2200	3850	5000	V	3000	3750
600	25		2800	4700	2400	4000				
700	27		2800	4700						
750	30		3300	5000						
900	36		4000	6000						

1 lb = 2.2 kg.

TABLE 21.16 Maximum Allowable Deflection for Pipe

Pipe size DN	Size, in	Maximum deflection, in
300	12	0.60
375	15	0.75
450	18	0.90
525	21	1.05
600	24	1.20
700	27	1.35
750	30	1.50
800	33	1.65
900	36	1.80

Deflection

A formula has been developed to calculate the deflection under the total calculated load transmitted to a buried flexible pipe. It is called the *modified Iowa formula.*

$$\mathrm{DEF} = \frac{\mathrm{LF} \times KL_T R^3}{EI + 0.061SR^3} \tag{21.8}$$

where DEF = deflection of pipe, in; refer to Table 21.16
LF = load factor
K = bedding constant
L_T = total load transmitted to pipe, lb/in
R = mean radius of pipe, in
E = modulus of elasticity, dimensionless
I = moment of inertia, dimensionless
S = modulus soil reaction, psi

Discussion:

1. LF is the load factor, which allows for continued deflection of pipe after the total load has been developed. It is an empirical number, with a recommended value of 1.25 for good backfill (85 percent density) or 1.50 for excellent backfill (95 percent density).
2. Here K is a factor depending on the width of bedding. An average value of 0.1 is recommended.
3. Here L_T is the total load (earth, superimposed, railroad), lb/in of pipe length.
4. Here R is the radius of pipe, in, measured from the center of the corrugation. For this formula, one-half of the nominal size should be used.
5. To determine I, refer to Table 21.17 for the moment of inertia of corrugated pipe, based on the gauge of steel selected. For plastic pipe, individual manufacturers should be consulted because of the great diversity of materials available.
6. Here E is 10^6 for all pipe sizes.
7. Here S is soil reaction modulus, in psi. S is dependent on the degree of compaction of backfill around the pipe. This number has not been fully correlated for backfilled soil. Estimates by the Bureau of Public Roads are 700 for good backfill at 85 percent

TABLE 21.17 Moment of Inertia for Corrugated Steel Pipe

Corrugation pitch and depth	16 gauge	14 gauge	12 gauge
2 × ½	0.00194	0.00246	0.00354
2 2/3 × ½	0.00189	0.00239	0.00343

compaction and 1400 for excellent backfill at 95 percent compaction. The value established for good or excellent backfill should be used in Eq. (21.8).

In certain cases, other structural failures for corrugated steel pipe may have to be considered during design. Seam, buckling, handling, and installation strengths and conduit wall compression strength may have to be determined. Refer to *Handbook of Steel Drainage and Highway Construction Products,* published by the American Iron and Steel Institute, for a description of the method of calculating all the above design parameters, if they are required.

BIBLIOGRAPHY

Cast Iron Soil Pipe Engineering Manual, Cast Iron Soil Pipe Institute.
Handbook of Steel Drainage and Highway Construction Products, American Iron and Steel Institute.
Handbook for Ductile Iron Pipe, Cast Iron Pipe, Cast Iron Pipe Research Association.
Moser, A. P., *Buried Pipe Design*, McGraw-Hill, New York.

CHAPTER 22
THE FREEZING OF WATER

GENERAL

The design engineer often must work outside the building on the site where temperatures are below freezing. This chapter explains the theory and methods to prevent freezing of water and wastewater in exterior pipes and tanks.

CODES AND STANDARDS

1. NFPA 22, Standard for Water Tanks for Private Fire Protection

FREEZING OF WATER

Much time, effort, and money must be spent to restore service when pipes and mains freeze. In addition, many times these efforts must be made under very difficult working conditions and with great urgency. Because of this, it is preferable to design systems that will not freeze.

Studies at the U.S. Army Cold Region Research and Engineering Laboratory (CRREL) have shown that the dynamics of the freezing process of water in pipes are much more complicated than originally thought, and that this process is different for both static and flowing water. It was found that when water starts to freeze, flow in a piping system can become blocked much earlier than previously believed. It was also established that the actual mechanics of the freezing process are less predictable in terms of both time and heat loss than was also previously believed.

Prior to these tests, it was believed that when ice formed inside a pipe or vessel, it started on the pipe wall and grew uniformly inward until the entire pipe was blocked. It was also thought that the only apparent difference between the freezing that occurred in static and flowing water was a difference in the rate due to the heat created by the flowing water. This was not correct.

Mechanics of the Freezing Process

Static Water. When the temperature outside a pipe falls below freezing, before any ice can form, the water inside must be cooled below the freezing point. This is called *supercooling*, and in the CRREL studies the temperature fell as low as 29°F (–1.6°C) before ice started to form. The initial ice formation is called *nucleation*. As the process continues, the ice takes the shape of thin feathery crystals interspersed with water, similar in shape to a Christmas tree. This new formation is called *dendritic ice*. As the dendritic crystals rapidly grow larger, they release latent heat of fusion due to the change of state of the water. Since the surrounding soil or air cannot absorb all this heat, the temperature of the water is then raised back to the 32°F (0°C) level. This brings the dendritic phase to an end. Only after all the heat is absorbed by the surrounding medium does the annular growth of ice actually start. The dendrils become more and more dense and are gradually incorporated into the growing annulus. Eventually, the annulus formed by the ice increases in size, becomes solid, and occupies the entire cross section of pipe. The pipe is then blocked. The pipe and its contents will continue to cool until the ambient temperature is reached.

Flowing Water. Prior to the CRREL studies, it was believed that flowing water began to freeze in a pipe with the formation of annular ice along the inside of the pipe perimeter, assuming a tapered cross section and having the smaller end at the downstream side. Blockage was thought to occur when the smaller cross section thickened and eventually closed, filling the pipe at that point. As the ambient temperature dropped, the remainder of the water in the pipe froze until the entire pipe became filled.

In actuality, it was found that as the ambient temperature fell, an ice/water mixture was produced in the shape of rippled ice surfaces that moved along with the flow of water. This mixture was found to be in the shape of a taper in cross section with the smaller end pointed downstream. At regular intervals there was sudden enlargement. A further lowering of the ambient temperature did not lead to a thickening of the ice, but rather, to a progressively closer spacing of the narrow ice bands. The friction factor of the flowing water was found to be greatly affected by this type of ice formation. Since the friction head loss of pressure consists primarily of wall drag and the nozzle losses that occur at each band, the hydraulics become very unstable. Any further decrease in the discharge rate of the system or an increase in the friction loss due to more ice will quickly create a condition where the total head requirements for flow are greater than the source can provide. It is at this point that all flow would stop and all the remaining liquid water would start to freeze immediately.

Pipe Damage Due to Freezing

Water falls into the category of substances that have a temperature of maximum density. Above that temperature it has a positive coefficient of expansion, which means that water will expand upon heating. However, if water is cooled below that temperature, it also has a positive coefficient of expansion since it will expand when cooled. The relationship between temperature and volume is shown in Table 22.1. Because of this expansion, water (or ice) will exert considerable pressure when confined within any enclosed space. The relationship between temperature and pressure is shown in Table 22.2.

A mathematical analysis of the actual stress produced by the freezing of water in an enclosed space has shown that the circumferential stress exceeds the longitudinal stress in all cases. What this means in practical terms is that any pipe can be expected to fail along its length in the absence of a weak joint or a flaw in the piping material. In tests conducted at the National Bureau of Standards, it was observed that in all cases of failure due to freezing, liquid water was released from the pipe at the point of failure.

Applying these facts to a piping system reveals that differences in heat transfer, caused by variations present in all piping systems, will result in the freezing of pipe in some locations but not in others along the same run. This will cause a rupture in one or more of the unfrozen sections of the piping network that are the last to freeze.

TABLE 22.1 Relationship between Water Temperature and Volume

Temperature		Specific volume, mL/g	
°C	°F	Water	Ice
−10	14	1.002069	
−5	23	1.000825	
0	32	1.0001324	1.0908
3	37.4	1.0000078	
4	39.2	1.0000000	
5	41	1.0000081	
10	50	1.0002720	

Source: Data of J. F. Mohler, at 1-atm pressure.

TABLE 22.2 Relationship between Temperature and Pressure for Water

Freezing point of water		Pressure		
°C	°F	atm	psi	kPa
0	32	1	0	0
−5	23	590	8,660	58,000
−10	14	1,090	16,000	107,200
−22	−7.6	2,047	30,000	201,000

Freezing of Water in Atmospheric Vessels

General. The most common method used to store water on a site is in an on-ground or elevated water storage tank made of either wood or steel. If enough heat cannot be added by means of new water flowing into the tank to keep the water from freezing, heat must be added by some other means. In addition to the tank itself, the riser from grade up to the bottom of an elevated tank must be protected. It is generally thought that adequate heating of any tank is almost as important as proper structural design.

Mechanics of Ice Formation in Tank. Water in an open tank or vessel loses heat on all the sides exposed to the atmosphere. For a tank on grade, the loss of heat into the surrounding soil is not as great due to the insulating factor of that soil. However, the greatest amount of heat is lost through the surface of the water. For water in a still tank, as the water cools, internal circulation takes place because the surface layer of water cools faster than the deeper water. This causes the surface layer to become denser, resulting in that water migrating to the bottom of the tank, causing the warmer water to migrate to the top. This circulation will continue to take place until the temperature of most of the water reaches about 32°F (0°C). When this occurs, the water is at its maximum density. At this point, an inversion takes place. The coldest water, at a lesser density, now rises to the surface. Continued cooling of the surface water further reduces its density, and the temperature rapidly falls to the point of freezing.

Ice generally appears first at the sides of the tank, then quickly forms a continuous layer over the surface. This same progression is true for similar installations such as pools and lagoons. A formula for the determination of ice formation on the surface of an open vessel has been developed. It considers the insulating effect that the forming layer of ice will have on the transfer of heat from the surface of the water to the air. One assumption is that the time is constant. The following formula calculates the thickness of ice formation in any period of time.

$$X = \frac{25I}{32 - T} \tag{22.1}$$

where X = time, h
I = ice thickness, in [A generally accepted figure of 1/16 in. (2 mm) is required to stop operation of float valves, etc.]
T = design air temperature, °F

A design air temperature value based on the mean of the high and low readings is considered adequate for design of tanks storing potable water. For tanks used to store fire protection water only, it would be advisable to use the lowest 1-day mean temperature. The lowest 1-day mean temperature can be found in Fig. 22.1.

Selecting the Heating System Type. Selecting the most economical type of heating system depends on the tank height, amount of heat required, and availability and cost of any particular fuel or heating medium. The basic methods or devices are

1. Direct discharge of steam into the water
2. Steam coils inside the tank
3. Hot water coils inside the tank
4. Electric immersion heating elements inside the tank

The direct discharge of steam into a tank is the method used most often for nonpotable water where steam is available. A steam supply line of adequate size is piped directly into the tank. This line ends in a tee placed about one-third of the height of the tank from the bottom.

When there is a potable water supply in combination with fire reserve water in a tank, the possibility of cross-contamination from an outside source must be minimized to the greatest extent possible. As a result, the following three methods should be considered.

The use of *steam* in coils at the bottom of a tank is mostly limited to tanks that have a flat bottom, are not elevated to a great height, and do not need too much heat transferred. Based on past experience, this method is not considered very reliable because a large number of problems have been reported.

Gravity circulation of hot water is a common solution if steam is not practical. This requires a heat exchanger or a hot water generator to be placed in close proximity to the tank, usually under the tank in the valve pit, if provided, or in a separate building adjacent to the tank. Cold water is taken from the discharge pipe and run through the heater. Since the now heated water is lighter than the colder supply, a natural circulation is obtained. Long runs of heater piping are not practical due to the heat loss from the piping and the expense of insulation needed to keep the heat from being lost. In addition, the hot water generator takes space in what might be a small valve pit under the tank.

The use of *electric immersion heaters* is practical only when the cost of the electricity is low. An advantage is the lowest initial cost of the heating methods discussed when used for potable water.

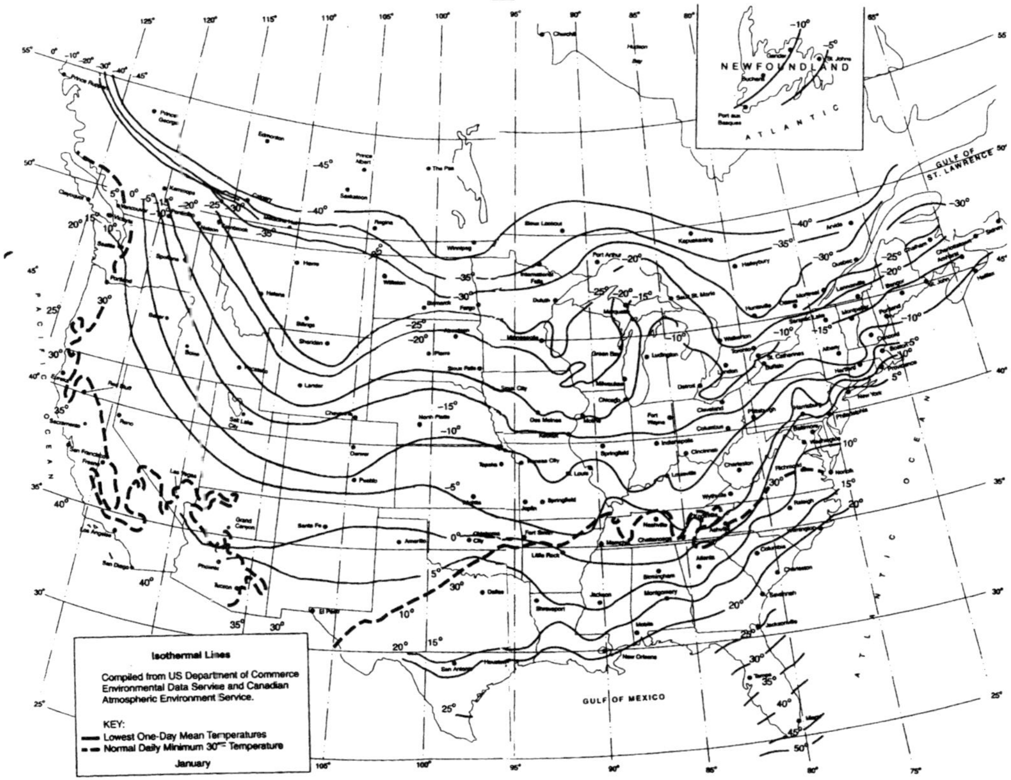

Isothermal lines for lowest one-day mean temperatures and normal daily minimum 30°F (−1°C) temperature line for January, United States and southern Canada.

FIGURE 22.1 Lowest 1-cay mean temperature. (NFPA)

The more warm water that is available to cycle inside a tank, the less ice buildup that will occur. It is strongly suggested by many authorities that a ladder inside a tank not be provided in an area that might be prone to freezing conditions, because of any ice that may exert pressure on the connections from the ladder to the tank.

Calculating Heat Loss from Elevated Tanks. A method to determine the amount of heat lost from water tanks has been developed by the National Fire Protection Association (NFPA) and has become the standard for design. It is very conservative in its approach. The basic consideration is not to allow the temperature of the water to fall below 42°F (5.6°C). This is a safe temperature above freezing that makes an allowance for tolerances in various equipment and for variations in weather design data.

The procedures required to calculate the amount of heat necessary to replace the heat that was lost to the atmosphere are as follows:

1. Determine the lowest 1-day mean temperature at the proposed site. Refer to Fig. 22.1 for this information.
2. Determine the number of gallons the tank will hold. Use the nominal, not actual, capacity.
3. Using the lowest temperature and the capacity from steps 1 and 2, refer to either Table 22.3 (for elevated wood or steel tanks or wood tanks on grade) or Table 22.4 (for steel tanks on grade) for the heat loss in thousands of Btu/per hour.
4. Refer to Table 22.5 to find the heat loss in Btu/per hour from the riser for each foot of length from grade to the bottom of the tank.
5. Add the two figures obtained in steps 3 and 4 to calculate the total heat loss for the entire tank. This figure is the capacity in thousands of Btu/per hour that the heating system must be capable of providing to the tank. Add a 10 percent safety factor to allow for unexpected variations in temperature.

SEWER AND WATER SUPPLY PIPING DESIGN

Designing a complex piping system to prevent freezing is a very involved procedure that requires a thermal analysis of the piping network. In current practice, precise analysis is not necessary and therefore is not done. The empirical methods discussed have been verified under actual field conditions and are considered sufficiently accurate for engineering design calculations. Although the following discussions are for a flowing sewage system, the same calculations can be made for flowing clear water piping with similar hydraulic characteristics.

The following assumptions have been made to simplify the solution of the heat-transfer process:

1. The thermal characteristics of sewage are essentially the same as those of clear water.
2. The sewer is assumed to be flowing full and at a minimum velocity of 2 ft/s (0.6 m/s).
3. There are critical flow periods that will last about 6 to 8 h, during which the flow will be approximately 25 percent of the 24-h average flow, and a 1- to 2-h period when the flow is about 5 percent of the 24-h flow rate.
4. The losses through manholes, catch basins, cleanouts, and the like are the same as those through the pipe run itself.
5. Heat is transferred mainly through conduction.

TABLE 22.3 Heat Loss from Typical Elevated Wood or Steel Tanks or Wood Tank on Grade
Thousands of British thermal units lost per hour when the temperature of the coldest water is 42°F (4°C)

Temperature		Wooden tanks—capacities, thousands of gallons								
°F	°C	10	15	20	25	30	40	50	75	100
35	2	8	10	11	13	14	19	21	28	33
30	–1	11	14	16	19	21	27	31	40	49
25	–4	15	20	21	25	28	36	42	54	65
20	–7	19	25	27	32	35	46	54	69	83
15	–10	24	31	34	39	44	57	66	85	102
10	–13	28	36	40	46	51	68	78	100	121
5	–15	33	43	47	54	60	78	92	117	142
0	–18	38	49	53	62	69	90	106	135	164
–5	–20	43	56	61	71	79	103	120	154	187
–10	–23	49	63	69	80	89	116	136	174	211
–15	–26	54	71	77	89	100	130	153	195	236
–20	–29	61	79	86	99	111	145	169	217	262
–25	–33	68	87	95	110	123	160	188	240	291
–30	–34	74	96	104	121	135	176	206	264	319
–35	–37	18	105	115	133	148	193	226	289	350
–40	–40	88	114	125	144	162	210	246	317	382
–50	–45	104	135	147	170	190	246	290	372	450
–60	–51	122	157	171	197	222	266	307	407	490

Temperature		Steel tanks—capacities, thousands of gallons							
°F	°C	30	40	50	75	100	150	200	250
35	2	43	51	59	77	92	120	145	168
30	–1	62	72	83	110	132	171	207	242
25	–4	82	96	111	146	175	228	275	323
20	–7	103	120	139	183	220	287	346	405
15	–10	145	146	169	222	267	267	347	419
10	–13	147	172	200	263	316	411	496	582
5	–15	171	200	233	306	367	478	577	676
0	–18	197	231	268	352	423	551	664	779
5	–20	224	262	304	400	480	626	755	884
–10	–23	253	296	344	452	543	707	853	1000
–15	–26	283	331	384	506	607	790	954	1118
–20	–29	314	368	427	562	674	878	1059	1241
–25	–33	348	407	473	622	747	972	1173	1375
–30	–34	382	447	519	683	820	1068	1288	1510
–35	–37	419	490	569	749	900	1171	1413	1656
–40	–40	456	534	620	816	979	1275	1538	1803
–50	–45	538	629	731	962	1154	1503	1814	2126
–60	–51	624	730	848	1116	1340	1745	2105	2467

TABLE 22.4 Heat Loss from Typical Steel Tank on Grade

Temperature		Tank capacities, thousands of gallons							
°F	°C	100	150	200	300	400	500	750	1000
35	2	85	114	135	175	206	238	312	380
30	–1	121	162	193	248	294	340	445	543
25	–4	161	216	257	330	393	453	594	722
20	–7	202	271	323	414	493	568	745	907
15	–10	245	329	391	502	537	689	904	1099
10	–13	290	389	463	595	707	816	1071	1302
5	–15	337	452	539	691	822	949	1244	1514
0	–18	388	521	620	796	947	1093	1434	1744
–5	–20	441	592	705	905	1076	1241	1628	1981
–10	–23	498	669	797	1023	1216	1403	1841	2239
–15	–26	557	748	891	1143	1360	1569	2058	2503
–20	–29	619	830	989	1270	1510	1742	2286	2781
–25	–33	685	920	1096	1406	1673	1930	2532	3080
–30	–34	752	1010	1203	1545	1837	2119	2781	3383
–35	–37	825	1108	1320	1694	2015	2325	3050	3710
–40	–40	898	1206	1437	1844	2193	2531	3320	4039
–50	–45	1059	1422	1694	2175	2586	2984	3915	4762
–60	51	1229	1651	1966	2524	3002	3463	4544	5528

TABLE 22.5 Heat Loss in Riser of Elevated Storage Tanks
Btu loss per hour through 4-ft-diameter (3.3-m) riser per foot-length

Atmospheric temperature, °F	°C	Btu/h
35	2	69
30	–1	192
25	–4	340
20	–7	506
15	–10	692
10	–13	893
5	–15	1092
0	–18	1309
–5	–20	1536
–10	–23	1771
–15	–26	2020
–20	–29	2291
–25	–33	2568

A basic equation of the steady-state heat flow by conduction is used to calculate the heat loss from sewers. It is based on the Fourier law and is

$$Q = \frac{K}{X} \times A(T_1 - T_2) \tag{22.2}$$

where Q = heat loss from pipe, Btu/h/ft
K = mean thermal conductivity, Btu/ft^2/h/°F
T_1 = temperature outside of pipe, °F
T_2 = ambient temperature at a distance from pipe that would not be affected by the heat of the pipe (This would be either the ground temperature for buried pipe or the design temperature of the air.)
A = area, in square feet of a pipe, normal to the direction of heat flow of a 1-ft-long section (Where there is insulation, the cylinder of insulation should be used to calculate the arithmetic mean of the area between the pipe wall and the outside of the insulation.)
X = distance, in inches, from the outside of the pipe wall where T_1 is measured to where T_2 is taken

Equation (22.2) has been used to create Fig. 22.2, which is a nomogram that shows the temperature drop of flowing water in a pipeline. Before Fig. 22.2 can be entered, the following values must be known:

V = velocity of flow, ft/s

H = heat-transfer coefficient (By field experience, 6 is average value.)

S = length of pipe run, ft

r = radius of pipe, ft

TI = inlet water temperature, °F

TO = outlet water temperature, °F (It is recommended that minimum temperature of 35°F be used to provide a safety factor to allow for variations in thermal properties of soil along the run, intermittent flow in pipe, and temperature fluctuations of both air and ground.)

TS = ambient temperature of either air or soil around pipe (Soil temperature can be approximated by adding 5°F to the lowest monthly mean temperature.)

Example. This example illustrates the use of the nomograph in Fig. 22.2 to find the temperature drop in flowing water in a sewer, given the following conditions:

Pipe diameter = 12 in

Inlet water temperature = 40°F

Proposed velocity = 2 ft/s

Length of run = 2.2 mi

Estimated ground temperature = 25°F (T_2)

$$S = 2.2 \times 5280$$

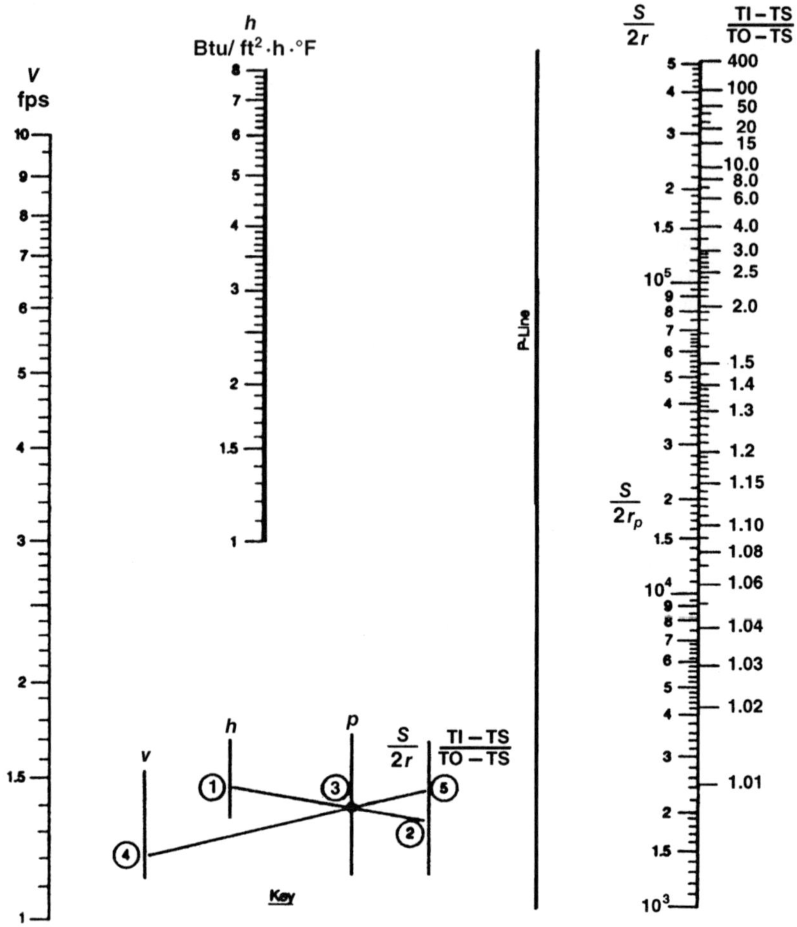

FIGURE 22.2 Temperature drop of flowing water in a pipeline.

Step 1. Find point 2:

$$\frac{S}{2r} = \frac{2.2 \times 5280}{2 \times 0.5} = 1.16 \times 10^4$$

Step 2. 6.0 value of H from previous discussion.

Step 3. Enter Fig. 22.2 and connect points 1 and 2 (6.0 and 1.16×10^4).

Step 4. Using pivot point 3 as the anchor, connect points 3 and 4. On the other side of the line, read a value of 1.77. We will call this value *Z*.

Step 5. Use the following formula to find TO.

$$Z = \frac{\text{TI} - \text{TS}}{\text{TS} - \text{TO}} \qquad 1.77 = \frac{40 - 25}{\text{TO} - 25} \qquad \text{TO} = 35.5°\text{F} \tag{22.4}$$

The recommended minimum temperature of water flowing in a sewer pipe is 35°F, which allows a small safety factor. The calculated temperature of 35.5°F is therefore considered unsafe. Using Fig. 22.2, and a temperature of 35°F as a guide, the velocity can be raised to 3 ft/s, or perhaps the inlet temperature raised to about 43°F (6°C) by adding water.

To prevent the freezing of water in a sewer line, the following methods have proved successful:

1. Provide insulation around the pipe to limit heat loss. The addition of insulation will not prevent the freezing of the water in time, but will considerably delay it.
2. Provide an enclosure around the pipe in such a manner that the enclosure does not touch the pipe walls. This will provide additional insulation to the system.
3. Heat-trace the pipe with electric cable, steam, or hot water.
4. Provide sufficient velocity to the liquid that it will reach its terminus without freezing.
5. Add warm water at the origin of the sewer, or at several places along the run (if possible), to counteract the heat loss of the water to the surrounding soil.

This formula will give the minimum flow required to keep water from freezing:

$$\text{GPM} == \frac{A \times \text{PS}(0.5\,tw - ta + 16)}{440.1\,d^2(tw - 32)} \tag{22.5}$$

where GPM = gallons per minute
A = area of water in pipe, ft^2 (m^2)
PS = pipe surface area, ft^2 (m^2)
t_w = temperature of water, °F (°C)
t_a = temperature of air, °F (°C)
d = pipe diameter, ft

DEPTH OF FREEZE IN SOILS

To prevent the freezing of water in underground piping systems, the pipes must be buried far enough below grade, that the soil used as backfill provides enough insulation from the air temperature to prevent the freezing process from starting. This depth is called the *frost depth*, which is the level to which the 32°F (0°C) isotherm will penetrate.

Various methods are used to determine the frost depth in a particular area: Consult the local authorities, such as the building department or fire marshal, for information and advice based on past experience; ask for recommendations of fire insurance carriers; or use maps such as the one shown in Fig. 22.3 which gives the approximate frost depth in feet as a starting point.

The following discussions will show how to calculate the frost depth and how to obtain the required criteria in order to perform the calculations with an acceptable degree of accuracy.

Derivation of the Basic Formula for Frost Depth

CRREL has developed a practical method for calculating the frost depth of soil. The method used to calculate frost penetration is based on heat-transfer principles involving a phase change (from water to ice) of pore water, or water held in the voids of soil. The actual freezing process under these conditions is very complex and does not lend itself to a fixed mathematical solution. To reduce the complexities involved for the exact determination of each value that follows, several assumptions have been made. These assumptions have the effect of simplifying the solution, but also introduce some slight error. All the following

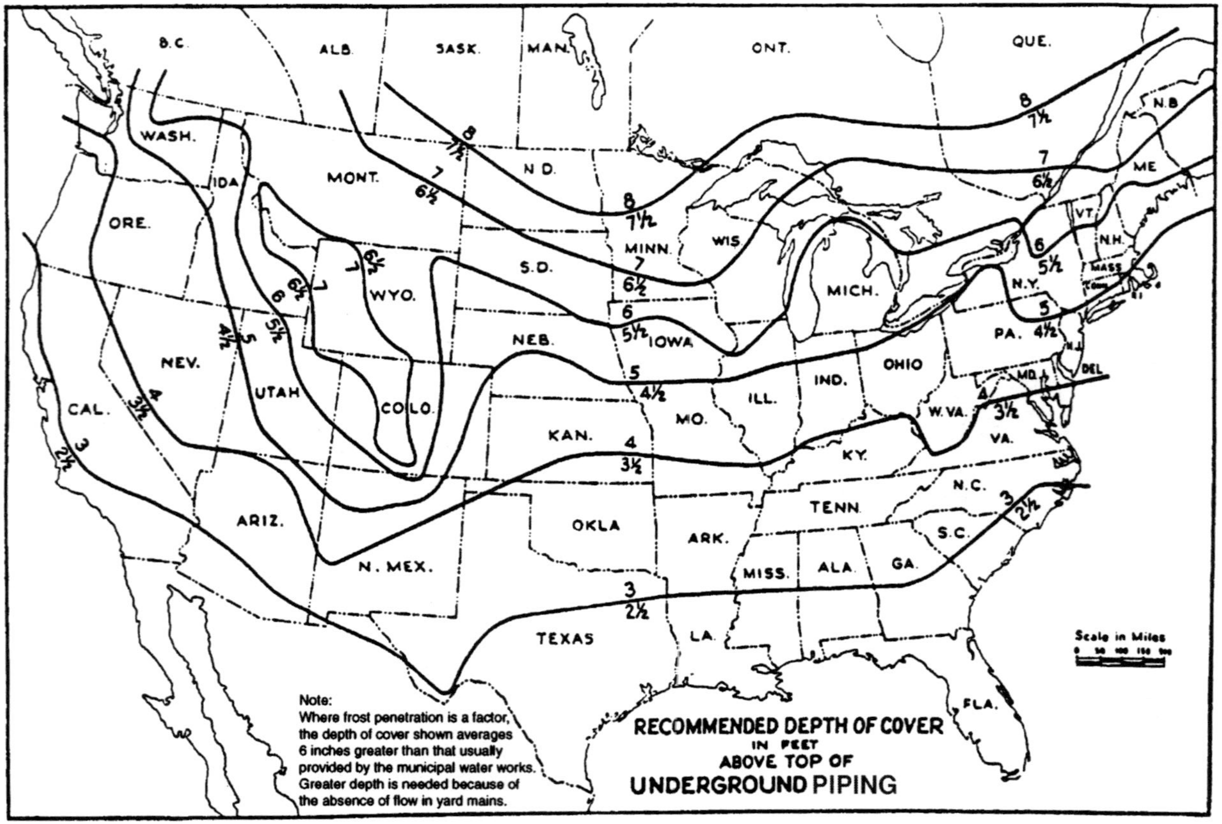

FIGURE 22.3 Approximate frost depth in feet. (*NFPA*.)

derivations have been verified under actual field conditions and found to be well within the accuracy necessary for engineering design purposes. These are the assumptions:

1. The soil is considered homogeneous throughout.
2. All pore water is converted to ice at 32°F (0.0°C).
3. Average thermal properties are applicable.

The depth to which the 32°F isobar will penetrate can be calculated by using the modified Berrigan equation

$$X = C \times \frac{\sqrt{48K - NF}}{L} \tag{22.6}$$

where X = penetration of frost into soil, ft
C = coefficient, dimensionless (This is a function of the freezing index, the mean winter temperature at the site, and the thermal properties of the soil.)
N = conversion factor, air index to surface index
SF = surface freezing index
K = thermal conductivity of soil, Btu/ft^2/h/°F
F = air freezing index, degree days Fahrenheit/yr
L = volumetric latent heat of fusion, Btu/ft^3

Now we define each item in the modified Berrigan formula.

1. Coefficient C. This is a general coefficient that considers the overall effect of the temperature change in the soil mass around a buried pipe. It is a function of the site freezing index, the mean annual site air temperature, and the thermal properties of the soil. To find this coefficient, two factors must be determined: the thermal ratio TR and the fusion parameter FP. Refer to Fig. 22.13.
2. The thermal ratio TR is calculated from the following formula:

$$\text{TR} = \frac{\text{ST}}{\text{AD}} \tag{22.7}$$

where TR = thermal ratio
ST = mean annual site air temperature, °F, minus 32 (The mean annual site air temperature can be obtained from the NOAA publication "Comparative Climatic Data." It is available only in tabular form and is much too large to be reproduced in this book. The 1-day lowest mean temperature could be used but will result in slightly lower temperature. Refer to Fig. 22.1.)
AD = average surface temperature differential (This is found by dividing the surface freezing index SF by the number of days in the freezing season.)

3. The fusion parameter FP is calculated from the following formula:

$$\text{FP} = \text{AD} \times \frac{\text{volumetric heat capacity}}{\text{latent heat of fusion in soil}} \tag{22.8}$$

4. The surface freezing index is derived from the formula

$$\text{SF} = \text{N} \times \text{AF} \tag{22.9}$$

The number of days in the freezing season, which is defined as the estimated number of days the temperature will fall to 32°F (0.0°C) or below, can be found in the mean air freezing index (Fig. 22.4).

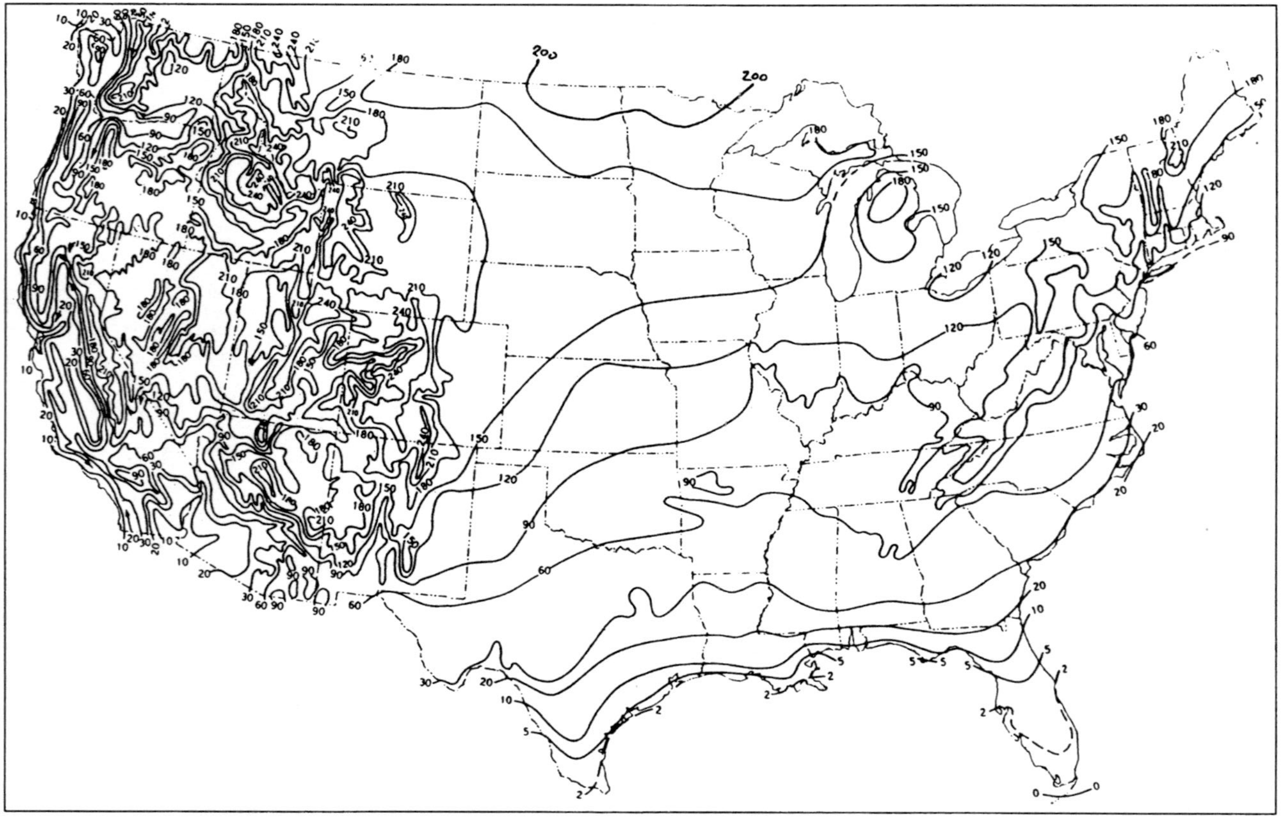

FIGURE 22.4 Mean air freezing index. Mean annual number of days with maximum temperature 32°F and below.

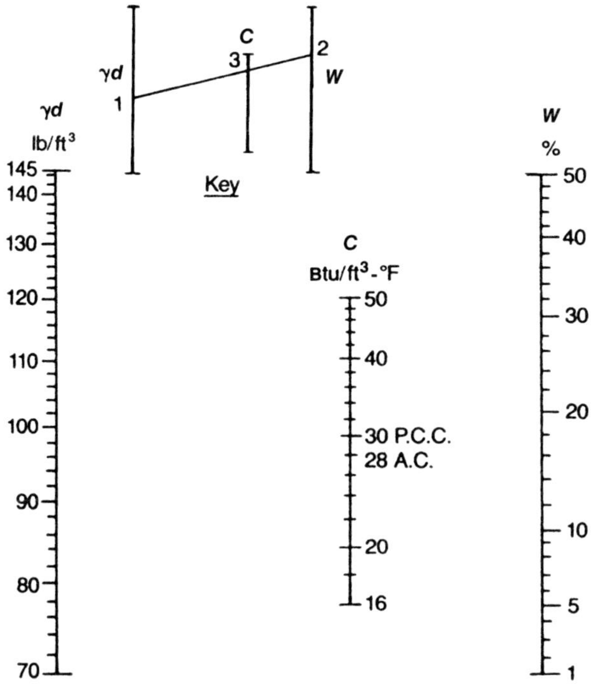

Note: Specific heat of soil solids assumed to be 0.17 Btu/lb-°F.

FIGURE 22.5 Volumetric heat capacity for soils.

AD is discussed in the explanation of Eq. (22.6). The volumetric specific heat capacity for soils is found in Fig. 22.5. The latent heat of fusion for soils is found in Fig. 22.6. Table 21.1 gives the soil weight and average value for moisture content.

5. The air–ground conversion factor is *N*. This factor is needed to approximate the temperature of the ground when only the air temperature is known. The combined effects of rediative, convective heat exchange at the air–ground junction have been considered in the determination of the actual value. Refer to Table 22.6 for values that have been established by experimentation under freezing conditions.

6. AF is the mean air freezing index (see Fig. 22.4). The penetration of freezing temperatures into the soil is partially dependent on the duration and magnitude of the temperature difference between the air and the ground. The lower air temperature and the longer freezing provide a cumulative increase in the penetration of frost. The method used to express the freezing temperature duration measurement is in degree days Fahrenheit. To find the value, the high and low temperatures for any 24-h period are averaged. Then 32 is subtracted from that number to obtain the freezing index for that day. If it is found to be 10°F (−12°C), this amounts to 12 degree days. If the number calculated came to 1°F (−17°C) below freezing and this continued for 10 days, the figure would also be 10 degree days. The freezing index should not be confused with the normal degree day measurement used by HVAC engineers for heating calculations.

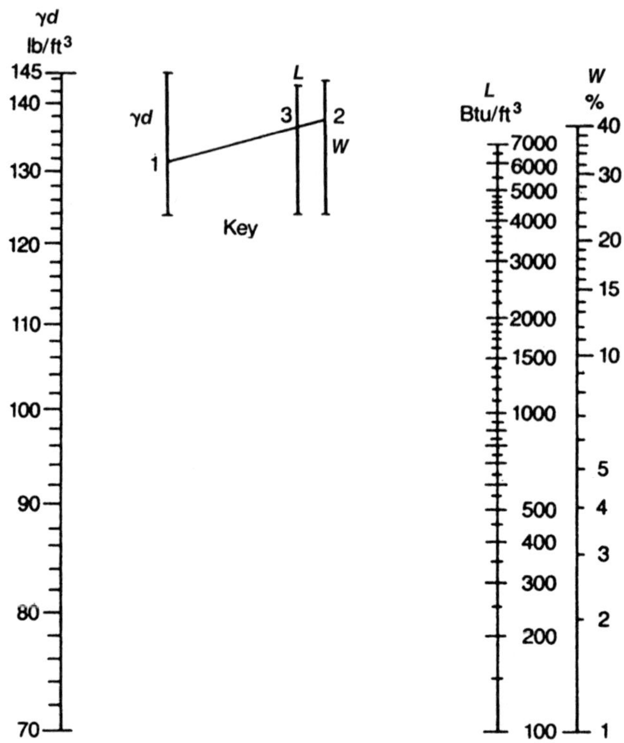

FIGURE 22.6 Latent heat of fusion for soils.

TABLE 22.6 Air–Ground Conversion Factor

Surface type	*N* Factor
Snow	1.0
Pavement free of snow and ice	0.9
Sand and gravel	0.9
Turf	0.5

Four separate methods have been established to determine the surface freezing index, each based on the number of years that weather information has been available. The first uses the single coldest year in a 30-yr period. The second uses an average of the 3 coldest years in a 30-yr period. The third method calculates the average of the 5 coldest years in a 30-yr period. The fourth method, called the *design freezing index*, uses the single coldest year in a 10-yr period. The design freezing index is illustrated in Fig. 22.7. The relationship between the mean freezing index and the other three indexes appears in Fig. 22.8. An isobaric map of the North American continent showing

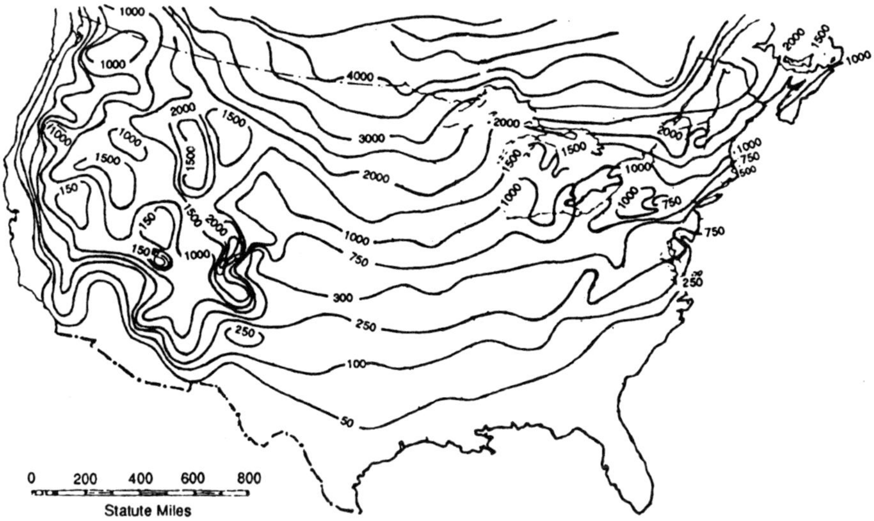

FIGURE 22.7 Design freezing index.

the mean air freezing index is illustrated in Fig. 22.4. With all this information available, what is the "correct" value? The best possible information is data obtained at the proposed site or the records of the nearest U.S. National Weather Service first-order weather station. If obtaining this information is not practical, use Fig. 22.8 to find the 30-day mean index, and then refer to Fig. 22.7 to obtain a comparative value. Generally accepted practice is to use the coldest index possible.

7. Thermal conductivity K is a measure of heat flow through a substance, given as Btu per foot of material per unit of time per degree Fahrenheit temperature difference. The thermal conductivity of soil is dependent on soil density, actual moisture content, temperature and state of pore water, and particle shape. Field tests have determined that the average values presented here are of sufficient accuracy for engineering design purposes. For this discussion only, soil groups are divided into three general categories: sand, silt with clay, and peat. Table 21.6 lists the soil weights used in the calculations following. Table 21.13 separates soils into unified soil groups and describes their composition. The average moisture content of these soils is about 15 percent for sand, 30 percent for clay, and 80 percent for peat. To find the value of K, use Fig. 22.9 for sand, Fig. 22.10 for clay, and Fig. 22.11 for loam, entering the weight of the soil and the moisture content. If the project under design has building construction associated with it, the structural engineer will have the soil tested for bearing strength. Part of this test is the determination of moisture content. A copy of the engineer's report should provide the necessary information.

8. The volumetric latent heat of fusion is L. This is a measure of the additional heat that must be absorbed by the surrounding medium in order for the pore water to change its phase state from water to ice. The more moisture a soil contains, the higher the L value. To find L, multiply the latent heat of water (144) by the weight of the soil (Table 22.6) and by the moisture content of the soil. This calculation is presented in graphic form in Fig. 22.12, where L (the volumetric latent heat of fusion) can be read directly.

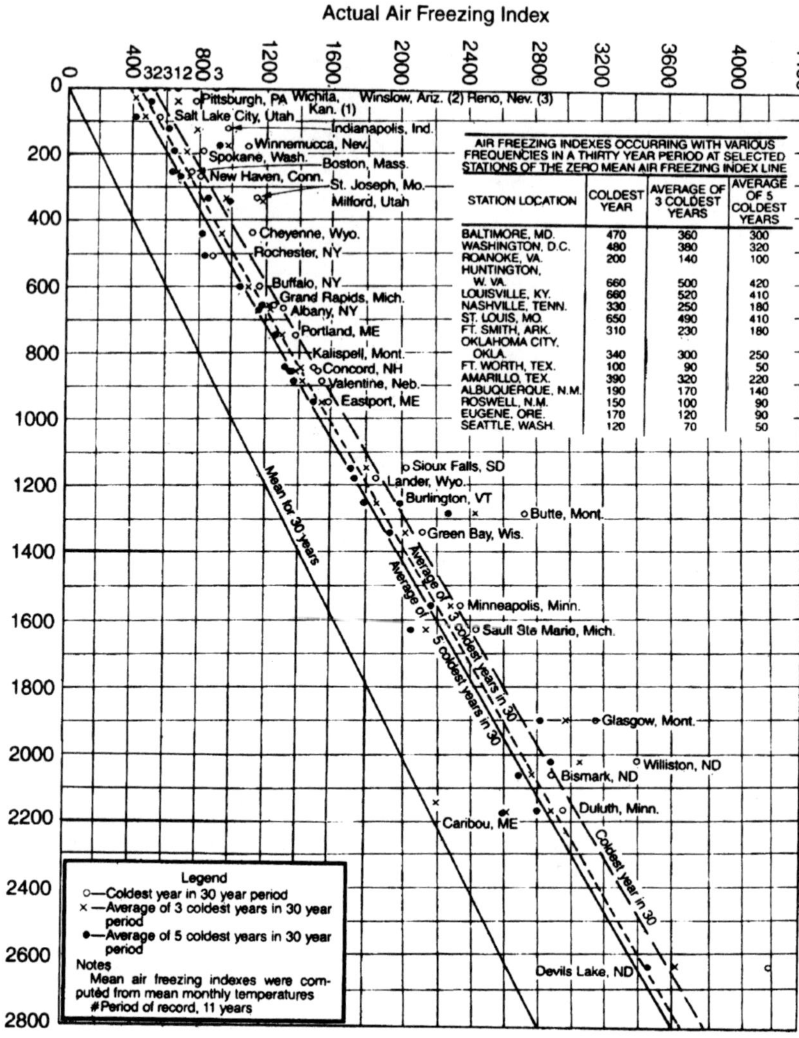

AIR FREEZING INDEXES OCCURRING WITH VARIOUS FREQUENCIES IN A THIRTY YEAR PERIOD AT SELECTED STATIONS OF THE ZERO MEAN AIR FREEZING INDEX LINE

STATION LOCATION	COLDEST YEAR	AVERAGE OF 3 COLDEST YEARS	AVERAGE OF 5 COLDEST YEARS
BALTIMORE, MD.	470	360	300
WASHINGTON, D.C.	480	380	320
ROANOKE, VA.	200	140	100
HUNTINGTON, W. VA.	660	500	420
LOUISVILLE, KY.	660	520	410
NASHVILLE, TENN.	330	250	180
ST. LOUIS, MO.	650	490	410
FT. SMITH, ARK.	310	230	180
OKLAHOMA CITY, OKLA.	340	300	250
FT. WORTH, TEX.	100	90	50
AMARILLO, TEX.	390	320	220
ALBUQUERQUE, N.M.	190	170	140
ROSWELL, N.M.	150	100	90
EUGENE, ORE.	170	120	90
SEATTLE, WASH.	120	70	50

FIGURE 22.8 Relationship between mean and other freezing indexes.

To illustrate the use of the modified Berrigan formula [Eq. (22.6)], the following example will determine the frost depth for a project in Green Bay, Wisconsin, with the following conditions:

Mean annual temperature	43.7°F (from NOAA)
Length of freezing season	200 days (obtained from Fig. 5.10)
Soil properties	Peat, snow cover

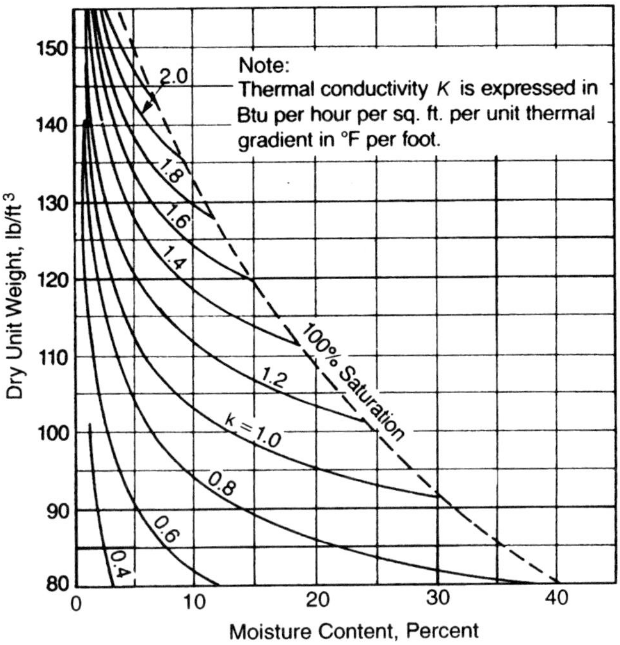

FIGURE 22.9 Average thermal conductivity for sandy soils.

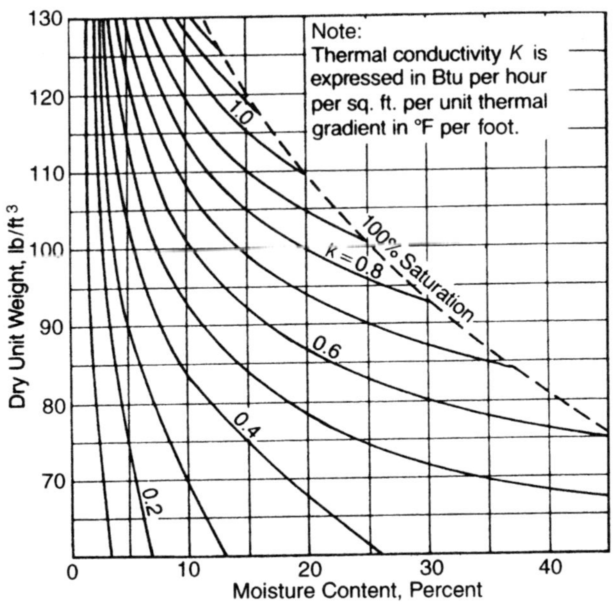

FIGURE 22.10 Average thermal conductivity for silt and clay soils.

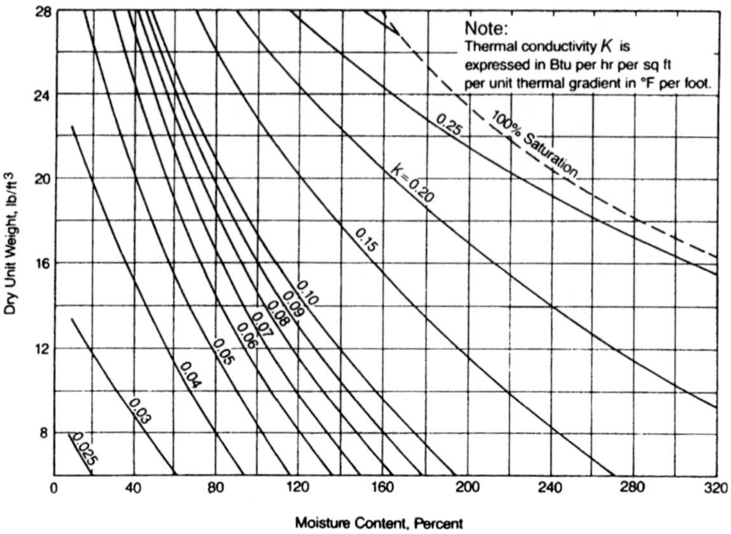

FIGURE 22.11 Average thermal conductivity for peat.

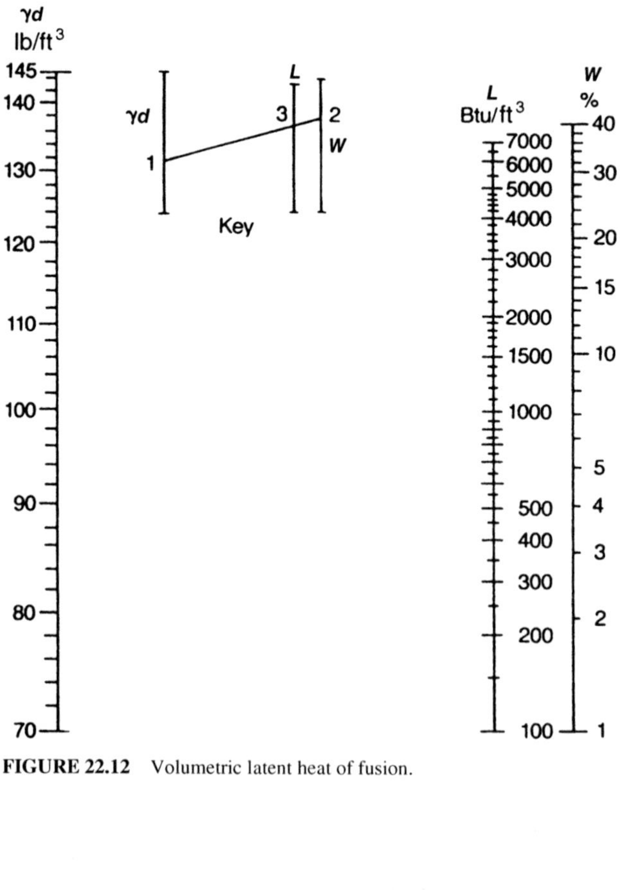

FIGURE 22.12 Volumetric latent heat of fusion.

Weight	100 lb/ft^3
Moisture content	15% (obtained from Table 22.6)
Latent heat of fusion	2150 Btu/ft^3
Thermal conductivity	1.01 Btu/h-ft^2-°F (obtained from Fig. 22.11)
Surface freezing index	2150 degree days (AF obtained from Fig. 22.4)
Coldest reading N	1, obtained from Fig. 22.1, snow cover
Coefficient	Found from Fig. 22.13
Design freezing index	2160, from Fig. 22.7

To obtain the coefficient C for the Berrigan formula, refer to Fig. 22.13. Enter with the fusion parameter and thermal ratio. To find the fusion parameter [FP in Eq. (22.8)], first calculate AD:

$$\text{AD} = \frac{2150}{200} = 10.7$$

Therefore
$$\text{FP} = 10.7 \times \frac{28.2}{2150} = 10.7 \times 0.013 = 0.14$$

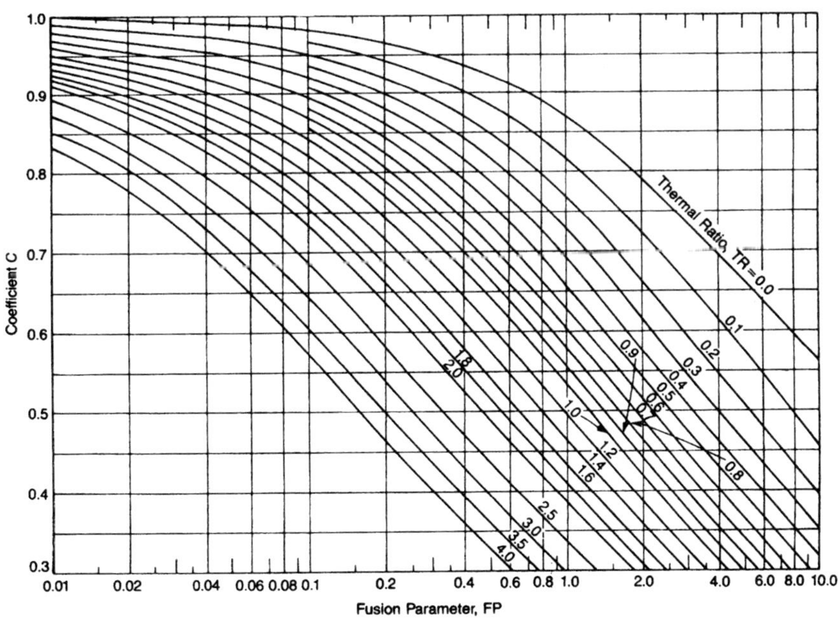

FIGURE 22.13 Coefficient in the modified Berrigan formula.

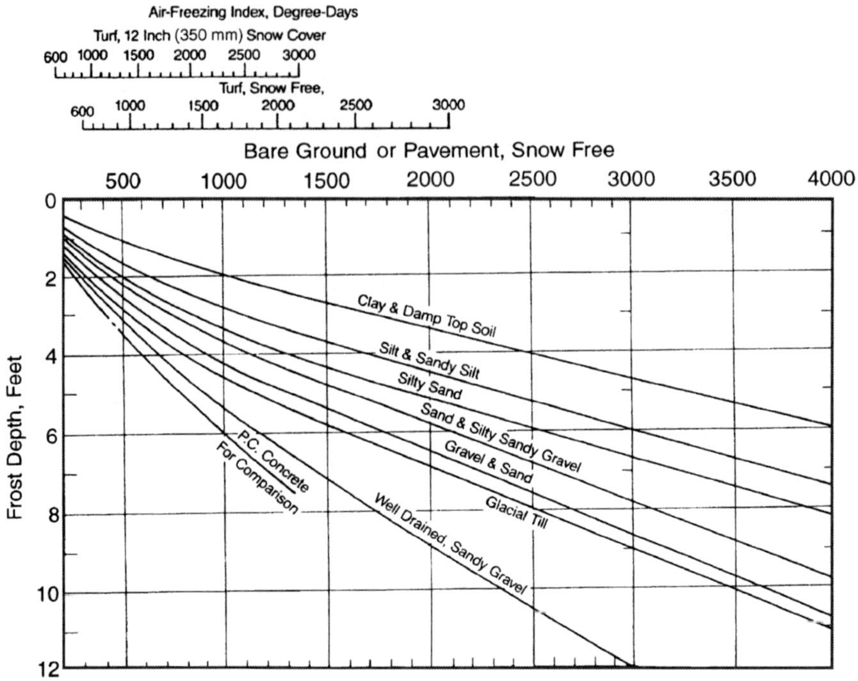

FIGURE 22.14 Nomograph to determine frost depth using modified Berrigan formula.

To find the thermal ratio [TR in Eq. (22.6)], calculate the following:

$$\text{TR} = \frac{\text{ST}}{\text{AD}} = \frac{43.7-32}{10.7} = \frac{11.7}{10.7} = 1.09$$

It is now possible to enter Fig. 22.13 with 0.14 and 1.09. Read $C = 0.8$. This gives us all the information needed to solve the modified Berrigan formula, Eq. (22.5).

$$X = 0.8\sqrt{\frac{48\times 1.01\times 2150}{2160}} = 0.8\sqrt{\frac{104{,}232}{2160}} = 0.8\times\sqrt{48.26} = 5.55\text{ ft}$$

The estimated depth of frost penetration is 5.55 ft.

Several nomographs have been prepared based on the modified Berrigan formula that are easier to use than the original formula. The first is shown in Fig. 22.14, giving the frost depth for various conditions as a direct reading.

In some cases, a pipe may have been buried under a road and a nonsusceptible fill used. This means that the fill will not be affected by frost, and so the condition known as frost heave will not occur. Use Fig. 22.15 to find the depth of fill that will prevent frost penetration below the fill. For a direct reading of the actual frost penetration under pavements, refer

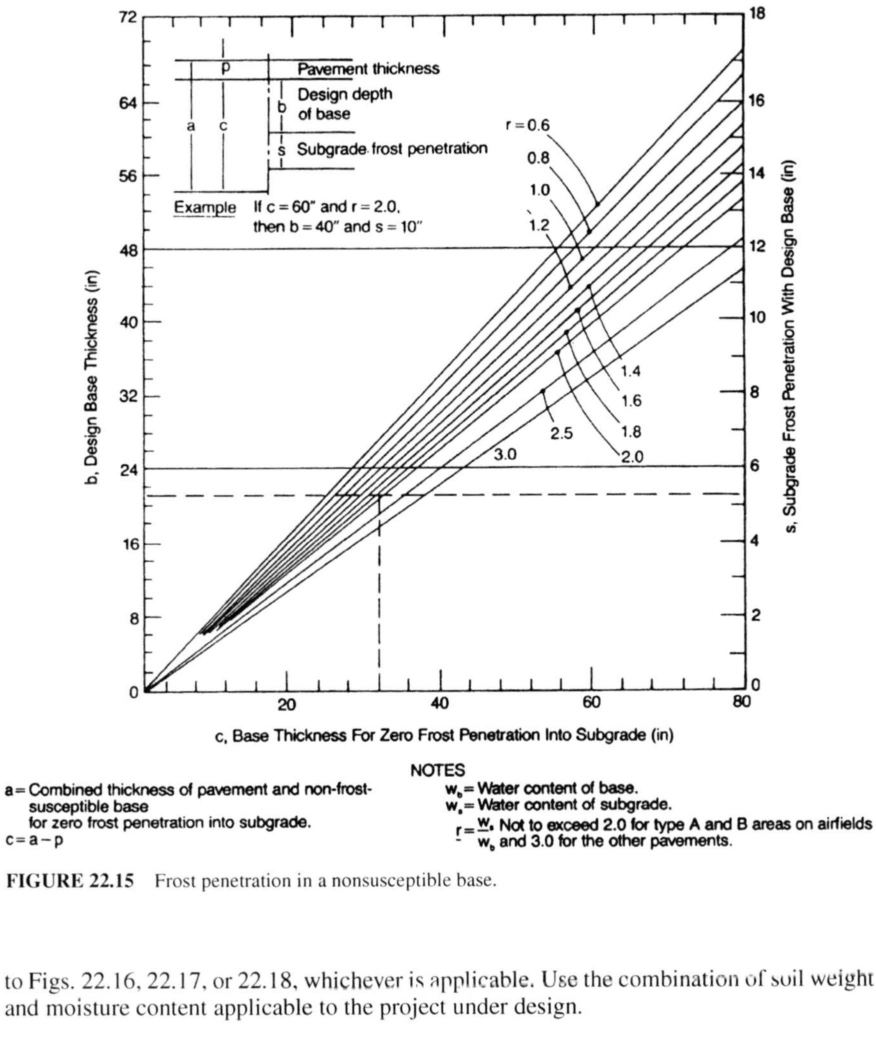

FIGURE 22.15 Frost penetration in a nonsusceptible base.

to Figs. 22.16, 22.17, or 22.18, whichever is applicable. Use the combination of soil weight and moisture content applicable to the project under design.

FROST CLOSURE OF VENTS

During cold weather, exposed vents on a roof may become wholly or partially blocked by frost on the inside portion of the exposed pipe. This is due to the fact that in cold weather, a current of warm, moist air rises through the plumbing piping when there is little or no flow through the system. This upward flow of air is caused by the temperature difference between the air outside the building and the air inside the pipe. Since the air inside the building is warmer than the free air, the inside air is lighter, causing the upward current. This is the *chimney effect*. When this warm, moist air reaches the chilled surface of the exposed pipe, moisture condenses on the colder surface of the pipe in the form of droplets.

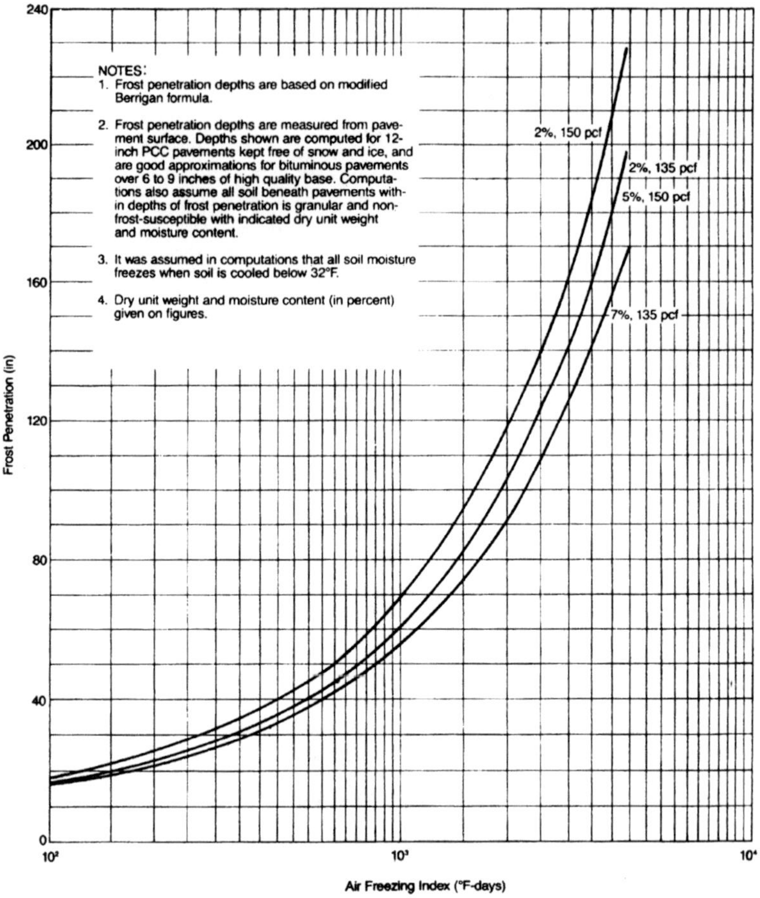

FIGURE 22.16 Frost penetration beneath pavement: 135 to 150 pcf base material.

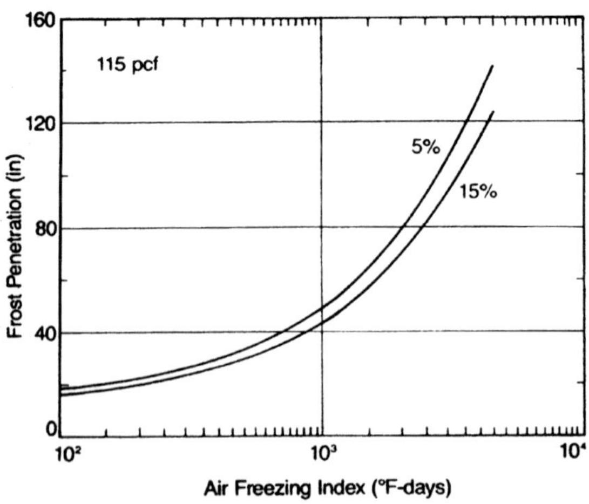

FIGURE 22.17 Frost penetration beneath pavement: 115 pcf base material.

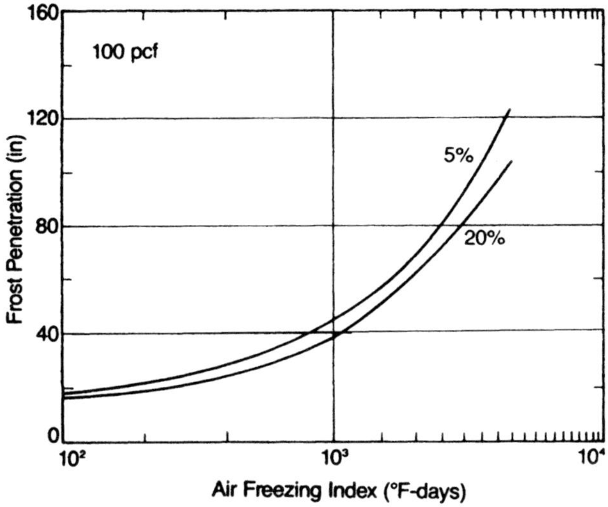

FIGURE 22.18 Frost penetration beneath pavement: 100 pcf base material.

If the correct conditions exist, these droplets will freeze. If this continues long enough, the pipe will become blocked.

There are actually two phenomena that can occur. The more common is the formation of ice in the form of an annular ring in the pipe interior. Another kind of blockage takes the form of a frost cap on top of the pipe. This problem of frost closure was the subject of a study at the National Bureau of Standards in 1922. The conclusions resulting from that study are still valid today.

The best method to prevent frost closure is to increase the size of the vent pipe just below the roof level, allowing the warm air to bypass the side of the exposed pipe as much as possible. It was shown that a 4-in (DN 100) pipe would not close up except under the most adverse conditions. One of these conditions was observed in the vicinity of Niagara Falls, New York, where the spray from the falls solidified on the pipes regardless of their size. The solution was to either run an electric heat trace line or circulate hot water around the perimeter of the pipe to keep it warm. For more unusual conditions, it was found that keeping the vent pipe as low to the roof as possible was acceptable. Blocking by snow at that low height never occurred, because the snow was porous enough to pass air and had a tendency to melt very quickly.

PIPING CONNECTIONS TO WATER TANKS

The piping connections to a water tank on the site shall include the necessary piping to keep both the water warm in the riser that reaches from underground to the tank itself and the water in the tank from freezing. Figures 22.19, 22.20, and 22.21 show typical connections and piping arrangements. Conditions will change based on available space under the tank.

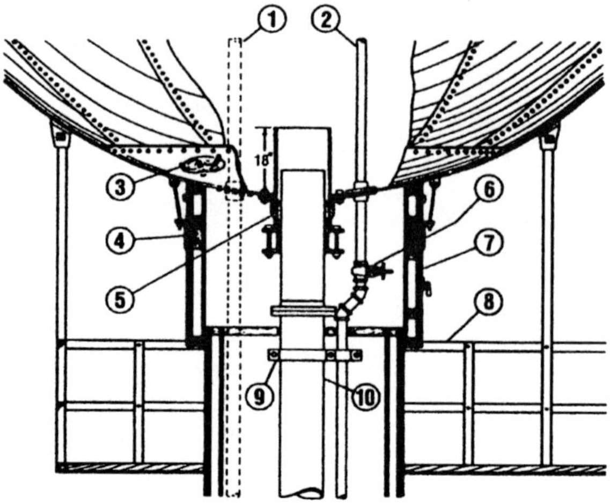

1. Inside brass overflow pipe, if used
2. Hot water circulating pipe
3. Handhole for removing sludge
4. Frostproof casing when required
5. Expansion joint
6. Approved OS&Y gate valve
7. Door in frostproof casing
8. Walkway
9. Brace for hot water circulating pipe
10. Pipe riser

FIGURE 22.19 Details of typical pipe connections to bottom of steel gravity tank with pipe riser.

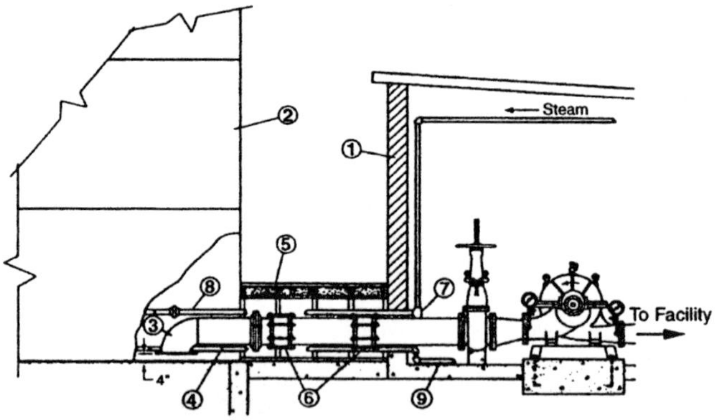

1. Pump room
2. Approved aboveground suction tank
3. Entrance elbow with vortex plate or flange 4 in (102 mm) above bottom of tank
4. Suction pipe
5. Frostproof casing, about 4 ft (1.2 m) high and 4 ft (1.2 m) wide
6. Flexible couplings
7. Steam trap on steam supply
8. Heating coil inside tank
9. Condensate return line to steam trap

FIGURE 22.20 Typical discharge pipe connected to side of aboveground suction tank.

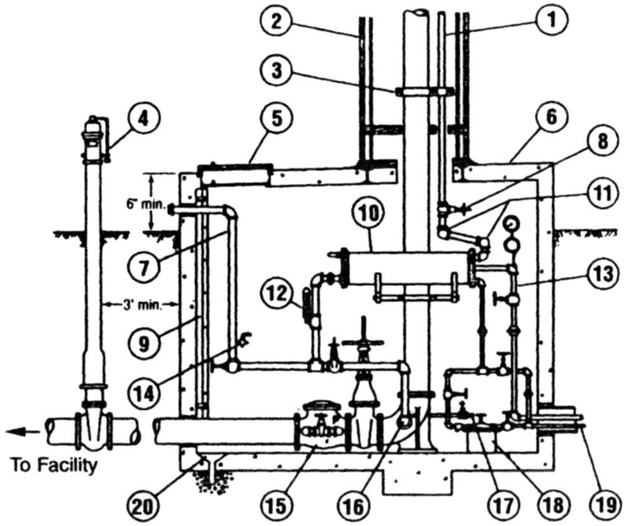

1. Hot water circulating pipe
2. Frostproof casing
3. Pipe clamps. Locate at about 25 ft (7.6 m) intervals. Loose fit around hot water circulating pipe.
4. Indicator-post valve. May be replaced with indicating-type valve in pit on yard side of check valve if space is not available for indicator post.
5. Hatch cover
6. Valve pit
7. Drain pipe
8. OS&Y gate valve
9. Ladder
10. Tank heater with relief valve set at 120 psi (825 kPa)
11. Four-elbow swing joint
12. Thermometer
13. Steam supply pipe
14. Drain cock, 1/2 in (13 mm)
15. Approved check valve with bypass
16. Cold water circulating pipe
17. Pipe to mercury gauge
18. Steam trap
19. Condensate return
20. Valve-pit drain

FIGURE 22.21 Typical valve pit and pipe connections at base of tank on independent tower. Tank has a pipe riser and steam-heated gravity circulating heating system.

BIBLIOGRAPHY

Alter, Amos, "Sewage and Sewage Disposal in Cold Regions," CRREL, U.S. Army Corps of Engineers.
Care, Kevin L., "The Freezing and Blocking of Water Pipes," CRREL.
Depth of Freeze and Thaw in Soils, U.S. Army Technical Manual, TM 5-852-6.
"Design of Fluid Systems Hook-ups," Spirix Sarco Inc., Allentown, Pa.
Dill, Richard S., "The Freezing of Pipes and Vessels," National Bureau of Standards.
Fire Protection Handbook, 16th ed., National Fire Protection Association.
Frankel, M., "Cold Weather Design of Plumbing Systems," parts 1 and 2, *Plumbing Engineer*.
Frankel, M., "Thermal Insulation for Plumbing Systems," parts 1 and 2, *Plumbing Engineer.*
Kenny, Thomas M., "Steam Tracing: Do It Right," *Chemical Engineering Progress.*
Sanger, F. J., "Foundations for Structures in Cold Weather," CRREL.

CHAPTER 23
BUILDING RATING AND ASSESSMENT SYSTEMS

GENERAL

The purpose of this chapter is to inform the design professional of what nationally accepted building rating and assessment organizations are available, and the goals they are intended to accomplish. This chapter is not intended as a complete design guide; rather, it will give the design engineer information to consider for the design of green buildings. It is included here because of the site utilities that should be a part of the overall building design. The primary subjects will be water conservation and facility rainwater and wastewater reuse. There are many other subjects concerning building rating that are included in the complete rating of buildings, but these two are the only subjects that concern a plumbing design professional.

A majority of states and many hundreds of local authorities have incorporated building rating and assessment systems into legislation suggesting their use. The incentives include tax breaks, expedited plumbing and building permits, grants and marketing assistance to the owner, to say nothing of reduced building operating costs. Whether building owners choose to abide by any particular energy efficiency standard, they will achieve as much energy efficiency as possible which will help to save money and limit natural resource consumption.

The design, construction, and occupation of commercial and industrial facilities use resources such as water, energy, and many types of raw materials. The building designers, architects, and construction workers face a challenge to meet the demand for new facilities designed for new and renovated buildings that are operated to minimize the impact on the environment, maximizing the return on investment and providing environmental benefits. The use of these facilities generates waste and often harmful atmospheric emissions. To minimize, reduce, or eliminate this impact, nationally accepted rating and assessment organizations have been established, called *building green rating systems*. There is no doubt that other green building and sustainable design will be recognized in time and will grow in importance and popularity over the next several years. Consultation by building teams with the applicable authorities is mandatory to select which rating method will be used. It is important to note that the standards are intended to only be a guide to energy efficiency. Some of the more popular standards are as follows:

1. One program is called Leadership in Energy and Environmental Design, or LEED. It is a certification program offering a method of rating systems that recognize projects which implement methods to achieve recognizable better environmental and health performance. Developed by the U.S. Green Building Council (GBC), LEED provides building owners and operators with general instructions for identifying and implementing a practical and measurable design.

2. Another program has been developed by the Green Building Initiative (GBI). This program originated in England and was brought to the United States in 2004. It is a green building guidance and assessment system that is intended to advance the overall environmental performance and sustainability of commercial buildings. This certification program, the exclusive provider of the Green Globe building certification, has two listings: one for new construction and another for existing buildings. As of this writing, this is the first and only building certification program that has become an American National Standards Institute (ANSI) national standard.
3. International Standards Organization (ISO) 5001 is intended to provide organizations with a recognized framework for integrating energy performance into their construction and management practices. Multinational organizations will have access to a single complete standard for implementation across the organization with a logical and consistent methodology for identifying and implementing improvements.

CODES AND STANDARDS

1. American Recovery and Reinvestment Act, 2009
2. Clean Water Act
3. U.S. Environmental Protection Agency rules and regulations
4. U.S. Green Building Council
5. Green Building Initiative supplement
6. Green Building LEED for Healthcare
7. International Code Council (ICC), "Safe and Sustainable By the Book"
8. International Association of Plumbing and Mechanical Officials (IAPMO), "Green Plumbing and Mechanical Code Supplement"
9. ISO 5001
10. ICC Fire Code, 2012

DEFINITIONS

Addition is the extension or increase in floor area or height of a building.

Alteration is any construction or renovation of an existing structure other than repair or addition.

Alternative transportation is intended to apply primarily to bicycle riders, although consideration is given to available public transportation. Along with bike racks, shower facilities must be provided for 1 percent of all full-time employees (FTEs),

Approved is any work acceptable to the code official or the authority having jurisdiction.

Blackwater is wastewater from toilets, urinals, and sinks containing organic, food, and toxic materials.

Building is any structure used or intended to support or shelter any use or occupancy.

Carbon footprint is a measure of the impact that human activities have on the environment, in terms of the amount of greenhouse gases produced, measured in units of carbon dioxide. These excessive emissions contribute to accelerating global warming and climate change. The carbon footprint indicates the effect on the climate and is measurable. Reducing the carbon footprint reverses the negative impact of greenhouse gases on the global environment.

Commercial building for this chapter is any building not used as a residence.

Conservation is generally defined as the practice of protecting against loss and waste by management of our natural resources, both renewable and nonrenewable. It is based on the premise of taking no more than what you need and the use of renewable resources whenever possible.

Graywater is untreated wastewater from lavatories, showers, washing machines, sinks, precipitation, and other sources not used for food preparation that has not come into contact with wastewater from toilets and urinals.

Green is a commonly used phrase to describe the prudent use and application of natural materials found in the environment and having an impact on the environment. When one is discussing materials, green basically means products that deliver the same or better performance, energy efficiency, function, or other benefits than comparable lesser products. It is safer, contains or uses less toxic material, and has a longer useful life cycle. It includes reducing energy cost, water use, elimination of waste; improves the environmental performance of a building; and reuses material in the production process. Government regulations and consumer demand for increased product usability have a main objective of reducing the use of fossil fuels.

Occupancy is the human presence and use of a facility as calculated by an applicable code. A large number of credits utilize the building occupancy number, which is established by the architect. All credits of the project must use the same number. The most common user groups for commercial projects are the full-time employees (FTEs), part-time employees such as transients, and residents converted to FTEs. If the building has a number of transients, such as a college or airport terminal, they must be converted to an FTE number.

Nonpotable water is water that is not safe for human consumption.

Rainwater is water from natural precipitation that has not been contaminated by use.

Rainwater collection and conveyance system has components extending between the collection surface and the storage tank that conveys collected precipitation.

Sustainability has a different definition depending on the individual or organization asked. For this handbook, it is the making use of renewable materials that grow naturally and are readily replaceable. Sustainable design, also called *green design*, is the art of designing physical objects and the built environment to accept the principles of economic, social, and ecological material reuse or replacement. It is thought of by many as the relentless pursuit of waste elimination, but there is more to it than that.

APPROVAL OF AUTHORITIES HAVING JURISDICTION

Due to alternative systems proposed, the design team should coordinate all engineered systems with the authorities having jurisdiction (AHJ). Most codes used have provisions for green building design and other types of engineered systems that are not intended to be code-compliant. Presentations and meetings shall be arranged to show the AHJ all the proposed systems for the purpose of answering all questions and obtaining approval.

It is generally conceded that the building codes are the greatest hindrance to the adaption of green building codes and construction for the green building at this time. This has partially been changed with the addition of the "Green Plumbing and Mechanical Code Supplement" into the National Standard Plumbing Code (NSPC) and the acceptance of the Green Building Initiative by ANSI. The provisions will become enforceable only where the approval of the AHJ has been obtained or has been adopted by the code officials.

LEED

LEED is an acronym for Leadership in Energy and Environmental Design. It is a major tool for guiding project teams to a greener building effort. It is a point-based process or rating method developed and administrated by the nonprofit U.S. Green Building Council (USGBC) to measure the steps and achievement in reducing energy consumption while preserving and/or improving the environment. It is a scoring system that includes several mandatory elements and other optional credits that can earn points for the project. The process of development and maintaining the LEED rating system is consensus-based along with additional voluntary committees for individual subjects. The rainwater management for LEED credits would include rainwater harvesting. This subject is discussed in Chap. 16.

Leed Building Certification

Certified buildings must meet LEED standards in six areas; health and comfort, material, location quality, water, energy use, and innovation.

To become LEED-certified, a building is rated on credits for six categorizes

1. Sustainable sites (SS)
2. Water efficiency (WE)
3. Energy and atmosphere (EA)
4. Materials and resources (MR)
5. Indoor environmental quality (IEQ)
6. Innovation and design (ID)

Each one of the categories is in turn divided into subdivisions. Items 4, 5 and 6 are not considered part of a plumbing design engineers scope of work and are outside the scope of this handbook.

The LEED credits for the new construction rating system have the following four award levels based on a maximum of 69 points:

1. Certified, 26 to 32 credits
2. Silver, 33 to 38 credits
3. Gold, 39 to 51 credits
4. Platinum, 52 or more credits

LEED points can be potentially earned with the installation of a rainwater harvesting:

1 point for measures that result in a 20 percent water saving

1 point for measures that result in a 30 percent water saving

TABLE 23.1 Typical LEED Fees

	Less than 50,000 ft^2 (fixed rate)	50,000–500,000 ft^2 (based on ft^2)	More than 500,000 ft^2 (fixed rate)
Design Review			
Members	$1,250	$0.025	$12,500
Nonmembers	1,500	0.03	15,000
Construction Review			
Members	500	0.01	5,000
Nonmembers	750	0.015	7,500
Combined Desiqn and Construction Review			
Members	1,750	0.035	17,500
Nonmembers	2,250	0.045	22,500
Initial Certification Review			
Members	1,250	0.025	12,500
Nonmembers	1,500	0.030	15,000

1 point for measures that result in a 40 percent water saving (innovation credit)

Up to 2 points for stormwater reduction

Up to 2 points for water-efficient irrigation

1 point for reducing the project's sewer and potable water usage by 50 percent

1 point for conservation of natural areas

Typical LEED fees for project review are given in Table 23.1.

Earning LEED Building Credits

The construction industry as a whole and design engineers in particular want to try to achieve the USGBC green building certification. To achieve what the USGBC considers an environmentally friendly building, many factors should be considered. Following are discussions of various strategies used to add LEED credits to different new buildings. Water efficiency points for various levels of water reduction are given in Table 23.2. The list is by no means complete and gives only a limited insight into the reduction possible. Not every

TABLE 23.2 Water Efficiency Points

Percentage reduction	Points
20	Prerequisite
30	2
35	3
40	4
45	Innovative design – 1

Designers can obtain up to 5 points for meeting the U.S. Green Building Council LEED Water Efficiency (WE) Credit 3—Water Use Reduction.

Source: U.S. Green Building Council.

strategy is applicable to every building, and the cost of implementing them might be excessive. Environmental policies do have an economic impact on the design of a project.

LEED credits for existing buildings are available and consider such items as building access, surroundings, efficient use of water, waste, and use of changing rooms and showers for bicycle users. It is not reasonable, nor is it always desirable, to have every building eligible for a LEED certification. It would not be possible for a major renovation consisting mostly of HVAC and architectural revisions to be eligible for LEED credits.

LEED—New Facility Building Construction

At the time of this writing, the International Association of Plumbing and Mechanical Officials (IAPMO), authors of model building codes, issued the LEED "Green Plumbing and Mechanical Code Supplement." It is regarded as a very comprehensive document that standardizes sustainable residential and commercial plumbing and mechanical systems. It is thought that there is very little information in existing codes as how to design green systems properly and to safely operate buildings within existing codes. Because of this, the Green Technical Committee developed the supplement.

In 2009, the LEED Building Council developed LEED version 3, which is the reference for this chapter. The following are the itemized different LEED reference guides. Only those reference guides that affect plumbing engineers are listed. LEED is targeting the top 25 percent of the market, which is the commercial buildings, with the intent of eventually including the remainder of the facilities. As the demand for green products and services is driven by these projects, the entire market will benefit from the greater availability of these products, which will lead to higher building quality and lower operating cost. The following are the different LEED reference guides in version LEED 3.

The Green Building Council has also developed an additional LEED-HC for Healthcare, incorporating a rating system written especially for the special needs of hospitals and other health care facilities. The building credits for new designs that concern the plumbing engineer are all in the water efficiency category.

In response to problems that have arisen with some design professionals who specify fixtures and methods that meet LEED-HC requirements but do not communicate these solutions to other team members, the U.S. Green Building Council is now requiring that an integrated team be selected to review and solve those problems.

Building Design and Construction (BD+C), Formally New Construction

1. New construction
2. Core and shell
3. Schools
4. Health care—in development
5. Retail—in development

Operation and Maintenance (O+M), Formally Existing Buildings

1. Existing schools—in development

Potable Water Use

Suggested water recycling treatment levels, if possible, are given in Table 23.3. One goal in green building design is to find ways to reduce water consumption. This can be accomplished

TABLE 23.3 Suggested Water Recycling Treatment and Uses*

Water Collection System	Increasing level of treatment →	
Primary Treatment: Sedimentation →	Secondary Treatment: Biological Oxidation, Disinfection →	Tertiary/Advanced Treatment: Chemical Coagulation, Filtration, Disinfection
• No uses recommended at this level	• Surface irrigation of orchards and vineyards • Non-food crop irrigation • Restricted landscape impoundments • Groundwater recharge of nonpotable aquifer† • Wetlands, wildlife habitat, stream augmentation† • Industrial cooling processes†	• Landscape and golf course irrigation • Toilet flushing • Vehicle washing • Food crop irrigation • Unrestricted recreational impoundment • Indirect potable reuse: Groundwater recharge of potable aquifer and surface water reservoir augmentation†
	Increasing level of human exposure →	

*Suggested uses are based on Guidelines for Water Reuse, developed by U.S. EPA.
†Recommended level of treatment is site-specific.
Source: U.S. Environmental Protection Agency.

by reducing or eliminating potable water use, reducing the use of material such as piping and insulation, and reducing energy use by using efficient piping design. One requirement of LEED certification is that a prerequisite reduction of 20 percent be obtained.

The reduction in water use can be accomplished by using systemwide technologies:

1. Non-potable water for flushing toilets
2. Use of automatic fixtures
3. Use of low-flow fixtures and waterless urinals
4. Use of small instantaneous water heaters near showers and sinks
5. Vacuum assist systems
6. Sewage conveyance and cooling tower water

The following potential water-saving reduction features for individual fixtures include

1. Infrared faucet sensors
2. Delayed-close shutoff faucets
3. Flow restrictors
4. Metering faucets (0.25 gal/cycle)
5. Dual-flush fixtures

TABLE 23.4 Flow Fixture Baseline Information

Flow fixture	Baseline flow rate	Uses per person per day FTE*	Transient/ visitor/student	Retail customer	Residential
Lavatory faucet	Commercial: 0.5 gpm, 15 s (12 s for automatic controls) Commercial Private Use (hotel, patient rooms): 2.2 gpm Residential: 2,2 gpm, 60 s All based on 60 psi.	3.0	0.5	0.2	5.0
Kitchen/break room sink	Commercial: 2.2 gpm Residential: 2.5 gpm All based on 60 psi	1.0	0.0	0.0	4.0
Shower	Commercial: 2.5 gpm, 300 s Residential: 2.5 gpm, 480 s All based on 80 psi.	0.1	0.0	0.0	1.0

*Full-time employees.

To judge the typical use of potable water, refer to Table 23.4. Table 23.5 gives the typical use of flush fixtures.

Fixture Groups

LEED version 3 has implemented the use of fixture groups. The difference between fixture groups is defined in Table 23.6 of the new version. This was done to eliminate the confusion of male restrooms with urinals versus male restrooms with two different types of fixtures, namely, no water urinals with low-flow water closets. Fixture groups require the plumbing engineer to break down the usage of each fixture into specific categories based on the user and fixture type. The baseline water usage flow is given in Tables 23.4 and 23.5. More detailed categories could be assigned in mixed-use buildings where there is a restaurant, offices, and hotel rooms, each with a different type and usage of fixtures.

When specifying a plumbing fixture for a LEED project, the design engineer must, as a minimum, specify low-flow urinals and water closets.

TABLE 23.5 Flush Fixture Baseline Information

Flush fixture	Gender	Flush rate, gpf	Uses per person per day FTE*	Transient/ visitor/student	Retail customer	Residential
Water closet	Female	1.6	3.0	0.5	0.2	0.5
	Male	1.6	1.0	0.1	0.1	5.0
Urinal	Male	1.0	2.0	0.4	0.1	—

*Full-time employees.

TABLE 23.6 Fixture Group Definitions

Group name	Annual days of operation	Full-time employees	Transients	Retail	Percent female	Percent male
Female	240	85	25	150	50	0
Male	240	85	25	150	0	50
Total lavatory use	240	85	25	150	50	50

The fixture groups are defined as male, female, and total lavatory use.
The annual days of operation are derived from an average of 50 workweeks of 5 days/week.
Source: Advanced Engineering Consultants.

LEED Certification Process for Engineers

The LEED accreditation program recognizes an individual as being knowledgeable about green design and a person who understands the LEED system and process. Anyone with an in-depth knowledge of the construction industry and green practices qualifies for accreditation. There are no prerequisites to take the exam.

Accrediation offers competitive advantages for the individual and companies. As LEED is a response to a growing market trend, accreditation is a strong indicator of modern practices. In addition, accreditation offers market value as well as shows commitment to sustainability. The name of the individual and company is listed on the USGBC website, which serves as a resource to owners. It helps reverse the downward pressure on professional scope and fees. The LEED process requires documentation of design services often unfunded in typical processes. A project could earn one point for including a LEED-accredited professional on the design team.

To become a LEED-accredited professional, one must pass a USGBC computerized exam offered at various sites throughout the United States. Various sources are available to assist the applicant to learn the LEED rating structure and all other information. The LEED reference guide is particularly valuable. There is a cost to take the test, and the amount depends upon membership in the USGBC. The test is divided into four categories:

1. Knowledge of LEED credit requirements
2. Coordinating projects and teams
3. Implementing the LEED process
4. Verifying, participating in, and performing technical analysis

The project must then be registered. There is a cost payable to the Green Building Council necessary to register a project, which is delineated in Table 23.1. This requires the project team to provide the basic details and create an online project listing. This provides access to LEED letter templates, credit inquiries, and technical support throughout the project. The second step is to certify the building. There is also a cost for certification. The certification has two phases. There is initially an optional design phase submission and a required final submission.

The focus is on a whole-building concept, including minimum consumption of resources (energy, water, raw materials, etc), providing a healthier and more comfortable indoor and outdoor environment, reducing operating costs, and improving maintenance. A coordinated, multidiscipline approach by all team members is necessary. This chapter focuses on the plumbing aspects of a LEED building. HVAC and electrical work are outside the scope of this chapter.

Construction Pollution Prevention

The LEED rating system has minimum requirements that must be met. This requirement is one that contractors have been meeting for years. The responsibility for design is the primary responsibility of the landscape architect, and the implementation is the responsibility of the contractor. The two parts of this prerequisite are to prevent loss of soil by rainwater runoff and to prevent sedimentation of stormwater.

Stormwater design credits have several aspects. Where the imperviousness of the site is less than 50 percent, the engineer may prevent the postconstruction runoff rate not to exceed the preconstruction rate. Another option would be to consider protecting any stream channels from receiving erosion material from the site. Where imperviousness is equal to 50 percent or more, runoff volume shall be limited to 25 percent of the preconstruction volume. LEED also allows diversion of rainwater for reuse by building functions or infiltrating it into the ground to help recharge aquifers. This shall receive approval of the applicable codes and the AHJ. Other credits are available. Calculations are based on a 2-year, 24-h storm event using the rational formula for stormwater calculations.

GREEN GLOBES

Green Globes is a software-based rating system for both new and existing construction. It is a tool for guiding project teams to a greener building effort. It is a point-based process or rating method developed and administrated by the nonprofit Green Globe rating system. The Green Globe rating system has minimum requirements that must be met. The eventual ratings are the number of globes (points) awarded based on the number of points accumulated in all assessment categories. To become certified, a first building is self-evaluated based on the rating of points for eight categories. It is a software scoring system that includes environmental assessment area elements. This is spelled out in Table 23.7. A 10 percent reduction is allowed for nonapplicable categories. Not every strategy is applicable to every building, and the cost of implementing them might be excessive. This has to be evaluated by the design team. Environmental policies do have an economic impact on the design of a project. More information can be obtained from www://theGBI.org.

Earning GBI Building Credits

The above assessment protocol is made using the software tools and certification system provided by GBI that ensures each facility is comprehensively rated. After achieving a baseline of a minimum of 35 percent of the 1000 points, new and existing buildings can be certified for their environmental achievements. Final certification is assigned by giving from one to four globes based on the percentage obtained from the final score.

Green Globe rating/certification is both suitable and obtainable for a wide range of industrial and commercial buildings, and it enables the owner to have the necessary tools for marketability to shareholders, tenants, and the broader community. First, the stage I subscription and assessment is requested. This is a self-evaluation of the project. The next preliminary rating is for a certification based on the size. The stage II assessment follows where GBI will assign a third-party assessor. The assessor will inspect the project from the contract document and a site survey. Following is the awarding of the appropriate

TABLE 23.7 Green Globe Environmental Assessment Areas

The Green Globes for New Construction tool delivers a comprehensive, thorough, and proven means for evaluating, improving, and rating the environmental impact of new commercial buildings and major renovations. It is based on a recognized environmental assessment protocol that comprehensively assesses building environmental impacts on a 1,000-point scale in multiple categories. The survey protocol covers seven different areas, with each area having an assigned number of points that are utilized to quantify overall building performance.

Assessment category	Points	Description
Energy	380	Performance, efficiency, demand reduction, energy-efficient features, use of renewable energy, transportation
Indoor environment	200	Ventilation, lighting, thermal and acoustical comfort, ventilation system
Site	115	Ecological impact, development area, watershed features, enhancement
Resources	100	Low-impact materials (LCA), reuse, demolition, durability, recycling
Water	85	Performance, conservation, treatment
Emissions and effluents	70	Air emissions (boilers), ozone depletion, water and sewer protection, pollution controls
Project management	50	Design process, environmental purchasing, commissioning
Total	1000	

number of globes, from one to four. This is given in Table 23.8. Typical GBI fees are given in Table 23.9. Expenses are billed separately.

Highly qualified third-party assessors with green building expertise in design, engineering, construction, and facility operation are assigned to inspect the facility during stage II. They interface with the design team and the owner during the third-party assessment via a review of the construction documents and on-site walk-throughs. The third-party assessors must go through a strict and complete training program in order to qualify.

There will be additional costs for miscellaneous items such as per diem. The prices shown are approximate and depend on the nature of the construction, with existing buildings having lower cost.

TABLE 23.8 Green Globe Ratings

Percent	Green Globe Rating	LEED Equivalent
35 to 54	1 Globe	Bronze
55 to 69	2 Globes	Silver
70 to 84	3 Globes	Gold
85 & above	4 Globes	Platinum

TABLE 23.9 Green Globe New Construction and Major Renovations Pricing and Certification

Software subscriptions	Price each
Green Globes® NC New Construction Single assessment/certification use (5-yr subscription)	$500/Building
Third-Party Assessment/Certification Price per building based on building square footage (enclosed or occupied)	

Service type		< 50,000 ft²	50,000 to < 100,000 ft²	100,000 to < 200,000 ft²	200,000 to < 500,000 ft²	> 500,000 ft²
NC Stage I Assessment		$3,000	$3,500	$4,000	$0.025/ft²	$12,500
NC Stage II Assessment & Rating/Certification		3,000	3,500	4,000	0.020/ft²	10,000
NC Stage I & Stage II Assessment & Rating/Certification	GBI Member	5,500	6,000	7,000	0.04/ft²	20,000
NC Stage I & Stage II Assessment & Rating/Certification	Nonmember	6,000	7,000	8,000	0.45/ft²	22,500
Assessor Travel Expenses (Mandatory for all on-site assessments)		Option 1: $1500 upfront flat fee Option 2: Actual expenses +20% overhead billed after the assessment is completed				

BACKGROUND FOR SUSTAINABILITY

The idea of sustainability is not new. For homeowners and in many industries, the recycling of materials has been a strategy since the late half of the twentieth century. Some manufacturers have been involved with cogeneration, alternate energy, water and chemical reuse, and recycling for many years. What is new, is the involvement and interest of the general public, the government, and the economic community. One of the more important considerations is the effect it would have on the profitability of a manufacturer. It has become a commonsense approach for first-class manufacturing facility, customers, and the environment.

To understand why building green is important, it's important to know the probable impact that any building will have on the environment. The knowledge of energy consumption that commercial and residential buildings are responsible for will help in the design of any particular facility. Typical numbers for the United States are

1. 63 percent of total energy consumption
2. 30 percent of total greenhouse gas emissions
3. 136 million tons of construction and demolition waste
4. 12 percent of potable water use
5. 40 percent (3 billion tons) of raw materials used globally

Benefits of Sustainable Construction

Such designs are not only beneficial to people and the planet but also profitable to the owner in the long run. Some of these benefits are tangible, and others are not. For example, reducing water consumption and associated waste removal costs saves on water and sewer bills. Energy-efficient buildings result in reduced gas costs. Developers procure better financing arrangements for sustainable buildings and lease their properties faster. Property values of sustainable buildings increase faster than those of conventional buildings. Sustainable buildings provide many benefits to the client and environment, such as

1. Reduced destruction of natural areas and habitats
2. Reduced air and water pollution and cost
3. Reduced depletion of finite resources
4. Healthier outdoor environments
5. Reduced operating costs

PLUMBING ENGINEERING OPPORTUNITIES

One of the premier objects of the green building is to find a way to reduce the building's environmental impact. Such designs are not only beneficial but also profitable to the owner in the long run. Some of the profits are tangible, and others are intangible.

An obvious focus for plumbing engineers is water conservation. Potable water is often scarce and expensive, and it is getting more so. Municipal water treatment is also costly and requires chemicals, and capacity is restricted. We need to help conserve this resource while avoiding contaminating the supply.

The codes we work with today mandate water conservation by limiting the use of water by means of low-flow water closets, urinals, showers, and lavatories. These fixtures impact both the amount of water consumed and the amount of waste discharged. Much of that clean water is used for waste removal, turf irrigation, and equipment washing. Such activities could use a lower quality than potable for these uses.

Methods to reduce the potable water demand can be implemented in the building's interior and exterior. Technology includes automatic fixtures, high-efficiency irrigation, native and adapted low-water-use plants. These types of technologies are common and often have little or no impact on the construction budget. They are considered good long-term investments because they reduce water cost throughout the life of the building. In addition, they are mostly low in maintenance.

Another option is the point-of-use water heater, located near or under sinks and showers. The advantage of using this type is that you use less water waiting for it to get hot, it leaves more space in the mechanical room, and energy is saved. The downside is that it requires more maintenance because it is located throughout the building instead of in one place.

Additional points can be obtained by reducing potable water use for irrigation, and further points gained from eliminating this water from public sources completely.

Potential Green Plumbing Features

At the start of a project, a list of potential plumbing features that would influence the LEED or GBI programs should be considered by the design team. Such items would include

1. Waterless urinals
2. Sensor-controlled faucets

3. Ultra-low-flow toilets
4. Graywater for toilet flushing
5. Rainwater collection and reuse
6. Solar heating hot water
7. Point-of-use water heaters
8. Vacuum plumbing drainage system
9. Heat recovery from electric motors for preheating hot water
10. Flow restrictors, metering faucets
11. Dual-flush toilets

This list is by no means complete. Further sustainable design features will ultimately be proposed and further evaluated by the design team. Many other features are possible because of the unique aspects of the project, and all must be brought up. Each must be analyzed regarding initial cost, construction schedule, and potential code restrictions. These technologies are common and often have little or no impact on the budget.

MECHANICAL CONTRACTOR AND CONSTRUCTION TEAM

Due to the unique nature of some of the systems selected and approved, the mechanical contractor shall be selected based on the quality of work and who has knowledgeable subcontractors. This is often impossible to achieve locally. This will require investigation of contractors as close to the project's location as possible, who have the ability to build the mechanical systems. A Request for Proposal for mechanical work should be issued which would bring replies from interested contractors. A meeting must be held to inform all contractors who replied about the project and to inform them about the operational features of the engineered systems. This meeting should include manufacturers who build any of the engineered systems and that would enable them to explain at length any special features. This should produce several mechanical contractors willing to bid on the project.

Based on the reputation and experience of all proposed contractors, one general contractor and the mechanical contractor would be selected. They should have a strong commitment to quality and past performance to deliver the project on time.

GREEN ROOFS

It is thought that the addition of grass and landscaping to a roof would provide an addition to an energy reduction. With the issuance of the 2012 ICC Fire Code, now specific code requirements must be followed. Nothing appears in the Plumbing Code. The roof drains and design are outside the scope of this handbook.

The ICC Building Code added a few requirements for loading, live load, and separation. The majority of regulations fall under the fire code, but some affect the plumbing designers. The concerns that arise from a building's energy savings are that the "green roofs" are not as green as thought.

One prime concern is the fact that the roof will have to be irrigated and the frequency of it. If there is not enough rain, the water used must come from somewhere. The size of the cultivated area is limited to 15,625 ft^2 (13,484 m^2). Other considerations include these questions: How will the roof grass be cut? Will it be a brush fire hazard? How will maintenance be performed and how often? All things considered, is it worth it?

LEGAL ASPECTS OF GREEN BUILDING DESIGN

In a lawsuit brought by an owner against the U.S. Green Building Council, it was alleged that the LEED building rating system made false promises as to the performance of a LEED-certified building and manipulated data showing that LEED-constructed buildings were more efficient than standard constructed buildings. The Federal Trade Commission (FTC) has issued guidelines restricting the use of "green" and other marketing assumptions that claim a producer or service benefits. This is an example of overpromising.

With any decline in building construction, green prominence grows. There are many opportunities that new or existing buildings could implement, such as installing new, low-use plumbing fixtures. Some owners decided to retrofit plumbing when they realized the savings in the quantity of water that actually occurred.

As more government bodies add incentives for green buildings, such as tax breaks, grants, and marketing assistance, the stakes for green engineers grow higher. Installing a system that does not meet whatever standards are applicable may result in large costs to retrofit such noncompliant systems. This may also result in losses in tax incentives, and grants and economic losses related to delays in achieving the desired certification.

Steps could be taken to protect yourself from major litigation. It is now possible to obtain professional liability insurance that protects against false green building claims. Also available is a "Green Reputation Coverage" that would reimburse you for the costs involved in protecting you from adverse green claims. There is also available insurance with broader property and environmental endorsements that could be added to your regular insurance policy.

While an insurance policy can protect you after a green building client files suit against you, there are steps that could be taken. The best rule to follow is to limit, as much as possible, the promises that are made. Make certain that all parties agree on reasonable expectations and that clients are aware of what is expected to be achieved.

BIBLIOGRAPHY

Altwies, J.,"A Plumbing Guide to LEED," *Plumbing Systems & Design.*

Balz, S., "LEED for Plumbing Professionals," *Plumbing Systems & Design.*

Biada, B., "Earning LEED 2009 Water Efficiency Standards," *Consulting Specifying Engineer.*

Cavazos, J., "Green Engineering Designs," *Evaluation Engineering.*

Green Building Initiative, Green Globes Web site.

Huff, W., "Alternate Building Water Systems," *Plumbing Systems & Design.*

Lynam, D., C. Sorock, "Green Litigation," *Plumbing Systems & Design.*

Natarajan, S., and D. Dexter, "Understanding LEED," *Plumbing Systems & Design.*

Phillips, N., "Let Profits Rain," *Water Technology.*

"Sustainability—It's Not Easy Being Clean," *Global Control Magazine.*

Williams, J., and S. Larocque, "Justifying Sustainable Projects," PW Magazine

APPENDIX
METRIC CONVERSIONS

The modern metric system is given the official designation *SI*, translated from the French title *Systeme International d'Unites* given by the 11th General Conference on Weights and Measures. Its purpose has been to originate a coherent system of units selected from metric systems. It was finalized in its current form by an international agreement in 1960 and is the standard international language of measurement. The metric system has been mandated as the preferred system of measurement in the United States since the Metric Conversion Act was passed by Congress in 1975. This was further enhanced by an Executive Order in 1991 that required all government publications to be revised, and that a transition plan for conversion be adapted. All federal government construction and most states now require metric units to be used for their projects, but not much progress has been made in the United States for total implementation of metric units for material and design used only in this country.

The United States, using the United States Customary System (USCS) units of weights and measures, is the last industrialized country that does not completely use the metric system (SI), but rather uses a system called inch-pound (IP) units. This appendix provides the most currently accepted means of expressing the various metric measurement units and their abbreviations. They have been obtained from experience with various code bodies and from vendors who supply equipment to both the United States and foreign markets. In addition, a limited table of conversion factors is provided for the most often referred to units used by engineers in the construction industry. For a complete list of conversion factors, refer to standard engineering references. The abbreviations used will refer to either IP for the inch-pound units currently in use in the United States or SI for the International System of Units.

BASIC METRIC FUNDAMENTALS

There are three classes of SI units: base units, derived units, and supplementary units.

Base units consist of seven well-defined units which, by convention, are regarded as dimensionally independent:

Length	meter (m)
Mass	kilogram (kg)
Time	second (s)
Electric current	ampere (A)
Thermodynamic temperature	kelvin (K)
Amount of a substance	mole (m)
Luminous intensity	candela (cd)

Derived units are formed by combining base units and other derived units.

Supplementary units consist of only two units which are not used in general engineering practice.

Pronunciation. The following are the correct pronunciation for often mispronounced words:

Joule rhymes with jewel.

Pascal rhymes with rascal.

Pico rhymes with peek-oh (not pie-ko)

Kilo rhymes with kill-oh (not keel-oh)

Basic Conventions

1. All symbols shall be lowercase except for the following:
   a. The liter is always capitalized when it is abbreviated. When it is written out, no capitalization is used. This is done to avoid any confusion with the number 1 or the lowercase l in other unit designations.
   b. All abbreviated units that refer to, or are derived from, a person's name are always capitalized (such as Pa for pascal or °C for degrees Celsius).
2. No periods are used after any symbol or abbreviation, except at the end of a sentence.
3. For technical writing, always use abbreviations in conjunction with numerals (such as 10 m). For nontechnical writing, the use of numerals could be combined with the written out unit names (such as 10 meters), if desired.
4. No plurals are used for any abbreviation. Plurals may be used for written out units.
5. The use of square and cubic designations in SI units is always a superscript, such as m^3 or m^2. In IP units, abbreviations are generally used, such as sq ft or cu ft, although it is not incorrect to use superscripts.
6. When a unit is written out, no capitalization is used on any measurement unit except where standard punctuation or proper grammar requires a capital letter.
7. Do not mix written out units names and abbreviations.
8. A soft conversion uses an exact equivalent. A hard conversion uses a rationalized, rounded number that is created to be easy to work with and remember.
9. Insert a space between the numerical value and the abbreviation (23 m). Do not insert a space between a prefix and an abbreviation (km).
10. When writing formulas, use a center dot (·) instead of a second slash for multiplication; for example, use $L/min \cdot m^2$, not $L/min/m^2$.

Accuracy

When converting from IP to SI units or vice versa, it is important that the precision of the most accurate figure be retained. Use the least significant figure (or the same number of digits) as a basis for accuracy. As an example, if a conversion from 50 kPa to pounds per square in (psi) is desired, a conversion factor of 6.894 is found. If the soft conversion factor of 7.0 is used, the degree of precision falls well within the accuracy required of most calculations and does not exceed the least significant digit of one. When extreme accuracy is required, hard conversions should be used.

The identification of the significant digit is possible only through a knowledge of the circumstances for which the conversion is used. Numbers that are intended to be exact are treated as though there are an infinite number of significant digits.

ABBREVIATIONS

The following are abbreviations for the most often used units:

second	s	liter	L	milli	m
minute	min	meter	m	centi	c
hour	h	gram	g	kilo	k
day	d	pascal	Pa	pound	lb
week	wk	newton	N	watt	W
month	mo	inch	in	joule	J
year	yr	foot	ft		
degree Celsius	°C	yard	yd		

PREFERRED USAGE

Although the base unit of temperature is the kelvin (K), the degree Celsius (°C) is used much more extensively and is the preferred unit of measurement. The °C is equal to K except that the starting point is different.

The preferred unit of area is the square meter (m^2) when converting square feet. Large areas should be measured in square kilometers (km^2), and for smaller areas use square millimeters (mm^2).

The term weight is commonly used as a synonym for mass. These terms can be used interchangeably.

IP customary units express reference to gauge or absolute pressure as either psig or psia. No such conventions are officially permitted in SI units. For expressing a pressure differential, the use of kilopascals (kPa) is preferred. Common use of the term kPa is generally understood and indicates the equivalent of psig. If kPa is intended to indicate absolute pressure, it should be written either as an absolute pressure of 456 kPa or 456 kPa abs.

The standard designation for pressure is the pascal (Pa) which is the equivalent of newtons per square meter (N/m^2). Other designations, such as kilograms per square centimeter (kg/cm^2) are no longer preferred. The bar and torr are still commonly used in vacuum and pressure calculations.

The preferred designation for velocity in SI units is meters per second (m/s). In the United States, foot per minute is the preferred unit of measure in IP units.

The commonly accepted practice of expressing standard conditions for gas flow rate in both IP and SI units is to use the prefix "s" and thus is written as "scfm" to indicate standard cubic feet per minute and "sLpm" to indicate standard liters per minute. Some foreign manufacturers are also using the prefix "n" (for normal) to indicate standard conditions.

The use of centimeters is discouraged, except for measurements of the human body. The use of millimeters is preferred for linear measurements where practical, generally up to 9 ft 0 in. As an example, a 4 ft 0 in × 8 ft 0 in sheet of plywood is written as 1200 × 2400 mm. The use of meters and a decimal add-on is also common practice for larger dimensions, such as 4.6 m.

For water and fire protection flow-rate calculations within facilities, the use of liters per minute (L/min) is preferred when converting SI units in the United States, rather than per second. This is to conform with the U.S. standard of gallons per minute. In other countries, liters per second may be the preferred unit of measure.

When more exact calculations are desired, it is common practice to round off to three places to the right of the decimal point.

CONVERSION FACTORS

Where applicable, all IP references are in U.S. customary units.

To calculate the reverse of the units, simply divide the conversion factor by 1. For example, feet per second multiplied by 0.305 equals meters per second. To convert meters per second to feet per second, divide 0.305 into 1 and get 3.28, which is the conversion factor.

PIPE SIZES

The pipe sizes currently used in the United States are nominal sizes roughly based on inches and fractions of an inch, referred to as "Iron Pipe Size" or IPS. The current standard most often used for SI units are "Nominal Dimension" or DN, roughly based on the pipe size in millimeters. The conversions from one to the other are as follows:

IPS	DN	IPS	DN
1/8	6	8	200
3/16	7	10	250
¼	8	12	300
3/8	10	14	350
½	15	15	375
¾	20	18	400
1	25	20	500
1¼	32	24	600
1½	40	30	750
2	50	36	900
2½	65	40	1000
3	80	48	1200
4	100	52	1300
5	125	56	1400
6	150	60	1500

PREFERRED SCALE

1 in to 16 ft = 1:200	1 in to 40 ft = 1:500
1 in to 8 ft = 1:100	1 in to 80 ft = 1:1000
1 in to 4 ft = 1:50	*Note*: SI is approximate

PITCH

in/ft	1/32	1/16	1/8	¼	½
cm/m	0.25	0.5	1	2	4

COMMONLY USED IP CONVERSIONS

To convert from	To	Multiply by
ft.hd of water	psi	0.433
psi	ft.hd	2.31
cfs	gal/s	450
gal of water	lb	8.33
lb/cu.ft	lb	62.36
gal of water	cu.ft	7.48
cu.ft of water	lb	62.3

CONVERSION FACTORS

To convert from	To	Multiply by

Pressure

Atmosphere	psi	14.7
ft lb	N · m	1.36
inHg	kPa	3.4
ft head	kPa	3.0
in water	Pa	250
psi	kg/cm^2	0.07
lb/ft^3	kg/m^3	16
psi	bar	0.07
psi	kPa	6.9
lb/ft^2	kg/m^2	5.0
psf	Pa	50

To convert from	To	Multiply by

Length

in	mm	25.4
yd	m	0.91
km	yd	1093
km	ft	3333.3
ft	m	0.3
mi	km	1.6
ft^3	m^3	0.3

Area and Spatial Volume

cu ft	m^3	0.03
cu ft	L	28.32
sq ft	m^2	0.836
sq mi	km^2	2.6
sq yd	m^2	0.836
sq in	mm^2	6.5
cu yd	m^3	0.764
cu in	mm^2	16.39
acre	m^2	4047
cu in	cm^3	16.4

Liquid Volume

gal	L	3.79
quart	L	0.95
m^3	gal	264

Fluid Flow Rate

gpm	L/min	3.8
gpm	L/sec	0.063
gpm	L/h	227
gpm	m^3/h	0.227
gpm/sq ft	L/m^2	0.68
gph	L/sec	0.0013
gph	L/min	0.08
gph	L/h	4.8
ft/sec	m/sec	0.3
ft/min	km/h	0.018

To convert from	To	Multiply by
cfm	m^3/sec	0.0005
cfm	m^3/h	0.03
efm	L/sec	0.47
efm	L/min	28.3
cfm	m^3/sec	2119
cfm	m^3/min	1.7
cfh	L/sec	0.008
cfh	m^3/h	0.03
cfs	m^3/sec	0.03
cfs	L/min	1700
L/min	cfs	0.0006
L/min	gal/sec	0.0044
L/sec	gal/sec	0.26412
fps	km/h	1.1
mph	km/h	1.6
mph	m/sec	0.45

Weight

kg	lb	2.2
lb	kg	0.433
lb/ft^3	kg/m^3	16
lb/ft^2	kg/m^2	5.0

Heat and Temperature

°F (measured temperature)	°C	°F − 32 × 0.56
°F (temperature difference)	°C	°F × 0.56
Btu/h	mJ/h	0.0010
Btu/ft^3	mJ/m^3	0.00.7
Btu/h	W/h	0.3
W/m^3 · h	3.15	Btu/sq ft · h

Velocity

fps	m/s	0.305
mph	m/s	0.45
mph	km/h	1.6
fpm	km/h	0.018
fps	km/h	1.1

INDEX

CPSIA information can be obtained at www.ICGtesting.com
Printed in the USA
BVOW042048240712

296088BV00001B/1/P

9 780071 760270